Elliott Coues

Key to North American Birds

Containing a concise account of every species of living and fossil bird at present

known from the continent north of the Mexican and United States boundary

Elliott Coues

Key to North American Birds
Containing a concise account of every species of living and fossil bird at present known from the continent north of the Mexican and United States boundary

ISBN/EAN: 9783337409616

Printed in Europe, USA, Canada, Australia, Japan

Cover: Foto ©Andreas Hilbeck / pixelio.de

More available books at **www.hansebooks.com**

KEY

TO

NORTH AMERICAN BIRDS

CONTAINING A CONCISE ACCOUNT OF EVERY SPECIES OF

LIVING AND FOSSIL BIRD

AT PRESENT KNOWN FROM THE CONTINENT NORTH OF THE MEXICAN
AND UNITED STATES BOUNDARY.

ILLUSTRATED BY 6 STEEL PLATES, AND UPWARDS OF 250 WOODCUTS.

BY

ELLIOTT COUES,

ASSISTANT SURGEON UNITED STATES ARMY.

SALEM: NATURALISTS' AGENCY.
NEW YORK: DODD AND MEAD.
BOSTON: ESTES AND LAURIAT.
1872.

PREFACE.

A PREFACE is indispensable in this instance, simply because I have no other opportunity of properly acknowledging the assistance I have received in preparing this work. I am particularly indebted to Mr. J. A. ALLEN, of Cambridge, Mass., who has diligently revised nearly all the proofsheets, and whose critical suggestions have proved invaluable. Mr. ROBERT RIDGWAY, of Illinois, has given me the benefit of his still unpublished studies of the *Raptores* and some other groups, besides rendering, as Mr. ALLEN also has, various essential services.

Prof. BAIRD kindly offered me the use of all the illustrations of his late Review, while Prof. AGASSIZ generously placed at my disposal the plates accompanying Mr. ALLEN's Memoir on the Birds of Florida. Several of the woodcuts have been taken from Prof. TENNEY's Manual of Zoology, with the author's permission; and a few others have been contributed by Messrs. LEE and SHEPARD. With a few exceptions, the rest of the illustrations have been drawn from nature by the author, and engraved by Mr. C. A. WALKER.

I have spoken elsewhere of Prof. MARSH's almost indispensable coöperation in one part of the work.

While material for the greater part of the descriptions has been furnished by the author's private cabinet, the Synopsis could hardly have been prepared without that free access to the collection of the Smithsonian Institution, of which I have been permitted to avail myself.

The only word of explanation that seems to be required is with regard to the large number of genera I have admitted. I have been led into this—unnecessarily, perhaps, and certainly against my judgment—partly by my desire to disturb a current nomenclature as little as possible, and partly because it is still uncertain what value should be attached to a generic name. Among wading and swimming birds—the groups of which are, on the whole, more precisely limited than those of *Insessores*—I have, however, indicated what I consider to be a reasonable reduction; and on another occasion I should probably extend a like practice, if not one even more "conservative," to the remaining groups. I will only add, that I consider that several of the admitted families of *Oscines* will require to be merged in one. These are the *Turdidæ, Saxicolidæ* and *Sylviidæ*, if not also the *Troglodytidæ* and *Motacillidæ;* while the same may prove true of the current Sylvicoline, Tanagrine and Fringilline groups.

E. C.

WASHINGTON, D. C., September 9th, 1872.

CONTENTS.

—

INTRODUCTION.

SECT. I. Ornithology Defined — Birds Defined — Brief Description
of their Peculiar Covering.

§ 1. Science (Lat. *scire*, to know) is knowledge set in order; knowledge disposed after the rational method that best shows, or tends to show, the mutual relations of observed facts. Such orderly knowledge of any particular class of facts — such methodical disposition of observations upon any particular set of objects — constitutes a Special Science. Thus, Ornithology (Gr. *ornithos*, of a bird, *logos*, a discourse) is the Science of Birds. Ornithology consists in the rational arrangement and exposition of all that is known of birds. Ornithology treats of the physical structure, physiological processes, and mental attributes of birds; of their habits and manners; of their geographical distribution; of their relations to each other and to other animals. The first business of Ornithology is to define its ground; to answer the question

§ 2. What is a Bird? A Bird is an air-breathing, egg-laying, warm-blooded, feathered vertebrate, with two limbs (legs) for walking or swimming, two limbs (wings) for flying or swimming, fixed lungs in a cavity communicating with other air-cavities, and one outlet of genito-urinary and digestive organs; with (*negative characters*) no teats, no teeth, no fleshy lips, no external fleshy ears, no (perfect) epiglottis nor diaphragm; no bladder, no scrotum, no corpus callosum; and with the following collateral characters, mostly shared by more or fewer other animals: — Under jaw hinged with the rest of the skull by means of an interposed movable bone, that is also movably jointed with two bones of the upper jaw; head jointed with neck by only one hinge; shoulder-joints connected with each other by a curved bone, the clavicle (with rare exceptions), and with breast-bone by a straight stout bone, the coracoid; ribs all bony, most of them jointed in the middle as well as with back-bone and breast-bone, and having bony offsets; less than three *separate* wrist and hand-bones; two fingers, of one or two bones; head of thigh-bone hinged in a ring, not in a cup; one of the two leg-bones not forming the ankle-joint; no *separate* ankle-bones; less than three *separate* foot-bones; two to four toes, of two to five bones, always ending in claws; both jaws horny-sheathed and nostrils in the upper one; feet and toes (when not feathered) horny-sheathed; three eyelids; eyeball with hard plates in it, eight muscles on it, and a peculiar vascular organ inside; two larynges, or "Adam's-apples"; two bronchi; two lungs, perforated to send air into various airsacs and even the inside of bones; four-chambered heart, with perfect double blood-circulation; tongue with several bones; two or three stomachs; one liver, forked to receive the heart in its cleft; gall-bladder or none; more or less diffuse pancreas, or "sweetbread"; a spleen; intestines of much the same size throughout; cœca, or none; two lóbulated, fixed kidneys; two testicles fixed in the small of the back, and subject to periodical enlargement and decrease; one functional ovary and oviduct; outlets of these last three organs in an enlargement at end of intestine, and their products, with refuse of digestion, all discharged through a common orifice. But of all these, and other characters, that come under the head of description rather than of definition, one is peculiarly characteristic of birds; for every bird has FEATHERS, and no other animal has feathers. Naturally, then, we look with special interest upon

FEATHERS :

§ 3. *a.* THEIR STRUCTURE. A perfect feather consists of a main stem, or scape (*scapus;* pl. 1, fig. 7, *a*), and a supplementary stem or after-shaft (*hyporhachis;* pl. 1, fig. 7, *b*), each bearing two webs or vanes (*vex-illum,* pl. *vexilla;* pl. 1, fig. 7, *c*), one on either side. The scape is divided into two parts; one, the tube or barrel, or "quill" proper (*calamus;* pl. 1, fig. 7, *d*) is hard, horny, hollow, cylindrical and semitransparent; one end tapers to be inserted into the skin; the other ends, at a point marked by a little pit (*umbilicus*), in the shaft (*rhachis*), or second part of the stem; the rhachis is squarish, and tapers to a point; is less horny, is opaque, and filled with white pith; it alone bears the vexilla. The after-shaft has the same structure, and likewise bears vexilla; it springs from the stem, at junction of calamus and rhachis, close by the umbilicus. It is generally very small compared with the rest of the feather; but in a few birds is quite as large; it is wanting in many; and is never developed on the principal wing and tail feathers. The vane consists of a series of appressed, flat, narrowly lance-shaped or linear laminæ, set obliquely on the rhachis, and divaricating outward from it at a varying angle; each lamina is called a barb (*barba;* pl. 1, fig. 6, *a, a*). Now just as the rhachis bears barbs, so does each barb bear its vanes (*barbules;* pl. 1, fig. 6, *b, b, c*); it is these last that make a vane truly a *web*, that is, they connect the barbs together, so that some force is required to pull them apart. They are to the barbs exactly what the barbs are to the shaft, and are similarly given off on both sides of the barbs, from the upper edge of the latter. They are variously shaped, but generally flat sideways, with upper and lower border at base, rapidly tapering to a slender thread-like end; and are long enough to reach over several barbules of the next barb, crossing the latter obliquely. All the foregoing structures are seen with the eye or a simple pocket lens, but the next two require a microscope; they are barbicels (or *cilia:* pl. 1, fig.

8), and hooklets (*hamuli;* pl. I, fig. 8). These are simply a sort of fringe to the barbules, just as if the lower edge of the barbule were *frayed out,* and only differ from each other in one being plain, hair-like processes, and the other being *hooked* at the end. Barbicels do occur on both anterior and posterior rows of barbules; but rarely on the latter; hooklets are confined to barbules of the anterior series, which, as we have seen, *overlie* the posterior rows of barbules, diagonally forming a meshwork. The beautiful design of this structure is evident; by it, the barbules are *interlocked,* and the vane of the feather made a web; for each hooklet of one barbule catches hold of a barbule from the next barb in front, —any barbule thus holding on to as many of the barbules of the next barb as it has hooklets. To facilitate this interlocking, the barbules have a thickened upper edge of such size that the hooklets can just grasp it. This is clearly illustrated in pl. I. fig. 2, where *a, a, a, a* are four barbs in transverse section, viewed from the cut surfaces; with their anterior (*b, b, b, b*), and posterior (*c, c, c, c*) barbules, the former bearing the hooklets which catch over the edge of the latter.

b. But all feathers do not answer the above description. First, the after-shaft may be wanting, as we have seen; then, as frequently happens, hooklets may not be developed, and barbicels may be few or wanting; barbules may be few or wanting, or so transformed as to be only recognized by position, and even barbs themselves may be wanting on one side of the shaft, as in some tail feathers of the famous Lyre-bird, or on both sides, as in certain bristly feathers about the mouth and eyelids of various birds. (Certain unusual styles of feathers are shown in figs. 1, 2, 3.) Consideration of these and other modifications has led to the recognition of *three*

§ 4. TYPES OF STRUCTURE. 1. The feathery (*pennacea*), characterized as above. 2. The downy (*plumulacea*), when the stem is short and weak, with soft rhachis and barbs, with long, extremely slender, mostly thread-like barbules, with little knotty dilatations in place of barbicels, and no hooklets. 3. The hairy (*filoplumacea*) with a thin, stiff calamus, usually no pith in the rhachis, fine cylindrical stiff barbs

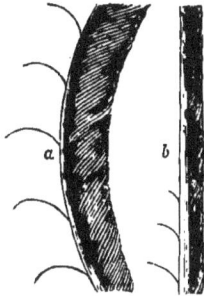
FIG. 2. Sections of a central rigid feather of the Lyre-bird. Natural size; *a,* from terminal curve; *b,* middle portion.

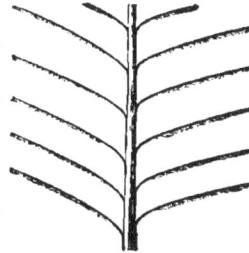
FIG. 1. Section from loosely barbed feather of Lyre-bird. Natural size.

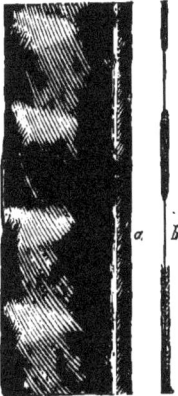
FIG. 3. *a;* section, 1-2 size, from one of the external feathers of Lyre-bird; *b,* single barb.

and barbules, the latter wanting barbicels, knots and hooklets. The first two types may be found in different parts of the same feather, as in pl. I, fig. 7, which is partly pennaceous, partly plumulaceous. All feathers are built upon one of these three plans; and, though seemingly endless in diversity, may be reduced to *four*

§ 5. DIFFERENT KINDS OF FEATHERS. 1. Contour-feathers (*pennæ*) have a perfect stem composed of barrel or shaft, and vanes of pennaceous structure at least in part, usually with downy structure toward the base. They form the great bulk of the plumage, that is upon the surface of a bird, exposed to light; their tints give the bird's colors; they are the most variously modified of all, from the fishlike scales of the penguin, to the glittering plates of the humming-bird, and all the endless array of tufts, crests, ruffs and other ornaments of the feathered tribe; even the imperfect bristle-like feathers above-mentioned belong here. Another feature is, that they are usually individually moved by cutaneous muscles, of which there may be several to each feather, passing to be inserted into the sheath of the tube, inside the skin, in which the stem is inserted; it is estimated that some birds have twelve thousand of these little feather muscles. Every one has seen their operation when a hen shakes herself after a sand-bath; and any one may see them plainly under the skin of a goose. 2. Down-feathers (*plumulæ*), characterized by the plumulaceous structure throughout. These form a more or less complete investment of the body; they are almost always hidden from view beneath the contour-feathers, like padding about the bases of the latter; occasionally they come to light, as in the ruff about a condor's neck, and then usually occur where there are no other feathers; they have an after-shaft or none, and sometimes no rhachis at all, when the barbs are sessile in a tuft on the end of the barrel. They often, but not always, stand in a regular quincunx between four contour-feathers. 3. The semiplumes (*semiplumæ*), which may be said to unite the characters of the last two, possessing the pennaceous stem of one and the plumulaceous vanes of the other. They stand among pennæ, like the plumulæ, about the edges of patches of the former, or in parcels by themselves, but are always covered over by contour-feathers. They are with or without an after-shaft. 4. Filoplumes (*filoplumæ*), or thread-feathers; these have an extremely slender, almost invisible, stem, not well distinguished into barrel and shaft, and no vanes (with rare exceptions), unless a few barbs near the end of the rhachis may be held for such. Long as they are, they are usually hidden by the contour-feathers, close to which they stand as accessories, one or more seeming to issue out of the very sac in which the larger feathers are implanted. They are the nearest approach to *hairs* that birds have.

§ 6. PECULIAR FEATHERS. Certain down-feathers are remarkable for continuing to grow indefinitely, and with this growth there is constant breaking off of the ends of the barbs. These feathers, from being always dusted over with the dry, scurfy exudation or exfoliation from the follicle in which they grow, are called *powder down-feathers*. They occur in the hawk, parrot, and gallinaceous tribes, but especially in the heron family, where they are always present, and readily seen as two large patches of greasy or dusty, whitish, matted feathers over the hips and in front of the breast. Their use is not known.

§ 7. FEATHER OIL-GLAND. With comparatively few and irregular exceptions, birds have a singular apparatus for secreting oil with which to lubricate and polish their feathers. It is a two-lobed, or rather heart-shaped, gland, saddled upon the root of the tail; consisting essentially of numerous slender secreting tubes or follicles, the ducts of which successively unite in larger tubes, and finally perforate the skin at one or more little nipple-like eminences. Birds press out a drop of oil with their beak, and then dress the feathers with it. The gland is largest in water-birds, which have most need of an impervious coating of feathers, and always present among them; very large in the fish-hawk; smaller in other land-birds, and wanting (it is said), among the ostriches, bustards, parrots and some others. (In pl. I, fig. 4, the line 6 points to the oil-gland.)

§ 8. DEVELOPMENT OF FEATHERS. In a manner analogous to that of hair, a feather grows in a little pit or pouch formed by inversion of the dermal layer, and is formed in a closed oval follicle consisting of an inner and outer coat separated by a layer of fine granular substance. The outer layer, or "outer follicle" is composed of several thin strata of nucleated epithelial cells; the inner is thicker, spongy and filled with gelatinous fluid; a little artery and vein furnish the blood-circulation. The inner is the true *matrix* of the feather, evolving from the blood-supply the gelatinous matter, and resolving this into cell nuclei; the granular layer is the formative material. The outer grows a little beyond the cutaneous sac that holds it, and opens at the end; from this orifice the future feather protrudes as a little, fine-rayed pencil point. During subsequent growth the follicular layers undergo little further change; it is the granular that becomes the feather.

§ 9. All a bird's feathers, of whatever kind and structure, taken together, constitute its *ptilosis* or

PLUMAGE.

(*a.*) FEATHERED TRACTS AND UNFEATHERED SPACES. With the exception of certain birds that have obviously naked spaces, as about the head, etc., all would be taken to be fully feathered. So they are fully *covered with* feathers; but it does not follow from this, that feathers are implanted everywhere upon the skin. On the contrary, this is the rarest of all kinds of feathering, though it occurs, almost or quite perfectly, among the penguins and toucans. Let us compare a bird's skin to a well-kept park, part woodland, part lawn; then where the feathers grow is the woodland; where they do not grow, the lawn; the former places are called *tracts* (*pterylæ*); the latter *spaces* (*apteria*); they mutually distinguish each other into certain definite areas. Not only are the tracts and spaces thus definite, but their size, form and arrangement mark whole families or orders of birds, and so are important for purposes of classification. They have been specially studied, named and classified by the celebrated Nitzsch, who has laid down the following as the general plan obtaining in the vast majority of birds : —

(*b.*) 1. The spinal or dorsal tract (*pteryla spinalis*, pl. I, fig. 4, I), running along the middle of the bird above from nape of the neck to the tail; subject to great variation in width, to dilation and contraction, to forking, to sending out branches, to interruption, etc. 2. The humeral tracts (*pt. humerales*, pl. I, fig. 4, 2), always present, one on each wing; narrow bands running from the shoulder obliquely backward upon the upper arm-bone, parallel with the shoulder-blade. 3. The femoral tracts (*pt. femorales*, pl. I, fig. 4, 3), a similar oblique band upon the outside of each thigh, but, unlike the last, subject to great variation. 4. The ventral tract (*pt. gastræi*, pl. I, fig. 3, 8), which forms most of the plumage on the under part of a bird; commencing at or near the throat, and continued to the anus; it is very variable like the dorsal tract, is usually bifurcate, or divided into right and left halves with a central apterium, is broad or narrow, branched, etc. ; thus, Nitzsch enumerates *seventeen* distinct modifications ! The foregoing are mostly isolated tracts, that is, bands nearly surrounded by apteria that are complementary to them; the following are continuously, uniformly feathered, and therefore, in general, equivalent to the part of the body they represent. Thus, 5, the head tract (*pt. capitis*, pl. I, figs. 3, 4; 4, I), clothes the head and generally runs into the beginning of both dorsal and ventral tracts. 6. The wing tract (*pt. alaris*, pl. I, figs. 3, 5; 4, 5), represents all the feathers that grow upon the wing, except those of the humeral tract. 7. The tail tract (*pt. caudalis*, pl. I, figs. 3, 6; 4, 6), includes the tail feathers and their coverts, those surrounding the oil-gland, and usually receives the termination of the dorsal, ventral, and femoral tracts. 8. The leg tract (*pt. cruralis*, pl. I, figs. 3, 7; 4, 7), clothes the legs as far as these are feathered, which is sometimes to the toes, generally only to the heel. I need not give the *spaces*, as these are merely the complements of the tracts ; and the highly important special feathering of the wings and tail will be examined in describing those members for purposes of classification.

§ 10. PROGRESS AND CHANGE. Newly hatched birds are covered with a kind of down, entirely different from the feathers they ultimately acquire. It is scanty, leaving much of the body naked, in *Altrices*, or those birds that are reared by the parent in the nest ; but thick and puffy in a few of these, and in all *Præcoces*, that run about at birth. But true feathers are soon gained, in some days or weeks, those of wings and tail being the first to sprout. The first plumage is usually only worn for a short time — then another is gained, and frequently several more changes ensue before the bird attains its mature covering. Feathers are of such rapid growth, that we can easily understand how exhaustive of vital energies the growth must be, and how critical a period the change is. The renewal of plumage is a process familiar to all under the term " moult " (*ecdysis*). It commonly occurs at least once a year, and generally twice, in spring and fall; when old, faded and worn out feathers are shed, and fresh ones take their place, either over a part or the whole of the body. The change frequently or generally results in considerable differences of color, constituting the "seasonal plumages" of so many birds, which, in the same bird, may change from black to white even, from plain to variegated, from dull to brilliant. But birds also change colors, by actual alteration in the tints of the feathers themselves, and by gaining new ones without losing any old ones. The generalization may be made, that when the sexes are strikingly different in color, the young at first resemble the female; but when the old birds are alike, the young are different from either. When the seasonal changes are great, the young resemble the fall plumage of the old. When the old birds of two different species of the same genus are strikingly alike, the young of both are usually intermediate between them, and different from either.

Besides being the most highly developed, most complex, wonderfully perfect and beautiful kind of tegumentary outgrowth; besides fulfilling the obvious design of covering and protecting the body, the plumage has its

§ 11. PECULIAR OFFICE: that of accomplishing the act of flying. For all vertebrates, except birds, that progress through the air — the flying-fish with its enlarged pectoral fins; the flying reptile (*Draco volans*) with its skinny parachute; the flying mammal (bat) with its great webbed fingers — accomplish aërial locomotion by means of tegumentary *expansions*. Birds, alone, fly with tegumentary *outgrowths*, or appendages.

SECT. II. AN ALLUSION TO THE CLASSIFICATION OF BIRDS — TAXONOMY — STRUCTURE — CHARACTERS — GROUPS OF DIFFERENT GRADES — TYPES AND ABERRATIONS — EQUIVALENCY — ANALOGY AND AFFINITY — EXAMPLE.

SEEING what a bird is, and how distinguished from other animals, our next business is to find out how birds are distinguished from each other; when we shall have the material for

§ 12. CLASSIFICATION, a prime object of ornithology, without which, birds, however pleasing they are to the senses, do not satisfy the mind, which always strives to make orderly disposition of things, and so discover their mutual relations and dependencies. Classification presupposes that there are such relations, as results of the operation of fixed inevitable law; it is, therefore,

§ 13. TAXONOMY (Gr. *taxis*, arrangement, and *nomos*, law), or the rational, *lawful* disposition of observed facts. Just as taxidermy is the art of fixing a bird's skin in a natural manner, so taxonomy is the science of arranging birds themselves in a natural manner, according to the rules that, to the best of our knowledge and belief, are deducible from examination of their

§ 14. STRUCTURE: The physical constitution of a bird; all the material constituents of a bird, and the way its parts or organs are put together.

Internal structure, or *anatomical* structure (*ana*, and *temnein*, to cut), so called because we have to cut into a bird to see it, comprehends all the parts of a bird that are ordinarily hidden from view; *external* structure, those that lie exposed to view upon the surface. Much time has been wasted in arguing the superiority of one or the other of these for purposes of classification; as if a natural classification must not be based upon *all* points of structure! as if internal and external points of structure were not reciprocal and the mutual exponents of each other! External points of structure stand to internal somewhat in the relation of interest and capital; it is legitimate and wise enough to use interest only unless we need to draw upon capital. In our greater taxonomic enterprises — in the founding of our higher groups — we require all the capital we can get; in our lesser undertakings the interest alone is sufficient. Moreover, birds are so much alike in their anatomical structure, that this answers taxonomic purpose only for higher groups; and practically, at any rate, we make our lesser divisions so readily from external structure, that this may be said to furnish most of our

§ 15. Zoological Characters. A "character" is any point of structure whatsoever that is susceptible of being perceived and described for the purpose of distinguishing birds from each other. Characters are of all *grades*, or values, from the trivial ones that separate two species, to the fundamental ones that mark off primary divisions. The more characters, of whatever grade, that birds have in common, the more closely they are allied to each other, and conversely. The possession of more or fewer characters in common, results in

§ 16. Degrees of Likeness. Were all birds alike, or did all birds differ by the same characters to the same degree, no classification would be possible. But we find that they vary within wide limits — from the almost imperceptible difference between two hatched in the same nest, to the extreme unlikeness between a thrush and a penguin. This is the arena of classification; this gives us both the room and the material to divide birds into groups, and subdivide these into other groups, of greater or lesser "value," or *grade*, according to the more or fewer characters shared in common. We saw that (in addition to other characters), *all* birds have feathers, which no other animals possess; birds can be separated from other animals, but not from each other, by this feature; it is therefore a CLASS character. Even the

§ 17. Primary Division of birds must be made from a character of less value than this. A broad generalization upon the sum total of all the exhibitions that (recent — geologically) birds make in their modes of life, shows that these are of three sorts. Either birds habitually live above the earth, in the air or on trees; or they habitually live on the ground; or they habitually live on the water; and in each case, their structure was designed and fitted for such particular end. We have, therefore, at the outset three *types of structure* correspondent with, and equivalent to, *three plans of life;* and, if our observations are correct, and our reasoning not fallacious, these types or plans, seemingly an abstract induction of ours, are as real as the birds themselves. It is natural then to divide birds into three primary groups: Aërial Birds (*Aves Aëreæ*), Terrestrial Birds (*Aves Terrestres*) and Aquatic Birds (*Aves Aquaticæ*). An illustration will make this clear. Men build machines to transport themselves and their goods; the only known media of transportation are the air, the earth and the water; and we do not imagine any sort of vehicles more unlike than a balloon, a buggy, and a brig; these, therefore, exemplify the most fundamental division of machines for transportation.

§ 18. ORDERS. Taking any one of these types of structure, we find that it may be unfolded, or carried out, in different ways. Studying all known aquatic birds, for example, we see that their plan of life is fulfilled in four different ways; it is exhibited under four aspects, or modes of execution, each distinguished by some particular combination of aquatic characters with certain other characters that we did not take into account in framing our *Aves Aquaticæ*. Thus a goose, a gannet, a gull and a guillemot, all agree in aquatic characters, but differ from each other by each having certain characters that the other three lack. Characters marking such modes of exhibition are called *ordinal;* and the groups so organized, *Orders.* In our illustration, there are likewise four plans of aquatic machines; diving bells, sailing vessels, steamships and rowboats, clearly distinguished by the way in which motion (the prime function of all vehicles) is effected; in this case it is by weight, by wind, by steam, by muscle; therefore the machinery by which these forces are applied, furnishes ordinal characters of aquatic vehicles.

§ 19. FAMILIES. But all the birds of an order are not alike; some resemble each other more than they do the rest; so another set of groups must be made. These groups are called *Families;* they consist in a certain combination of all ordinal characters with special sets of characters of the next lower grade or value. Let x represent the sum total of strictly ordinal characters, and suppose we find these variously combined with a certain number of the next lower grade of characters, as $a, b, —f$ for instance; then the particular combination x (abc) is one family; x (bef) another; x (cde) another, etc., and we shall have as many families under an order as there actually are such combinations. Sometimes an order may be represented by x ($a. . .f$); then there is but one family, as, for example, in the aquatic order *Lamellirostres* where the *Anatidæ* alone furnish every one of the ordinal features, and are equivalent to the order; that is to say $(a . . .f) = x$, because no character from a to f is wanting in any member of the order. In our order *sailing vessels*, of aquatic machines, *masts* and *sails* are ordinal characters, because they are essential apparatus to catch the wind. But these may be of a varying number, etc., upon which we might found families of sailing-vessels, as the ship family, represented by x (three masts + square sails); the schooner family x (two masts + fore-and-aft sails); the sloop family x (one mast + fore-and-aft sails), etc. Diving bells, I suppose, are so much alike, that they might be called an order of aquatic machines of but one family.

§ 20. GENERA. After family manifestations of ordinal characters, we come to the modifications of families themselves, enquiring how many *kinds of difference (genus,* a kind, pl. *genera)* there are in the birds composing a family. The mode of determining genera in a family is precisely like that of determining families in an order; it is *x* again (this time representing family characters) into a varying number or combination of characters of the next lower grade, *a—f.* A genus is the last definite grouping of birds that is usually recognized; it may be defined as the ultimate essential modification of structure (*ultimate,* because there is none lower; *essential,* because trivial features do not constitute a genus; *of structure,* because mere size, color, etc., are only specific characters). In the ship family, the three-masted vessel, full-rigged, with square sails, is a genus (ship-proper); one with square sails on two masts only, and fore-and-aft sails on the mizzen, is another genus (bark), and so on. Genera are composed of one or more

§ 21. SPECIES. The definition of a species has become difficult of late years, but for present purposes we may assume that it is any one of the *constant exponents* of a genus, comprehending all the birds that bear to each other the relation of parent and offspring; the latter capable of reproducing 'each after its kind' and maintaining certain characters to an evident degree peculiar to itself. Resting, then, upon this, we have little else to consider before we reach that most unquestionable fact, an individual bird. Species, however, are not absolutely constant; they vary in size, color, etc., within certain limits, under influences not always comprehended as yet, but which seem a part of that universal tendency in nature toward the production of essential unity in diversity; the operation of which, if completely effective, would level distinctions and abolish difference in sameness.

§ 22. A VARIETY is a step in this direction; for, although it may seem an opposite step, yet departure from any given point or standard must be approach toward some other. A variety is (*generally*) distinguished from a species by its tendency to revert to its original stock, or, diverging further from that, to approach some other type. The former case is constantly being demonstrated, and the latter is probably susceptible of being proven; but in either case, *inconstancy* is a marked feature of varieties. Varieties apparently produced by difference in food, climate, etc., are called *local races,* when restricted to a small area in or around the general distribution of the parent stock; *geographical races,* when more widely separated over large areas. A *hybrid* is a cross between two species, almost always of the same natural genus. Hybrids are generally infertile, while crosses between mere varieties are capable of reproduction, so that hybridism becomes in some measure a test; nevertheless, exceptions are not wanting.

§ 23. INTERMEDIATE GROUPS. Having arrived at the individual bird, we will retrace our steps for a moment, for the student must sooner or later learn, that, easy as it seems to theoretically determine the foregoing groups, there are many difficulties in the way of their practical definition. This is partly because all birds are singularly inter-related, presenting few broad, unequivocal, unexceptional characters in the midst of numberless minor modifications, and partly because the higher groups, no less than species and varieties, shade into each other. In our illustration, for example, we find exactly intermediate aquatic machines; thus, it would be difficult for a landsman to say whether an hermaphrodite brig belonged to the ship family, or the schooner family; he would have to decide according as he considered number of masts, or shape of sails, the more essential family character. But the *intermediate groups* which remain to be examined are not of this ambiguous nature; they are unequivocally referable to some particular group of the next higher grade, and, being subordinate divisions, they are distinguished by the prefix *sub*, as sub-order, sub-family. Though somewhat difficult to define, they are, I think, susceptible of intelligible, if not always precise, definition. A sub-group of any grade is framed, without taking into consideration any new or additional characters, upon the *varying prominence* of one or more of the characters just used to form the group next above. In our formula above x (abc) for a certain family of the order x, suppose the family character a to be *emphasized*, as it were, and to predominate over b and c, to the partial suppression of these last: then a sub-family of x (abc) might be expressed thus: — x (Abc); and it is further evident, that there will be as many sub-families as there are groups of birds in the family representing varying emphasis of a, or b, or c; as x (a B c), x (A b C), etc. While we take account of *new* characters of another grade, in forming our successive main groups, in our sub-groups, then, we recognize only *more or less* of the same characters. But the distinction is not always evident; nor is it observed so often as, perhaps, it should be.

§ 24. TYPICAL AND ABERRANT GROUPS. Waiving what might be reasonably argued against considering any group specially "typical" of the next higher, we may define a convenient and frequent term: —The *typical* genus of a family, or family of an order, is that one which develops most strongly, or displays most clearly, the more essential characters of the next higher group, of which it is one member. And in proportion as it fails to express these in the most marked manner, either by bearing their stamp more lightly, or by having it obscured or defaced by admixture of the characters of a neighboring group, does it become less and less typical ("subtypical") and finally *aberrant*. Suppose the ordinal symbol x, as before, to represent the sum of various ordinal characters, more or less essential to the integrity of the order; then obviously, the family characters abc, or def may be combined with a varying value of x; thus, x^1 (abc) or x^2 (def) and the formula of the *typical* family would be x^n ($a—f$). Thus, it is characteristic of most thrushes (*Turdidæ*) to have the tarsus booted, but all do not have it so; therefore, in subdividing the family, we properly make a division into thrushes with booted tarsi, and thrushes with scutellated tarsi; the former are *typical* of the family, the latter sub-typical or even aberrant.

§ 25. EQUIVALENCE OF GROUPS. It may sound like a truism to say, that groups of the same grade, bearing the same name, whatever that may be, from sub-class to sub-genus, must be of the same value; must be distinguished by characters of equal or equivalent importance. *Equivalence of groups* is necessary to the stability and harmony of any classificatory system. It will not do to frame an order upon one set of characters here, and a family upon a similar set of characters there; but order must differ from order, and family from family, by an equal or corresponding amount of difference. Let a group called a family differ as much from the other families in its own order as it does from *some* other order, and it is by this very fact *not* a family, but an order itself. Let the orders of birds stand apart a yard, say; if, then, any families, so-called, stand as far apart, they are not families. It seems a simple proposition, yet it is too often ignored, and always with ill result. Two points should be remembered here: first, that the absolute size or bulk of a group has nothing to do with its grade; one order might contain a thousand species, and another only one, without having its ordinal value disturbed. Secondly, any given character may be of different value in its application to different groups. Thus, number of primaries, whether nine or ten, is a family character almost throughout *Oscines;* but in one Oscine family, *Vireonidæ*, it is scarcely a generic feature. It is difficult, however, to determine such a point as this last without faithful training in ornithology.

§ 26. AFFINITY AND ANALOGY. Birds are allied, or *affined*, according to the number of like characters they employ for like purposes; they are *analogically related* according to the number of unlike characters that they use for similar purposes. A loon and a cormorant, for instance, are closely affined, because they are both fitted in the same way for the pursuit of their prey under water. A dipper (family *Cinclidæ*), and a loon (family *Colymbidæ*), are analogous, because they both pursue their prey under water; but they stand almost at the extremes of the ornithological system; they have almost no affinity beyond their common birdhood; totally different structure is only modified for the same ends, that are thus brought about by totally different means. So the wings of a butterfly, a bat, and a bird are analogical, because they subserve the same purpose in each case; needless to add, these creatures have no affinity.

§ 27. With this cursory glance* at some taxonomic principles I pass to a brief explanation of modifications of external characters alone; some knowledge of which is necessary to the slightest appreciation of ornithological definitions and descriptions. I shall confine myself mainly to consideration of those that the student will need to understand in order to use the present

---

* As the present occasion obviously affords no opportunity for an adequate discussion of the classification of birds, it is hardly necessary to say to ornithologists, that here I simply assume a class *Aves* composed of recent birds, as an initial step, without considering the broader generalizations deducible from extinct forms; and that I speak of species and varieties, in the sense in which these terms are commonly used, waiving the biological questions involved.

volume easily and successfully. Here, however, I will insert a tabular illustration of the foregoing remarks:—

Class AVES: — Birds.

  (Sub-class* *Insessores:* — Perching Birds.)

    Order PASSERES: — Passerine Perchers.

      (Sub-order *Oscines:* — Singing Passerines.)

        Family † TURDIDÆ: — Thrushes.

          (Sub-family † *Miminæ:* — Mocking Thrushes.)

            Genus ‡ MIMUS: — Mockers.

              (Sub-genus ‡ *Mimus:* — Typical Mockers.)

                Species ‡ POLYGLOTTUS: — Many-tongued.

                  (Variety *caudatus:* — Long-tailed.)

SECT. III. DEFINITION AND BRIEF DESCRIPTION OF THE EXTERIOR OF A BIRD. — PARTS AND ORGANS — I. THE BODY: HEAD, NECK AND BODY PROPER. — II. THE MEMBERS: BILL, WINGS, TAIL, FEET.

§ 28. THE CONTOUR of a bird with the feathers on, is spindle-shaped, or *fusiform*, tapering at both ends; it represents two cones, joined base to base at the middle, or greatest girth of body, tapering in front to the tip of the bill, behind to the end of the tail. *Obvious design:* easiest cleavage of air in front, and lessening of drag or wash behind. But this shape is largely produced by the lay of the plumage; a

§ 29. 'NAKED BIRD presents several prominences and depressions; this irregular contour is reducible, in general terms, to *two* double cones. The head tapers to a point in front, at the tip of the bill; and nearly to a point behind, towards the middle of the neck, in consequence of the swelling muscles by which it is slung on the neck; from the middle of the somewhat contracted or hour-glass shaped neck, this last enlarges toward the body, by the swelling of the muscles by which it is slung to the body; the body then tapers to the tail. The

§ 30. EXTERIOR OF A BIRD is divided into seven parts: 1, head (*caput*), 2, neck (*collum*), 3, body (*truncus*), 4, bill (*rostrum*), 5, wings (*alæ*), 6, tail (*cauda*), 7, feet (*pedes*): 1, 2, 3, are collectively called "body," in distinction to 4, 5, 6, 7, which are *members.* The

---

* Intermediate groups are in italics and parentheses.

† *Families* now always end in *-idæ*, and sub-families in *-inæ*, a very convenient distinction, since we thus always know the rank designated by words so ending.

‡ A bird's scientific name now INVARIABLY consists of two words — the genus and the species, the former first, the latter last: thus, *Mimus polyglottus;* but we may, if we wish, interpolate the sub-genus in parentheses, and affix the variety with sign var.: thus, *Mimus (Mimus) polyglottus,* var. *caudatus.* Generic names are *always* written with a capital; specific names, according to the rules of the British Association, now generally followed, should never be, though it is customary to so write those that are derived from the names of persons and places, as well as all substantive appellations.

§ 31. HEAD has the general shape of a 4-sided pyramid; of which the base is applied to the end of the neck, and does not appear from the exterior; the uppermost side is more or less convex or vaulted, sloping in every direction, and tapering in front; the sides proper are flatter, more or less perpendicular, and taper in front; the bottom is likewise flattish and similarly tapering. The departures from this typical shape are endless in degree, and variable in kind; they give rise to numerous *general* descriptive terms, as "head flattened," "head globular," etc., but these are not susceptible of precise definition. The sides present each two openings, *eyes* and *ears;* their position is variable, both absolutely and in respect to each other. But in the vast majority of birds, the eyes are strictly *lateral*, and near the middle of the side of the head, while the ears are behind and a little below. Exceptions:—owls have eyes "anterior;" woodcock and snipe have ears below and not behind the eyes. The *mouth* is always a horizontal fissure in the apex of the cone; there are no other openings in the head proper, for the nostrils are always in the bill. The

§ 32. NECK, in effect, is a simple cylinder: rendered somewhat hourglass-shaped as above stated. Its length is variable, as is the number of bones it has. Bearing the head with the bill, which is a bird's true *hand*, it is unusually *flexible*, to permit the necessarily varied motions of this important organ. Its least length may be said to be that which allows the point of the bird's beak to touch the oil-gland on the rump: its length is usually in direct proportion to length of legs, in obvious design of allowing the beak to touch the ground easily to pick up food. Its habitual shape is a double curve like the letter S; the lower belly of the curve fits in the space between the legs of the merry-thought (*furcula*); the upper limb of the curve holds the head horizontal. This sigmoid flexure (*sigma*, Greek S) is produced by the shape of the jointing surfaces of the several bones: it may be increased, so that the upper end touches the lower belly; may be decreased to a straight line, but is scarcely carried beyond this in the opposite direction. As a generalization, the neck may be called longest in wading birds; shortest in perching birds; intermediate in swimming birds; but some waders, as plovers, have short necks; and some swimmers, as swans, extremely long ones; a very long neck, however, among perching birds is rare, and confined mainly to a crane-like African hawk, and certain of the lowest perchers that stand on the confines of the waders. The shape of the

§ 33. BODY PROPER or trunk (L. *truncus*), is obviously referable to that of the egg; it is *ovate*, (L. *ovum*, an egg). The swelling breast muscles represent the but of the egg, which tapers backwards. But this shape is never perfectly expressed, and its variations are unnumbered. In general, perching birds have a body the nearest to an oval; among waders, the oval is usually *compressed*, or flattened perpendicularly, as is well seen in the heron family, and still better in the rail family, where the narrowing is at an extreme; among swimmers, the body is *always* more or less *depressed*, or flattened horizontally, and especially underneath, to enable these birds to rest with stability on the water; a duck or a diver shows this well. Speaking of shape of body, I must allude to the

§ 34. CENTRE OF GRAVITY of a bird, and show the admirable provision by which this is kept beneath the centre of the body. The enormous breast-muscles of a bird are its heaviest parts; sometimes they weigh, to speak roundly, as much as one-sixth of the whole bird. Now these are they that effect all the movements of the wings at the shoulder-joint, lifting as well as lowering the wings; did they all pull straight, the lifters would have to be *above* the shoulder; but they all lie below, and the lifters accomplish their office by running through a pulley, which changes their line of traction; they work, in short, like men hoisting sails from the deck of a vessel; and thus, like a ship's cargo, a bird's chief weight is kept below the centre of motion. Topheaviness is further obviated by the fact that birds with a long, heavy neck and head draw this in upon the breast, and extend the legs behind, as is well shown in a heron flying. The nice adjustment of balance by the variable extension of the head and legs is exactly like that produced by shifting the weight along the bar of a steel-yard; this, with the slinging of the chief weight under the wings instead of over or even between them, enables a bird to keep right side up in flight, without exertion.

Sub-sect. 1.   Of the Body; its Topography, etc.

§ 35. BESIDES being divided as above into body and members, the exterior of a bird is further subdivided; the body being mapped out, mainly for purposes of description, into *regions*, and the members being similarly resolved into their component parts or organs. We have first to notice, as the most general, the

§ 36. UPPER AND UNDER PARTS. Draw a line from the corner of the mouth along the side of the neck to and through the shoulder-joint and thence along the side of the body to the root of the tail; all above this line, including upper surface of wings and tail, are *upper parts;* all below, including under surfaces of wings and tail, are *under parts;* called respectively, "above" and "below." The distinction is purely arbitrary, but so convenient that it is practically indispensable; for it will be seen in a moment, how an otherwise lengthy description can be compressed into, for example, four words: "above, green; below, yellow:" and these terms are often used because many birds' colors have some such simple *general* character. The "upper parts" of the body proper (§ 33) have, also, received the general name of *notæum* (Gr. *notos*, back: fig. 4, 12): the "under parts," similarly restricted, that of *gastræum* (Gr. *gaster*, belly; fig. 4, 20). These two are

§ 37. NEVER NAKED, while both head and neck may be variously bare of feathers. The only exception is the transient condition of certain birds during incubation: when, either, like the eider duck, they pull feathers off the belly to cover the eggs or even to build the nest, or, like several other birds, the plumage below is worn off in setting. The gastræum is rarely pecu-

liarly ornamented with feathers of different texture or structure from those
of the general plumage; but an instance of this is seen in our Lewis' woodpecker. The notæum, on the contrary, is often the seat of extraordinary
development of feathers, either in size, shape or texture; as the singularly
elegant plumes of the herons. Individual feathers of the notæum are
generally pennaceous (§ 4), in greatest part straight and lanceolate; and

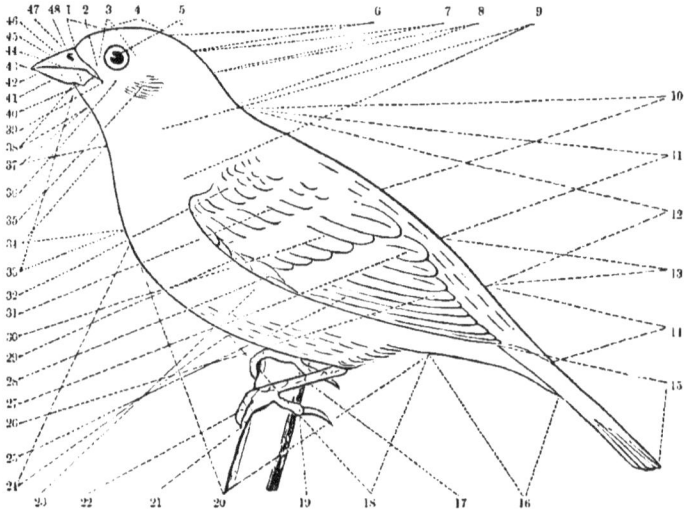

FIG. 4. — Topography of a Bird.

1, forehead (*frons*). 2, lore. 3, circumocular region.
4, crown (*vertex*). 5, eye. 6, hind head (*occiput*). 7,
nape (*nucha*). 8, hind neck (*cervix*). 9, side of neck.
10, interscapular region. 11, *dorsum*, or back proper, including 10. 12, *notæum*, or upper part of body proper,
including 10, 11, and 13. 13, rump (*uropygium*). 14, upper
tail coverts. 15, tail. 16, under tail coverts. 17, tarsus.
18, abdomen. 19, hind toe (*hallux*). 20, *gastræum*, including 18 and 24. 21, outer or fourth toe. 22, middle or third
toe. 23, side of the body. 24, breast (*pectus*). 25, primaries. 26, secondaries. 27, tertiaries; nos. 25, 26, 27 are all
remiges. 28, primary coverts. 29, *alula*, or bastard wing.
30, greater coverts, 31, median coverts. 32, lesser coverts.
33, the "throat," including 34, 37, 38. 34, *jugulum* or lower
throat. 35, auriculars. 36, malar region. 37, *gula*, or middle throat. 38, *mentum*, or chin. 39, angle of commissure, or corner of mouth. 40, ramus of under mandible.
41, side of under mandible. 42, *gonys*. 43, *apex*, or tip of
bill. 44, *tomia*, or cutting edges of the bill. 45, *culmen*,
or ridge of upper mandible, corresponding to gonys. 46,
side of upper mandible. 47, nostril. 48 passes across the
bill a little in front of its *base*.

as a whole they lie smoothly *imbricated* (like shingles on a roof). The
gastræal feathers are more largely plumulaceous (§ 4), less flat and imbricated, but even more compact, that is, thicker, than those of the upper
parts; especially among water birds, where they are all more or less curly,
and very thickset. There are subdivisions of the

§ 38. NOTÆUM. Beginning where the neck ends, and ending where the
tail coverts begin, this part of the bird is divided into back (Lat. *dorsum;*
fig. 4, 11) and rump (L. *uropygium* fig. 4, 13). These are direct continuations
of each other, and their limits are not precisely defined. The feathers of
both are on the *pteryla dorsalis* (§ 8, *b*). In general, we may say that the
anterior two-thirds or three-quarters of notæum is back, and the rest rump.
With the former are generally included the scapular feathers, or *scapulars:*
these are they that grow on the *pterylæ humerales* (§ 8, *b*): the region of

notæum that they form is called *scapulare* (L. *scapula*, shoulder-blade) ;
that part of notæum strictly between them is called *interscapulare* (fig. 4, 10) ;
it is often marked, as in the chipping sparrow, with streaks or some other
distinguishing coloration. A part of dorsum, lying between interscapulare
and uropygium, is sometimes recognized as the "lower back" (L. *tergum*),
but the distinction is not practically useful. To uropygium probably also
belong the feathers of the *pterylæ femorales* (§ 8, *b*), and at any rate they
are practically included there in descriptions; but these properly represent
the *flanks* (L. *hypochondria*), that is, the sides of the rump. They are
sometimes the seat of peculiarly developed or otherwise modified feathers.
The whole of notæum, taken with the upper surfaces of the folded wings, is
called the "mantle" (L. *stragulum*), and is often a convenient term, especially in describing gulls. In like manner, the

§ 39. GASTRÆUM is subdivided into regions, called, in general terms,
"breast" (*pectus;* fig. 4, 21), "belly" (*abdomen;* fig. 4, 18) and "sides of the
body" (fig. 4, 22). The latter belong really as much to back, of course, as to
belly ; but in consequence of the underneath freighted shape of a bird's body,
the line we drew (§ 36) passes so high up along the sides, that these last are
almost entirely given to gastræum. The *breast* begins over the merry-
thought, where *jugulum* (§ 40) ends; on either hand it slopes up into
"sides :" behind, its extension is indefinite. Properly, it should reach as far
as the breast-bone (*sternum*) does; but this would leave, in many birds,
almost nothing for abdomen, and the limit would, moreover, fluctuate with
almost every family of birds, the sternum is so variable in length and shape.
Practically, therefore, we restrict pectus to the *swelling* anterior part of
gastræum, which we call abdomen as soon as it begins to straighten out and
flatten. Abdomen, like breast, rounds up on either hand into *sides;* behind,
it ends in a transverse line that passes across the anus. It has been unnecessarily divided into *epigastrium*, or "pit of the stomach," and *venter*,
or "lower belly;" but these terms are rarely used. ("*Crissum*" is a word
constantly employed for a region immediately about the anus; but it is
loosely used, sometimes including the hypochondria, and oftener meaning
simply the under tail coverts; I refer to it again in speaking of these last.)
Although these various boundaries seem fluctuating and not perfectly definite, yet a little practice will enable the student to appreciate their proper
use in descriptions, and then use them himself with sufficient accuracy.
The anterior continuation of body in general, or the

§ 40. NECK, is likewise subdivided into regions. Its lateral aspects (except in a few birds that have lateral neck tracts of feathers) are formed
by the meeting over its sides of the feathers that grow on the dorsal and
ventral pterylæ; the skin is really not planted with feathers; and partly on
this account, perhaps, a distinctively named region is not often expressed;
we say simply "sides of the neck" (*parauchenia*, fig. 4, 9). Behind, it is
divided into two portions : a lower, the "hind neck," or "scruff of the neck,"
cervix (fig. 4, 8), adjoining the back; and an upper, the "nape of the ·

neck" (*nucha*; fig. 4, [7]), adjoining the hind head; these are otherwise known
as the *cervical region*, and the *nuchal region*, respectively, and both together
as "the neck behind." The front of the neck has been, perhaps, unnecessarily subdivided, and the divisions vary with almost every writer. It will be
sufficient for us in the present connection to call it *throat* (Lat. *gula*, fig.
4, 37), and *jugulum* (fig. 4, 34), remembering that the jugular portion is
lower, vanishing in breast, and the gular higher, running into chin along the
under surface of the head. *Guttur* is a term used to signify gula and
jugulum together; it is simply equivalent to "throat" as just defined.
Though generally fully covered with feathers, the neck, unlike the body
proper, is frequently in part naked. When naked *behind*, it is almost invariably *cervix* that is bare, from interruption of the upward extension of the
pteryla dorsalis; as exemplified in many herons. *Nucha* is rarely, if ever,
naked except in continuation of general nakedness of the head. Similarly,
gula is naked from above downwards, as is especially illustrated in nearly
all the order *Steganopodes*, as pelicans, cormorants, etc., that have a naked
throat-pouch; or some vultures, whose nakedness of head extends over
nucha, and along gula, as if the feathers were killed by over-manuring with
the filthy substances these birds eat. The condor has a singular ruffle all
around the neck, of close, *downy* feathers, as if to defend the roots of the
other feathers from such consequence. Jugulum becomes naked in a few
birds, where a distended crop or craw protrudes, pushing apart the feathers
of two branches of the pteryla ventralis as these ascend the throat. The
neck is not ordinarily the place of remarkably modified feathers; they might
restrict freedom of motion in the neck; to this rule, however, there are
signal exceptions. Among these may be mentioned here, the grouse family,
among our representatives of which, the "ruffed" has singular tufts on the
sides of the neck; the "pinnated" little wing-like feathers there, covering
bare, distensible skin, and the "cock of the plains" has curious, stiff, scaly
feathers; unless these rather belong to pectus. Cervix proper almost never
has modified feathers, but often a transverse coloration different from that of
the rest of the upper parts; when conspicuous, this is called "cervical collar,"
to distinguish it from the guttural or jugular "collars" or rings of color.
Nucha is frequently similarly marked with a "nuchal band;" often, special
developments there take the form of *lengthening* of the feathers, and we
have a "nuchal crest." More particularly in birds of largely variegated
colors, guttur and jugulum are marked *lengthwise* with stripes and streaks,
of which those on the sides are apt to be different from those along the
middle line in front. Jugulum occasionally has lengthened feathers, as in
many herons. Higher up, the neck in front may have variously lengthened or otherwise modified feathers. Conspicuous among these are the
ruffs, or tippets, of some birds, especially of the grebe family, and, above
all our other birds, of the male ruff (*Machetes pugnax*). But these, and a
few other modifications of the feathers of the upper neck, are more conveniently considered with those of the

§ 41. HEAD. Though smaller than either of the parts already considered, the head has been more minutely mapped out, and such detail is necessary from the number of recognizable parts or regions it includes. Without professing to give all that have been named, I describe what will be needed for our present purposes.

(a). "Top of the head" is a collective expression for all the superior surface, from base of the bill to nucha, and on the sides nearly or quite to the level of the upper border of the eyes. This is *pileum* (fig. 4, 1, 4, 6); it is divided into three portions. *Forehead*, or frontal region, or, simply, "the front" (L. *frons;* fig. 4, 1) includes all that slopes upward from the bill — generally to about opposite the anterior border of the eyes. *Middle head* or crown (L. *corona*), or vertex (L. *vertex;* fig. 4, 4), includes the top of the head proper, extending from forehead to the downward slope towards nucha. This last slope is hind head, or *occiput* (fig. 4, 6). The lateral border of all three together constitutes the "superciliary line," that is, line over the eye (Lat. *super*, over, *cilia*, hairs [of the brows in particular]).

(b). "Side of the head" is a general term defining itself. It presents for consideration the following regions: *orbital*, or circumorbital (L. *orbis*, an orb, properly, here, the circular hole in the skull itself that contains the eyeball; fig. 4, 3) is the small space forming a ring around the eyes; it embraces these organs, with the upper and under *lids* (L. *palpebræ*); where these meet in front and behind respectively, is the *anterior canthus* and *posterior canthus*. The region is also subdivided into supra-orbital, infra-orbital, ante-orbital and post-orbital, according as its upper, under, front or back portion is specially meant. The position of the circumorbital varies in different families; generally, it is midway, as stated, but may be higher or lower, crowded forward toward the base of the bill, or removed to the back upper corner of the side of the head, as strikingly shown in the woodcock. The *aural* or *auricular* (fig. 4, 35) region is the part lying over the external ear-opening; its position varies in heads of different shape; but in the vast majority of cases it is situated a little behind and below the eye. Wherever located it may be known at a glance, by the texture of the auricular feathers (shortly, the *auriculars*) covering the opening. Doubtless to offer least obstacle to passage of sound, these are a tuft of feathers with loose vexilla (§ 3) from greater or less disconnection of the barbs (§ 3); and they may collectively be raised and turned forward, exposing the ear-opening; they are extremely large and conspicuous in most owls. "Temporal region," or the temples (L. *tempora*, times, or age, because an elderly man's hair whitens there first) is a term not often used; it designates the part between eyes and ears, not well distinguished from the post-orbital space. At the lowermost posterior corner of the head a protuberance is seen, or may be felt; it is where the lower jaw is hinged to the skull, and is called the "angle of the jaw;" it is generally just below and behind the ear. The *lore* (L. *lorum*, strap or thong; hence, reins or bridle; hence, place where the main strap of a bridle passes; fig. 4, 2) is an important region. It is generally pretty

much all the space betwixt the eye and the sides of the base of the upper
mandible (§ 44). Thus, we say of a hawk, "lores bristly;" and examination of a bird of that kind will show how large a space is covered by the
term. Lore, however, should properly be restricted to a narrow line
between the eye and bill in the direction of the nostrils. It is excellently
shown in the heron and grebe families, where "naked lores" is a distinctive
family character. The lore is an important place, not only from being thus
marked in many birds, but from being frequently the seat of specially
modified or specially colored feathers. The rest of the side of the head,
including the space between angle of jaw and bill, has the name of *cheek*
(L. *gena*, firstly eyelid, then, and generally, the prominence under the eye
formed by the cheek-bones; fig. 4, 36). It is bounded above by lore, infra-
orbital, and auricular; below, by a more or less straight line, representing
the lower edge of the bony prong of the under mandible (§ 44). It is
cleft in front for a varying distance by the backward extension of the gape
of the mouth; above this gape is more properly *gena*, or *malar region* in
strictness; below it is *jaw* (*maxilla*), or rather "side of the jaw." The
lower edge of the jaw definitely separates the side of the head from the

(c). "Under surface" of the head; properly bounded behind by an imag-
inary line drawn straight across from one angle of the jaw to the other, and
running forward to a point between the forks of the under mandible (§ 44).
As already hinted, "throat" (*gula*; fig. 4, 37) extends upward and forward
into this space without obvious dividing line; it runs into *chin* (L. *mentum*;
fig. 4, 38) of which it is only to be said, that it is the (varying in extent)
anterior part of the under surface of the head. Anteriorly, it may be conveniently marked off, opposite the point where the feathers end on the side
of the lower jaw, from the feathery space (when any) *between* the branches
of the under mandible itself; this latter space is called the *interramal* (L.
inter, between, *ramus*, fork).

(d). The head is so often marked lengthwise with different colors, apt to
take such definite position, that these lines have received special names.
Median vertical line is one along the middle of pileum, from base of culmen
(§ 50) to nucha; *lateral vertical lines* bound it on either side. *Superciliary
line* has just been noticed; below it runs the *lateral stripe;* that part of it
before the eye, is loral or ante-orbital; behind the eye, post-orbital; when
these are continuous through the eye, they form a *transocular line;* below
this is *malar line*, cheek-stripe, or *frenum;* below this, on the under jaw,
maxillary line; in the middle below, *mental* or *gular lines.* The lines are
stripes (L. *plagae*) when narrow and distinct, like the welt of a whip-lash;
streaks (*striae*) when narrow and somewhat erratic; and *vittae* or *fasciae* when
quite broad, as is particularly likely to be the case with the eye-line.*

---

* I had thought of a section on *patterns of coloration* (*pictura*), but the attempt to reduce birds' infinitely
varied colors to generalized formulas would take too much space. I may add, however, conveniently in
this connection, the following: Considerable areas of color take name from the parts they occupy, down
to what may be called variegations. These are produced in two ways: (1) by insensible change of colors, either
in fading into lighter, or shading into darker tints of the same; as an indefinite brown into black, gray or

(e). No part of the body has so variable a ptilosis (§ 9) as the head. In the vast majority of birds, it is wholly and densely feathered; it ranges from this to wholly naked; but nakedness, it should be observed, means only absence of perfect feathers, for most birds with unfeathered heads have a hair-like growth on the skin. Our samples of naked headed birds, are the turkey, the vultures, the cranes, and some few birds of the heron tribe. Associated with more or less complete "baldness," is frequently the presence of various fleshy outgrowths, as *combs*, *wattles*, *caruncles* (warty excrescences), *lobes* and *flaps* of all sorts, even to enumerate which would exceed our limits. The parts of the barn-yard cock exemplify the whole; among North American birds they are very rare, being confined, in evident development at any rate, to the wild turkey. Sometimes *horny plates* take the place of feathers on part of the head; as in the coots and gallinules. A very common form of head nakedness marks one whole order of birds, the *Steganopodes*, which have mentum and more or less of gula naked, and transformed into a sort of pouch, extremely developed in the pelicans, and well seen in the cormorants. The next commonest is definite bareness of the lores, as in all herons and grebes. A little orbital space is bare in many birds, as the vulturine hawks, and some pigeons. Among water birds particularly more or less of the interramal space is almost always unfeathered; the nakedness always proceeds from before backwards. With the rare exceptions of a narrow frontal line, and a little space about the angle of the mouth, no other special parts of the head than those above given are naked in any North American bird, unless associated with general baldness.

(f). The opposite condition, that of redundant feathering, gives rise to all the various CRESTS (L., pl. *cristæ*) that form such striking ornaments of many birds. Crests proper belong to the top of the head, but may be also held to include those growths on its side; these together being called crests in distinction to the ruffs, ruffles, beard, etc., of gula or mentum. Crests may be divided into two kinds:—1, where the feathers are simply lengthened or otherwise enlarged, and 2, where the texture, and sometimes even the structure (§ 4) is altered. Nearly all birds possess the power of moving and elevating the feathers on the head, simulating a slight crest in moments of excitement. The general form of a crest is a full soft elongation of the coronal feathers collectively; when perfect such a crest is *globular*, as in the *Pyrocephalus* (genus 111); generally, however, the feathers lengthen on

white; or by unmarked change of a secondary color, as green into blue or yellow. (2) by obvious *markings*. Markings are all reducible to two kinds, *streaking* and *spotting*. Streaking, as a generic term, is sharply divisible into *lengthwise* and *crosswise*. *Lengthwise streaking* comprehends all kinds of streaks, stripes, vittæ, fasciæ, with the distinctions above given in the text. *Crosswise streaking* is called *barring*, and always runs transverse to the axis of a bird; if the lines are straight, it is *banding*; if undulating, it is *waving*; if very fine and irregular, it is *vermiculation* (L. *vermiculus*, a little worm). *Spotting* is graded according to size of the markings, from dotting or pointing, to blotching or splashing; and spots are also designated according to their shape, as round, square, U-shaped, V-shaped, hastate, sagittate. etc. Very fine spotting mixed with streaking, is called *marbling*; when indistinct, *nebulation* or *clouding*; intermediate special marks have particular names, as crescents. Distinct round spots are *ocelli* ("little eyes"). Indistinct variegations of any sort are called *obsolete*. Washes of color over a definite color, are called *tinges* or tints. Color is *glossy* when it shines; *metallic*, when it glitters; *iridescent* when it changes with different lights. Colors are also *bright*, *dull*, *dead* (said of white), *opaque*, or *returty* (said of deep colors, chiefly black), etc.

the occiput more than on the vertex or front, and this gives us the simplest and commonest form. Such crests, when more particularly occipital, are usually connected with lengthening of nuchal feathers, and are likely to be of a thin, pointed shape, as well shown in the kingfisher. Coronal or vertical crests proper, are apt to be rather different in coloration than in specially marked elongation of the feathers; they are perfectly illustrated in the kingbird, and other species of that genus. Frontal crests are the most elegant of all; they generally rise as a pyramid from the forehead, as excellently shown in the blue jay, cardinal bird, tufted titmouse and others. All the foregoing crests are generally single, but sometimes double; as shown in the two lateral occipital tufts of the "horned" lark, in all the tufted or "horned" owls, and in a few cormorants. Lateral crests are, of course, always double, one on each side of the head; they are of various shapes, but need not be particularized here, especially since they mostly belong to the second class of crests — those consisting of texturally modified feathers. It is a general — though not exclusive — character of these last, that they are *temporary;* while the other kind is only changed with the general moult, these are assumed for a short season only — the breeding season; and furthermore, they are often distinctive of *sex.* Occurring on the top of the head they furnish the most remarkable ornaments of birds. I need only instance the elegant helmet-like plumes of the partridges of the genus *Lophortyx* (186); the graceful flowing train of the *Oreortyx* (gen. 185); the somewhat similar plumes of the night and other herons. The majority of the cormorants, and many of the auks, possess lateral plumes of similar description; these, and those of the herons are probably — in most cases certainly — *deciduous;* while those of the partridges above mentioned last as long as the general plumage. These lateral plumes, in many birds, especially among grebes, are associated with, and, in fact, coalesce with, the ruffs, which are singular lengthening and modifying in different ways of feathers of auriculars, genæ and gula; and are almost always temporary. *Beards,* or special lengthening of the mental feathers alone, are comparatively rare; we have no good example among our birds, but a European vulture, *Gypaëtos barbatus,* is one. The feathers sometimes become *scaly* (*squamous*) forming, for instance, the exquisite gorglets or frontlets of humming-birds. They are often *bristly* (*setaceous*), as about the lores of nearly all hawks, the forehead of the dabchick, meadowlark, etc. While usually all the unlengthened head feathers point backward, they are sometimes *erect,* forming a velvety pile, or they may radiate in a circle from a given point, as from the eye in most owls, where they form a *disk.*

In the foregoing, I only mention a few types, chiefly needed to be known in the study of our birds; but should add that there are many others, with endless modifications, among exotic birds; to these, however, I cannot even allude by name. Peculiarities of nasal feathers, and others around the base of the bill, are noticed below. Forms of crests are illustrated in figs. 21, 22, 23, 24, 32, 56, 95, 96, 107, 109, 114, 117, 125, 127, 135, 136, 152, 153, 154, 177, 191, 202.

Sub-sect. 2. *Of the Members; their parts and organs.*

1. THE BILL.

§ 42. THE BILL is hand and mouth in one : the instrument of *prehension.* As hand, it takes, holds and carries food or other substances, and in many instances, *feels;* as mouth, it tears, cuts, or crushes, according to the nature of the substances taken ; assuming the functions of both lips and teeth, neither of which birds possess. An organ thus essential to the prime functions of birds, one directly related to their various modes of life is of the utmost consequence in a taxonomic point of view ; yet, its structural modifications are so various and so variously interrelated, that it is more important in framing families and genera than orders ; more *constant* characters must be employed for the higher groups. The general

§ 43. SHAPE of the bill is referable to the *cone;* it is the anterior part of the general cone that we have seen to reach from its point to the base of the skull. This shape confers the greatest strength combined with the greatest delicacy ; the end is fine to apprehend the smallest objects, while the base is stout to manipulate the largest. But in no bird is the cone expressed with entire precision ; and in most, the departure from this figure is great. The bill ALWAYS consists of two, the upper and the lower

§ 44. MANDIBLES (fig. 5), which lie, as their names indicate, above and below, and are separated by a horizontal fissure — the mouth. Each mandible ALWAYS consists of certain projecting skull-bones, sheathed with more or less *horny* integument in lieu of true skin. The frame-work of the Upper mandible is (chiefly) a bone called the *intermaxillary*, or better, in this case, the *premaxillary.* In general, this is a three-pronged or tripodal bone running to a point in front, with one, the uppermost prong, or foot, implanted upon the forehead, and the other two, lower and horizontal, running into the sides of the front of the skull. The scaffold of the Under mandible is a compound bone called *inferior maxillary;* it is U-, or V-shaped, with the point or convexity in front, and the prongs running to either side of the base of the skull behind, to be there movably hinged. These two

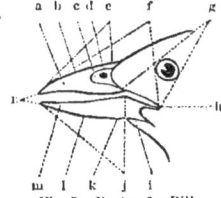

Fig. 5. Parts of a Bill.

a, side of upper mandible; *b*, culmen; *c*, nasal fossa; *d*, nostril; *e*, (see below); *f*, gape, or whole commissural line; *g*, rictus; *h*, commissural point or angle of the mouth; *i*, ramus of under jaw; *j*, tomia of under mandibles (the reference lines *e* should have been drawn to indicate the corresponding tomia of upper mandible); *k*, angle of gonys; *l*, gonys; *m*, side of under mandible; *n*, tips of mandibles.

bones, with certain accessory bones of the upper mandible, as the *palate* bones, etc., together with the horny investment, constitute the JAWS. *Both* jaws, in birds, are *movable;* the under, by the joint just mentioned ; the upper, either by a joint at, or by the elasticity of the bones of, the forehead ; it is moved by a singular muscular and bony apparatus in the palate, further notice of which would involve anatomical details. When closed, the jaws meet and fit along their opposed-edges or surfaces, in the same manner, and for the same purposes, as the lips and teeth of man or other vertebrate animals.   All bills, thus similarly constituted, have been divided * into

§ 45. FOUR CLASSES, representing as many ways in which the two mandibles close upon each other at the end. 1. The *epignathous* (Gr. *epi*, upon, *gnathos*, jaw) way, plan, or type, in which the upper mandible is longer than the under, and its tip is evidently bent down over the tip of the lower. 2. The *hypognathous* (Gr. *hypo*, under), in which the lower mandible is longer than the other. 3. The *paragnathous* (Gr. *para*, at or by), in which both are of about equal length, and neither is evidently bent over the other. 4. The *metagnathous* (Gr. *meta*, with, beside, etc.), in which the points of the mandibles cross each other. The second and fourth of these are extremely rare; they are exemplified, respectively, by the skimmer and the cross-bill (genera 295 and 60). The first is common, occurring throughout the birds of prey, the parrots, and among the petrels, gulls, etc., etc. The great majority of birds exhibit the third; and among them, there is such evident gradation into epignathism, that it is necessary to restrict the latter to its complete development, exhibited in the intermaxillary bone divested of its horny sheath, which often, as among flycatchers, etc., forms a little overhanging point, but does not constitute epignathism. These classes, it should be added, though always applicable, and very convenient in descriptions, are purely arbitrary, that is, they by no means correspond to any four primary groups of birds, but on the contrary, usually only mark families and the subdivisions of families; and the four types may be seen in contiguous genera. The general shape of the bill has also furnished

§ 46. OTHER CLASSES, for many years used as a large basis for ornithological classification; but which the progress of the science has shown to be merely as convenient as, and only less arbitrary than, the foregoing. The principal of these are represented by the following types : — A, among land birds. 1. The *fissirostral*, or cleft, in which the bill is small, *short*, and with a very large gap running down the side of the head, as in the swallow, chimney-swift, whippoorwill. 2. The *tenuirostral*, or slender, in which the bill is small, *long*, and with a short cleft; as in the humming-bird, creeper, nuthatch. 3. The *dentirostral*, or toothed, in which, with a various general shape, there is present a nick, tooth, or evident lobe in the opposed edges of one or both mandibles near the end; as in the shrike, vireo, and some wrens, thrushes and warblers. 4. The *conirostral*, or conical, sufficiently defined by its name, and illustrated by the great finch family and some allied ones. B, among water birds. 5. The *longirostral*, or long, an aquatic style of the tenuirostral, best exhibited in the great snipe family. 6. The *pressirostral*, or the compact, illustrated by the plovers, etc., and quite likely analogous to the conirostral. 7. The *cultrirostral*, cutting, perhaps analogous to the dentirostral, exemplified in the heron group. None of these are now used to express natural groups, in strict definitions; all are.

---

* By the writer: Proc. Acad. Nat. Sci. Phila., Dec. 1869, p. 213.

convenient incidental terms in general descriptions. Various other lesser terms, expressing special modifications, as *lamellirostral*, *acutirostral*, etc., are employed; but all are best used, now, as common, not as proper names, simply descriptive of

§ 47. OTHER FORMS. A bill is called *long*, when notably longer than the head proper; *short*, when notably shorter; *medium*, in neither of these conditions. It is *compressed*, when higher than wide, at the base at least, and generally for some portion of its length; *depressed*, when wider than high; *terete*, under neither of these conditions. It is *recurved*, when curved upward; *decurved*, when curved downward; *bent*, when the variation in either direction is at an angle; *straight*, when not out of line with axis of the head. A bill is *obtuse* (said chiefly of the paragnathous sort) when it rapidly comes to an end that therefore is not fine; or when the end is knobby; it is *acute* when it runs to a sharp point; *acuminate*, when equally sharp and slenderer; *attenuate*, when still slenderer; *subulate* (awl-shaped), when slenderer still; *acicular* (needle-shaped), when slenderest possible, as in some humming-birds. A bill is *arched*, *vaulted*, *turgid*, *tumid*, *inflated*, etc., when its outlines, both crosswise and lengthwise, are notably more or less convex; and *contracted*, when some, or the principal, outlines are concave (said chiefly of depressions about the base of the upper mandible, or of concavity along the sides of both mandibles). A bill is *hamulate* (hooked) or *unguiculate* (clawed), when strongly epignathous, as in rapacious birds, where the upper mandible is like the talon of a carnivorous beast; it is *dentate*, when toothed (§ 46), as in a falçon; if there are a number of similar "teeth," it is *serrate* (like a saw); it is *cultrate* (knife-like) when extremely compressed and sharp-edged, as in the auk, skimmer; if much curved as well as cultrate, it is *falcate* (scythe-shaped); and each mandible may be oppositely falcate, as in the cross-bill. A bill much flattened and widened at the end (rare) is *spatulate;* examples: spoonbill, shoveller duck. One is called *lamellate*, when it has a series of plates or processes just inside the edges of the mandibles; as in all the duck order, and in a few petrels; the design is to furnish a sifter or strainer of water, just what is effected in the whale, by the "bone" in its mouth. Finally, the far end of the bill, of whatever shape, is called the *tip* or *apex* (fig. 5, *n*); the near end, joined to the rest of the skull, the *base;* the rest is the *continuity.* Some other features of the bill as a whole are best treated under separate head of

§ 48. THE COVERING OF THE BILL. (*a.*) In the great majority of birds, including nearly all perchers, many walkers and some swimmers, the sheathing of the mandibles is wholly *hard*, *horny* or *corneous;* it is integument modified much as in the case of the nails or claws of beasts. In nearly all waders and most swimmers, the sheath becomes, wholly or partly, softer, and is of a dense, leathery texture. But some swimmers, as among the auks, furnish bills as hard-covered as any, while some perchers have it partly quite soft, so that no unexceptional rule can be laid down; and, moreover, the gradations from one extreme to the other are insensible. Probably, the softest bill is found among the snipes, where it is skinny throughout, and in typical snipes vascular and nervous at the tip, becoming a true organ of touch, used to feel for worms out of sight in the mud.   In all the duck order, the bill is likewise soft; but there it is always terminated by a hard, horny "nail," more or less distinct; and such horny claw also occurs in other water birds with softish bills, as the pelican.   An interesting modification occurs in all, or nearly all, of the pigeon order; these birds have the bill hard or hardish at tip and through most of continuity, but towards and at the base of the upper mandible the sheath changes to a soft, tumid, skinny texture, overarching the nostrils; it is much the same with most plovers. But the most important feature in this connection is afforded by the parrots and all the birds of prey; one so remarkable that it has received a distinct name : — CERE.   The cere (L. *cera*, wax; because it looks waxy) is a dense membrane saddled on the upper mandible at base, so different from the rest of the bill, that it might be questioned rather it does not more properly belong to the head than to the bill, were it not for the fact that the nostrils open in it.   Moreover, the cere is often densely feathered, as in the Carolina parroquet, in the bill proper of which no nostrils are seen, these being hidden in the feathered cere, which, therefore, might be easily mistaken, at first sight, for the bird's forehead.   A sort of false cere occurs in some water birds, as the jaegers, or skua-gulls (genera 280 and 279).   The tumid nasal skin of pigeons is sometimes so called; but the term had better be restricted to the birds first above named.   The under mandible probably never presents softening except as a part of general skinniness of the bill.

(*b.*) The covering is either *entire* or *pieced*.   In most birds it is entire; that is, the sheath of either mandible may be pulled off whole, like the finger of a glove.   It is, however, in many birds divided into parts, by various lines of slight connection, and then comes off in pieces; as is the case with some water birds, particularly petrels, where the divisions are regular, and the pieces have received distinctive names.   The entire covering of both jaws together, is called *rhamphotheca;* of the upper alone, *rhinotheca;* of the under, *gnathotheca.*

(*c.*) The covering is otherwise variously marked; sometimes so strongly, that similar features are impressed upon the bones themselves beneath. The most frequent marks are various *ridges* (L. pl. *carinæ*, keels) of all lengths and degrees of expression, straight or curved, vertical, oblique, horizontal, lengthwise or transverse; a bill so marked is said to be *striate* or *carinate;* when numerous and irregular, they are called *rugæ* (L. *ruga*, a wrinkle) and the bill is said to be *corrugated* or *rugose.*   When the elevations are in points or spots instead of lines, they are called *punctæ;* a bill so furnished is *punctate*, but the last word is oftener employed to designate the presence of little *pits* or depressions, as in the dried bill of a snipe, towards the end.   Larger, softish, irregular knobs or elevations pass under the general name of *warts* or *papillæ*, and the bill so marked is *papillose;* when the processes are very large and soft, the bill is said to be *carunculate*

(L. *caro*, flesh, diminutive *carunculus*, little bit of flesh). Various linear *depressions*, often but not always associated with carinæ, are grooves or *sulci* (L. *sulcus*, a furrow) and the bill is then *sulcate*. Sulci, like carinæ, are of all shapes, sizes and positions; when very large and definite, they are sometimes called *canaliculi*, or channels. The various knobs, "horns," and large special features of the bill cannot be here particularized. Any of the foregoing features may occur on both mandibles, and they are exclusive of that special mark of the upper, in which the nostrils open, and which is considered below (§ 51). We have still to notice the special parts of either mandible; and will begin with the simplest, the

§ 49. UNDER MANDIBLE. In the majority of birds it is a little shorter and a little narrower and not nearly so deep as the upper; but sometimes quite as large, or even larger. The upper edge, double (*i. e.* there is an edge on both sides), is called the mandibular *tomium* (Gr. *temnein*, to cut; fig. 5, *j*), as far as it is hard; this is received against, and usually a little within, the corresponding edge of the upper mandible. The prongs already mentioned (§ 44) are the mandibular *rami* (pl. of L. *ramus*, a branch; fig. 5, *i*); these meet at some point in front, either at a short angle (like >) or with a rounded joining (like ⊐). At their point of union there is a prominence, more or less marked (fig. 5, *k*); this is the GONYS (corrupted from the Gr. *gonu*, a knee; hence, any similar protuberance). That is to say, this point is gonys proper; but the term is extended to apply to the whole line of union of the rami, from gonys proper to the tip of the under mandible; and in descriptions it means, then, the *under outline of the bill* for a corresponding distance (fig. 5, *l*). This important term must be constantly held in mind. The gonys is to the under mandible what the keel is to a boat. It varies greatly in length. Ordinarily, it forms, say, one-half to three-fourths of the under outline. Sometimes, as in conirostral birds, a sparrow for example, it represents nearly all this outline; while in a few birds it makes the whole, and in some, as the puffin, is actually longer than the lower mandible proper, because it extends backwards in a point. Other birds have almost no gonys at all: as a pelican, where the rami only meet at the extreme tip, or in the whole duck family, where there is hardly more. As the student must see, the length of the gonys is simply a matter of the early or late fusion of the rami, and that similarly, their mode of fusion, as in a sharp ridge, a flat surface, a straight line, a curve, etc., results in corresponding modifications of its special shape. The interramal space (§ 41, *c*) is complementary to length of gonys: sometimes it runs to the tip of the bill, as in a pelican, sometimes there is next to none, as in a puffin; while its width depends upon the degree of divergence, and the straightness or curvature of the rami. The surface between the tomium and the lower edge of rami and gonys together is the *side of the under mandible* (fig. 5, *m*). The most important feature of the

§ 50. UPPER MANDIBLE is the *culmen* (Lat. for top of anything; fig. 5, *b*). The culmen is to the upper mandible what the ridge is to the roof of a

house; it is the upper profile of the bill — the *highest middle lengthwise line
of the bill;* it begins where the feathers end on the forehead, and extends
to the tip of the upper mandible. According to the shape of the bill it may
be straight or convex, or concave, or even somewhat ∞-shaped; or double-
convex, as in the tufted puffin : but in the vast majority of cases it is convex, with increasing convexity towards the tip. Sometimes it rises up into
a thin elevated crest, as well shown in *Crotophaga* (gen. 126) and in the
puffins, when the upper mandible is said to be *keeled,* and the culmen itself to be *cultrate;* sometimes it is really a furrow instead of a ridge, as
toward the end of a snipe's bill; but generally it is simply the uppermost
line of union of the gently convex and sloping *sides of the upper mandible*
(fig. 5, *a*). In a great many birds, especially those with depressed bill, as
all the ducks, there is really no culmen; but then the *median lengthwise* line
of the surface of the upper mandible, takes the place and name of culmen.
The culmen generally stops short about opposite the proper base of the bill;
then the feathers sweep across its end, and downwards across the base of the
sides of the upper mandible, usually also obliquely backwards. Variations
in both directions from this standard are frequent; the feathers may run out
in a point on the culmen, shortening the latter, or the culmen may run
a way up the forehead parting the feathers; thus either in a point, as in the
rails and gallinaceous birds, or as a broad plate of horn, as in the coots
and gallinules. The lower edge (double) of the upper mandible is the
maxillary tomium, as far backward as it is hard and horny. The most conspicuous feature of the upper mandible in most birds is the

§ 51. NASAL FOSSA (L. *fossa,* a ditch), or *nasal groove* (fig. 5, *c*), in
which the nostrils open. The upper prong of the intermaxillary bone (§ 44)
is usually separated some ways from the two lateral ones; the skinny or
horny sheath that stretches betwixt them is usually sunken below the general
level of the bill, especially in those birds where the prongs are long or widely
separated; this "ditch" is what we are about. It is called *fossa* when short
and wide, with varying depth; *sulcus* or groove when long and narrow; the
former is well illustrated in the gallinaceous birds; the latter in nearly
all wading birds and many swimmers. When the prongs are soldered
throughout, or are very short and close together, there is no (or no evident)
nasal depression, and the nostrils open flush with the level of the bill. The

§ 52. NOSTRILS (fig. 5, *d*) vary in *position* as follows : — they are *lateral*
when on the sides of the upper mandible (almost always) ; *culminal* when
together on the ridge (rare) ; *superior* or *inferior* when evidently above or
below midway betwixt culmen and tomia; they are *basal,* when at the base
of the upper mandible ; *sub-basal* when near it (usual) ; *median* when at or
near the middle of the upper mandible (frequent, as in cranes, geese, etc.) ;
terminal when beyond this (very rare ; and probably there are now *no* birds
with nostrils at the end of the bill, except the *Apteryx*). The nostrils are
pervious, when open, as in nearly all birds ; *impervious,* when not visibly
open, as among cormorants and other birds of the same order; they are

perforate when there is no *septum* (partition) between them, so that you can see through them from one side of the bill to the other, as in the turkey-buzzard, crane, etc.; *imperforate* when partitioned off from each other, as in most birds; but different ornithologists use these terms interchangeably. The principal *shapes* of the nostrils may be thus exhibited:—a line, *linear* nostrils; a line variously enlarged at either end, *clavate*, *club-shaped*, *oblong*, *ovate* nostrils; a line, enlarged in the middle, *oval* or *elliptic*, nostrils; this passing insensibly into the circle, *round* or *circular* nostrils; and the various kinds of more or less linear nostrils may be either longitudinal, as in most birds, or oblique, as in a few; almost never directly transverse (up and down). Rounded nostrils may have a raised border or *rim;* when this is prolonged they are called *tubular*, as in some of the goatsucker family, and in all the petrels. Usually, the nostrils are formed entirely by the substance surrounding them, thus, of cere, in a hawk, of softish skin, in a pigeon, plover or snipe, or of horn, in most birds; but often their contour is partly formed by a special development somewhat distinct either in form or texture, and this is called the *nasal scale*. Generally, it forms a sort of overhanging arch or portico, as well shown in all the gallinaceous birds, among the wrens, etc. A very curious case of this is seen in the European wryneck (*Iynx torquilla*), where the scale forms the floor instead of the roof of the nostrils. The nostrils also vary in being *feathered* or *naked;* the nasal fossa being a place where the frontal feathers are apt to run out in points (called *antiæ*) embracing the root of the culmen. This extension may completely fill and hide the fossa, as in many grouse and ptarmigan; but it oftener runs for a varying distance toward, or *above* and beyond the nostrils; sometimes, similarly below them, as in a chimney-swift; and the nostrils may be densely feathered when there is no evident fossa, as in an auk. When thus truly feathered in varying degree, they are still open to view; another condition is, their being covered over and hidden by modified feathers. These are usually bristle-like (*setaceous*), and form two tufts, close-pressed, and directed forwards, as is perfectly shown in a crow; or the feathers may be less modified in texture, and form either two *tufts*, one over each nostril, or a single *ruff*, embracing the whole base of the upper mandible; as in nuthatches, titmice, redpoll linnets, snow buntings and other northern *Fringillidæ*. Bristles or feathers thus growing forwards are called *retrorse* (L. *retrorsum*, backward; here used in the sense of *in an opposite direction from* the lay of the general plumage; but they should properly be called *antrorse*, *i. e.*, forward). The nostrils, whether culminal or lateral, are, like the eyes and ears, *always two* in number, though they may be united in one tube, as in the petrels.

§ 53. THE GAPE. It only remains to consider what results from the relations of the two mandibles to each other. When the bill is opened, there is a cleft, or fissure between them; this is the *gape* or *rictus* (L. *rictus*, mouth in the act of grinning); but, while thus really meaning the open *space* between the mandibles, it is generally used to signify the *line of their*

closure. Commissure (L. *committere,* to put or join together) means the
point where the gape ends behind, that is, the *angle of the mouth,* where
the opposed edges of the mandibles join each other; but as in the last case,
it is loosely applied to the whole line of closure, from true commissure
to tip of the bill. So we say, "commissure straight," or "commissure
curved;" also "commissural edge" of either mandible (equivalent to
"tomial edge") in distinction from culmen or gonys. But it would be
well to have more precision in this matter. Let, then, *tomia* (fig. 5, *j*) be
the true cutting edges of either mandible from tip to opposite base of bill
proper, *rictus* (fig. 5, *g*) be their edges thence to the ᴘᴏɪɴᴛ *commissure*
(fig. 5, *h*) where they join when the bill is open; the ʟɪɴᴇ commissure (fig.
5, *f*) to include both when the bill is closed. The gape is *straight,* when
rictus and tomia are both straight and lie in the same line; *curved, sinuate,*
when they lie in the same curved or waved line; *angulated,* when they are
straight, or nearly so, but do not lie in the same line, and therefore meet at
an angle. (An important distinction. See under family *Fringillidæ* in the
Synopsis.)

THE WINGS.

§ 54. Dᴇꜰɪɴɪᴛɪᴏɴ. Pair of anterior or *pectoral* limbs organized for
flight by means of dermal outgrowths. Used for this purpose by birds in
general; but by ostriches and their allies only as outriggers to aid running;
by penguins as fins for swimming under water; used also in the latter
capacity by some birds that fly too, as divers. Wanting in no recent birds,
but imperfect in a few. To understand their structure we must notice

§ 55. Tʜᴇɪʀ Bᴏɴʏ Fʀᴀᴍᴇᴡᴏʀᴋ. (Fig. 6.) This ordinarily consists
of *nine* actually separate bones; but there are several more that fuse
together. The arm-bone, *humerus,* a single bone, reaches from shoulder to
elbow; it is succeeded by two parallel bones, *ulna* and *radius,* of about
equal lengths, reaching from elbow to wrist, forming the forearm, cubit or
antibrachium. The wrist (*carpus*) has two little knobby carpal bones, called
scapholunar and *cuneiform;* very early in life there is another, the *magnum,* that soon fuses with the hand-bone, or *metacarpal.* At first, this last
is of three bones, corresponding to those of our hand that support our fore,
middle and ring finger respectively; afterwards they all run together. The
one corresponding to the *middle* finger is much the largest of the three, and
it supports two finger-bones (*phalanges*) placed end to end, just as our
three similar finger-bones are placed one after the other at the end of their
own hand-bone. The forefinger hand-bone sticks out a little from the side
of the principal one, and bears on its end one finger-bone (sometimes two),
which is commonly, but wrongly, called the bird's "thumb." For although
on the extreme border of the hand, it is *homological* with the forefinger;
birds have no thumb (exc. *Archæopteryx, Struthio, Rhea*); and no little
finger. The third hand-bone is joined to the second, and bears no finger-
bone.

§ 56. THE MECHANISM of these bones is admirable. The shoulder-joint is loose, much like ours, and allows the humerus to swing all about, though chiefly up and down. The elbow-joint is tight, permitting only bending and unbending in a horizontal line. The finger bones have scarcely any motion. But it is in the wrist that the singular mechanism exists. In the first place, the two forearm bones are fixed with relation to each other so that they cannot roll over each other, like ours. Stretch your arm out on the table ; without moving the elbow, you can turn the hand over so that either its palm or its back lies flat on the table. It is a motion (*rotation*) of the bones of the forearm, resulting in what is called *pronation* and *supination*. This is absent from the bird's arm, necessarily ; for if the hand could thus roll over, the air striking the pinion-feathers, when the bird is flying, would throw them up, and render flight difficult or impossible. Next, the hinging of the hand upon the wrist is such, that the hand does not move up and down, like ours, in a plane perpendicular to the plane of the elbow-bend, but back and forwards, in a plane horizontal to the elbow ; it is as if we could bring our little finger and its side of the hand around to touch the corresponding border of the forearm. Thus, evidently, extension of the hand upon the wrist-joint increases and completes the unfolding of the wing that commenced by straightening out the forearm at the elbow. There is another essential feature in a bird's wing. In the figure, 6, ABC represents a deep angle formed by the bones, but none such is seen upon the outside of the wing. This is because this triangular space is filled up by a fold of skin stretched over a cord that passes straight from near A to C. But A and C approach or recede as the wing is folded or unfolded, and a simple cord long enough to reach the full distance A—C would be *slack* in the folded wing; so the cord is made *elastic*, like an india rubber band ; it stretches when the wing is unfolded, and contracts when the wing is shut ; it is thus always hauled taut. The cord makes the always straightish and smooth anterior border of the wing. *The carpus* C, or the always prominent point of the anterior border, is a highly important landmark in descriptions, and should be thoroughly understood ; it is also called the "bend of the wing." (See under Directions for Measurement ; see also explanation of fig. 6.)

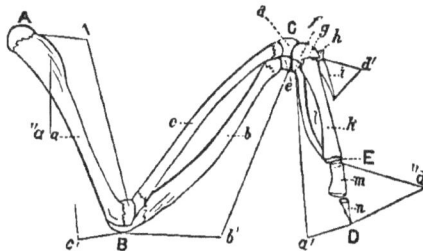

FIG. 6, taken from a young chicken (right wing, upper surface), shows the composition and mechanism of a bird's wing. A, shoulder ; B, elbow ; C, wrist or *carpus* ; D, tip of principal (the third) finger; AB, arm; BC, forearm; CD, pinion, or hand, composed of c, carpus, thence to E, *metacarpus* or hand proper, except the bone *i*, this, and ED, being *digits* or fingers. *a*, shaft of humerus; *b*, ulna; *c*, radius; *d*, scapholunar bone; *e*, cuneiform bone; these last two composing wrist or carpus proper. Now the figure (1) marks two lines that run to the two ends of the humerus, designating a sort of cap on either end of that bone; this cap is an EPIPHYSIS;* both ends of ulna and radius show similar epiphyses, connected in the figure, as in case of the humerus, with the shaft by *waved* lines. Then, of the metacarpus, *g* and *f* are the epiphyses of, respectively, the two principal metacarpal bones *k*, the third, and *l*, the fourth; *k* and *l* have not yet coalesced together, but lie simply opposed to each other, whereas their epiphyses themselves, *g* and *f*, are seen *nearly* fused together. *h*, which seems to be the epiphysis of *i*, is not; it is a metacarpal itself (the second), bearing the *digit*, *i*; it is nearly soldered with *g*, in which its epiphysis is already absorbed. Later in life, *k* sends a plate-like process towards *l; l* and *k* grow together; *h* grows into *k* and *g; f* and *g* grow into *lk*, with the compound result *fghlk*, forming a single bone, THE METACARPAL, bearing the "thumb" phalanx *i* and the two finger phalanges *m*, *n*, all three of which remain permanently separate. (Observe, that *k* is called the THIRD metacarpal, because it represents that bone in the hand of man and beasts; that in actual position it is second, *h* being first and *l* third; that ordinary birds have *no* first and *no* fifth metacarpals; and that the bone *i*, though called "thumb," corresponds to the first joint of our forefinger.) *d'*, first finger, or thumb, the seat of the *bastard wing-feathers* (alula, § 58); *d''*, actually the second finger, but morphologically the third finger, composed of two movable bones *m*, *n*. *a'*, seat of primaries (upon whole pinion); *b'*, seat of secondaries (upon forearm); *c'*, seat of tertiaries (about and above elbow); *a''*, seat of scapularies (upon *pteryla humeralis*). This wing is shown half-spread; in closing or folding, c approaches A, and D approaches B; all *nearly* in the plane of the paper; and in unfolding, the elbow-joint B is such a perfect hinge that c cannot sink down below the level of the paper, and c is similarly so hinged that D cannot fly up from the same level, as the air, pressing upon the quill feathers a' and b', would tend to make it do. Observe also; *b* and *c* are two rods connecting B and c, and the construction of their jointing at B and c, and of their jointings with each other at their ends, is such, that they can *slide along* each other a little way. Now when the point c, revolving about B, approaches A in the arc of a circle, the rod *c pushes on* towards *d, f, g,* etc., while the rod *b pulls back c, l,* etc.; so that the point D is brought nearer B. Conversely, in opening the wing, when c recedes from A, c pulls back, and *b* pushes on, effecting recedence of D from B. So the angle ABC cannot be increased or diminished without similarly increasing or diminishing the angle BCD. In other words, you cannot open or shut one part of the wing, without opening or shutting the other; it is like killing two birds with one stone, this wonderful bony mechanism for economizing muscular power.†

We are now ready to examine the

§ 57. WING-FEATHERS. These all grow upon the pteryla alaris (§ 9, *b*, and Pl. I, fig. 4, 5). They are of two main sorts; the *remiges* (L. *remex*, a rower) or long quills collectively, and the coverts, *tectrices* (L. *tectrix*, arbitrary feminine corruption of *tector*, a coverer) ; to which may be added as a third distinct group the *bastard quills* (*alula*, or *ala spuria*). The

§ 58. ALULA (L. diminutive of *ala*, a wing, Pl. I, fig. 1, *al*), or little wing, is simply the bunch of feathers that grow upon the "thumb." Highly

---

*Epiphysis (Gr. *epi*, upon, *phusis*, growth). Young bones are wholly cartilaginous, or gristly; they harden at length by deposition in the cartilage of bone-earth. This deposit begins at certain points called *ossific centres*. Now in what are called "long" bones, that is, bones like a humerus, etc., there may be one such centre for the shaft and one upon each end of the bone. The shaft ossifies first; the ends later; and before the bone has completed its growth these ends remain distinct from the shaft with which they afterwards solder. These cartilaginous or gristly *caps* on the ends are called epiphyses.

†See BERGMANN, *Arch. f. Anat.*, 1839, 296; COUES, *Amer. Nat.* v, 1870, 513.

important as it is in a morphological point of view, it is taken into little
account in practical ornithology, unless when largely modified in form, conspicuous in color, or bearing special organs, as claws, spurs, etc.   It
strengthens, and defends, and adds to the symmetry of the anterior outer
border of the wing.   (The student must carefully distinguish the use of
the word *spurious* in this connection from its application to a certain state
of the first primary — see § 62.)

§ 59. THE WING-COVERTS are conveniently divided into the *upper* (*tectrices superiores*) and *under* (*tect. inferiores*); they include all the small
feathers that clothe the wings, extending a varying distance along the bases
of the remiges (§ 60). The ordinary disposition and division of the upper
coverts is as follows : — There is one set, rather long and stiffish, close-pressed
over the bases of the outer nine or ten remiges, covering these, in general,
about as far as their structure is plumulaceous. These spring from the hand
or pinion (§ 55) and are the *upper* PRIMARY *coverts* (Pl. I. fig. 1, *pc*) : they
are ordinarily the least conspicuous of any. All the rest of the upper coverts
are SECONDARY, and spring mostly from the forearm ; they are considered in
three groups, or *rows*. The *greater* coverts (Pl. I, fig. 1, *gsc*) are the first,
outermost, longest row, covering the bases of most of the remiges except
the first nine or ten ; the *median coverts* (Pl. I, fig. 1, *msc*), are a next
row, shorter, but still almost always forming a conspicuous series. All the
rest of the secondary coverts pass under the general name of *lesser coverts*
(Pl. I, fig. 1, *bc*). The greater coverts have furnished a very important zoö-
logical character : for in all *Passeres* they are not more than half as long as
the remiges they cover, while the reverse is believed to be the case in nearly
all other birds. The under coverts have the same general disposition as the
upper : but they are all like each other, have less distinction into rows or
series, and for practical purposes generally pass under the common name of
under wing-coverts; and since, when the wing is strikingly colored underneath, it is these feathers, and not the remiges, that are highly or variously
tinted, the expression " wing below," or "under surface of the wing" generally refers to them more particularly. We should distinguish, however,
from the under wing-coverts in general, the *axillary feathers*, or *axillars* (L.
axilla, arm-pit). These are the innermost of the under wing-coverts ; almost always longer, stiffer, and otherwise distinguishable from the rest ; in
ducks, for example, and many waders, they take on remarkable development.

§ 60. (a.) THE REMIGES (Pl. I, fig. 1, *b, s,* and *t*) mainly give the size,
shape, and general character to the wing, and are its most important features ; they represent the whole of its posterior outline, most of its surface,
and most of its outer and inner borders. Taken collectively, they form a
flattened surface for striking the air ; this surface may be quite flat, as in
birds with long pointed wings that cut the air like oar-blades ; generally it
is a little concave underneath, and correspondingly convex above ; this con-
cavo-convexity varying insensibly within certain limits. It is usually greatest in birds with a short rounded wing, as in the gallinaceous order. Two

extremes of the mode of flight result. The short, round wing confers a
heavy, powerful, cutting flight, for short distances, with a whirring noise,
produced by quick vibrations of the wing : birds that fly thus are almost
always thickset and heavy. The long, pointed wing gives a light, airy,
skimming flight, indefinitely prolonged, with little or no noise, as the wing
beats are more deliberate : birds of this style of wing are generally trim
and elegant. These, of course, are merely generalizations, mixed and obscured in every degree in actual bird-life. Thus the humming-bird, with
long pointed wings, whirs them fastest of all birds ; so fast that the eye cannot follow the strokes, and merely perceives a mist on each side of the bird.
The combination of a pointed with a somewhat concave shape of wing is a
remarkably strong one ; it results in a rapid, vigorous, *whistling* flight, as in
a pigeon or duck. An *ample* wing, as it is called, that is, one long as well
as broad, without being pointed, is seen in the herons ; it confers a slow and
somewhat lumbering, but still strong, flight. The longest winged birds are
found among the swimmers, as albatrosses ; but here the extreme length is
largely produced by the length of the humerus ; some land birds, as swallows,
swifts, humming-birds, and other fissirostral birds, would have a still longer
wing, were not the humerus extraordinarily short. The shortest wings
(among birds with perfect remiges), occur in the lowest swimmers, as among
the auks and divers, and in the gallinaceous birds. The various special
shapes of wings are too numerous and too insensibly gradated to be mentioned here. The mechanics of ordinary flying are probably now understood,* though the "way of an eagle in the air" was an enigma to the wise
man of old. But the sailing of some birds for an indefinite period through
the air, up as well as down, without visible motion of the wings, remains a
stumbling-block ; the flight of the turkey vulture is yet unexplained, I venture to affirm.

(b.) The *number* of remiges ranges from sixteen, in the humming-bird,
to upwards of fifty, in the albatross. This statement is exclusive of the
penguins, in which there are no true remiges. The remiges subserve flight
in nearly all existing birds except these last, the ostriches and their allies,
and the great auk, *Alca impennis*—if indeed this bird still lives.

(c.) Of the *shape* of remiges there is little to be said, they are, with few
exceptions, so uniform. They are the stiffest, strongest, most truly pennaceous (§ 4) of a bird's feathers ; they have no evident hyporhachis (§ 3, a) ;
they are generally *lanceolate*, that is, taper regularly and gradually to a
rounded point. Sometimes one or both webs are *incised* or *attenuate*
towards the end, that is, they narrow abruptly ; this is also called *emargination.* (See fig. 110.) The tips of the remiges may be squarely or obliquely
cut off, as it were, or nicked in various ways. Except in the case of a few
of the innermost remiges, their outer vexillum (§ 3, a) is always narrower

---

* The student should not fail to consult, in this connection, M. Marey's " Lectures on the Phenomena of
Flight," *Smithsonian Report* for 1869, p. 226. (Translated from *Revue des Cours Scientifiques.*)

than the inner, and its barbs stand out less from the rhachis (§3, a). Remiges are divided into three classes, according to their seat; and in this is involved one of the most important considerations in practical ornithology, of which the student must make himself master. The three classes are 1, the *primaries*; 2, the *secondaries*; 3, the *tertiaries*.

§ 61. THE PRIMARIES (Pl. I, fig. 1, *b*) are those remiges which grow upon the pinion, or hand- and finger-bones (fig. 6, CD). Whatever the total number of remiges may be, *in all birds with remiges the primaries are either* NINE *or* TEN *in number*, as far as is known. The albatross and the humming-bird (§ 60, b) both have ten. All birds, probably, below the highest, the oscine *Passeres*, have ten. Among *Oscines*, there are nine or ten indifferently; and just this difference of one primary more or less forms one of the most marked distinctions between some families of that suborder. So the tenth feather in a bird's wing, counting from the outside, is a sort of crucial test in many cases; if it be first secondary, the bird is one thing; if it be last primary, the bird is another; the necessity, therefore, of determining which it is, becomes evident. It is, of course, always possible to settle the question by striking at the roots of the remiges and seeing how many are seated on the pinion; but this generally involves some defacing of a specimen, and ordinarily there is an easier way of determining. Hold the wing half spread; then, in nearly all *Oscines*, the primaries come sloping down on one side, and the secondaries similarly on the other, to form, where they meet, a reëntrant angle in the general contour of the posterior border of the wing; the feather that occupies this notch is the one we are after, and unluckily is sometimes last primary, and sometimes first secondary. But primaries are, so to speak, *emphatic, self-asserting, italicized* remiges, stiff, strong, obstinate; while secondaries are *whispering, retiring* remiges *in brevier*, limber, weak, and yielding. This difference in character is almost always shown by *something* in their general shape, impossible to describe, but which the student will soon learn to detect. Let the reader examine plate I, fig. 1, where *b* marks the 9 primaries of a sparrow's wing, and *s* indicates the secondaries; he will see a difference at once. The primaries express themselves, though with constantly diminishing force, to the last; then the secondaries immediately begin to tell a different tale. Among North American birds, the only ones with NINE primaries are the families *Motacillidæ, Alaudidæ, Sylvicolidæ, Hirundinidæ, Fringillidæ, Icteridæ*, part of the *Vireonidæ*, and the genus *Ampelis*.* The condition of the *first* primary, whether

§ 62. SPURIOUS or not, is often of great help in this determination. The first primary is said to be *spurious* (compare § 58) when it is very short; say a third, or less than a third, of the length of the second primary. A

---

* This really has ten; but the first is so small and so out of position that it is only theoretically accounted as such, and would not be so considered by the student. I should add, that recent researches of Prof. Baird's tend to show that *all* supposed nine-primaried birds have really ten; but only an expert ornithologist could find the additional one in question; and it need not be taken into account for present purposes. (See explanation of Pl. I. fig. 1, *pu*.) Nitzsch says the grebes have 11 primaries; this may be confirmed.

spurious first primary only occurs in certain ten-primaried *Oscines*. It is evident, therefore, that the finding of this short primary is equivalent to determining the presence of ten primaries; but, on the other hand, not finding it does not prove nine primaries; the count must be made in all cases where the first primary is more than one-third as long as the second.

§ 63. THE SECONDARIES (Pl. I, fig. 1, *s*) are those remiges that are seated on the forearm (fig. 6, BC); they vary in number from six upward, the precise greatest number probably not ascertained, unless it be the forty of the albatross. They have the peculiarity of being actually attached to one of the bones of the forearm (*ulna*) which the other remiges are not. If you examine an ulna, you will see a row of little points showing the attachment. The secondaries present no special features necessary to describe in the present connection. They are enormously developed in the argus pheasant.

§ 64. (a.) THE TERTIARIES (Pl. I, fig. 1, *t*) are, properly, the remiges that grow upon the upper arm (*humerus*); but they are not evident in most birds, and the two or three innermost secondaries, that grow upon the very elbow, and are commonly different from the rest, in form or color, or both, pass under the name of tertiaries. So also some of the scapular feathers (§ 38, and Pl. I, fig. 1, *scp*), when long or otherwise conspicuous, are called tertiaries. But there is an evident and proper distinction. Scapulars are feathers of the *pteryla humeralis* (§ 9, *b*); while tertiaries, whether seated on the elbow or higher up, are the innermost remiges of the *pteryla alaris* (§ 9, *b*). They are oftener called "tertials," for short, though the other name is more correct, besides being formed in analogy with the names of the other remiges. Tertiaries do not often afford conspicuous or important characters; but in many birds they are very long and flowing. This is particularly the case in most sharp-winged wading birds; and, in fact, is mainly confined to birds with such a wing.

(b.) Occasionally, any of the wing feathers take on remarkable special developments, and such is particularly the case with the tertials and secondary upper coverts; but it would be superfluous to particularize these here. The wing rarely produces anything but feathers; sometimes, however, offensive weapons are found, as in the horny spur-like process of the pinion of the spur-winged thrush, *Turdus dactylopterus*, the spur-winged goose (*Plectropterus*), spur-winged pigeon (*Didunculus*), several plovers (*Chettusia*, etc.), the jacanas (*Parra*), etc., and the one or two claws of the ostriches and their allies, as well as of the extinct *Archæopteryx*. But we have no illustration of these outgrowths among North American birds.

THE TAIL.

§ 65. TIME was when birds flew about with long bony and fleshy tails, with the feathers inserted in a row on either side (*distichous*) like the hairs of a squirrel's. But we have changed all that. Now the bones are few (generally about nine in number), and short, not projecting beyond the general plumage, and the last one, called *coccyx* or *vomer* (L. *vomer*, a plough-share), is large and singularly shaped, and the feathers are stuck around this like the blades upon a lady's fan. The whole bony and muscular apparatus is familiar to every one as the "pope's nose" of the Christmas turkey; and in descriptive ornithology the word "tail" refers solely to the feathers, all of which grow upon the pteryla caudalis (§9, b). The tail feathers, like those of the wings, are of two sorts; *coverts* (*tectrices*) and *rectrices* (L. *rectrix*, a female ruler or governess; here in the sense of a *steerer* or rudder, because they guide the bird's flight); these correspond precisely to the wing-coverts (§ 59) and the remiges (§ 60, *a*). The

§ 66. TAIL-COVERTS are the numerous, generally rather small, in compar-ison with rectrices, feathers that overlie and underlie the rectrices, defending their bases, and contributing to the firmness and symmetry of the tail. An obvious division of them is into an *upper* (*tect. superiores*) and *under* (*tect. inferiores*) set. Neither set is EVER wholly wanting; but sometimes one or the other, and particularly the upper, is very short, and not distinguishable from the general plumage of notæum (§ 38), as in the ruddy duck (genus 270). The upper coverts are the most variable in size, shape and texture. While usually shorter than the under, and reaching only from a fourth to a half of the length of the rectrices, sometimes they take an extraordinary development, project far beyond the rectrices, and form the bird's chiefest ornament. The gorgeous argus-eyed train of the peacock is upper tail coverts, not rectrices; the elegant plumes of the paradise trogon (*Pharo-macrus mocinno*), several times longer than the bird itself, are likewise coverts. The under tail coverts are more uniform in development, and very rarely, as in some of the storks, become plumes of any considerable pretensions. Ordinarily, they are about half as long as the tail, but fre-quently reach its whole length, and form a dense tuft, as in the ducks. I do not now recall an instance of their projecting noticeably beyond the tail. It is to this bundle of under tail-coverts that the word *crissum* (§ 39) prop-erly applies. The

§ 67. RECTRICES or true tail feathers can almost never be confounded with the coverts: they are, like the remiges, stiff, well-pronounced feathers, pennaceous to the very base of the vexilla, wanting after-shafts (at least evident after-shafts, in the great majority of cases), and have one vexillum wider than the other, except, sometimes, the central pair. They are always *in pairs:* that is, there is the same number on each side of the middle line of the tail, and their number, consequently, is always an even one. The ex-ceptions to this rule are so few (and then only among birds with the higher numbers of tail feathers) that they are probably to be regarded as simple anomalies, from accidental arrest of a feather. They are imbricated over each other in this way :—The central pair are highest, and lie with both their webs over the next feather on either side (the inner web of either of these middle two underlying or overlying the inner web of the other); and they all thus successively overlie each other, so that they would form a pyramid were they thick, not flat. This disposition is perceived at once in the accompanying diagram, where it will also be seen that *spreading* of the tail is simply the greater divergence of *a* from *b*, while *closing* the tail is bringing *a* and *b* together directly under *c*. The act is accomplished by certain muscles that pull on either side at the bases of the quills collectively: they are the same that pull the whole tail to one side or the other, just as tiller-ropes of a boat's rudder work on that instrument. The *general*

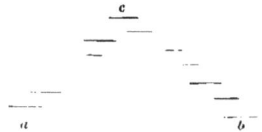

§ 68. SHAPE of a rectrix, is shown in Pl. I, fig. 5. The feather is somewhat clubbed, or oblong, widening gradually and nearly regularly towards the tip, where it is gently rounded. But the obvious departures from this are various. A rectrix broad to the very tip, and there cut squarely off, is *truncate;* one such cut diagonally off is *incised*, especially when, as usually happens, the outline of the cut portion is concave. A *linear* rectrix is very narrow, with parallel sides; a *lanceolate* one is broader at the base, and tapers regularly and gradually to a point. A noticeably pointed rectrix is *acute;* when the pointing is produced by *abrupt* contraction towards the tip it is called *acuminate*, as in woodpeckers generally. A very long, slender, more or less linear feather is said to be *filamentous*, as the lateral one of a barn-swallow or of most terns, the middle one of a tropic bird (gen. 278), etc. When such protrude suddenly and far beyond all the rest, I call them *long-exserted*, after an analogous term in botany. An unusually stiff feather is called *rigid*, as in woodpeckers and other birds that use the tail as a prop or support. When the rhachis projects beyond the vexilla, the feather is *spinose*, or better, *mucronate* (L. *spina*, a prickle, or *mucro*, a point; *e. g.*, chimney-swift, fig. 123). The bob-o'-link (gen. 87) and sharp-tailed finch (fig. 84) both approximate towards this condition. When the vexilla are wavy-edged, the feather is *crenulate* (fine example in *Plotus*, gen. 276). While the great majority of rectrices are *straight*, some are *curved*, either outwards or inwards, in the horizontal plane; those curved in a perpendicular plane are *arched* or *vaulted*—the latter particularly when the vanes are concavo-convex in transverse section. The typical

§ 69. NUMBER of rectrices is TWELVE. This holds in the vast majority of birds. It is so uniform throughout the great group *Oscines*, that the rare exceptions are perfectly anomalous; in the other group of *Passeres* (*Clamatores*) it is usually twelve, but sometimes *ten*. Among *Strisores* there are never more than *ten* rectrices. In *Scansores*, the number varies from *eight* to *twelve;* eight is rare, as in the genus *Crotophaga* (no. 126); other cuckoos have *ten;* the woodpeckers have APPARENTLY *ten*, but there are really *twelve*, of which the outer pair on each side are very small, almost rudimentary, hidden betwixt the bases of the second and third pair (see Key, III). Birds of prey have about twelve. Pigeons (all ours at least) have twelve or fourteen. In birds below these the number begins to increase; thus directly, among the grouse, we may find up to twenty, as in the great cock of the plains; but in a few singular types (*Tinamidæ*) of the order *Gallinæ*, there are *none*, or only rudimentary ones. Among water birds the numbers vary so that they are usually of only generic, and sometimes only specific, importance. Those swimmers with long, well-formed tails, as the *Longipennes*, and particularly the gull family, and some of the ducks, have the fewest; here there are twelve, sometimes fourteen, rarely sixteen; while those with short, soft tails have the most, as sixteen, eighteen, twenty; and, as in the pelicans, twenty-two, or even twenty-four—the last being *about* the maximum, although in one genus of penguins (*Aptenodytes*) there are thirty-two or more. Swimmers again, furnish birds with *no* rectrices, the whole grebe family (*Podicipidæ*) being thus distinguished. So rectrices run among birds from none to over thirty. The *typical*

§ 70. SHAPE OF THE TAIL, as a whole, is the FAN. The modifications, however, are as many as, and greater and more varied than, those of the wing, at the same time that they are susceptible of better definition, and have received special names that must be learned. Taking the simplest case, where the rectrices are all of the same length, we have what is called the *even, square* or *truncate* tail, from which nearly all the others are simple departures in one way or another. A square, or nearly so, tail with the two central feathers long-exserted (§ 68) is common: we see it in all jaegers (gen. 280), in *Momotus* (gen. 112) and especially in *Phaëthon* (gen. 278). The most frequent departure from the even tail is by gradual successive shortening of the rectrices from the pair next the middle to the exterior ones; and this shortening is called *gradation*. Gradation is a generic term, implying such shortening in any degree. Precisely, it should mean shortening each successive pair of rectrices by the *same* amount; say, each pair being half an inch shorter than the next. But this exactness is not often preserved. When the feathers shorten by *more and more*, we have the true *rounded* tail, probably the commonest form among birds: thus, let the gradation between the middle and next pair be just appreciable, and then increase regularly, to half an inch between the next to the outermost and the lateral pair. The opposite gradation, by *less and less* shortening, gives the *wedge-shaped* or *cuneate* tail; it is well shown in the magpie, where, as in many other birds, the central feathers would be called long-exserted, were all the rest of the same length as the outer. A cuneate tail, especially with narrow acute feathers, is also called *pointed*, in contradistinction to rounded, as in the sprig-tailed duck (gen. 253). The generic opposite of the gradated tail is the *forked;* where the lateral feathers *increase* in length from the central to the outer pair. The least appreciable forking is called *emargination*, and such a tail is *emarginate;* when it is more marked, as for instance, say an inch of forking in a tail six inches long, the tail is truly *forked.* The degrees of forking are so various and intimately connected, that they are usually expressed by qualified terms: as, "slightly forked," "deeply forked," etc. The deeper forkings are *usually* accompanied by a more or less *filamentous* elongation of the outer pair of rectrices: as in the barn swallow, some flycatchers, most of the terns, etc., etc.   It would be advisable to have a term to express such extreme condition, which I shall call *forficate*, when the depth of the fork is equal to, or greater than, the length of the shortest (middle) pair of feathers; it occurs among our birds in the genera *Milvulus* (no. 104), *Sterna* (291), and elsewhere.   *Double*-forked or *double*-rounded tails are not uncommon; they result from combination of both gradation and forking, in this way: — Let the middle feathers remain constant, and the next two or three pairs progressively increase in length, then the rest successively decrease; evidently, the tail is forked centrally, gradated externally: this is the double rounded form; it is shown in the genera *Myiadestes* (no. 52) and *Anous* (294).   Now with middle feathers as before, let the next pair or two decrease in length, and the rest progressively increase to the outermost: then we have the double-forked, a common shape among sandpipers.   In the latter case, the forking rarely amounts to more than simple emargination, and generally is really little more than simple protrusion of the middle pair of rectrices in an otherwise slightly forked tail; and in neither case is the gradation either way often great.

Various shapes of tails, which the student will readily name from the foregoing paragraph, are illustrated in figs. 17, 19, 29, 30, 32, 54, 57, 68, 73, 76, 84, 98, 106, 117, 120, 121, 126, 133, 135, 137, 144, 145, 147–52, 177, 206, 214.   I should also allude to the *folded* tail of the barn-yard fowl (*Gallus bankivi*, var.) a very familiar but rare form.   One of the most beautiful and wonderful of all the shapes of the tail is illustrated by the male of the famous lyre-bird (*Menura superba*), shown in the figure at the end of this Introduction.

It should be remembered that to determine the shape, the tail should be viewed *nearly* closed; for spreading will obviously make a square tail round, an emarginate one square, etc.   I append a diagram of the principal forms.

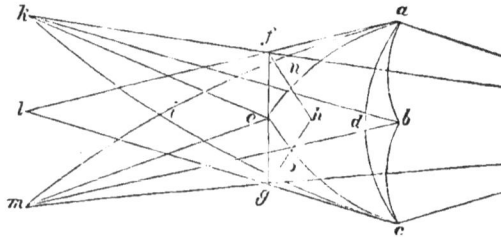

FIG. 7.—Diagram of shapes of tail.

FIG. 7. *ade*, rounded; *acc*, gradate; *aic*, cuneate-gradate; *alc*, cuneate; *abc*, double-rounded; *feg*, square; *fhg*, emarginate; *faeog*, double-emarginate; *kim*, forked; *kem*, deeply forked; *kbm*, forficate.

THE FEET.

§ 71. IN ALL BIRDS, the posterior extremities are organized for progression; for walking, hopping, or running on land, in all; but a few of the lowest birds can scarcely walk; for perching on trees, etc., in the vast majority, most of which hop about there, and many of which climb or scramble in every imaginable way, with or without the aid of the tail; for swimming on the water, or diving, in a great many; for grasping and holding detached objects in some, as the parrots, birds of prey, and a few others. The modifications of the leg and foot are more numerous, more diverse, and more important, in their bearing upon taxonomy, than those of either bill, wing or tail.

§ 72. (a.) THE BONY FRAMEWORK. (Fig. 8, somewhat diagrammatic illustration, taken from a loon's right leg.) This ordinarily consists of twenty bones, of which fourteen are toe-bones, one is a little bone connecting the hind toe with the rest of the foot, one a little bone in front of the knee-joint, and four are the principal bones from the hip-joint down to the roots of the toes. The first is the *femur* or thigh-bone, *a*, reaching from hip A, to knee B; a large terete bone, corresponding to the humerus of the wing. Then come two bones, *b*, the *tibia*, or principal (and inner) leg-bone, and *c*, the *fibula*, or lesser (and outer) leg-bone; both these joint with the femur above, and in front of this, the knee-joint, there is in many or most birds a

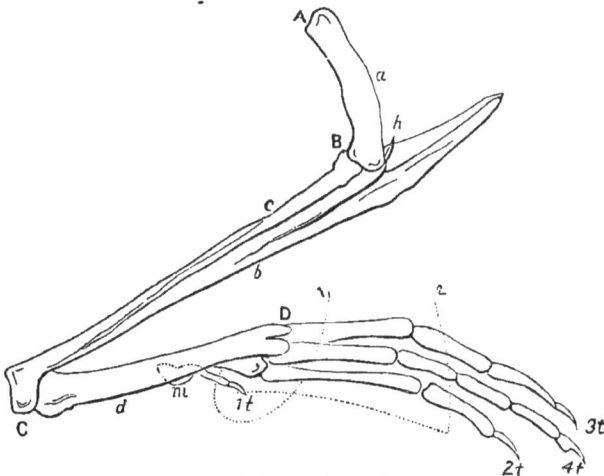

FIG. 8. Bones of leg and foot.

little knee-pan, or knee-cap: the *patella*, *p*. The tibia runs to the *heel*, c, and there has an enlarged extremity to joint with the next bone: but the fibula is only a slender spicula not reaching the heel, but ending in a sharp point part way down the leg, and partly soldered with the tibia. It is only in a few of the lowest birds, that the tibia runs up to a point *above* the knee-joint, as shown in this figure: ordinarily, it ends at the knee itself. The portion of the leg represented by the femur, or from A to B, is the THIGH; that represented by tibia and fibula is the LEG or CRUS; leg proper, therefore, is from knee to heel, or B to C only.

(b.) Now a bird's legs are not like ours, separate from the body from the hip downward, but are for a variable distance inclosed within the general skin of the body. The freedom is greatest among the higher birds, and especially rapacious birds, that use the feet for grasping, and least in the lowest swimming birds : the entire range of enclosure of the leg, is from part way up the thigh down almost to the very point c, as in the case of the loon and other diving swimmers. And in no birds, is the knee, B, seen outside the general contour of the *plumage;* it must be looked or felt for among the feathers, and in most prepared skins will not be found at all. Practically, it is a landmark of no consequence in determining genera and species, though of the utmost importance in primary classification ; the student may for awhile ignore its existence if he chooses. The first *joint* that sticks out from the plumage is the HEEL, c ; and this is what, in loose popular terms, is called "knee," upon the same erroneous notion that the wrist of a horse's foreleg is called "knee." Just so people call a bird's crus the "thigh," and disregard the thigh altogether. There is no need of this confusion ; and even without the slightest anatomical knowledge, any one can tell knee from heel at a glance, whatever their position relative to the body ; for knees ALWAYS bend *forward,* and heels ALWAYS bend *backward.*

(c.) This point c corresponds to the point c in fig. 6 of the wing. There we found two little carpal bones, or wrist-bones, intervening between forearm and hand, or metacarpus ; but adult birds have no such actual bones intervening between tibia and the next bone, *d,* the METATARSUS. So there is no tarsus proper ; metatarsus hinges directly upon tibia, or foot upon leg, without true ankle-bones ; that is, the foot-bone itself makes the ankle-joint, with the leg, at the point c, heel. (Theoretically, however, there are tarsal bones : for there is an epiphysis (§ 56*) at the lower end of the tibia, and an epiphysis at the upper end of the metatarsal bone ; afterwards fused with these bones respectively. One or the other, or both of these are held by different anatomists to be tarsal bones ; more particularly, the one that fuses with the metatarsus ; which last, therefore, represents both tarsus and metatarsus, and is on this account called *tarso-metatarsus.**)

---

*This is as usually taught. But Gegenbaur has shown that these so-called epiphyses are true tarsal bones. He represents, in the chick at the ninth day of embryonic life, two bones, an upper and an under, the former afterward anchylosing with the tibia, the latter with the metacarpus, leaving the ankle-joint between them, as in reptiles. Morse, who has studied the embryos of several species, goes still further ; he shows that the upper tarsal bone of Gegenbaur is really two bones, corresponding to the tibiale and fibulare, or astragalus and calcaneum ; these subsequently co-ossify to form the upper one seen by Gegenbaur, and finally co-ossify with the tibia to form the bitrochlear condyle characteristic of this bone in *Aves.* The distal tarsal ossicle he believes to be the centrale of reptiles. Wyman discovers that the so-called process of the astragalus has a distinct ossification, and Morse interprets it as the intermedium. (*Am. Nat.* v. 1871, 524.) In the light of these late discoveries, the homologies of the bird's carpus and metacarpus become clearer. We have seen (§ 55, 56, fig. 6) that birds retain throughout life two distinct proximal carpal bones (called scapholunar and cuneiform, but better named simply *radiale* and *ulnare*), and that in early life they have a distal bone, that was mentioned as the magnum, but appears to be centrale, corresponding to the distal tarsal ossicle, just as the ulnare and radiale do to the proximal tarsal ossicles. Morse has even found in the carpus of birds, two more ossicles, the homology of which remains undetermined. But what we now know, renders it almost certain, that the so-called epiphyses upon the proximal ends of the metacarpals, are not epiphyses, any more than the so-called tarsal epiphyses ; and that the metacarpus of birds is really carpo-metacarpus, just as the metatarsus is actually tarso-metatarsus. This view is strengthened by the fact that the metacarpal bones of higher vertebrates, except the first, ordinarily lack epiphyses.

(d.) The principal metatarsal bone, *d*, representing the distance C D, between the lower end of the leg and the roots of the toes, really consists of three bones fused in one; these are partly distinct only in the penguins, among recent birds; but in all birds except ostriches, the original distinction is indicated by three prongs or claws at the lower end of the bone : for jointing with the three principal toes. The other toe, almost always the hinder one, when it is present, is hinged on the metatarsus in an entirely different way; by means of a separate little rudimentary bone, the ACCESSORY METATARSAL, *m*, in the figure, in dotted outline. It is of various shapes and sizes, and variable in position up and down the lower part of the metatarsus. Ordinarily it is too small, or too flat, to be seen from the outside of the foot at all; it has no true jointing with the main metatarsal, but is simply pressed flat against it, and more or less soldered, much as the lower part of the fibula is with the tibia. It may be wanting in some birds with no hind toe; in others, without hind toe, it still persists.

(e.) In spite of the anatomical proprieties involved, this part of the leg, from heel to bases of toes,—from C to D—represented really by the metatarsal bone and its accessory, has gained a name now so firmly established, that it would be finical to attempt to change it in ordinary descriptive writings. *This is* THE TARSUS; we shall soon see how important a thing it is.

(f.) The toes or *digits* consist of a certain number of bones placed end to end, all jointed upon each other, and the first series upon the metatarsal or its accessory. Each of these individual bones is called a *phalanx* (pl. *phalanges*) or *internode* (because intervening between the joints or *nodes* of the toes). The furthermost one of each toe almost invariably bears a claw. They are of various lengths relative to each other, and of variable number in the same or different toes; but these points, and others, are fully considered farther on. We may here glance at the

§ 73. (a.) MECHANISM involved. The hip is a ball-and-socket joint, permitting roundabout as well as fore-and-aft movements of the thigh. The knee is usually a hinge-joint only, allowing back and forward motion of the leg; so constructed that the forward movement is *never* carried beyond a right line with the leg, while the backward is so free that the leg may be completely doubled under the thigh. In some birds there are also rotatory movements at the knee, very evident in certain swimmers. The ankle or heel-joint is a strict hinge, and sometimes a wonderful one, too, taken in connection with the action of certain muscles that move the tarsus. For in some birds the interior structure of the joint is such that it locks the tarsus, when straightened out upon the leg, in that position, so firmly that some voluntary muscular effort is needed to overcome the resistance; such birds can sleep standing up on one leg, and this is the design of the mechanism. The ankle permits just the opposite bendings to those of the knee; the tarsus cannot pass backward out of a straight line with the leg; but can come forward until the toes nearly touch the knee. The jointing of the toes on the metatarsal bone is peculiar; for the hinge-surfaces of the metatarsal prongs have such mutual obliquity, that when the toes are brought forward, at right angles or thereabouts with the tarsus, they spread themselves in the action, and the open foot, with its diverging toes, are pressed on the ground or against the water; and when the toes are bent around in the other direction, they close together more or less parallel with each other, besides being bent or flexed, each one at its several nodes. The mechanism is best illustrated in the swimmers, which must present a broad surface to the water in giving the backward stroke, and bring the foot forward closed with only an edge opposed to the water. It is carried to such extreme in the loon, that the digit marked $2t$ in the figure lies below and behind $3t$, as there shown; in most birds with the foot in much the same position relative to the tarsus, $2t$ would appear above $3t$ (compare other figures of feet). It is probably least marked in birds of prey, that clutch with all the toes spread. The individual toe joints are all simple hinges.

(b.) In ordinary hopping, walking, perching, etc., only the toes rest upon or grasp the support, and c is more or less perpendicularly above D. This resting of the toes is complete for all the anterior ones; for the hind toe it varies according to the position and length of the latter from complete resting like the others, to mere touching of the tip, and finally to not even this; the hind toe is then said to be *functionless*. But the lowest birds cannot stand upright on their toes at all; these rest with the tarsus horizontal, and the heel c touching the ground; moreover, in all these birds, the tail affords additional support, making a tripod with the legs, as in the kangaroo. These birds might be called *plantigrade*, in strict anatomical analogy with the beasts so called; the others are *digitigrade*, quite as analogously; but there are no birds, that, like horses and cows, walk on the *ends* of their toes, or toe-nails. A bird's ordinary walking or running, corresponds exactly with ours, as far as the mechanics of motion are concerned; but its hopping, as it is called, is really leaping, both legs being brought forward at once. Nearly all birds down to *Gallinæ, leap* when on the ground; all others walk or run, advancing one leg after the other. Leaping is thus really distinctive of the *Insessores;* though many of them, as titlarks, shore larks, meadow larks, many terrestrial sparrows, blackbirds, crows, turkey buzzards, and others, including all the pigeon family, walk instead of leaping.

§ 74. THE PLUMAGE of the legs varies within wide limits. In general, the leg is feathered to the heel, and the tarsus and toes are naked. The thigh is ALWAYS feathered. The crus is feathered in all *Insessores* (with rare exceptions), and in all *Natatores* without exception; in the loon family the feathering extends on as well as to the heel-joint. It is among the *Cursores*, or walkers, and especially wading birds, that the crus is most naked; here it may be denuded half way up. A few waders—among ours, chiefly in the snipe family—have the crus apparently clothed to the joint, but this is in most if not all cases due to the length of the feathers, for probably no one of them has the crural pteryla itself extended to the joint. The crural feathers are almost always short and inconspicuous; sometimes long and flowing, as in nearly all the hawks, our tree-cuckoos, etc. The *tarsus* in the vast majority of birds is naked of feathers; it is so in all the higher *Insessores*, with very few exceptions (as in the swift family, for instance), in all waders, without exception, and in all swimmers with the single exception of the frigate bird (*Tachypetes*, gen. 277), and here the feathering is not complete. The *Raptores* and the *Gallinæ* give us the most feathered tarsi. Thus feathering is the rule, among the owls (*Strigidæ*); frequent (either partial or complete) in hawks and eagles, as the genera *Aquila* (161) *Archibuteo* (160) and *Buteo* (159). All our grouse, as distinguished from the turkeys and partridges of the same order, have the tarsus more or less feathered. The *toes* are feathered in few birds; but we have fine examples of this, in the snowy owl, and all the ptarmigan. Partial feathering of the tarsus is often continued further down to or on the toes by sparse modified bristly feathers; this is well illustrated in the barn owl. When incomplete, the feathering is usually wanting *behind* and *below*; being almost invariably continuous above with the crural feathering. But, in that spirit of delight that birds show in proving every rule we make about them by furnishing exceptions to it, the tarsus is sometimes partly feathered without connection with the general plumage above. A curious example is afforded by the bank swallow, with its little tuft of feathers at the base of the hind toe; and some varieties of the barnyard fowl sprout monstrous leggings of feathers from the side of the tarsus.

§ 75. THE LENGTH OF THE LEG, compared with the size of the bird, is extremely variable. A thrush or a sparrow probably represents about an average in this respect. The shortest-legged known bird is probably the frigate, just mentioned; a yard long, more or less, it has a tibia not half as long as the skull, and a tarsus under an inch. The leg is very short in the order *Strisores*, as among humming-birds, swifts, goatsuckers, kingfishers, trogons, etc.; while the swallows, of *Oscines*, are like swifts in this respect. It is likewise pretty short among *Scansores*. The leg is also "short" in all swimmers; the femur especially being very short, and the tarsus likewise; while the toes, bearing their broad webs, are longer. The leg lengthens in lower *Insessores*, as most hawks, and especially among some of the terrestrial pigeons. It is still longer among the walkers; and reaches its maximum among the waders, especially the larger kinds, as flamingoes, cranes, storks and herons, among all of which it is accompanied by corresponding increase in length of the neck. Probably the longest legged of all birds for its size is the stilt (*Himantopus*, 197). It is seen from the above, that, taking the tarsus alone, as an index of the whole comparative length of the leg, this is in the frigate bird under one thirty-sixth of the total length; a flamingo, four feet long, has a tarsus one foot; a stilt, fourteen inches long, a tarsus four inches; so the maximum and minimum of length of tarsus are represented by nearly thirty, and under three, per cent. of the bird's whole length.

§ 76. THE NAKED PART of the leg is covered, like the bill, by a hard-
ened, thickened, modified integument, which varies in texture between corneous and leathery. This is called the PODOTHECA (Gr. *podos*, of a foot,
theke, sheath). Land birds have the most horny covering, and water birds
the most skinny; in general this is distinctive of these two great divisions
of birds, and the exceptions are few. The perfectly horny envelope is
tight and immovably fixed, or nearly so, while the skinny is looser, and may
usually be slipped round about a little. The covering may also differ on
different parts of the same leg; in fact, such is usually the case to a degree.
Unlike the covering of the bill, that of the legs is NEVER simple and continuous throughout; it is divided and subdivided in various ways. The lower
part of the crus, when naked, and the tarsus and toes, are variously cut up
into scales, plates, tubercles, etc.; these have all received special names;
and moreover, the mode of this division becomes, especially among higher
birds, a matter of the utmost consequence, for purposes of classification,
since it is fixed and definite in the same groups.

§ 77. SCUTELLA (pl. of L. *scutellum*, a little shield; figs. 10, 11, *b*) are
scales, generally of large comparative size, arranged in definite up and down
lines, and apt to be imbricated, or fixed shingle-wise, with the lower edge of
one overlapping the upper edge of the next below. The great majority of
birds have them. They generally occur on the front of the tarsus (which is
called *acrotarsium*, and corresponds to our "instep"), and almost invariably on the top of the toes (called *acropodium*); frequently on the back
of the tarsus; not so often on the tibia, sides of the tarsus, sides and under
surfaces of the toes (if ever in the latter situation). A tarsus so furnished
is said to be *scutellate*, before or behind, or both, as the case may be;
the term is equally applicable to the acropodium, but the expression is
rarely used because the scutella are so commonly there.

§ 78. PLATES, or *reticulations* (L. *reticulum*, a little net or web; fig.
11, *a*), result from the cutting up of the envelope by cross lines in various
ways. Plates are of various shapes and sizes; but however they may be, in
these respects, they are distinguished from scutella by not appearing *imbricated;* their edges simply meet, but do not overlap. They are generally
smaller than scutella. The commonest shape is the six-sided, or hexagonal;
a form best adapted to close packing, as strikingly shown, and long ago
mathematically proven, in case of the cells of bees' honey-comb. They are
sometimes five-sided, or even four-sided; but are more likely to have more
sides, becoming irregularly polygonal, or even circular; when crowded in
one direction and loosened in another, this develops into the oval, or even
somewhat linear. A leg so furnished is called *reticulate;* it may be wholly
so, but is generally partly scutellate. A particular case of reticulation is
called

§ 79. GRANULATION (L. *granum*, a grain); when the plates become elevated into little tubercles, roughened or not. Such a leg is said to be *granulated* or *rugose;* it is well seen in the parroquet and fish hawk.

§ 80. WHEN the harder sorts of either scutella or plates are roughened without obvious elevation, the leg is said to be *scabrous* or *scarious*. But *scabrous* is also said of the under surfaces of the toes, when these develop special *pads*, or wart-like bulbs (called *tylari*); excellently shown in most hawks. The softer sorts of legs, and especially the webs of swimming birds, are often crosswise or otherwise marked by lines, without these being strong enough to produce plates; this is a condition analogous to the little raised lines and depressions seen on our own palms, and especially our finger-tips. Occasionally, the plates of a part of the leg become so developed as to form actual *serration;* seen on the hinder edge of the tarsus of grebes.

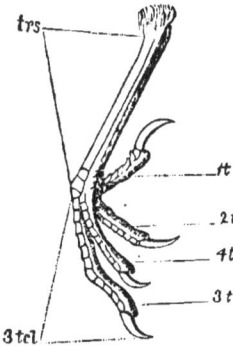

FIG. 9. "Booted" tarsus, of a robin.　　FIG. 10.　Scutellate tarsus, of a cat-bird.　　FIG. 11. *b.* Scutellate tarsus, of a pigeon; *a.* reticulate tarsus, of a plover.

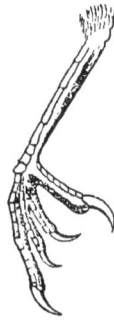

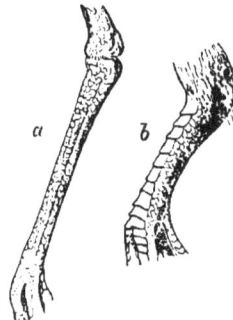

§ 81. WHEN an unfeathered tarsus shows on its front surface no divisions of the podotheca, or only two or three divisions close by the toes, it is said to be *booted*, and the podotheca is said to be *fused*. (Fig. 9.) This condition chiefly occurs in higher *Oscines*, and is supposed by many, particularly German ornithologists, to indicate the highest type of structure; but it is also found in some water birds, as Wilson's stormy petrel. It is not a very common modification. Among North American birds it only occurs in the following cases: — Genera *Turdus* (1), *Cinclus* (5), *Saxicola* (6), *Sialia* (7), *Regulus* (9), *Chamæa?* (11), *Myiadestes* (52) and *Oceanites* (307); and even these birds, *when young*, show scutella, which disappear with age, by progressive fusion of the acrotarsial podotheca.

§ 82. THE CRUS, when bare below, may present scutellation either before or behind, or both, as is seen in many waders where the crus is largely naked; often again, the crural podotheca may consist of loose, softish, movable skin, not obviously subdivided: sometimes it is truly reticulate, as in the genus *Heteroscelus* (221).

§ 83. THE TARSUS, in general, may be called subcylindrical; it is often quite circular in transverse section; very rarely thicker across than fore-and-aft (as in penguins); but very often thicker in the reverse direction. When this transverse thinness becomes noticeable, the tarsus is said to be

compressed: the form is seen in its highest development in the loon, where
the tarsus is almost like a knife-blade. Cylindrical tarsi occur chiefly when
there are scutella before and behind; it occurs in our shore lark (*Eremophila,* gen. 26), but is a rare modification among land birds, though very
common among waders. The tarsus of the vast majority of land birds is
seen, on close inspection, to be sharp-ridged behind, and gently rounded in
front. This generally results from the presence, in front, of a series of
scutella, associated, on the sides and hinder edge of the tarsus, with fusion,
or with a few large plates variously arranged. The meeting of these two
kinds of envelope on the sides of the tarsus is generally in a more or less
complete straight up and down line; either a mere flush trace of union, or
a ridge, oftener a groove (well seen in the crows) that may or may not be
filled in with a few small linear plates. But further consideration of special
states of the tarsal envelope, however important and interesting, would be
part of a systematic treatise, rather than of an outline sketch like this.

§ 84. THE TOES (individually, *digiti;* collectively, *podium*). Their normal number is FOUR : there are *never* more. The ostrich alone has only *two.*
There are *three* in all the auks (fam. *Alcidæ*) and albatrosses (subfam. *Dio-
medeinæ*) : in all struthious birds, except the ostrich and *Apteryx;* and in a
large number of waders (*Grallæ*). Three toes only occur as an anomaly
among Insessores, as in the cases of the exotic genus *Ceyx* of kingfishers,
and the genus *Picoides* of woodpeckers. North American three-toed birds
are only these : — the woodpeckers just named ; auks and albatrosses ; plovers
(except one, *Squatarola,* 189) ; the oystercatchers (*Hæmatopus,* 194) ; the
sanderling (*Calidris,* 211) ; the stilt (*Himantopus,* 197). In the vast
majority of cases, there are *three* toes in front, and *one* behind ; occasionally,
either the hind one, or the outermost front one, is *versatile,* that is, capable of
being turned either way ; the outermost one is mostly so in the owls, the fish
hawk (gen. 153), and a few other birds. We have no case of true versatility
of the hind toe among North American birds, but several cases of its *lateral*
stationary position (goatsuckers, some Western swifts, loons, and all the totipalmate swimmers) ; nor have we any example of that rarest condition
(seen in the European swifts, *Cypselus,* and in the *Coliidæ*) where *all four*
toes are turned forward. This only occurs in the order *Strisores.* The arrangement of toes *two* in front, and *two* behind, or *in pairs,* characterizes the
whole order *Scansores,* or climbers ; such birds are said to be *zygodactylous*
(yoke-toed ; see fig. 128). Our examples are the parrot, woodpeckers and
cuckoos, to which some add the trogons ; in all these, except the last named,
it is the outer anterior toe that is reversed. In nearly every three-toed bird,
all three are anterior; our single exception is the genus *Picoides* (132),
where the hind toe is *wanting,* the outer anterior reversed to take its place,
and only two left in front. No bird has more toes behind than in front.
All birds' toes are

§ 85. NUMBERED, in a certain definite order, as follows (see figs. 8, 9) :—
hind toe (1t) = *first* toe ; inner anterior toe (2t) = *second* toe ; middle an-

terior toe (3t) = *third* toe ; outer anterior toe (4t) = *fourth* toe. In birds.
with the hind toe reversed, the same order is obvious : only, inner anterior
toe = 1t, etc. In *zygodactyli* (except *Trogonidæ*), inner hind toe =1t ;
inner front toe = 2t ; outer front toe = 3t ; outer hind toe = 4t. Now
when the number of toes decreases, the toes are always reduced in the
same order : thus, in all three-toed birds, 1t is wanting : in the two-toed
birds 1t and 2t are wanting. This is proven by the

§ 86. NUMBER OF JOINTS, or number of *phalanges* (§ 72, *f*) of the toes.
The constancy of the joints in birds' toes is remarkable, one of the strongest expressions we have of the highly monomorphic character of the class
Aves. In all birds, 1t has *two* joints (not counting the accessory metatarsal). In all birds, 2t has *three* joints. In nearly all birds, 3t has *four* joints.
In nearly all birds, 4t has *five* joints. The only exceptions to this, consist
in the lessening of the joints of 3t by *one*, and the lessening of the joints
of 4t by *one* or *two*. So in all cases, where the joints do not run 2, 3, 4, 5,
for the toes from 1st to 4th, they run either 2, 3, 4, 4, or 2, 3, 3, 4, or 2,
3, 3, 3. This variability in number of the internodes is confined (wholly?)
to the order *Strisores*. Our examples are in the sub-families *Cypselinæ* and
Caprimulginæ (which see ; see also figs. 119 and 122). This admirable
conservatism enables us to always determine what toes are missing, in birds
with less than four ; thus, in *Picoides*, the hind toe, though seemingly 1t,
is evidently 4t, because 5-jointed ; in the ostrich, with only two toes, 3t
and 4t are seen to be preserved, because they are respectively 4- and 5-
jointed. (In fig. 8, the dotted line 1 indicates the first series of phalanges
of all the toes ; dot-line 2, the second ; the correspondence of the remaining
phalanges is seen at a glance.)

§ 87. THE POSITION of the toes, other than in respect of their direction,
is important. In ALL birds the front toes are on the same level, or so
nearly so, that the difference is not notable. And the same may be said of
the hind toes, when there are *two*, as in *Scansores*. But the hind toe, when
present and single, varies remarkably in position, and must have special
notice, as this character is important in taxonomy. The insertion of this
toe varies, from the very bottom of the tarsus, where it is on a level with
the front toes, to some distance up the tarsus. When flush with the bases
of the other toes, so that its whole under surface touches the ground, it is
said to be *incumbent*. When just so much raised that its tip only touches
the ground, it is called *insistent*. When so high up that it does not reach
the ground at all, it is termed *remote* (*amotus*). But as the precise position
varies insensibly, so that the foregoing distinctions are not readily perceived, it is practically best to recognize only two of these three conditions,
and say simply, "hind toe elevated," when it is inserted appreciably above
the rest, or "hind toe not elevated," when its insertion is flush with that of
the other toes. In round terms : it is characteristic of all *Insessores* to have
the hind toe DOWN ; it is characteristic of all other birds to have the hind toe
UP (when present). The exceptions to the first statement are extremely

rare; they are confined, among our birds, in any marked degree, to the two
genera of *Caprimulgidæ* (gen. 114, 115) and the turkey buzzard (*Cathartes*,
gen. 166); but among other *Raptores* besides *Cathartes*, such as certain
owls, and in some pigeons (lowest of *Insessores*, it will be remembered),
the toe is not *quite* down, or is even perceptibly uplifted. Technically, however, I take all these but the three first named, as having the toe down. It is
elevated in all our *Rasores* or *Gallinæ* (gen. 177 to 188); elevated in all our
waders except the *herons, ibises,* and *spoonbill;* the elevation is least marked
in the rail family, but still plain enough there. It is elevated in ALL
swimming birds, whether lobe-footed, or partly or wholly web-footed; but
in the *Totipalmate* order (*Steganopodes*, gen. 273 to 278) where it is lateral
and webbed with the inner toe, the elevation is slight. Now since, curiously
enough, the only three of our insessorial genera above mentioned (two of
Caprimulgidæ, and *Cathartes*) that have the hind toe well up, have also
little webs connecting the anterior toes; and since some *Raptores* are our
only other Insessores with any such true webs; and since herons, ibises
and spoonbills are our only birds with such true webs, that have the hind
toe down, the following rule is infallible for all our birds : *Consider the
hind toe* up *in every bird with any true webbing or lobing of the front toes,*
except herons and their allies and some birds of prey. The converse,
also, holds nearly as well; for our only birds with fully-cleft anterior toes,
and hind toe up, are the rails and gallinules, the black-bellied plover
(our only 4-toed plover), the turnstone, the woodcock, Wilson's snipe, and
most of the true sandpipers. Besides its versatility of position the hind
toe has

§ 88. OTHER NOTABLE CHARACTERS. It is *free* and *simple*, in the vast
majority of birds; in *all Insessores*, nearly all *Cursores*, and most *Natatores*. In length, it may equal or surpass (with its claw included) the
longest anterior toe, and generally surpasses at least one or two of them.
It is never so long as when down on a level with the rest; here also, it
attains its greatest mobility, and among *Passeres* is virtually provided with a
special muscle for its apposition with the others in the act of grasping. In
general, it grows shorter as it gets higher up; and probably in no bird
where it is truly elevated, is it so long as the shortest anterior toe. It is
short and barely touches the ground in most waders; shorter still in some
swimmers, as the gulls, where probably it is functionless; rudimentary in
one genus of gulls, *Rissa* (284), where it bears no perfect claw; represented
only by an immovable sessile claw, liable to be overlooked unless carefully
sought for, in the petrels; it disappears in the birds above named (§ 84),
and some others. It is never actually joined by direct soldering to either of
the other toes, for any noticeable distance; but is united to the base of the
inner toe by a web in the loons, and to the whole length of the inner toe in
all the *Steganopodes* (fig. 183). But it may be, as it were, independently
webbed; that is, have a lobe or flap of membrane hanging from it; this condition is seen in all the sea-ducks (*Fuligina*, gen. 260 to 270), and in all

our truly lobe-footed birds. I may finally consider the modes of union of the anterior toes under the head of the .

§ 89. THREE MODIFICATIONS OF THE BIRD'S FOOT. All birds' feet are built upon one or the other of three plans, corresponding to the three subclasses *Insessores, Cursores* and *Natatores.* These are the *perching* plan, the *walking* or *wading* plan, and the *swimming* plan; and these are pretty sharply distinguished (independently of differences in the number and position of the toes) by the method of union. In the perching plan, the toes are only very exceptionally connected by true movable webbing; they are *cleft to the base,* or else joined, for a part, or the whole, of one joint, or a part also of the second joint, by *actual cohesion.* Our thrushes show about as complete cleavage as is ever seen; our wrens, titmice, creepers, etc., exhibit considerable basal cohesion. A remarkable exception is seen in the *syngnesious* foot; where the outer and middle toes fuse for nearly their whole length; the kingfisher (figs. 116, 117), illustrates this; and all such birds are called *syndactylous* (Gr. *sun* together, *dactylon* a finger). In the walking plan, the toes are never, probably, thus joined by fusion; and they are seldom cleft to the base; the union is generally by a movable basal web, of variable extent. This constitutes the *semipalmate* ($\frac{1}{2}$-webbed, that is,) *foot.* But the webs occasionally, in true wading birds, run out to the ends of the toes, as in the avocet (gen. 196), and in the flamingo (if indeed this bird really belongs among waders). Generally they run out to the end of the first, or along part of the second joint, constituting true semipalmation; shown in the semipalmated sandpiper and willet. (Figs. 166, 170.) Oftener the web is of about this size between the outer and middle toes, and slighter or wholly deficient between the middle and inner; this is shown in nearly all our larger waders, including herons. (It is also the usual state of webbing of those hawks that have semipalmation.) In the swimming plan, the foot is changed into a paddle by webbing or lobing; the former constitutes the *palmate,* and the latter the *lobate,* foot. In the palmate, the webbing is usually complete betwixt the three front toes; it is extended to the hind toe, likewise, in all *Steganopodes,* and partly in the loons. Sometimes the webbing is defective, from deep *incision,* or cutting away of the free anterior border of the webs for some distance : this is seen partly in the genus *Philacte* (249) and much more so in the short-tailed tern, *Hydrochelidon* (gen. 292; fig. 208), where it simulates semipalmation. But in such a case, if the fresh foot be carefully examined, the webbing will be seen running as a narrow border, quite to the claws, as usual. Frequently, one web is larger than the other, as in all our terns (fig. 207, for example) where the inner web is somewhat defective. In the lobate foot, instead of connecting webs, we have a series of broad lobes along each joint of the toes, as in the coot, and all the grebes : but it is almost always, if not always, associated with semipalmation. It occurs, again, in some wading birds, as the remarkable family of the phalaropes, which swim, in fact, better than they walk. Here the lobation may be either *scolloped,* or cut out at the joints, as in the coot, or *plain*, that is, straight-edged. (Fig. 162.) True lobation, occurring, among North American birds, only in the grebes, coots, and phalaropes, must be carefully distinguished from various

§ 90. MARGINAL FRINGES, or processes, that birds of the lower orders often exhibit. Thus, if a gallinule be examined in a fresh state, it will be found to have a margin of membrane running along the sides of the toes, and the same is the case, if less evident, in a great many waders. Palmate birds also show it, on the free borders of 2*t* and 4*t;* it is very conspicuous in the albatrosses, and plain enough in geese, &c. In the grouse family there is a remarkable development of horny substance, resembling a real fringe, being cut into a series of sharp teeth, or pectinations.

§ 91. THE CLAWS. With certain anomalous exceptions, as in case of a rudimentary hind toe, every toe bears a claw. The general shape of the claw is remarkably constant throughout birds: variations are in degree only, rather than in kind. A cat's claw represents nearly the usual shape, viz: *compressed, arched, acute.* The great talons of a bird of prey are only the extreme of this typical shape. Besides this general shape, the claws are usually dug out underneath, so that the transverse section, as well as lengthwise outline below, is concave, and the under surface is bounded on either side by a sharp edge. One of these edges, and particularly the inner edge of the middle claw, is somewhat dilated or expanded in a great many birds; and in some it becomes changed into a perfect *comb,* by having a regular series of teeth. This pectination occurs only on the inner edge of the middle claw; it is beautifully shown by all the true herons (*Ardeidæ*); by the whippoorwills and nighthawks, by the frigate pelican, and, to a less degree, by the barn owl. It is supposed to be used for cleaning out lice from parts that cannot be reached by the bill; but this is open to question, seeing that outside the herons, it chiefly occurs among very short-legged birds, that cannot possibly reach many parts of the plumage with the toes. Besides *Raptores*, most perching birds are very sharp-clawed; the claws are more obtuse among the pigeons and *Gallinæ* (scratchers) and still more so among most swimming birds. Obtuseness is generally associated with flatness, or *depression;* this is seen in Wilson's petrel, as distinguished from all our others, and carried to the extreme in the grebes, where the claws resemble human nails. The deviations from curvature occur principally in the hind claw; this is straight or nearly so, in the shore lark, and some terrestrial sparrows, as the genus *Plectrophanes* (63). All the claws are straight, and prodigiously long, in some exotic birds of the rail tribe—the jacanas (*Parra*); this enables the birds to run lightly over the floating leaves of aquatic plants, by so much increase of breadth of support that they do not slump in. Claws are also variously carinate, sulcate, etc. They are always horny. They take name from and are reckoned by the digits they belong to: thus, 1*cl.* = claw of 1*t*: 2*cl.* = claw of 2*t*, etc.

SECT. IV.   DIRECTIONS.—HOW TO USE THE KEY.—HOW TO MEASURE A SPECIMEN, ETC.

1.   HOW TO USE THE KEY.

§ 92.   WE have in hand a bird which we know nothing about, and desire to *identify;* that is, to discover its name and position in the system; and to learn whatever else the present volume may afford.   Let us suppose it to be that little black and white spotted bird which we often see climbing about our fruit trees, boring holes in the bark.

The Key opens with an arbitrary division of our birds, according to the number and position of their toes.   Our specimen, we see, has four toes, arranged in pairs; that is, two before and two behind.   It therefore comes under the third division (III).   Turning to III, we read :—

Bill with a cere, and strongly epignathous, etc.,
— not cered; inner hind toe with 3 joints, etc.,
        — only 2 joints.   (f)

We see that the bill of the specimen is neither cered nor hooked, and that the inner hind toe is 2-jointed.   Following, therefore, the reference-letter (f), we find three alternatives, viz.,

(f) Tail of 8 feathers, etc.,
    — 10 soft feathers, etc.,
    — 12 (apparently only 10) rigid acuminate feathers.   (g)

The tail feathers of the specimen are stiff and pointed, and we count ten perfect ones, besides a rudimentary pair concealed at the bases of the others. Evidently, then, we continue with the reference letter (g), as follows :—

(g) Birds > 14 inches long, etc.,
(g) Birds < 14 in.; ridges on upper mandible reaching tip, etc., ... Picus, 131.

The specimen is much less than fourteen inches long, and the sharp ridges on the sides of the upper mandible run quite to the end of the bill; and here, at last, instead of a reference-letter, we find a genus named; which is the one to which the specimen belongs.   The bird is a *Picus.*

§ 93.   THUS the key conducts to a genus, by presenting in succession, certain *alternatives,* on meeting with each of which, the student has only to determine which one of the two or more sets of characters agrees with those afforded by his specimen.   There will not, it is believed, be any trouble in determining whether a given character *is so,* or *is not so,* since only the most tangible, definite, and obvious features have been selected in framing the key.   After each determination, either the name of a genus is encountered, or else a reference-letter leads on to some new alternative, until by a gradual process of elimination the proper genus is reached.   After a few trials, with specimens representing different groups, the process will be shortened, for the main divisions will have been learned; still, the student must be careful how he strikes in any where except at the beginning, for a false start will soon set him hopelessly adrift. The Key has been tested [*] so thoroughly that there is little danger of his running off the track except through carelessness, or misconception of technical terms; but there is no excuse for the former, and the latter may be obviated by the Glossary and the Introduction, which should be consulted when any doubt arises. Time spent upon the Introduction will be time saved in the end.

§ 94. Now the genus *Picus* that we found has a number after it, which refers to the Systematic Synopsis, where the genera are numbered consecutively. The running numbers at the top of the pages catch the eye in a moment, and enable us to turn directly to *Picus*, 131. Here we find a few remarks, illustrative of the general character of all our species of the genus; and these we see, are six in number. We have now to find out which one of the six ours is; and to this end they are *analyzed*, that is, mapped out in groups, in such way that we perceive their most striking features, or diagnostic characters, almost at a glance : —

    [*] Body not banded, streaked nor spotted.
    [**] Spotted and crosswise banded, but not streaked.
  [***] Spotted and lengthwise streaked, but not banded.
      [†] Usually 9-10 long; outer tail-feathers wholly white.
     [††] Usually 6-7 long; outer tail-feathers barred with black and white.

The specimen has no transverse bars of color on the body, but a long white streak down the back, and a profusion of white spots on the wings and their coverts; it is not over seven inches long, and has the outer tail feathers black and white; so that we know it comes under [***][††]. As there is but one species given there, our bird is at last identified. It is the downy woodpecker, *Picus pubescens*. The term *pubescens*, at the end of the descriptive paragraph, is the specific name, which, joined with the generic name, *Picus*, constitutes the scientific designation of the species, as explained in the Introduction, p. 13. In this case of the downy woodpecker, no full description appears, merely because the bird "is exactly like *P. villosus*" (the preceding species) except in the diagnostic points of size and barred rectrices; but in general, a concise specific description will be found. These descriptions are not always, or even usually, full and complete; being designed simply to discriminate the several species of the same genus, or to certify that the student has discovered the right species, if there be but one under the genus. But since mere identification of a specimen is not all that we may desire, many other particulars are really given. Thus we discover that the downy woodpecker inhabits Eastern North America, and is replaced in the West by a variety closely resembling it. We discover its exact relations to its congener, *P. villosus*, and of both these to the other

---

[*] In the cases of over nine-tenths of the genera, by actual comparison with the specimens themselves, and found to give accurate results. It is just possible, that an occasional *immature* specimen, or one offering unusual deviation from the normal standard, cannot be determined by the Key.

species of the genus. We have a reference to several standard authors, which may be consulted if desired. Turning back a few pages, we find that the genus *Picus* belongs to the sub-family *Picinæ*, of the family *Picidæ*, of the order *Scansores;* and each of these groups is defined, illustrated, or otherwise noticed. In this way, it is believed, a single specimen may be made the means of imparting no inconsiderable amount of information.

2. HOW TO MEASURE A SPECIMEN.

§ 95. For large birds, a tape line showing inches and fourths will do : for small ones, a foot rule, graduated for inches and eighths, or better, decimals to hundredths, must be used ; and for all nice measurements the dividers are indispensable.

§ 96. In comparing measurements made with those given in the Synopsis, absolute agreement must not be expected ; individual specimens vary too much for this. It will generally be satisfactory enough, if the discrepancy is not beyond certain bounds. A variation of, say, five per cent., may be safely allowed on birds not larger than a robin : from this size up to that of a crow or hawk, ten per cent. ; for larger birds even more. Some birds vary up to twenty or twenty-five per cent., in their total length at least. So if I say of a sparrow for instance, "length five inches," and the specimen is found to be anywhere between four and three-fourths and five and one-fourth, it will be quite near enough. *But :* — the relative proportions of the different parts of a bird are much more constant, and here less discrepancy is allowable. Thus "tarsus longer than the middle toe," or the reverse, is often a matter of much less than a quarter of an inch : and as it is upon just such nice points as this that a great many of the generic analyses rest, the necessity of the utmost accuracy in measuring, for use of the key, becomes obvious. When I find it necessary to use the qualification "about" (as, "bill *about*=tarsus") I probably never mean to indicate a difference of more than five per cent. of the length of the part in question.

§ 97. "LENGTH." Distance between the tip of the bill and the end of the longest tail feather. Lay the bird on its back on the ruler on a table, take hold of the bill with one hand, and of both legs with the other ; pull with reasonable force, to get the curve all out of the neck ; hold the bird thus with tip of the bill flush with the end of the rule, and see how much the end of the tail points to. Put the tape line in place of the ruler, in the same way, for larger birds.

§ 98. "EXTENT." Distance between the tips of the outspread wings. They must be *fully* outstretched. With the bird on its back, crosswise on the ruler, its bill pointing to your breast, take hold of right and left metacarpus with thumb and forefinger of your right and left hand, respectively, *stretch* with reasonable force, getting one wing-tip flush with one end of the ruler, and see how much the other wing-tip points to. With large birds, pull away as hard as you please, and use the table, floor, or side of the room, as convenient ; mark the points and apply tape line.

As this measurement cannot be got at all from dried skins, I do not often use it in this book. But it is highly important, and for the very reason that it cannot be got afterwards, always *note it down* from fresh specimens. The first measurement, likewise, can only be got at approximately in skins, and the following details are really our chief data in all cases :—

§ 99. "LENGTH OF WING." Distance from the angle formed at the (carpus) bend of the wing, to the end of the longest primary. Get it with compasses for small birds. In birds with a convex wing, do not lay the tape line over the curve, but under the wing, stretching in a straight line from the carpal angle, to end of longest primary. This measurement is the one called, for short, "the wing ;" thus when I say, simply, "wg. 12," I mean that this distance is twelve inches ; so, also, "wg. $= \frac{1}{2}$ tl.," means that this distance is half as great as the length of the tail.

§ 100. "LENGTH OF TAIL." Distance from the roots of the rectrices, to the end of the longest one, whichever one that may be. Feel for the pope's nose ; in either a fresh or dried specimen, there is more or less of a palpable lump into which the tail feathers stick. Guess as near as you can to the middle of this lump ; place the end of the ruler opposite the point, and see how much the tip of the longest tail feather points to. "Depth of fork " and "amount of gradation," in a tail, is the difference between the shortest and the longest tail feathers ; in the one case the outer, in the other the middle, pair of rectrices is the longest.

§ 101. "LENGTH OF BILL." Exactly what this is, depends upon the writer. Some take the curve of the upper mandible ; others the side of the upper mandible from the feathers ; others the gape, etc. I take *the chord of the culmen.* Place one foot of the dividers on the culmen just where the feathers end — no matter whether the culmen runs up on the forehead, or the frontal feathers run out on the culmen, and no matter whether the culmen is straight or curved. Then with me the *length of the bill* is the shortest distance from the point just indicated to the tip of the upper mandible. Measure it with the dividers. In a straight bill, of course it is the length of the culmen itself ; in a curved bill, however, it is quite another thing. The "depth of bill" is determined opposite the same point ; it is a perpendicular transverse dimension : the "width of the bill" is determined at the same point ; it is the horizontal transverse dimension. "The gape" is the shortest distance between the commissure proper (see § 53, and fig. 5, *h*) and the tip of the upper mandible.

§ 102. "LENGTH OF TARSUS." This is the most important measurement for the purposes of this volume. Measure it *always* with dividers, and *in front of* the leg. It is the distance between the joint of the tarsus with the leg above, and that with the first phalanx of the middle toe below. Place one foot of the dividers exactly upon the middle of the tibio-tarsal joint in front. The front of this joint is rounded on either side by two little semi-circular rims, or lateral elevations, more or less evident in different birds ; you want to get just between them. In the softer-legged wading, or water birds, there is a slight elevated point right in the middle; this, or the position of it in other birds, is the precise place. Place the other foot of the dividers over the transverse line of jointing of the base of the middle toe. This latter point, in all birds, when the toes are bent backward, becomes a more or less salient angle easily determined. In hard-legged birds it is usually indicated by the termination of last tarsal scutellum : in water birds, there will be seen a little crosswise nick, showing just where the skin has shrunk into the crack between the end of the metatarsus and the base of the toe. It will be evident that a measurement taken as here directed will not always be the same as one taken behind, up over the convexity of the heel, and down to the level of the sole; but there are behind no other tangible points of termination. (See fig. 9, *trs.*) What, now, is the meaning of the expression—"$b. = \frac{1}{2}$ *trs.*"?

§ 103. "LENGTH OF TOES." Distance in a straight line along the upper surface of a toe, from the point last indicated, to the root of the claw on top. Observe that, as the claws are inserted upon the ends of the toes, somewhat as the nails are on our fingers, this measurement is a different thing from one taken along the under surface of the toes. Always make it with the dividers. Length of toe is always taken *without* the claw unless otherwise specified. When no particular toe is specified, $3l$ is always meant. (See fig. 9, *3tcl.*) Define this expression : —"*trs.* > $3l$."

. § 104. "LENGTH OF THE CLAWS." Distance in a *straight line* from the point last indicated to the tip of the claw. (See § 101.) When this measurement is meant to be included in the length of toe, I say *tcl*. Determine this : —"*trs.* < *3tcl*."

§ 105. "LENGTH OF HEAD" is an often convenient dimension for comparison with the bill. Set one foot of the dividers on the base of culmen (determined as above), and allow the other to just slip snugly down over the arch of the occiput. This is the required measurement. What does this mean : —"*hd.* = b."?

§ 106. ALL MEASUREMENTS are in the English inch and vulgar fractions or decimals, unless otherwise specified.

§ 107. FINALLY, it may be well to call attention to the fact, that most persons unaccustomed to handling birds are liable to be deceived in attempting to *estimate* a given dimension; they generally make it out *less* than measurement shows it to be. This seems to be an optical effect connected with the solidarity of the object, as is well illustrated in drawing plates of birds, which, when made exactly of life-size. always look larger than the original, on account of the flatness of the paper. The ruler or tape-line, therefore, should always be used, and are more particularly necessary in those cases where analyses in the Key rest upon dimensions. It is hardly necessary to add, that in taking, approximately, the total length from a prepared specimen, regard should be had for the "make-up" of the skin. A little practice will enable one to determine pretty accurately how much a skin is stretched or shrunken, and to make the due allowance in either case.

ABBREVIATIONS USED.

abd. Abdomen.
ad. Adult.
aut. Autumn.
axill. Axillaries.
b. Bill.
bl. Blue.
blk. Black.
Br. Am. British America.
brn. Brown.
brst. Breast.
cl. Claw, claws.
col. Color.
col'd. Colored.
comm. Commissure.
culm. Culmen.
Eur. Europe.
fthr. Feather.
fthr'd. Feathered.
fthrs. Feathers.
hd. Head.
gon. Gonys.
gr. Gray.
grn. Green.
intersc. Interscapularies.

lgth. Length.
mand. Mandible.
max. Maxilla.
Mex. Mexico.
N. Am. North America (at large).
nost. Nostrils.
obs. Observation.
occip. Occiput.
ole. Olive.
plmg. Plumage.
prim. Primary — ies.
purp. Purple.
rect. Rectrices.
rem. Remiges.
retic'tions. Reticulations.
retic. Reticulate.
rmp. Rump.
scap. Scapularies.
scut. Scutella.
scutl. Scutellate.
sec. Secondary — ies.
'sh. Diminishing suffix; as *blk'sh*, blackish.
spr. Spring.

sum. Summer.
superc. Superciliary.
wg. wgs. Wing, wings.
win. Winter.
W. I. West Indies.
t. Toe, toes.
tcl. Toe and claw together.
tert. Tertiary — ies.
tib. Tibia.
thrt. Throat.
tl. Tail.
trs. Tarsus.
un. mand. Under mandible.
un.-tl.-cov. Under tail coverts.
un.-wg.-cov. Under wing coverts.
up. mand. Upper mandible.
up.-tl.-cov. Upper tail coverts.
up.-wg.-cov. Upper wing coverts.
U. S. United States, except Alaska; usual abbreviations for names of States and Territories.
wht. White.
yell. Yellow.
yg. Young.

A few contractions, not given above, are self-explanatory.

SIGNS USED.

♂. Male.
♀. Female.
O♂, or *yg.* ♂. Young male.
O♀, or *yg.* ♀. Young female.
= Sign of equality; *generally,* as long as.
> More; *generally,* longer than; *also,* greater than, or more than.
< Less; *generally,* shorter than; *also,* smaller than, or less than.

| Certainty, with personal responsibility (not exclamation or surprise). All other punctuation as usual.

*, †, ‡, etc. Refer as usual to foot notes, when at the end of a word: when before a word or paragraph, they are used to point off sections in a manner that will be evident.

⁎ Interpolated sentences.

§ Complete paragraphs (in the introduction only).

WORKS REFERRED TO.

I quote throughout the following standard American works when they notice the species in question : —

" *Wils.*" WILSON, ALEXANDER. American Ornithology. 9 vols., 4to. 1808-14. (The original ed., and Ord's continuation.)

" *Nutt.*" NUTTALL, THOMAS. Manual of the Ornithology of the United States and of Canada. 2 vols. 12mo. (The first ed., of 1832-3f. unless the 2d (of 1840) is specified.)

" *Aud.*" AUDUBON, JOHN JAMES. Birds of America. 7 vols., 8vo. 1840-44. (Octavo reprint of the " Ornithological Biography," repaged and with systematic arrangement and renumbering of the plates of the folio edition.)

" *Cass.*" CASSIN, JOHN. Illustrations of the Birds of California, Texas, etc. 8vo, 1 vol. 1853-55.

" *Bd.*" " *Cass.* in Bd." " *Lawr.* in Bd." BAIRD, SPENCER F., with the coöperation of JOHN CASSIN and GEORGE N. LAWRENCE. Birds of North America; constituting the ninth vol. of the Pacific Railroad Explorations and Surveys. 1858. (Also republished separately, with a few additions and a 2d vol. of 100 plates, by the Naturalists' Agency. 1870.)

" *Ell.*" ELLIOT, D. G. Birds of North America. 2 vols., folio. (Plates and descriptions of many species recently introduced to our fauna, or before unfigured.)

" *Coop.*" Birds of California. From the MSS. notes of J. G. COOPER. Edited by S. F. Baird. 1 vol. 8vo, 1870.

I also quote, in particular cases, papers from the proceedings of different societies, etc., by various writers. The references in these instances are sufficiently explicit.

The Roman numerals immediately after the italicized author's name, refer to the *volume* ; the next figure, to the *page* ; " pl." with figures after it, to the number of the *plate* ; " fig.," to the number of the *figure*.

Lyre Bird of Australia. (See § 70.)

TOES 3, — 2 IN FRONT, 1 BEHIND. . . . . PICOIDES 132
TOES 3, — 3 IN FRONT. (**II.**)
TOES 4, — 2 IN FRONT, 2 BEHIND. (**III.**)
TOES 4, — 3 IN FRONT, 1 BEHIND. (**IV.**)

II. [TOES 3. — 3 IN FRONT.]

Toes incompletely, or not webbed. (a)
Toes completely webbed. (d)

(a) Naked leg and foot together about wing. Bill subulate; one basal web. . . . . . HIMANTOPUS 197
    — much < the wing. (b)
(b) Bill much > tarsus, truncate at tip; trs. reticulate. Birds over 12 inches long, . . . HÆMATOPUS 194
    — much <, or about = trs. Birds under 12 inches long. (c)
(c) Tarsus in front scutellate, about = bill. . . . . CALIDRIS 212
    — reticulate, > bill; plumage speckled. . . . . CHARADRIUS 190
      — not speckled; trs. nearly twice = 3t., ... PODASOCYS 192
        — not nearly twice = 3t., ... ÆGIALITIS 191
(d) Nostrils tubular; sides of under mandible not sulcate, ... DIOMEDEA 296
    — with a long colored groove, ... PHŒBETRIA 297
    — not tubular (linear, oval, etc.) (e)
(e) Nostrils naked; eyelids horny; both mandibles sulcate. Not crested . . . . FRATERCULA 318
    — simple. Birds > 12 long; up. mand. sulcate. Crested, ... LUNDA 319
      — not sulcate. Crested, . . . CERATORHINA 320
      — < 12 long; un. mand. falcate, up. mand. oval, . . PHALERIS 321
        — not falcate; up. mand. wrinkled, . PTYCHORHAMPHUS 323
          — smooth, . . . SIMORHYNCHUS 322
(e) Nostrils incompletely feathered; tail nearly even; b. and trs. compressed. . SYNTHLIBORHAMPHUS 325
    — gradated; bill and tarsus not compressed, ... MERGULUS 324
(e) Nostrils completely feathered; tarsus in front reticulate. Birds under 12 long, . BRACHYRHAMPHUS 326
    — 12 or more long. . . . URIA 327
    — scutellate; b. not sulcate or cultrate, ... LOMVIA 328
      — sulcate. Bird < 24. . . . UTAMANIA 317
        Bird > 24, ... ALCA 316

III. [TOES 4, IN PAIRS. 2 IN FRONT, 2 BEHIND.]

Bill with a cere, and strongly epignathous; tarsus granulated. . . . . CONURUS 138
    — not cered; inner hind toe with 3 joints; plumage iridescent, ... TROGON 125
    — only 2 joints. (f)
(f) Tail of 8 feathers; upper mandible sulcate; sides of head partly naked, ... CROTOPHAGA 126
    — 10 soft feathers; tarsus > middle toe and claw; lores bristly; birds about 2 feet long, GEOCOCCYX 127
      — < middle toe and claw; lores soft; birds about 1 foot long, . COCCYZUS 128
    — 12 (*apparently* only 10) rigid acuminate feathers. (g)
(g) Birds > 14 inches long, conspicuously crested; bill and nasal feathers not dark. . . . CAMPEPHILUS 129
      — dark, . . . HYLOTOMUS 130
(g) Birds < 14 in.; ridges on up. mand. reaching tip; tongue acute, barbed. No yellow, . . . PICUS 131
    — ridges running into tomium; tongue obtuse, brushy. Some yellow, . SPHYRAPICUS 133
    — ridges wanting, or indistinct and not reaching tip or tomia. (g²)
    (g²) plumage of belly bristly, of back with metallic iridescence, ... ASYNDESMUS 136
      — normal, with many round black spots, ... COLAPTES 137
        — not spotted; not white, ... CENTURUS 134
          — white, ... MELANERPES 135

IV. [Toes 4, — 3 in front, 1 behind.]

§. Hind toe inserted above the level of the rest (and always shorter than the shortest anterior toe). (**A**)

§. Hind toe not inserted above the level of the rest (and *generally but not always* not shorter than the shortest anterior toe). (**B**)

A. (*The hind toe elevated.*)

1. Feet TOTIPALMATE; (*all 4 toes webbed; hind toe semilateral and barely elevated.*) (A)

2. Feet PALMATE; (*3 front toes completely webbed, hind toe well up, simple or lobed, free or connected by slight webbing with base only of inner toe.*) (B)

3. Feet LOBATE; (*3 front toes (partly webbed, or not, and)* CONSPICUOUSLY *bordered with plain or scolloped membranes; hind toe free, and simple or lobed.*) (C)

4. Feet SEMIPALMATE; (*2, or 3, front toes webbed at base only by small yet evident membrane; hind toe well up, simple.*) (D)

5. Feet SIMPLE; *front toes with no evident membranes; hind toe well up, simple.* (E)

(A) Tarsus feathered, partly; tail deeply forked; bill epignathous, ... TACHYPETES 277
— naked; bill > tail, hooked at tip, furnished with an enormous pouch, ... PELECANUS 274
—< tail; throat feathered; middle tail feathers filamentous, ... PHAËTHON 278
— naked; tail pointed, soft; tomia subserrate, ... SULA 273
— rounded, stiff; bill paragnathous, ... PLOTUS 276
— epignathous, ... GRACULUS 275
(B) Hind toe somewhat lateral, and joined by slight web to base only of inner toe, ... COLYMBUS 311
— directly posterior, free, and simple or lobed. (h)
(h) Bill — recurved, depressed at base, subulate, extremely acute, ... RECURVIROSTRA 196
— bent abruptly downward near its middle, and lamellate, ... PHOENICOPTERUS 246
— neither recurved, nor abruptly bent. (i)
(i) Bill — *hypognathous*, corneous, cultrate, sulcate, ... RHYNCHOPS 295
— paragnathous, corneous, *not* lamellate; nostrils *not* tubular; tail *not* even. (k)
— epignathous (or paragn. and tl. even), corneous, *not* lamellate; nostrils *not* tubular. (l)
— epignathous, corneous, *not* lamellate; *nostrils tubular.* (m)
— paragnathous, mostly membranous, *lamellate,* nostrils not tubular. (n)
(k) Tail graduated, and middle feathers shorter than next pair. Plumage sombre brown . . . ANOUS 294
— forked; toes almost semipalmate. Black, brown or ashy, and white, ... HYDROCHELIDON 292
— well webbed; feet not black; back pale; no crest, ... STERNA 291
— blk.; wht. crescent on forehead (Sterna, 291, or) . HALIPLANA 293
— no crescent; not cre-ted; b. barely > trs., GELOCHELIDON 289
— crested; b. much > trs., . THALASSEUS 290
l) Bill with a sort of cere; middle tail feathers exserted; tarsus < 3tcl., ... BUPHAGUS 279
— = 3tcl., ... STERCORARIUS 280
— not cered; hind toe rudimentary, not bearing a perfect claw, ... RISSA 284
— perfect; tail wedge-shaped; a dark collar round neck, . . RHODOSTETHIA 286
— forked; bill black, tipped with yellow, ... XEMA 287
— reddish, not tipped with yellow, . . . CREAGRUS 288
— even; tarsus black, rough; webs inci-ed, ... PAGOPHILA 285
— not black. Under plumage — (l²)
(l²) dark, head white, tail black, bill and feet reddish, ... BLASIPUS 282
(l²) white, head dark (if dark, head whitish), ... CHROCOCEPHALUS 283
(l²) white, head white (if dark, head not whitish), ... LARUS 281
(m) Tarsus not < 3t.; claws depressed, obtuse; tarsal scutella fused; webs with yellow, . OCEANITES 307
— distinct; webs black, ... FREGETTA 308
— compressed; tail cuneate; no white anywhere, ... HALOCYPTENA 303
— nearly even. Blk. or smoky brown, and white, PROCELLARIA 304
— forked. Blk. or smoky brown, and white, . CYMOCHOREA 305
— Not black and white, ... OCEANODROMA 306
m) Trs. < 3tcl.; tail of 12 feathers; nasal tube obliquely truncate, septum thick, ... PUFFINUS 310
— vertically truncate, septum thin, ... PROFINUS 309
— > 12 flirs.; plmg. conspicuously spotted, ... DAPTION 301
— unspotted; tl. cuneate. . . . . . AESTRELATA 302
— not cuneate; 16-fthr'd. OSSIFRAGA 298
— 14-fthrd. b. < trs., FULMARUS 299
— = trs., PRIOCELLA 300
(n) Lamellae acute, like saw-teeth, retrorse; bill terete, black; trs. = ¾ 3t., ... LOPHODYTES 272
— not black; trs. > ¾ 3t., ... MERGUS 271
— simple; bill depressed toward end; lores naked. Adult entirely white, ... CYGNUS 247
— feathered; trs. in front — reticulate. (o)
— scutellate. (p)
(o) Trs. not > 3tcl. Plumage partly lavender-colored, head white, throat black, ... PHILACTE 249
— > 3tcl.; bill and legs not black. White or gray, bluish, speckled, etc., ... ANSER 248
— black; neck all black; nostrils median, ... BRANTA 250
— not all black; nostrils subbasal, . . . DENDROCYGNA 251

(p) Hind toe *simple*; head crested, and narrow tip of bill formed wholly by the nail, ... AIX 259
— not crested; bill *much* wider at end than at base, ... SPATULA 258
— not wider; tail cuneate, ⅓ or more of the wing, . . DAFILA 253
— not cuneate, not ½ the wing. (p²)
(p²) Bill < head; crown streaked; tl. fthrs lance-acute; ♀ and yg. of . DAFILA 253
— creamy or white; speculum green, ... MARECA 255
— about = hd.; speculum white; wing coverts chestnut, CHAULELASMUS 254
— little > hd. speculum violet, black and white bordered, ... ANAS 252
— green; wing coverts sky blue, . QUERQUEDULA 257
— not blue, . . . NETTION 256
(p) Hind toe *lobed*; cheeks bristly. Colors black and white. . . . . CAMPTOLÆMUS 264
tail *pointed*; in the *adult* = or > wing; bill black and orange, ... HARELDA 273
— rounded, the feathers stiff, lance-linear, exposed to their bases, . . ERISMATURA 270
— Ducks with none of the foregoing characters. (p³)
(p³) Up. mand. *gibbous* at its unfeathered base. Black or brown, ... ŒDEMIA 269
— not gibbous where unfeathered; nail narrow, distinct. (p⁴)
— broad, fused. (p⁵)
(p⁴) Head black or gray, with white; nost. nearly median; b. about = rs., BUCEPHALA 262
— reddish or brownish, no white; nost. nearly median; b. > trs., . AYTHYA 261
— black or brown; nost. subbasal; b. > trs., ... FULIX 259
(p⁵) Feathers not extending on culmen; bill barely tapering to tip, . . POLYSTICTA 266
— much tapering to tip, . . HISTRIONICUS 265
— extending on culm., and partly on sides of upper mand., . SOMATERIA 268
— entirely on sides of up. mand., . LAMPRONETTA 267

(C) Forehead naked, with a large horny plate formed by extension of culmen. . . . . FULICA 245
— feathered; lores feathered; tail perfect; bill flattened, membranes scolloped, PHALAROPUS 200
— subulate, membranes scolloped, . . LOBIPES 199
— subulate, membranes plain, . . STEGANOPUS 198
— naked; tail, *none*; forehead bristly; bill epignathous, . . . PODILYMBUS 315
— soft; bill paragnathous. (q)
(q) Tarsus = middle toe and claw. Birds 29 inches, or more, long, ... ÆCHMOPHORUS 312
— < middle toe and claw. Birds from 12 to 19 inches long, ... PODICEPS 313
Birds under 12 inches long, ... SYLBEOCYCLUS 314

(D) Middle claw *pectinate*; lt. 4-jointed; lt. lateral; tail rounded; long rictal bristles. . ANTROSTOMUS 111
— forked; short rictal bristles, . . CHORDEILES 115
(D) Mid. claw *not* pectinate; head *naked*; nostrils imperforate; naked leg and foot < tail, . MELEAGRIS 177
— perforate; naked leg and foot < tail, . . CATHARTES 166
— perforate; naked leg and foot > tail, ... GRUS 223
— feathered; nostrils feathered, or overhung by a scale, in deep
fossa of stout, hard bill. (r)
— not feathered nor scaled, in long groove of
slender softish bill. (s)
(r) Toes feathered; tarsi and nasal fossæ feathered. Plumage pure white in winter. . . . . LAGOPUS 183
— naked; — tarsi feathered, part way down; tail of 18 soft broad feathers, ... BONASA 182
— to the toes; tail of — 20 stiff acuminate feathers, . . . CENTROCERCUS 179
— 20 or 16 soft broad feathers, ... TETRAO 178
— 18 fthrs; neck with lanceolate feathers, CUPIDONIA 181
— without such feathers, PEDIŒCETES 180
— naked; tail nearly = wing; crest — slender, clubbed, recurved, . . LOPHORTYX 186
— full, soft, depressed, ... CALLIPEPLA 187
— ⅓ to ⅔ the wing; crest — long, straight, filamentous, . . . OREORTYX 185
— full, soft, depressed, ... CYRTONYX 188
— rudimentary or none, ... ORTYX 184
(s) Trs. entirely reticulate; hind toe minute; bill straight, not > head, ... SQUATAROLA 189
— scutellate in front only; bill much > hd. very slender, decurved, ... NUMENIUS 222
— barely > hd., comparatively stout, straight,. . . HETEROSCELUS 221
— and behind; tl. not barred; one minute web; primaries mottled, TRYNGITES 220
— 2 plain webs; b. <. or about = hd., EREUNETES 206
— much > hd., MICROPALAMA 205
— tl. barred crosswise with light and dark colors. (s²)
(s²) Gape not reaching beyond base of — furrowed culmen. Under a foot long, . MACRORHAMPHUS 204
— unfurrowed culmen. Over a foot long, ... LIMOSA 213
(s²) Gape longer. Length < 9 in.; 2t. unwebbed; bill grooved nearly to tip, ... TRINGOIDES 217
— about half-way to tip, RHYACOPHILUS 216
— > 9 in.; b. not > hd., grooved ⅔ its length; tl. about = ⅓ the wg., ACTITURUS 219
— not = ½ the wing, PHILOMACHUS 218
— > head; 2t. webbed; legs not green or yellow, SYMPHEMIA 214
— barely or not webbed; legs green or yell., GLOTTIS 215

(E) Forehead covered with a broad horny plate; nostrils linear, trs. < 2 in. long, ... GALLINULA 243
— nearly circular; trs. 2 in. long. . . PORPHYRULA 244
— feathered; first primary attenuate; bill straight, > hd., culm. grooved, ... SCOLOPAX 202
—3 outer primaries attenuate; bill same: tibiæ feathered; trs. < 3t. . PHILOHELA 201
— not attenuate; first primary much < second. (t)
— =, or >, second. (u)
(t) Length 2 feet or more: bill much > head, decurved; tibia half bare; trs. not < 3tcl., . . . ARAMUS 239
— < 2 feet; bill > head, decurved; tibia little bare; trs. < 3tcl., ... RALLUS 240
— < head. straight: feet as before. Length 10 inches or more, ... CREX 242
— less than 10 inches. . . . . PORZANA 241
(u) Trs. evidently < 3tcl.; tibiæ naked below: bill about twice = head, culm. furrowed, . GALLINAGO 203
— feathered; b. little > head. culm. unfurrowed, ... ARQUATELLA 210
— about =, or >. 3tcl.; trs. in front—reticulate. . . . . APHRIZA 193
scutellate, legs reddish; bill acute, < head, . STREPSILAS 195
legs dark. (v)
(v) Bill slightly curved. much > hd.; tarsus evidently > middle toe and claw, . . . ANCYLOCHEILUS 209
— = or barely > middle toe and claw, ... TRINGA 211
— straight, much < head. Primaries mottled with black, ... TRYNGITES 220
— about =, or > hd; tarsus much > middle toe and claw, ... TRINGA 211
— about = middle toe and claw, ... ACTODROMAS 207

B. (*The hind toe not elevated.*)

1. TIBIÆ NAKED BELOW. (w)
2. NOSTRILS OPENING BENEATH SOFT SWOLLEN MEMBRANE. (x)*
3. BILL HOOKED AND FURNISHED WITH A CERE. (y)
4. BIRDS WITHOUT THE ABOVE CHARACTERS. (z)
(w) Middle claw simple; tarsus reticulate; bill flat, spoonshaped at end, ... PLATALEA 227
— not flat, very stout, tapering, decurved, . . TANTALUS 224
— scutellate; bill grooved, curved; claws — straightish, . FALCINELLUS 225
— curved. . . . . IBIS 226
(w) Mid. claw *pectinate*; tail of 10 feathers; lower neck bare behind, BOTAURUS 237
—Length under 18 inches, ARDETTA 238
—12 feathers; lateral toes not more than ¼ as long as tarsus, HYDRANASSA 233
— more than ¼ the tarsus. (w²)
(w²) Tibiæ bare 1 inch or less; trs. > 3tcl.; bill over ½ inch deep at base, . . NYCTHERODIUS 236
— < 3tcl.; b. not thrice as long as high, . . . NYCTIARDEA 235
b. more than thrice as long as high. . . BUTORIDES 234
— 2 in. or more. Lgth. 2 feet or less. Blue (or white). legs blk. and blue. FLORIDA 230
White; legs black and yellow, GARZETTA 231
—3 feet or more. Bluish, ashy. brown, &c., ... ARDEA 228
White; trs. < 7 in. b. < 6, . HERODIAS 232
White; trs. > 7 in. b. > 6, . . AUDUBONIA 229

(x) Tail-feathers — 12. *Greenish*; bird over 18 inches long, ... ORTALIDA 176
— 14. long, tapering, much graduated; circumorbital space naked, ... ZENAIDURA 169
— 12; trs. flat'd above; tail broad, rounded, much < wings, ... COLUMBA 167
— narrow, pointed, about = wings, . . . ECTOPISTES 168
trs. wholly naked, — reticulate, ... STARNŒNAS 175
— scutellate in front; tail pointed, ... SCARDAFELLA 173
— rounded, (x²)
(x²) Trs. about = 3t., without claw; wing rounded, 1st primary < 4th, ... GEOTRYGON 174
— evidently < 3t.; wing pointed, and — under 4 inches long, ... CHAMÆPELIA 172
— over 4; lores — naked, ... MELOPELIA 171
— feathered, ... ZENÆDA 170

---

* This membrane (not scale). which distinguishes the *pigeons*, shrinks in drying, when it may be recognized by its closing up the nostrils, or at least making them *irregular*; but if still in doubt, observe tarsi *reticulate on sides and behind*, and (generally) *scutellate in front*. See fig. 11, b.

(y) Nostrils *at edge of* the cere; eyes *anterior*, surrounded by *radiating* feathers, the anterior of which are *bristly* and hide the base of the bill; outer anterior toe *shorter* than inner anterior toe. (y^1)

y) Nostrils *in* the cere; eyes *lateral*, *not* surrounded by a disc; outer anterior toe (generally) *not* shorter than inner anterior toe. (y^2)

(y^1) Trs. naked or scant-feathered. Facial disc perfect; 3cl. somewhat pectinate, ... STRIX 139

—imperfect; 3cl. simple; trs.>3tcl., ... SPEOTYTO 150

—<3tcl., . . . MICRATHENE 149

—full-feathered; head tufted; tail about ⅔ the wing. Over 18 inches long, ... BUBO 140

—½ the wing. Under 12 inches long, ... SCOPS 141

—Over 12, under 18; tufts—of 8 to 12 fthrs., . OTUS 142

—of 3 to 6 fthrs., BRACHYOTUS 143

—not tufted; tail about ½ the wing. Length under 12 inches, ... NYCTALE 147

—⅔ the wg. Lgth. 18 or more. Pure wht., spotted, NYCTEA 145

—Not pure white, SYRNIUM 144

—¾ the wg. Length over 12, under 18 inches, . SURNIA 146

Length much under 12, . . . GLAUCIDIUM 148

(y^2) Trs. feathered to the toes—all around; tail a foot or more long, ... AQUILA 163

—except a narrow strip behind: tail not a foot long, ... ARCHIBUTEO 160

—reticulate—upper mandible toothed, under mandible notched, nostrils circular, . . FALCO 158

—Claws all of same length, rounded underneath: tibial feathers close, . . PANDION 162

—Tail emarginate, *and* outer feathers not longer than middle, ... ELANUS 154

—forked, outer feather about twice as long as middle, ... NAUCLERUS 155

—scutellate in front; no web at base of toes; tail a foot or more long, ... HALIÆTUS 164

—a web; nostrils circular; tail not ⅔ as long as wing, ... ICTINIA 153

—oval; bill not ½ as deep at base as long, . . . ROSTRHAMUS 152

—oval; tarsus feathered about ½ way down in front, ASTUR 157

—hardly ½ way down, . . . ACCIPITER 156

—and behind—Tibial feathers not reaching below the joint, ... ONYCHOTES 161

—Bill yellowish; nost. linear, oblique, near up. edge of cere, POLYBORUS 165

—Face with a ruff; trs. twice 3t.; up. tail-coverts white, ... CIRCUS 151

—Hawks without these characters, ... BUTEO 159

(z) PRIMARIES,—10; the 1st (never spurious) *always more than ⅔ as long as the longest.* (a)

(z) PRIMARIES,—10; the 1st (spurious or) *at most not ⅔ as long as the longest.* (b)

(z) PRIMARIES,— 9; the 1st (*never spurious*) of variable length. (e)

(a) Feet *syndactylous*; bill serrate: middle tail feathers long-exserted, ... MOMOTUS 112

—not serrate, middle tail feathers not exserted, ... CERYLE 113

(a) Feet *normal*; *tail of* 10 *feathers*; more than 6 *secondaries*; trs. feathered; 4t. 3-jointed, . PANYPTILA 116

—naked; tail not spiny, . NEPHŒCETES 117

—spiny, . . CHÆTURA 118

—only 6 *secondaries*; bill subulate, = or > head. (a^2)

(a^2) Trs. feathered. Grass-green, head striped with black and white, ... HELIOPÆDICA 119

—naked; b. serrate, twice = head. Black below, throat not scaly, ... LAMPORNIS 120

—not serrate; 1st primary rigid; tail truncate. Green, ... STELLULA 123

—not rigid; attenuate; or—rufous on sides; or crown scaly, SELASPHORUS 122

—not attenuate; no rufous; throat green, . ARGYRTRIA 124

—not green, . . . TROCHILUS 121

(a) Feet *normal*; *tail of* 12 *feathers*: 1st primary—attenuate; tail > wings, forficate, ... MILVULUS 104

—not > wings: forked or not, TYRANNUS 105

—not attenuate; crown plain, or full-crested. (a^3)

(a^3) Tail = or little < wing, not forked, *edged with chestnut*; trs. = or > 3tcl., . MYIARCHUS 106

—slightly or not forked; not edged; trs. > 3tcl., ... SAYORNIS 107

—*much* < wing., a little forked; trs. < 3tcl. Length 6¼ or *more*. . . . CONTOPUS 108

—< or nearly = wg., barely or not forked: trs. = or > 3tcl. Length 6¼ in. *or less*. (a^4)

(a^4) Colors greenish, olive, etc.; no buff, red or *pure* brown, . . . EMPIDONAX 109

—brownish olive, etc.; buffy below. Subcrested, . MITREPHORUS 110

—fiery (or rosy) red, and deep brown; ♂ full-crested, PYROCEPHALUS 111

(b) *Tarsus "booted;"* wings < tail, both much rounded; plumage remarkably lax, ... CHAMÆA 11

wings > tail; nostrils linear; no rictal bristles; plumage close. Aquatic, CINCLUS 7

—not linear: tail double-rounded, ... MYIADESTES 52

—not double-rounded. Under 5 in. long. REGULUS 9

Over 5 in. long. (b^2)

(b^2) Tarsus not > mid. toe and claw. *Blue* the chief color, ... SIALIA 6

—> mid. toe and claw. No blue. Tail only ⅔ the wing, . SAXICOLA 5

—more than ⅔ the wing, TURDUS 1

(b) *Tarsus scutellate*; nostrils covered with tufts of antrorse bristly feathers. (c)

—nostrils exposed; base of bill with few such feathers, or none. (d)

(c) Bill— strongly epignathous, toothed and notched near tip. Gray, wings and tail black, . COLLURIO 54
— paragnathous. Not 7 in. long; b. nearly = hd., wg. much > tl., trs. not > 3tcl., ... SITTA 16
— barely or not ⅓ as long as hd.—Crested, . . LOPHOPHANES 12
— Not crested. (c²)
(c²) Head yellow; bend of wing chestnut, ... AURIPARUS 15
— not yell.; crown and throat blk. or dark, ... PARUS 13
— crown ashy or light brown, . PSALTREPARUS 14
— Over 7 long. Crested. Blue, with black bars on wings and tail, . CYANURUS 100
No crest. Iridescent blk. and wht.; wgs. much < tl., . . PICA 99
— Uniform glossy blk.; wgs. much > tl., . CORVUS 95
— Gray; blk. wgs. > blk. and wht. tail, PICICORVUS 96
— Gray; no blue; wgs. about = tail, . PERISOREUS 103
— Blue, &c., no green or yellow, . APHELOCOMA 101
— Blue, black, green and yellow, . . . XANTHOURA 102
(d) Length over 14 inches; color dark brown; rounded tail not < wings, ... PSILORHINUS 98
— 10 — 12 inches; color all blue, square tail < wings, ... GYMNOKITTA 97
— 7 — 8 inches; glossy black (♀ brown) with large white wing patch. Crested, . PHÆNOPEPLA 51
— 4½ — 5½ inches; brown, streaked, below white, tail feathers rigid, acuminate, . . . CERTHIA 17
— 4 — 5 inches; bluish gray, unstreaked, below wht., tail soft, blk. and white, . POLIOPTILA 10
— 4 — 5 in.; olive-green, below yellowish, tail like back, bill not hooked, . . . PHYLLOPNEUSTE 8
— 4½ — 6½; greenish or grayish olive, wht'sh or yell'sh below, bill distinctly hooked at tip, VIREO 53
(d) Birds presenting no one of the foregoing combinations of characters. (d²)
(d²) Rictus bristled; inner toe cleft to base; wg. not < tail; b. little < head, ... OREOSCOPTES 2
— < tail; bill much < head, ... MIMUS 3
— little <, =, or > hd.. HARPORHYNCHUS 4
— unbristled; breast — with distinct round black spots; b. < hd., . . . CAMPYLORHYNCHUS 18
— unspotted; back — uniform in color; wg. = or < tail; b. < head, . THRYOTHORUS 21
— speckled; throat pure white; b. nearly = hd., . . . CATHERPES 20
— streaked; b. much < hd., . . . SALPINCTES 19
— barred crosswise; tail nearly = wg.; . . . . . . TROGLODYTES 22
— much < wg., ... ANORTHURA 23
— streaked lengthwise; bill about ⅔ as long as head, TELMATODYTES 24
— hardly or not ½ = hd.. CISTOTHORUS 25
(e) Bill metagnathous; both mandibles falcate, their points crossed, ... CURVIROSTRA 59
(e) Tarsus scutellate behind; hind claw straight; nostrils concealed; little ear tufts, . . . EREMOPHILA 26
(e) Quills (usually) tipped with red horny appendages; tail tipped with yellow. Crested, . . AMPELIS 50
(e) Tomia of up. mand. toothed or lobed near middle. Bright red, or greenish and yellow, . PYRANGA 43
(e) Greenish or grayish-olive, below golden yellow, belly white, lores black. Length 7-8 inches, . ICTERIA 39
(e) Greenish or grayish-olive, below white or yellow; bill notched and hooked at tip. Length 5½—6½, VIREO 53
(e) Birds with no one of these special characters.— Commissure* straight or gently curved. (f)
— Commissure* abruptly angulated. (i)
(f) Bill triangular-depressed, about as wide at base as long, gape twice as long as culmen, reaching
to about opposite eyes, trs. not > outer lateral toe and claw; 1st primary = or > 2d. (g)
(f) Bill not nearly so wide as long, gape not twice as long as culmen; trs. > lateral toes. (h)
(g) Outer web of 1st primary saw-like, with a series of minute recurved hooks, . . . STELGIDOPTERYX 48
(g) A little feathery tuft at base of hind toe. Plain gray, below white, breast like back, . . . COTYLE 47
(g) No hooks nor tuft. Tail deeply forked, outer feathers attenuate, or with white spots, . HIRUNDO 44
— forked. Below, pure white; above, lustrous or velvety. . . TACHYCINETA 45
— forked. Uniform lustrous blue-black, or partly white below, . . PROGNE 49
— barely or not forked. Rump not colored like back, . . PETROCHELIDON 46
(h) Longest secondary nearly = primaries in closed wing; hind claw slightly curved, twice as long
as middle claw.—Tail < wing. No spots or streaks below, ... BUDYTES 27
— < wing; trs. > hind toe and claw. Breast spotted, ... ANTHUS 28
— not > 1t'l. Back and breast spotted, ... NEOCORYS 29
(h) Longest secondary much < primaries in closed wing; hind claw well curved, not nearly twice
as long as middle claw. (h²)
(h²) Rictus with many conspicuous bristles reaching decidedly beyond nostrils. (h³)
(h²) Rictus with no evident bristles or a few short ones reaching little if any beyond nostrils. (h⁴)
(h³) Bill barely or not twice as long as wide at base; tail blk. and orange, or brn. and yell.. SETOPHAGA 41
(h³) Bill fully twice as long as wide; tail unmarked, or with white feathers, . . . MYIODIOCTES 40
(h⁴) Trs. <, or about = 3tcl.; hind toe much > its claw. Entirely blk. and wht., streaked, MNIOTILTA 30
— little > its claw; breast and rump yellow, . . . CERTHIOLA 42
(h⁴) Trs. > 3tcl.; tl. not < wgs. Olive, with yell. below, hd. of ♂ with blk. or ashy, . . GEOTHLYPIS 38
— < wgs.; tail feathers yellow on inner webs, dusky on outer webs, DENDRŒCA 35
— all unmarked, same color on both webs. (h⁵)
— (some or all) marked with white blotches. (h⁶)

---

* As this important distinction may not be perfectly plain to the student in some dried (especially if distorted) specimens, it may help him to be here told, that (f) will take him to the swallows and the great warbler group of little insectivorous birds; while (i) will carry him to the blackbirds, orioles, meadow starlings, cowbirds, and bobolinks, and the
great conirostral granivorous finch family, including grosbeaks, linnets, buntings, finches, and all the sparrows.

(h^5) Conspicuously *streaked* below; crown plain, or with 2 black stripes, ... SEIURUS 36
(h^5) No streaks below; b. at least $\frac{1}{2}$ inch long; 4 black stripes on head, or none, . . . HELMITHERUS 33
—not $\frac{1}{2}$ inch long; wg. > $2\frac{1}{2}$ in.; crown plain or with black, . . . OPORORNIS 37
— < $2\frac{1}{2}$ in.; crown plain, or with bright spot, HELMINTHOPHAGA 34
(h^6) Rictal bristles not evident; b. at least $\frac{1}{2}$ inch long; whole hd. and neck rich yellow, PROTONOTARIA 32
bill < $\frac{1}{2}$ in. long; whole head and neck not yellow. HELMINTHOPHAGA 34
(h^6) Rictal bristles evident; hind toe much > its claw. Length under 5 inches, ... PARULA 31
—little if any longer than its claw, ... DENDROECA 35
(i) Length *less than* 5 in., wing and tail 2 in. or less, tail feathers acute, ... SPERMOPHILA 82
(i) Hind claw slightly curved, twice as long as middle claw, ... PLECTROPHANES 63
(i) Feathers of crown *bristle-tipped*. Streaked: below yellow, with black breast patch, . . STURNELLA 91
(i) Longest secondary nearly = primary in closed wgs. *Black*, with white wing patch, . CALAMOSPIZA 78
(i) Conspicuously *crested*. Red the prevailing color. Bill reddish, face black, ... CARDINALIS 84
—not reddish, face not black, . PYRRHULOXIA 83
(i) Tail-feathers *acute*. Black, nape buff, ♂; or streaked yell'sh-brn., ♀; wg. > 3 inches, . DOLICHONYX 87
—Small streaked marsh-sparrows, wg. < 3 inches, its edge yell., . AMMODROMUS 68
(i) Colors *greenish* and white, with yell. on edge of wing; —rufous head-stripes, ... EMBERNAGRA 86
—crown chestnut, breast ashy }
(i) Length about 8 in.; *tl. > wgs.* Plain brown, &c., or black, white and chestnut, } . . PIPILO 65
(i) Inner claw reaching at least $\frac{1}{2}$ way to tip of 3cl. Black, white and chestnut, }
—Blk. (or brn.) hd. *yellow*; > 8 long, XANTHOCEPHALUS 90
—Spotted and streaked; < 8 long, . . . PASSERELLA 77
(i) Birds with none of the foregoing combinations of chars.—*Bill with a ruff of antrorse bristly feathers.* (k)
—*Bill without ruff; nostrils exposed.* (l)
(k) Length—8 or more. *Red*, or gray with brownish yellow on head and rump, ... PINICOLA 56
—under 8. *White*, with blk. on wgs. and tl.; or washed with clear brown, . PLECTROPHANES 63
—Bluish-gray, below reddish-gray, crown, face, wgs. & tl., blk. . . PYRRHULA 57
—Reddish-brown, blk.-streaked, crown ashy, throat blk. Imported, . PASSER 76
—Unstreaked, chocolate-brown, rosy-tinted, hd. with blk. and ash, LEUCOSTICTE 60
—*Streaked*: no yell., crown crimson, face and throat *dusky*, . . . AEGIOTHUS 61
—*Streaked*: no yell.; no red; or else crown, and throat too, red, . CARPODACUS 58
—Streaked *or not*: with some yellow, but no red, ... CHRYSOMITRIS 62
(l) Species at least over 7 inches long. (m)
(l) Species at most not over 7 inches long; plumage nowhere decidedly spotted or streaked. (n)
—somewhere or everywhere spotted or streaked. (o)
(m) Bill *jet-black*; plumage glossy blk., (plumage of ♀ plain brown) {
with head and neck rich *brown*, ... MOLOTHRUS 68
with head, neck and breast *yellow*, ... XANTHOCEPHALUS 90
with head black, bend of wing *red*, ... AGELÆUS 89
with *no* red or yell.; tail rounded, < wg., . . SCOLECOPHAGUS 93
—graduated, = or > wg., . . . QUISCALUS 94
(m) Bill dark horn-blue, very acute; plumage black, with orange, yellow, or white }
—plain olivaceous, yellowish below, ♀, } . . . . ICTERUS 92
—obtuse; plumage blk., white and brown, tl. with wht. spots, . . GONIAPHEA 80
(m) Bill greenish-yellow, as long as tarsus; wgs. black, many secondaries white. . . . HESPERIPHONA 55
(m) Bill not bluish nor greenish; tail with white spots; under wg.-coverts *rosy or yellow*, . GONIAPHEA 80
—not rosy or yell.; yg. of, . PIPILO 65
tail plain; entire plumage *streaked*. ♀ of . . . . AGELÆUS 89
(n) Black and chestnut, or orange, ♂, or olive yellowish below, ♀; b. acute, horn blue, or brn., ICTERUS 92
(n) Dusky grayish-brown, nearly uniform; bill blackish, obtuse. ♀ of . . . . MOLOTHRUS 68
(n) Blackish, or ashy, belly and 1 to 3 outer tail feathers white; bill flesh color, ... JUNCO 72
(n) Throat and tail black, latter with white spots: head with 2 white stripes, ... POOSPIZA 71
(n) *Blue*, with or without red, purplish, &c.; or greenish and yell.; or plain brown; < 6 long, CYANOSPIZA 81
(n) *Blue*, with chestnut on wings, ♂; or plain brown, ♀; > 6 long, ... GONIAPHEA 80
(o) Wings > tail; *breast* more or less *yellow*; throat patch, or maxillary streaks, black, ... EUSPIZA 79
—not yell.; wg. < $2\frac{1}{2}$ inches, its edge yellow; tail 2 in. or less, . . . COTURNICULUS 67
> $2\frac{1}{2}$ in.,—without yell.; lesser wg.-cov. chestnut, . . POOECETES 66
with yell. or not; longest sec. nearly = prim., PASSERCULUS 65
—without yell.; longest sec. much < primaries, CENTRONYX 64
(o) Wings not > tail; tail forked. Lgth. 5 to 6 in., wg. or tl. $2\frac{1}{2}$ to 3, trs. $\frac{5}{8}$ to $\frac{3}{4}$, ... SPIZELLA 73
—graduated, tipped with wht.; head striped with chestnut and wht., CHONDESTES 75
—little rounded, *black*, outer feather pale edged, ... POOSPIZA 71
—not black. Streaked below or crown chestnut, . . MELOSPIZA 69
Not streaked below. Lgth. 6 or more, ZONOTRICHIA 74
Length under 6 in., . PEUCÆA 70

SYSTEMATIC SYNOPSIS

OF

NORTH AMERICAN BIRDS.*

———◇◇◇◇———

Subclass I. AVES AËREÆ, or INSESSORES.

AËRIAL BIRDS, or PERCHERS.

THE first and highest one of three primary divisions of the class† *Aves*, embracing all existing birds down to the *Gallinæ*.

The knee and part of the thigh are free from the body, and the leg is almost always feathered to or beyond the tibio-tarsal joint. With rare exceptions, the toes are all on the same level, and touch the support throughout; being thus fitted for grasping or *perching*. In other respects the members of this great group are too various to be defined by external characters, unless it be negatively, in the absence of the special features of the other two groups. They are *Altrices*. They are now usually divided into *five* Orders, of which the first is the

Order PASSERES. Perchers Proper.

The feet are perfectly adapted for grasping by the length and low insertion of the hind toe, great power of opposing which to the front toes, and great mobility of which, are secured by separation of its principal muscle from that that bends the other toes collectively. The hind toe is always present, and never turned for-

---

* North of the present Mexican Boundary; inclusive of Lower California; exclusive of Greenland.

† As commonly received, without recognizing, however, the fossil *Archæopteryx* (see Introd. § p. 12) a mesozoic bird, which probably alone represents a primary group *Saururæ*; admitting which, some high authorities then divide all existing birds into two other primary groups, *Ratitæ* (Ostriches), in which the sternum has no keel, and *Carinatæ*, embracing all other birds. On this basis, our *Aves aëreæ* would represent a group of less value than a subclass; and I desire to be understood as using this term provisionally, in a conventional sense.

wards or even sideways; its claw is as long as, or longer than, the claw of the middle toe. The feet are never zygodactyle, nor syndactyle, nor semipalmate, though the front toes are usually immovably joined to each other at base, for a part, or the whole, of the basal joints. Various as are the shapes of the wings, these members agree in having the great row of coverts not longer than half the secondaries; the primaries either nine or ten in number, and the secondaries more than six. The tail, extremely variable in shape, has twelve rectrices (with certain anomalous exceptions). The bill is too variable to furnish characters of groups higher than families; but it is always córneous, either wholly or in part, is never largely membranous, as in many wading and swimming birds, nor cered, as in birds of prey. No Passeres are known to have

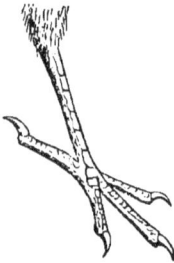

FIG. 12. Passerine foot.

more than one common carotid artery; and they all have the sternum cast in one particular mould, with slight minor modifications of shape. They are the typical *Insessores*, as such representing the highest grade of development, and the most complex organization, of the class. Their high physical irritability is coördinate with the rapidity of their respiration and circulation; they consume the most oxygen, and live the fastest, of all birds. They habitually reside above the earth, in the air that surrounds it, among the plants that with them adorn it; not on the ground, nor on "the waters under the earth."

Passeres, corresponding to the Insessores proper of most ornithologists, and comprising the great majority of birds, are divisible into two groups, commonly called suborders, mainly according to the structure of the lower larynx. In one, this organ is a complex muscular vocal apparatus; in the other the singing parts are less developed, rudimentary, or wanting. In the first, likewise, the tarsus is *normally* covered on either side with two entire horny plates, that meet behind in a sharp ridge; in the other, these plates are subdivided or otherwise differently arranged. This latter is about the only *external* feature that can be pointed out as of extensive applicability; and even this does not always hold good. For example, among our birds, the larks (*Alaudidæ*), held to be Oscine, and certainly to be called songsters, have the tarsus perfectly scutellate behind.

Suborder OSCINES.   Singing Birds.

The first and higher of the two suborders just indicated. All of the birds composing it have a more or less complex vocal apparatus, consisting of five pairs of muscles; but many of them do not sing.

It is a question, which one of the numerous Oscine families should be placed at the head of the series. Largely, perhaps, through the influence of those ornithologists who hold that fusion of the tarsal envelope into one continuous plate indicates the acmé of bird-structure, the place of honor has of late been usually assigned to the thrushes. But only a part of the thrushes themselves show this character; on which account, probably, the rest were associated by Cabanis with the wrens. It seems to me most probable that this character, though unquestionably of high import, should be taken as of less value than the reduction of the number of primaries from ten to nine; and I am at present inclined to believe that eventually some Oscine family with only nine primaries — as the finches or tanagers — will take the leading position. Here, however, I follow usage.

Family TURDIDÆ. Thrushes.

The oval nostrils are nearly or quite reached, but not covered, by feathers. There are bristles in all our genera about the rictus and base of upper mandible. The toes are deeply cleft, the inner one almost to its very base, the outer to the end of its basal joint. The bill is not conical, nor deeply fissured, and *usually* has a slight notch near the tip. There are 12 tail-feathers (in all our forms), and 10 primaries, of which the 1st is short or spurious, and the 2d is shorter than the 4th. Our two subfamilies are sharply defined by the character of the tarsus.

Subfamily TURDINÆ. Typical Thrushes.

With the tarsus, in the adult, enveloped in one continuous plate, or "boot," formed by fusion of all the scutella except two or three just above the toes. Thus easily distinguished; for our few other birds that show this feature are very different in other respects. The 1st quill is spurious or very short; the 2d is longer than the 6th; the 4th toe is longer than the 2d. Upwards of one hundred and fifty recorded species are now usually assigned to the *Turdinæ* proper, most of them being referable to the single genus *Turdus* with its subgenera. They are nearly cosmopolitan, and have a great development in the warmer parts of America, where the subfamily is, however, mainly represented by types closely allied to *Turdus* proper; more aberrant forms, constituting very distinct genera, occur in the old world. We have but one genus in the United States, of which the robin is the most familiar example, though several other species are common and well known birds. These are diffused over all the woodland parts of our country, and are all strictly migratory. They are insectivorous, but like many other insect-eating birds, feed much upon berries and other soft fruits. Although not truly gregarious, some, as the robin for instance, often collect in troops at favorite feeding places, or migrate in companies. They build rather rude nests, often plastered with mud, never pensile, but saddled on a bough, or fixed in a forked branch, or on the ground; and lay 4–6 greenish or bluish eggs, sometimes plain, sometimes spotted. They are all vocal, and some, like the woodthrush, are exquisitely melodious.

1. Genus TURDUS Linnæus.

* Not spotted nor banded below; throat streaked. (Subgenus *Planesticus*.)

FIG. 13. Robin; natural size.

Robin. Dark olive-gray, head and tail blackish; below reddish-brown, throat black and white, under tail coverts and crissum white with dark marks, eyelids and tips of outer tail-feathers with white spots, bill brown or yellow, feet dark. Very *young* birds are spotted above. 9–10 long, wing 5–5½, tail 4–4½. N. Am. Nest in trees and bushes; eggs plain. Wils., i, 35, pl. 2; Aud., iii, 14 pl., 142; Nutt., i, 338; Bd., 218.   . migratorius.

    Var. *confinis*, described from Cape St. Lucas, is paler, duller, &c. Bd., Rev. 29.

    ** Banded crosswise, not spotted, below. (Subgenus *Hesperocichla*.)

Varied Thrush. Slate-color, below orange-brown, with black pectoral band which runs up on sides of neck and head; crissum and under tail coverts whitish; eyelids, postocular stripe, 2 wing-bars and much edging of quills, orange-brown; bill dark, feet pale; ♀ and *young*, duller, browner, pectoral bar obscure, etc. Size of the last. Pacific slopes, N. Am.: accidental in Mass., N. J. and Long Island. Aud., iii, 22, pl. 143; Bd., 219. . . . nævius.

    *** Spotted, not banded, below. (Subgenus *Hylocichla*.)

    † Upper parts not uniform in color.*

    ‡ Upper parts tawny, shading into olive on rump.

Wood Thrush. Under parts white,

FIG. 11. Wood Thrush; natural size.

barely or not buff-tinted, marked with large distinct dusky spots, middle of throat and belly only immaculate; bill dusky and yellowish; legs flesh-color; 7–8 long; wing 4–4½, tail 3–3½. Eastern United States. Nest in bushes and low trees; eggs plain. Wils., i, 35, pl. 2; Nutt., i, 343; Aud., iii, 24, pl. 144; Bd., 212. mustelinus.

    ‡‡ Upper parts olive, shading into rufous on rump and tail.

Hermit Thrush. Under parts white, with slight buffy tint anteriorly and olive shade on sides, breast and sides of throat thickly marked with large distinct dusky spots; bill dusky and yellowish; legs pale. About 7 long; wing 3½, tail 2¾. Eastern (and Arctic) North America. Nest in bushes; eggs plain. Wils., v, 95, but *not* his fig. 2 of pl. 45; Nutt., i, 346; Aud., iii, 29, pl. 146; Bd., 212. . . . . pallasii.

    Var. *auduboni*, is entirely similar in color, but rather larger. South-western United States into Mexico. Bd., Rev. 16.

    Var. *nanus*, is entirely similar in color, but rather smaller. Rocky Mountains to Pacific. Aud., iii, 32, pl. 147; Bd., 223; Rev. 16; Coop., 4.

    †† Upper parts uniform in color.*

    → Upper parts olive.

Olive-backed Thrush. Under parts white, olive-shaded on sides, the fore parts and sides of head and eyelids strongly tinged with buff, the breast and throat thickly marked with large dusky-olive spots. 6¾–7¾ long; wing

---

* *Very young* birds of all the species of *Hylocichla* are spotted above; but these spots disappear the first autumn, and then the ground color is always as stated.

3½–4, tail 2¾–3. North America, except perhaps south-west U. S. Nest in bushes; eggs speckled. Wils., v, pl. 45, f. 2, but not his description on p. 95; Bd., 216. swainsoni.

Var. *aliciæ.* *Alice's Thrush.* Similar; but without any buffy tint about head, nor yellowish ring around eye; *averaging* a trifle larger, with longer, slenderer bill. Much the same distribution, but breeds further north. Nest and eggs similar. Bd., 217, and Rev. 21.

Var. *ustulatus.* Similar; but with the upper parts slightly suffused with tawny, and the spots below smaller, fewer and paler; thus approximating to the following species. Nest and eggs, however, as in *swainsoni.* Pacific Coast, U. S. Nutt., 2d ed. i, 400; Bd., 215; Coor., 5.

+ + Upper parts tawny.

Wilson's Thrush. *Veery.* Under parts white, with olive shade on sides, and strong fulvous (almost pinkish-brown) tint on breast; breast and sides of neck with very small, sparse, sometimes indistinct dusky spots. 7–7½ long; wing 4–4¼, tail 3¼. Eastern N. Am. Nest built on the ground; eggs plain. Wils., v, 98, pl. 43; Nutt., i, 349; Aud., iii, 27, pl. 145; Bd. 214. fuscescens.

Subfamily MIMINÆ. Mocking Thrushes.

Distinguished from the last by having the tarsus scutellate in front, the tail longer and rounder (usually longer than the wings, but not so in *Oreoscoptes*), the wings shorter and rounder, with 1st primary hardly to be called spurious. Birds very much like overgrown wrens (with which they used to be associated), but distinguished therefrom by more deeply cleft toes, different nostrils, and bristly rictus (compare diagnoses of the two families). The bill is usually longer, or at least slenderer, and more curved than in the typical thrushes: in some species of *Harporhynchus* it attains extraordinary length and curvature. As a group they are rather southern, hardly passing beyond the United States; and attaining their maximum development in Central and South America. The *Miminæ* may be properly restricted to these American birds, represented

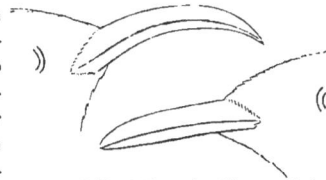

Fig. 15. Bills of *Harporhynchi;* natural size.

by the genera *Mimus, Harporhynchus,* and five or six other closely related forms. Upwards of forty species are recorded, about two-thirds of which are certainly genuine. About one-half of the current species fall in the genus *Mimus* alone; of *Harporhynchus,* all but one of the known species occur within our own limits. In their general habits they resemble the true thrushes; but they habitually reside nearer the ground, relying for self-preservation more upon the concealment of the shrubbery, than upon their own activity and vigilance. They are all melodious, and some, like the mockingbird, are celebrated songsters, famous for their powers of mimicry, and their brilliant vocal execution. In compensation, perhaps, for this great gift, they are plainly clad. grays and browns being the prevailing colors. The nest is generally placed in a bush; the eggs, four or five in number, are greenish-blue, plain or speckled.

2. Genus OREOSCOPTES Baird.

Mountain Mockingbird. Brownish ash, below whitish, shaded behind, thickly spotted with dusky; 8; wing nearly 5; tail 4. Rocky Mountain region of United States. Aud., ii, 194, pl. 139; Bd., 347. . montanus.

3. Genus MIMUS Boie.

Mockingbird. Ashy gray, below white, slightly shaded across breast and along sides, wings and tail blackish, former with two white bars, and much white at base of primaries, latter with 1-3 outer feathers partly or wholly white. The ♂ is known by the much greater extent of white on the primaries, which is the mark of a "singer," as he is called, the ♀ being songless, in captivity at any rate; *young* birds are spotted below the first autumn. 9-10 long; wing about 4; tail about 5 (nearly 6 in var. from California). Southern U. S. to Massachusetts, but not common north of 38°; thronging the groves of the South Atlantic and Gulf States. Two or three broods are generally reared each season. When taken from the nest, the mockingbird becomes a contented captive; and has been known to live many years in confinement. Naturally an accomplished songster, he proves an apt scholar, susceptible of improvement by education to an astonishing degree; but there is a great difference with individual birds in this respect. Wils., ii, 14, pl. 10, fig. 1; Aud., ii, 187, pl. 137; Bd., 344. polyglottus.

Fig. 16. Mockingbird; about ⅔ natural size.

Catbird. Blackish-ash, or dark slate; crown and tail black; under tail coverts chestnut. 8-9 long; wing 3?, tail 4. Eastern United States; also Washington Territory, Mexico, Central America and Bermuda. An abundant and familiar inhabitant of our groves and briery tracts, remarkable for its harsh cry, like the mewing of a cat (whence its name), but also possessed

of no mean vocal powers. *** The tarsal scutella are frequently obsolete.
WILS., ii, 90, pl. 14, f. 3; AUD., ii, 195, pl.140; BD., 346. CAROLINENSIS.

4. Genus HARPORHYNCHUS Cabanis.

* Bill equal to or shorter than head, not, or not much, curved; tail moderately
longer than wings. *Breast spotted.*

Brown Thrush.   Thrasher.   Sandy Mockingbird. Reddish-brown,
below white, with more or less tawny tinge, and thickly spotted with dark
brown, except on throat and middle of belly, the spots lengthening into
streaks on the sides; wings with two white bars; tail feathers with pale
tips; bill black, yellow below; feet pale; iris yellow; *about* 11 long;
wing 4, tail 5 or 6; bill nearly straight, 1 inch long. Eastern United
States; a delightful songster, abundant in thickets, etc. WILS., ii, 83, pl.
14; NUTT., i, 328; BD., 353. . . . . RUFUS.

Var. *longirostris.   Long-billed Thrush.* Somewhat similar; darker brown
above; the markings below blackish; bill longer and a little more curved. Mexico
to the Rio Grande. BD., 352, pl. 52; REV., 44.

Curve-billed Thrush. Dull grayish-brown, below whitish, breast, etc.,
spotted with color of the back, wing coverts and lateral tail feathers tipped
with white; size of the last; bill over an inch long, and decidedly curved.
Valley of Rio Grande and Colorado. BD. 351, pl. 51.   . CURVIROSTRIS.

Cinereous Thrush. Brownish-ash, below whitish, shaded with fulvous,
especially behind, and with brown spots; two wing-bars and tips of lateral
tail feathers obscurely white. Rather smaller than the foregoing. Cape St.
Lucas. BD., Pr. Acad. Phil. 1859, 303, and Rev. 46; ELLIOT, pl. 1;
COOP., 19. . . . . CINEREUS.

**Bill longer than head, arcuate! Tail much longer than wings. *Breast not
spotted.*

Sickle-billed Thrush.   Californian Mockingbird. Dark oily olive-
brown, paler below, deepening into rusty brown on belly and under tail
coverts; throat rusty whitish; auriculars streaked : bill black, at a maximum
of curvature, about 1½ long, but very variable in length and degree of
curve; tarsus about 1½; total length 11 or more; wing 4 or less, tail 5 or
6 inches long. Coast region of California. CASS., Ill. 260, pl. 43; BD.,
349; Rev. 48; COOP., 16. . . . . REDIVIVUS.

Var. *lecontei.   Leconte's Thrush.* Pale ash, still paler below, shading into
brownish-yellow on under tail coverts; throat whitish, with slight maxillary streaks;
bill black; no decided markings anywhere. A bleached desert race. Colorado
Valley; only two specimens known. (Ft. Yuma, *Leconte;* Ft. Mojave, *Coues.*)
BD., 350, pl. 50; Rev. 47; COOP., 17.

Crissal Thrush. Olive-brown, paler on throat and belly; throat whitish
with blackish maxillary streaks; under tail coverts chestnut in marked contrast; auriculars slightly streaked; bill black. Size of the last, or rather
larger; tail 6 or more. Valley of Rio Grande and Colorado. BD., 351,
pl. 82; Rev. 47; COOP., 18. . . . . CRISSALIS.

Family SAXICOLIDÆ.   Stone-chats and Bluebirds.

Chiefly Old World; represented in North America by one European straggler and the familiar bluebirds; authors assign different limits to it, and frequently transpose the genera; it might come under *Turdidæ* without violence. As usually constituted, it contains upwards of one hundred species, commonly referred to about a dozen genera. Like most other groups of *Passeres*, it has never been defined with precision, the family being known, conventionally, by the birds ornithologists put in it. The following birds have booted tarsi; oval nostrils; bristled rictus; rather short, square or emarginate tail; long, pointed wings,

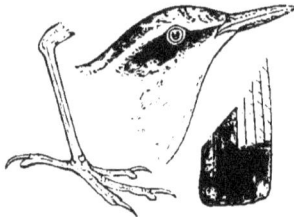

Fig. 17. Stone-chat; natural size.

with very short spurious 1st quill, and the tip formed by the 2d, 3d and 4th quills.

5. Genus SAXICOLA Bechstein.

Stone-chat.   Wheat-ear. Adult:—ashy gray; forehead, superciliary line and under parts white, latter often brownish-tinted; upper tail coverts white, wings and tail black, latter with most of the feathers white for half their length; line from nostril to eye, and broad band on side of head, black; bill and feet black; *young* everywhere cinnamon-brown, paler below; wing 3½, tail 2½, tarsus 1; middle toe and claw ¾. Atlantic coast, astray from Europe *via* Greenland; also, North Pacific Coast, from Asia. Cass., Ill., 208, pl. 34; Bd., 220, and Rev. 61. . . . . ŒNANTHE.

6. Genus SIALIA Swainson.

⁎⁎⁎ More or less *blue:* bill and feet black; ♀ grayish or brownish, with blue traces, especially on rump, wings and tail. *Young* like the ♀, but curiously *spotted.* 6–7 long, wing 3¾–4½, tail 2¾–3¼, bill ¼ or less, tarsus ¾ or less.

Eastern Bluebird. ♂ rich sky-blue, uniform above; below reddish-brown, belly white. Eastern U. S. Wils. i, 56, pl. 3; Nutt. i, 445; ii, 171, pl. 134; Bd., 222. . . . . SIALIS.

Western Bluebird. ♂ above, and the throat, deep blue, with a dorsal patch of same color as breast and sides, which are rather darker than in the last species; belly dull bluish-gray. Rocky Mountains to Pacific. Nutt., i, 513; Aud., ii, 176, pl. 135; Bd., 223; Coop., 28. . . . MEXICANA.

Arctic Bluebird. ♂ everywhere clear pale blue, with a faint green shade, paler below, fading into white on belly. Chiefly central region of N. A., from 65° into Mexico; also Pacific coast. Nutt., 514; Aud., ii, 178, pl. 136; Bd., 224; Coop., 29. . .   . . ARCTICA.

Family CINCLIDÆ.   Dippers.

Aquatic! thrush-like birds (formerly included under *Turdidæ*), with thickset bodies, compact plumage to resist water, short, stiff, concave wings, with 10 primaries of which the 1st is spurious, still shorter, square tail, almost hidden by the coverts, linear nostrils, slender bill, almost a little turned up (gonys convex, culmen slightly concave), with no trace of rictal bristles.   There is only one genus, with about a dozen species, all inhabiting clear mountain streams of most parts of the world, easily progressing *under* water; feed on aquatic animal substances; moderately vocal; our species builds a remarkable and elegant dome-like nest of moss, with a hole in one side.

7. Genus CINCLUS Bechstein.

Water Ouzel. Dipper. Lead-colored, nearly uniform, but apt to be brownish on head; 7; wing $3\frac{1}{2}$; tail $2\frac{1}{4}$. Rocky Mountain region of N. A. NUTT., ii, 569; AUD , ii, 182, pl. 137; BD., 229; COOP., 25.   . . . . MEXICANUS.

FIG. 18. Dipper; natural size.

Family SYLVIIDÆ. Sylvias.

A large family, chiefly Old World, sparingly represented in the New.   Primaries 10, the 1st short or spurious, about half the 2d, which is shorter than the 6th; bill slender, about straight, shorter than the head, usually slightly notched and hooked at tip; rictus bristly; nostrils exposed, or slightly overhung, but never densely hidden : part have booted tarsi, and these are difficult to distinguish technically from *Turdinæ* and *Saxicolidæ*, but here size is a good criterion, none of our *Sylviidæ* being over five inches long; the rest, with scutellate tarsi, are of course distinguishable on sight from the last mentioned families; from the *Certhiidæ*, by not having stiff acuminate tail-feathers; from the *Paridæ* and *Sittidæ* by not having densely feathered nostrils; from the *Troglodytidæ*, by the less cohesion of the toes at base; and from all the *Sylvicolidæ* by having more than nine primaries.   Three subfamilies occur in North America; one of them, *Polioptilinæ*, peculiar to this country, used to be associated with the *Paridæ*, with which, however, it has no special affinity; another, *Regulinæ*, is simply warblers with booted tarsi; a third, *Sylviinæ*, with its several not well defined groups, constitutes an immense assemblage of upwards of five hundred recorded species, among them the famous nightingale of Europe.

Subfamily SYLVIINÆ. Typical Old World Warblers.

Represented in North America by a single waif from Asia.

8. Genus PHYLLOPNEUSTE Meyer.

Kennicott's Sylvia.   Olive-green; below yellowish and white; superciliary line yellow; wings and tail dusky, olive-edged; wing coverts yellowish-tipped. $4\frac{3}{4}$; wing $2\frac{1}{2}$; tail 2. Alaska (*Dall*). BD., Trans. Chicago Acad., 1869, 313, pl. 30. f. 2. . .   . . BOREALIS.

Subfamily REGULINÆ. Kinglets.

Tarsus *booted;* wings longer than the emarginate tail. Elegant greenish-olive pigmies, with brilliant colors on the head when adult.   There are about ten species of the following genus, inhabiting Europe, Asia and America; two of them are very common in our woods, thickets and orchards. Migratory, insectivorous; have a sweet song.

9. Genus REGULUS Cuvier.

*** Greenish-olive, below whitish or yellowish; wings and tail dusky, edged with greenish or yellowish, wing coverts whitish-tipped. 4–4½ long, wing 2⅛–2¼; tail 1½–1¾.

Ruby-crowned Kinglet. Crown with a rich scarlet patch (in *both* sexes, but wanting in both the first year); no black about head; bill and feet black. North America. WILS., i, 83, pl. 5, f. i; NUTT., i, 415; AUD., ii, 168, pl. 133; BD., 227. . CALENDULUS.

Golden-crowned Kinglet. Crown bordered in front and on sides by black, inclosing a yellow and flame colored patch (in the ♂; in the ♀, the scarlet wanting); extreme forehead, and line over eye, whitish; *young,* if ever without traces of black and yellow on the head, may be told from the last species, by smaller size and presence of a tiny bristly feather overlying the nostrils; this is wanting in *calendulus.* North America. WILS., i, 126, pl. 8, f. 2; AUD., ii, 265, pl. 132; BD., 227. . . . . SATRAPA.

FIG. 19.   Golden-crowned Kinglet.

OBS. Cuvier's Kinglet (*R. cuvieri* AUD., ii, 163, pl. 131; NUTT., i, 416, Schuylkill River, June, 1812), not now known, is said to have *two* black stripes on each side of head. *R. tricolor* NUTT., i, 420, is *R. satrapa;* so is his *R. cristatus,* which is the name of the European species, not found in North America.

Subfamily POLIOPTILINÆ.   *Gnatcatchers.*

Tarsus not booted, and wings not longer than the rounded tail; bill slender (too thick in the figs.), depressed and well bristled at base; tip evidently overhanging (not in the figs.); tarsus long, slender; toes very short.   Delicate little woodland birds, peculiar to America; migratory, insectivorous, very active and sprightly, with sharp, squeaking notes.   There are about a dozen, chiefly Central and South American, species of the single

10. Genus POLIOPTILA Sclater.

*** Bluish-ash, paler or white below; tail black and white; wings dusky, edged with hoary white; bill and feet black; only 4–4½ long; wing scarcely 2, tail rather more.

Blue-gray Gnatcatcher. Clear ashy blue, bluer on head; forehead, and line over eye, black (wanting in ♀): outer tail feather white. United States to Mass.; Arizona; Mexico. WILS., ii, 164, pl. 18, f. 3; NUTT., i, 297; 2d ed., i, 327; AUD., i, 244, pl. 70; BD., 380; COOP., 35. CÆRULEA.

Black-headed Gnatcatcher. ♂ bluish-ash, with whole crown black. ♀ with crown like back; outer tail feather white-edged only. Southwest United States. Cass., Ill., 164, pl. 27; Bd., 382; Coop., 31. MELANURA.

Plumbeous Gnatcatcher. Duller leaden gray; crown like back; a white, and also a black (latter not in ♀) superciliary line; whole outer web of outer tail feather, and its tip for some distance, white. Arizona. Bd., 382; Coop., 37. . PLUMBEA.

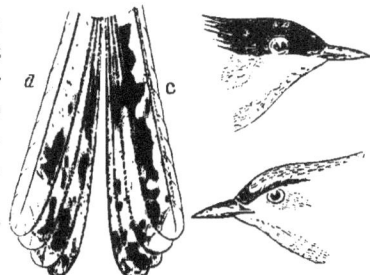

Family CHAMÆIDÆ. Wren-tits.

Recently framed for a single species, much like a titmouse in general appearance, but with the tarsus not evidently scutellate in front; rounded wings much shorter than the graduated tail; lores bristly, and plumage extraordinarily soft and lax. With the general habits of wrens, with which the species was formerly associated.

FIG. 20. Under fig., blue-gray gnatcatcher; upper fig., black-headed gnatcatcher; *c*, tail of the same; *d*, tail of plumbeous guatcatcher; all of natural size.

11. Genus CHAMÆA Gambel.

Fasciated Tit, or Ground Wren. Dull grayish or olivaceous-brown, below paler and more fulvous; throat and breast streaked with darker; wings and tail brown, obscurely waved with dusky; whitish ring round eye; iris white. $5\frac{1}{4}$–6; wing only $2\frac{1}{4}$–$2\frac{1}{2}$, tail $3\frac{1}{2}$; the graduation an inch. Coast region of California. A curious bird, with no special resemblance to any other species. Cass., Ill., 36, pl. 7; Bd., 370; Coop., 39. FASCIATA.

Family PARIDÆ. Titmice, or Chickadees.

Ours are all small (under 7 in. long) birds, at once distinguished by having 10 primaries, the 1st much shorter than the 2d; wings barely or not longer than the tail; tail-feathers not stiff nor acuminate; tarsi scutellate, longer than the middle toe, anterior toes much soldered at base; nostrils concealed by dense tufts, and bill compressed, stout, straight, unnotched and much shorter than the head: characters that readily marked them off from all their allies, as wrens, creepers, etc. Really, they are hard to distinguish, technically, from jays; but all our jays are much over 7 inches long.

They are distributed over North America, but the crested species are rather southern, and all but one of them western. Most of them are hardy birds, enduring the rigors of winter without inconvenience, and as a consequence, none of them are properly migratory. They are musical, after a fashion of their own, chirping a quaint ditty; are active, restless, and very heedless of man's presence; and eat everything. Some of the western species build astonishingly large and curiously shaped nests, pensile, like a bottle or purse with a hole in one side; others live in knotholes, and similar snuggeries that they are said to dig out for themselves. They are very prolific, laying numerous eggs, and raising more than one brood a season; the young closely resemble the parents, and there are no obvious seasonal or sexual changes of plumage. All but one of our species are plainly clad ; still they have a pleasing look, with their trim form and the tasteful colors of the head.

Subfamily PARINÆ.   True Titmice.

Exclusive of certain aberrant forms, usually allowed to constitute a separate subfamily, and sometimes altogether removed from Paridæ, the Titmice compose a natural and pretty well defined group, to which the foregoing diagnosis and remarks are particularly applicable. There may be about seventy-five good species of the Parinæ, thus restricted, most of them falling in the genus Parus, or in its immediate neighborhood. With few exceptions they are birds of the northern hemisphere. abounding in Europe, Asia and North America. The larger proportion of the genera and species inhabit the Old World ; all those of the New World occur within our limits, except two — Psaltriparus melanotis and Parus meridionalis, which are Mexican, though they have been lately included in our systematic works. The former is a very distinct and beautiful species ; the latter is perhaps only a southern variety of the common Chickadee.

12. Genus LOPHOPHANES Kaup.

***Conspicuously crested. Leaden-gray, often with a faint olivaceous shade, paler or whitish below ; wings and tail unmarked. (All the figures are of natural size.)

Tufted Titmouse.   Forehead alone black ; nearly white below ; sides washed with rusty-brown ; feet leaden-blue. Young birds have the crest plain, thus resembling the next species ; but they are nearly white below, the sides showing rusty traces. Largest of our species of the family, 6–6½ ; wing 3-3¼, tail about the same. Eastern United States, north to Long Island ; "Nova Scotia" (Aud.). WILS., i, 137, pl. 8, f. 5 ; Aud., ii, 143, pl. 125 ; Bd., 384.   BICOLOR.

FIG. 21. Tufted Titmouse.

Plain Titmouse.   Plain leaden gray with faint olive shade, merely paler below ; no markings anywhere. 5½–6 ; wing and tail about 2¾. New Mexico, Arizona and California. CASS., Ill., p. 19 ; Bd., 386 ; ELLIOT, pl. 3 ; COOP., 42. . . INORNATUS.

Black-crested Titmouse.   Size of the last, or rather less ; similar to the

FIG. 22. Plain Titmouse.

FIG. 23. Black-crested Titmouse.

FIG. 24. Bridled Titmouse.

first in color, but forehead whitish, and whole crest black. Valley of the Rio Grande. CASS., p. 13, pl. 3 ; Bd., 385 ; COOP., 43.   ATRICRISTATUS.

Bridled Titmouse.   Olivaceous-ash ; below soiled whitish ; chin and throat pure black; sides of head and neck white, commonly striped with black in two or three places; crest like back, margined with black; smallest; 5–5¼; wing and tail about 2½; *young* with the black head-markings obscure. New Mexico, Arizona, and southward. CASS., Ill., 19; BD., 386; COOP., 43. . . . . . . WOLLWEBERI.

FIG. 25. Black-capped Chickadee; nearly natural size.

13. Genus PARUS Linnæus.

* Crown and nape, with chin and throat, black, separated by ashy or whitish; above brownish or grayish ash, often with faint olivaceous tinge; below whitish or rusty or brownish shaded on sides; wings and tail plain, more or less whitish-edged.

† No white superciliary line.

Titmouse. Black-capped Chickadee. Average dimensions:—Length 5⅓; extent 8¼; wing and tail, each, 2½; tarsus 7-10. *Extremes:*—Length 4¾–5½; extent 7½–8½; wing and tail 2⅓–2⅝; tarsus ⅔–¾. North America. Everywhere abundant. WILS., i, 137, pl. 8, f. 4; AUD., ii, 146, pl. 126; NUTT., i, 244; BD., 390. . . . . ATRICAPILLUS.

Var. SEPTENTRIONALIS. *Long-tailed Chickadee.* Averaging larger; paler below, and less shaded on sides; wings and tail much edged with whitish; tail longer compared with the wings (nearly or quite 3). Missouri and Rocky Mountain region. CASS., Ill., 80, pl. 14; BD., 389.

Var. CAROLINENSIS. *Carolina Titmouse.* Averaging smaller than *P. atricapillus;* wings and tail less edged with whitish. Eastern United States, southerly. AUD., ii, 152, pl. 127; BD., 392.

Var. OCCIDENTALIS. *Western Titmouse.* Size of the first; said to be darker, with longer tarsi. Pacific Coast. BD., 391.

†† A distinct white line in the black over eyes and across forehead.

Mountain Chickadee. Otherwise exactly like *P. atricapillus.* Rocky Mountains to Pacific. BD., 394; ELLIOT, pl. 2; COOP., 46. . MONTANUS.

** Body with chestnut brown; chin and throat brownish-black.

Hudsonian Chickadee. Pale olive-brown; crown similar but browner; below on sides, and behind, pale chestnut. About 5; wing 2½, tail 2⅔. British America into Northern States (Alaska, *Dall*). AUD., ii, 155, pl. 123; BD., 395. Var. *littoralis* is described from Nova Scotia. HUDSONIUS.

Chestnut-backed Chickadee. Crown, nape and throat alike in color, sooty brown; back and sides chestnut. Under 5; wing 2⅓, tail less. Pacific coast. AUD., ii, 158, pl. 129; BD., 394; COOP., 47. . . . RUFESCENS.

14. Genus PSALTRIPARUS Bonaparte.

Dwarfs among pygmies! 3¾–4¼ long; wing 2 or less, tail 2 or more; ashy or olive gray; paler (whitish, etc.) below.   Both species are western; these and *Auriparus flaviceps* build the curious pensile nests above mentioned.

Least Titmouse.   Crown dark brown, unlike back.   Pacific coast to Sierra Nevada.   AUD., ii, 160, pl. 130; BD., 397; COOP., 48. .   MINIMUS.

Leaden Titmouse.   Crown like back.   Iris brown or yellow.   Arizona. BD., 398; COOP., 49. . . . .   PLUMBEUS.

15. Genus AURIPARUS Baird.

Yellow-headed Titmouse.   Ashy; paler below; head all yellow (this color wanting in the *young*); bend of wing chestnut; 4½; wing and tail about 2¼. Texas, New Mexico, Arizona, South and Lower California.   BD., 400, and Rev., 85; COOP., 51. . . . .   FLAVICEPS.

Family SITTIDÆ.   Nuthatches.

These birds differ in so many respects from either *Certhiidæ* or *Paridæ*, with both of which they have been associated, that I shall give them independent family rank. CHARS.—Bill subcylindrical, tapering, compressed, slender, acute, nearly or about as long as the head, culmen and commissure about straight, gonys long, convex, ascending (giving a sort of recurved look to a really straight bill).   Nostrils rounded, concealed by bristly tufts.   Wings long, pointed, with 10 primaries, the 1st very short or spurious; tail much shorter than wings, broad, soft, nearly even; tarsus shorter than the middle toe and claw, scutellate in front; toes all long, with large, much curved, compressed claws; 1st toe and claw about equal to the 3d; 2d and 4th toes, very unequal in length; plumage compact; body flattened; tongue horny, acute, barbed.   Nuthatches are amongst the most nimble and adroit of creepers; they scramble about and hang in every conceivable attitude, head downwards as often as otherwise.   This is done, too, without any help from the tail—the whole tarsus being often applied to the support.   They are chiefly insectivorous, but feed also on hard fruits; and get their English name from their habit of sticking nuts and seeds in cracks in bark, and hammering away with the bill till they break the shell.   They are very active and restless little birds, quite sociable, often going in troops, which keep up a continuous noise; lay 4–6 white, spotted eggs, in hollows of trees.   The family is a small one, of less than thirty species, among them a single remarkable Madagascan form, *Hypherpes*, a genus peculiar to Australia (*Sittella*), and another confined to New Zealand (*Acanthisitta*): but it is chiefly represented by the genus *Sitta*, with 12 or 14 species, 8 or 9 of Europe and Asia, and the following of our country:—

16. Genus SITTA Linnæus.

* *White* below, flanks and under tail coverts washed with rusty brown; ashy-blue above, middle tail feathers the same, other tail feathers black, spotted with white; *crown and nape glossy black, without stripes;* wings varied with black, white and the color of the back.   Large; 5½–6; wing 3½, tail 2.

FIG. 26. White-bellied Nuthatch; nat. size.   *White-bellied Nuthatch.*   As above; bill over 15-100 deep at base.   In the young and many ♀'s, black of head

restricted to nape, or altogether absent.   Eastern United States to the
Plains. WILS., i, p. 40; NUTT., i, 581; AUD., iv, 175, pl. 247; BD.,
374. CAROLINENSIS.

Var. ACULEATA. *Slender-billed Nuthatch.* Exactly like the last, except slenderer bill; not over 15-100 deep at base. Plains to Pacific. BD., 375; COOP., 54.

** Rusty brown below, nearly uniform; back bluer than in the last, *head with
white stripes*, crown black or like back; tail as in the last; wings plain; medium
in size; 4½ to nearly 5; wing 2¾, tail 1½.

Red-bellied Nuthatch.   ♂ with crown glossy black, bordered by white
stripes meeting across forehead, below these a black bar through eye to
hind nape, below this, and the chin, white. ♀ with crown like back, and the
lateral black stripe merely dusky; young with no
black on the crown and lateral stripes obscure.
North America, but rather northerly. WILS., i,
40, pl. 2; NUTT., i, 583; AUD., iv, 179, pl. 248;
BD., 376. . . . . CANADENSIS.

*** Pale rusty or brownish white below; wings,
tail and back, much as in the last; *crown and nape
brown to below eyes, the lower border darker; head without white stripes. Smallest; 4, or less, long; wing 2½, tail 1½.

FIG. 27.   Red-bellied Nuthatch;
natural size.

Brown-headed Nuthatch.   Crown clear hair-brown; a distinct little
whitish spot on nape; middle tail feathers like back, with no black, and
little or no white at base. South Atlantic States, strictly. WILS., ii, 105,
pl. 15; NUTT., i, 584; AUD., ii, 181, pl. 249; BD., 377, . . PUSILLA.

Pygmy Nuthatch.   Crown dull olive brown, its lateral borders blackish;
the nuchal whitish spot wanting or obscure; middle tail feathers white at
base, and there black-edged on outer web. This species is apt to be quite
brownish underneath, instead of merely muddy white, as in the last; but
both vary much in this respect. Rocky Mountains to Pacific, United States.
AUD., iv, 184, pl. 250; BD., 378; COOP., 55. . . . . . . PYGMÆA.

Family CERTHIIDÆ.   Creepers.

A very small, well-marked group, of about a dozen species, and four or five
genera, which fall in two sections, commonly called subfamilies; one of these,
Tichodrominæ, is represented by the well known European Creeper, *T. muraria*, and
several, chiefly Australian, species of the genus *Climacteris;* while the genus *Cer-
thea*, with five or six species or varieties, and one or two allied genera (all but one
Old World) constitutes the

Subfamily CERTHIINÆ.   Typical Creepers.

Our species may be known on sight, among North American Oscines, by its
rigid, acuminate tail-feathers, like a woodpecker's. Besides:—bill about equal to
head, extremely slender, sharp and decurved; nostrils exposed; tarsus shorter
than 3d toe and claw, which is connate for the whole of the 1st joint, with both
2d and 4th toe; 1st toe shorter than its claw; claws all much curved and very
sharp; tarsus scutellate; wings 10-primaried, 1st very short, not one-half the 2d,

which is less than the 3d; point of wing formed by 3d, 4th and 5th quills; tail rounded, equal to or longer than wing. Restless, active little forest birds that make a living by picking bugs out of cracks in bark. In scrambling about, they use the tail as woodpeckers do, and never hang head downwards, like the nut-hatches. Lay numerous eggs in knotholes; not migratory; no song; slight seasonal or sexual changes of plumage.

17. Genus CERTHIA Linnæus.

Brown Creeper. Plumage above singularly barred with dusky, whitish, tawny or fulvous brown, and bright

FIG. 28. Brown creeper; nat. size.

brown — latter chiefly on the rump; below, white, either pure or soiled, and generally slightly brownish-washed behind; wings dusky, oddly varied with tawny or whitish bars and spots; tail plain; about $5\frac{1}{2}$; wing and tail about $2\frac{3}{4}$. North America. WILS., i, 122, pl. 8; NUTT.; AUD., ii, 109, pl. 115; .BD., 372. . . . . FAMILIARIS.

Family TROGLODYTIDÆ. Wrens.

Embracing a number of forms assembled in considerable variety, and difficult to limit with precision. Closely related to the last two or three families; known from these by non-acuminate tail feathers and exposed nostrils. Very intimately resembling, in particular, the mocking group of thrushes—those with scutellate tarsi and not strictly spurious first primary; but all our wrens are *smaller* than any of the *Miminæ*, and otherwise distinguished by less deeply cleft toes, as stated on p. 73; "the inner toe is united by half its basal joint to the middle toe, sometimes by the whole of this joint; and the second joint of the outer toe enters wholly or partially into this union, instead of the basal only." Nostrils narrowly or broadly oval, exposed, overhung by a scale resembling that of the *Gallinæ;* bill rather or very slender, straight or slightly decurved, from half as long to about as long as the head, unnotched in all our genera; no evident rictal bristles: wings short, more or less rounded, primaries 10, the 1st short, but not strictly spurious; tail variable in length, much or little rounded: tarsus scutellate, hind toe very long.

Excluding certain Old World forms sometimes placed with the Wrens, but probably better assigned elsewhere; and excepting the European wren and its congeners, the *Troglodytidæ* are confined to America. If thus restricted, the family is susceptible of more exact limitation, as shown by Baird in his elaborate ' Review' (p. 91). There are about a hundred recognized species or varieties, usually referred to about sixteen genera or subgenera; most of these belong to tropical America, where the family reaches its maximum development; for instance, over twenty species of *Campylorhynchus* alone are described. Of the North American forms, genera 18, 19 and 20 are confined to the West, and represent a section distinguished by the breadth of the individual tail feathers, which widen noticeably towards the tip. Species of all our other genera are common and familiar eastern birds, much alike in disposition, manners and habits ; the house wren may be taken to typify these. They are sprightly, fearless and impudent little creatures, apt to show bad temper when they fancy themselves aggrieved by cats or people, or anything else that is big or unpleasant to them; they quarrel a good deal, and are particularly spiteful towards martins and swallows, whose homes they often invade and occupy. Their song is bright and hearty, and they are fond of their own music; when disturbed at it, they make a great ado with noisy scolding. Part of them live in reedy swamps and marshes, where they hang astonishingly big globular nests, with a little hole in one side, on tufts of rushes, and lay six or eight dark colored eggs; the others nest anywhere, in shrubbery, knotholes, hollow stumps and other odd nooks. Nearly all are migratory; one is stationary; one comes to us in fall from the north, the rest in spring from the south. Insectivorous, and very prolific, laying several sets of eggs each season. Plainly colored, the browns being the usual colors; no red, blue, yellow or green in any of our species.

18. Genus CAMPYLORHYNCHUS Spix.

Brown-headed Creeper Wren. Brown, conspicuously white-streaked, crown brown, plain; below whitish becoming pale brownish behind, with many very distinct round black spots, largest and closest on throat and breast; tail feathers black, only the outer and central pair with more than one white bar on the inner web. Largest of all, 8; wing and tail about 3½. South-western United States, and southward. CASS., Ill., 156, pl. 25; BD., 355; COOP., 61. . . . . BRUNNEICAPILLUS.

Allied Creeper Wren. Similar; smaller; fewer and smaller black spots on breast; tail feathers all with white bars or spots on both webs. Cape St. Lucas, XANTUS, Proc. Acad. Philada., 1859, p. 298; ELLIOT, pl. 3; BD., Rev. 100; COOP., 62. A variety of the last? . . . . AFFINIS.

19. Genus SALPINCTES Cabanis.

Rock Wren. Brownish-gray, often obsoletely waved with lighter and darker shades, becoming cinnamon or fulvous-brown on rump, everywhere speckled with black and white dots; below whitish, throat and breast obscurely streaked with dusky; belly and sides fulvous-brown-tinted, under tail coverts blackish-barred; wings dusky, obscurely waved with paler, chiefly on outer webs; middle tail feathers barred like wings, others with broad subterminal black bar and fulvous tip; outer feather often with several such markings. 5½-6; wings 2¾, tail 2⅛; all the markings are obscure and blended; the brown has often a slight pinkish shade. Central and Rocky Mountain region of the United States into Mexico. (*Myiothera obsoleta*, BONAP., Am. Orn., i, 6, pl. 1, f. 2.) NUTT., i, 435; AUD., ii, 113, pl. 116; BD., 357; COOP., 65, ... OBSOLETA.

20. Genus CATHERPES Baird.

Mexican or *White-throated Wren.* Brown, grayer towards and on head, becoming rich ferruginous or brownish-red behind, both above and below; chin and throat pure white; back and crown finely speckled with black and white dots; wings dusky, waved with brown; tail rich brown, like the rump and belly, with numerous narrow distinct black bars; belly waved or speckled with dusky and whitish; bill long (¾ or more), extremely

slender; 5–5½; wing 2¼, tail 2½. South-western United States and southward. Cass., Ill., 173, pl. 30; Bd., 356; Coop., 66.   . . mexicanus.

21. Genus THRYOTHORUS Vieillot.

* Tail not longer than wings, all its feathers reddish-brown with numerous fine
black bars.

Carolina Wren.   Clear reddish-brown, slightly grayer on head, brightest
on rump; below tawny of varying shade; long conspicuous superciliary
line white or tawny; wings edged
with color of back, and dusky waved;
wing coverts usually whitish spotted;
under tail coverts usually blackish
barred; sides of body unmarked.
5½ to nearly 6; wings 2¼, tail rather
less. Eastern United States, rather
southern; north to Connecticut, and
scarcely or not migratory; winters
at Washington, D. C.   A voluble
songster. Wils., ii, 61, pl. 12, f.
5; Nutt., 1, 429; Aud., ii, 116,
pl. 117; Bd., 361.   ludovicianus.

Var. berlandieri.   *Berlandier's
Wren.* Similar; rather smaller; bill
larger; darker, especially below; sides
dusky-barred. Near Mexican boundary.
A geographical race of the last, with
which it is perfectly connected, according to Mr. Allen, by intermediate Floridan
specimens. Bd., 362, pl. 83, f. 1; Rev. 124.

** Tail longer than wings; its feathers mostly black.

Bewick's Wren.   Grayish-brown; below ashy-white; superciliary line
white; wings dusky, faintly waved; under tail coverts dark-barred; two
middle tail feathers like back, with numerous fine black bars, others black,
several of the lateral with white or gray spots or tips. 5½; extent 6¾;
wings little or not over 2, tail 2½. United States, southern; in New Mexico
and Arizona, whiter below (var. *leucogaster*); on Pacific coast, grayer above
and bill longer (var. *spilurus*). Nutt., i, 434; Aud., ii, 120, pl. 118;
Bd., 303; Rev. 126; Coop., 69, ... bewickii.

22. Genus TROGLODYTES Vieillot.

House Wren. Brown, brighter behind; below rusty-brown, or grayish-
brown, or even grayish-white; everywhere waved with darker shade, very
plainly on wings, tail, flanks and under tail coverts; breast apt to be darker
than either throat or belly; bill less than head, about half an inch long;
wings and tail nearly equal, about 1⅞–2½; total length from 4½–5¼ (average 4⅜). Eastern United States, very abundant anywhere. Wils., ii, 129,

pl. 8; NUTT., i, 422; AUD., ii, 125, pl. 120; BD., 367. Very variable in precise tint, distinctness of the barring, etc.; old spring birds are apt to be grayer and clearer below; *young* fall specimens are usually browner. *T. americanus* AUD., as I have said (Proc. Essex Inst. v, 1867, 278; specimen in my cabinet, personally identified by Audubon; see also MAYNARD, Guide, p. 95), is not otherwise different, and I shall now drop it. . . . ÆDON.

Var. *parkmanni* AUD. On an average, grayer and paler. Western United States (see COUES, Proc. Acad. Phil., 1866, p. 43). BD., 367; COOP., 71.

23. Genus ANORTHURA Rennie.

Winter Wren. Deep brown, darkest on head, brightest on rump and tail, obscurely waved with dusky and sometimes with whitish also; tail like rump; wings dusky, edged with color of back, and dark barred; several outer primaries also whitish barred; a superciliary line, and obscure streaks on sides of head and neck, whitish; below pale brown; belly, flanks and under tail coverts strongly barred with dusky and whitish. Only 4–4½ long; extent 6¼–6½; wing 2 or *less*, tail 1½ or

FIG. 30. Winter Wren.

less—so short that the outstretched feet reach beyond it. Tarsus and middle toe and claw together about 1¼; bill ⅜. North America; United States in winter. *Sylvia hyemalis*, WILS., i, 139, pl. 8, f. 6; *Trog. hyemalis*, AUD., ii, 128, pl. 121; BD., 369; *Trog. europæus*, NUTT., i, 427. Var. *pacificus* is described; BD., Rev. 145. . . . TROGLODYTES.

Alaskan Wren. "Form like that of the winter wren;" size and colors nearly the same; darker; bill larger; culmen, gape and gonys almost perfectly straight—latter slightly ascending. St. George's Island, Bering's Sea. One specimen known. A variety of the last? BD., Trans. Chic. Acad., 1869, 315, pl. 30, f. 3. . . . . ALASCENSIS.

24. Genus TELMATODYTES Cabanis.

Long-billed Marsh Wren. Above clear brown, unbarred, back with a black patch containing distinct white streaks, crown brownish-black, superciliary line to nape white: wings not noticeably barred, but outer webs of inner secondaries blackish; tail brown, dusky barred; below dull white, often quite pure, the sides alone brownish-washed, and under tail coverts somewhat barred. 4¾–5½ long; wing about 2, tail less, tarsus ¾–⅞; bill ½ or more, barely curved. North America; particularly reedy swamps and marshes of United States, abundant. WILS., ii, 58, pl. 12, f. 4; NUTT., i, 439; AUD., ii, 135, pl. 123; BD., 364; var. *paludicola*, BD., Rev. 148. . . PALUSTRIS.

FIG. 31. Long-billed Marsh Wren; nat. size.

25. Genus CISTOTHORUS Cabanis.

Short-billed Marsh Wren. Dark brown above, crown and middle of back blackish, nearly *everywhere conspicuously streaked with white;* below buffy white, shading into pale brown on sides and behind; wings and tail barred with blackish and light brown; flanks barred with dusky; throat and middle of belly whitish : $4\frac{1}{2}$; wing and tail about $1\frac{3}{4}$; bill not $\frac{1}{2}$ long and very slender; tarsus and middle toe and claw together about $1\frac{1}{8}$. Eastern United States, in reedy swamps and marshes, not common. *Troglodytes brevirostris.* NUTT., i, 431 ; AUD., ii, 138, pl. 124 ; BD., 365. STELLARIS.

Family ALAUDIDÆ. Larks.

A rather small group, well defined by the character of the feet, in adaptation to terrestrial life. The subcylindrical tarsi are scutellate and blunt behind as in front, with a deep groove along the inner side, and a slight one, or none, on the outer face. Other characters (shared, however, with some *Motacillidæ*) are the very long, straight, hind claw, which equals or exceeds its digit in length ; the long, pointed wings, with the 1st primary spurious or wanting, and the inner secondaries ("tertiaries") lengthened and flowing. The nostrils are usually concealed by dense tufts of antrorse feathers. The shape of the bill is not diagnostic, being sometimes short, stout and conic, much as in some *Fringillidæ*, while in other genera it is slenderer, and more like that of insectivorous Passeres. The family is composed, nominally, of a hundred species ; with the exception of one genus and two or three species or varieties, it is confined to the Old World. Its systematic position is open to question ; Lilljeborg removes it from *Oscines* altogether, probably on account of the peculiarities of the podotheca ; authors generally place it near the *Fringillidæ*, perhaps from the resemblance of the bill of some species to that of the finches ; but it has many relationships with the *Motacillidæ*, and in the arrangement of this work I find no better place for it than here, though it has no special affinity with the preceding family. Moreover, the fact that it has indifferently nine or ten primaries may indicate a natural position between the sets of families in which number of primaries is among the diagnostic features. According to shape of bill, structure of nostrils, and number of primaries, the family may be divided into two subfamilies, the *Alaudinæ*, typified by the celebrated skylark of Europe, and the

Subfamily CALANDRITINÆ,

Represented in America by the single genus *Eremophila*, of which there are nominally ten, really four or five, species. The birds of this genus have the bill compressed-conoid, shorter than the head, the nostrils densely feathered, and apparently only nine primaries (though I suspect that a rudimentary 1st primary exists in the condition mentioned under *Ampelis* and *Vireo*) ; the point of the wing formed by the first three primaries ; the tail of medium length and nearly square ; and a peculiar little tuft of lengthened feathers over each ear, like the "horns" of certain owls. They frequent open places, are strictly terrestrial in habits, and never hop when on the ground, like most Passeres ; they are migratory in most localities, and gregarious, except when breeding ; nest on the ground, and lay 4–5 speckled eggs ; sing sweetly in the spring time.

26. Genus EREMOPHILA Boie.

Horned Lark. *Shore Lark.* In spring: — Pinkish-brown, brightest on rump, nape and wing coverts, thickly streaked with dusky; below, white, breast and sides shaded with the color of the back, chin, throat and superciliary line pale yellow, or yellowish-white; a pectoral crescent and curved stripe under the eye, black; tail black, outer feathers white-edged and middle ones like the back. Tints extremely variable; young birds, and fall and winter specimens of the Atlantic States are plain grayish-brown, streaked with darker, below soiled whitish, and with the black markings of the head and breast obscure or wanting, though the yellow is usually bright — even

Fig. 32. Horned Lark.

more so than in spring. Length 7–7½, wing 4⅓, tail 2¾–3, tarsus ⅞, hind claw ½–⅔, very slender and sharp. North America; in the east retires in spring beyond the United States, but in the west breeds on the plains much further south. WILS., i, 85, pl. 5, f. 4; NUTT., i, 455; AUD., iii, 44, pl. 151; BD., 403. . · . . . . ALPESTRIS.

Var. CHRYSOLÆMA. A rather smaller, brighter colored race, occurring in southwestern United States and Mexico. It looks quite different at first sight, but is not distinguishable as a species by any definite or constant characters. *Alauda rufa* AUD., vii, 353, pl. 497; BD., 403. The foregoing, with *E. peregrina*, a South American species or variety, are the only American *Alaudidæ*.

Family MOTACILLIDÆ. Wagtails.

Bill shorter than the head, very slender, straight, acute, notched at tip. Rictus not evidently bristled. Primaries nine, of which the 1st is about as long as the 2d, and the first three, four or five, form the point; inner secondaries enlarged, the longest one nearly, or quite, equalling the primaries in the closed wing. Tail lengthened, generally about equalling the wing. Feet large; tarsus scutellate, longer than the middle toe and claw; inner toe cleft to the very base, but basal joint of outer toe soldered with the middle one; hind toe usually bearing a long and little curved claw. A pretty well defined group of one hundred, chiefly Old World, species, which may be termed terrestrial Sylvias, all living mostly on the ground, where they run with facility, never hopping like most Oscines. They are usually gregarious; are insectivorous and migratory. They have gained their name from the characteristic habit of moving the tail with a peculiar see-saw motion, as if they were using it to balance themselves upon unsteady footing. They may be distinguished from all the foregoing birds, except *Alaudidæ*, by having only nine primaries; and from all the following birds by having long flowing inner secondaries; and from *Alaudidæ*, with which they agree in this respect, as well as in usually having a lengthened, straightish hind claw, by having the tarsal envelope as in Oscines generally, slender bill and exposed nostrils. Two subfamilies are generally recognized, though the distinctions are scarcely more than generic.

Subfamily MOTACILLINÆ.   True Wagtails.

Represented in America by a single species; in the Old World by nearly fifty species or varieties, chiefly belonging to the genus *Motacilla* and its subdivisions or immediate allies, of which *Budytes* is one.   In *Motacilla* itself, the hind claw is of about the ordinary length and curvature; in *Budytes*, the hind claw is longer and nearly straight, and the tail is about as long as the wing, the point of which is formed by only three quills.

27. Genus BUDYTES Cuvier.

FIG. 33. Yellow Wagtail; natural size.

Yellow Wagtail.   Greenish-olive, below yellow; crown and nape ashy, superciliary line white, wings and tail blackish, white-edged.   Length 6; wing and tail about 3.   Alaska; a well known, widely spread and extremely variable Old World species, unknown in America until the recent discovery by Dr. Bannister that it is abundant at St. Michael's.   DALL and BANN., Trans. Chicago Acad., 1869, 277, pl. 30, f. 2. . . . . FLAVA.

Subfamily ANTHINÆ.   Titlarks.

Consisting of the single genus *Anthus*, of which, however, there are several subdivisions.   In typical *Anthus*, the wing is longer than the tail, and its point is formed by the four outer primaries, the 5th being abruptly shorter; the hind claw is nearly straight, and nearly or quite equals its digit in length.   Here belong our species; in certain South American forms even five primaries enter into the tip of the wing; in several European subgenera only three primaries are abruptly longer than the succeeding ones.   Our *Anthus* is strictly congeneric with the European *A. spinoletta*, type of the genus: *Neocorys* only differs in having the feet larger and tail shorter.   About fifty species (among them six or eight Central and South American ones) are ascribed to *Anthinæ*, of which half may prove genuine.   They are terrestrial and more or less gregarious birds, migratory and insectivorous.

28. Genus ANTHUS Bechstein.

Brown Lark.   Titlark.   Wagtail.   Pipit.   Dark brown with a slight olive shade, and most of the feathers with dusky centres, giving a slightly streaked appearance; eyelids, superciliary line and all the under parts pale buffy or ochrey brown (very variable in shade), the breast and sides of the neck and body thickly streaked with dusky; wings and tail blackish, inner secondaries pale-edged, and one or more outer tail feathers wholly or partly white; 6¼–6¾, wing 3¼–3½, tail 2¾–3. North America, everywhere; an abundant and well known bird of fields and plains.   In the United States, seen chiefly in flocks, in the fall and winter: breeds in high latitudes, and in the Rocky Mountains, above the timber line, as far south as Park county, Colorado (*Allen*); lays 4–6 very dark colored eggs in a mossy nest on the ground; voice querulous, gait

FIG. 34. Brown Lark; natural size.

tremulous, flight vacillating. WILS., v, 89, pl. 89; NUTT., i, 450; AUD.,
iii, 40, pl. 140; Bn., 232. . . . . LUDOVICIANUS.

29. Genus NEOCORYS Sclater.

Missouri Skylark. Brown, the feathers with paler edges; below and a
superciliary line, whitish, the breast sharply speckled with dusky; wings
and tail dusky, inner secondaries pale-edged, outer tail feathers white; 5½;
wing 3, tail 2⅓. Region of the Upper Missouri and Saskatchewan, extremely rare; said to resemble closely the European skylark in habits.
AUD., vii, 335, pl. 486; BD., 232. . . . . SPRAGUEI.

Family SYLVICOLIDÆ. American Warblers.

Primaries, nine; inner secondaries not enlarged, nor hind toe lengthened and
straightened, as in the two preceding families; bill without a lobe or tooth near the
middle of the commissure, as in *Pyranga*, nor strongly toothed and hooked at end,
as in *Collurio* and *Vireo*, nor greatly flattened with gape reaching to eyes, as in
Hirundinidæ, nor strictly conical with angulated commissure, as in *Fringillidæ*.
The family presents such a number of minor modifications of form, that it seems
impossible to characterize it, except negatively; in fact, it has never been satis-
factorily defined. But doubtless the student will be able to assure himself that his
specimen is a sylvicoline, by its not showing the peculiarities of our other nine-
primaried Oscines.

All the sylvicolas are *small* birds; excepting *Icteria*, and perhaps a species of
Seiurus, not one is over six inches long, and they hardly average over five. With
few exceptions they are beautifully clothed in variegated colors; but the sexes are
generally unlike, and the changes of plumage, with age and season of the year, are
usually strongly marked, so that different specimens of the same species may bear
to each other but little resemblance; this of course renders careful discrimination
necessary. The usual shape of the bill may be called conoid-elongate (something
like a slender minié bullet in miniature), but the variations in precise shape are
endless. The rictus is usually bristled; the bristles sometimes have an extraordinary development, and are sometimes wanting. The wings are longer than the
tail, except in *Geothlypis*, *Icteria*, and one or two exotic genera; neither the wing
nor tail ever presents striking forms. The feet have no special peculiarities, though
they show some slight modifications corresponding to somewhat terrestrial, or more
strictly arboricole, habits. Some of the warblers have the habits of titmice or
wrens; others of creepers or nuthatches; the *Seiuri* closely resemble the titlarks in some respects, and have even been placed in the *Motacillidæ;* while the
Setophaginæ simulate the *Tyrannidæ* (of a different suborder) so perfectly that
they used to be classed with the true flycatchers. The warblers grade so perfectly
towards the tanagers that they have all been made a subfamily of *Tanagridæ*
(where possibly they belong). The affinity of some of them with the *Cœrebidæ*,
or honey-creepers of the tropics, is so close that the dividing line has not been
drawn. The position of *Icteria* and its two associate exotic genera, *Granatellus*
and *Teretristis*, is open to question; perhaps they come nearer *Vireonidæ*. It is
probable that final critical study will result in a remapping of the whole group;
meanwhile, the very diversity of forms included in it enables us to mark off sections with ease.

As at present constituted, the *Sylvicolidæ*, comprising upwards of a hundred genuine species, may be considered to represent, in America to which they are confined, the *Sylviidæ* or typical Old World warblers. I divide them into three subfamilies, uniting the *Geothlypinæ* of Baird with the true *Sylvicolinæ*. Their characters, mostly borrowed from Baird's excellent analysis, will be found in full beyond; here they may be shortly contrasted : —

Sylvicolinæ. — Wings longer than tail (except in *Geothlypis*); commissure slightly curved, with short bristles or none.

Icterinæ. — Wings shorter than tail; commissure much curved, unbristled.

Setophaginæ. — Wings longer than tail; commissure slightly curved, with bristles reaching beyond the nostrils.

Subfamily SYLVICOLINÆ. *Warblers.*

Bill conoid-elongate, shorter than the head, about as high as, or rather higher than, wide opposite the nostrils, not hooked, but with a slight notch, or none, at tip ; commissure straight or slightly curved ; a few rictal bristles, reaching little if any beyond the nostrils, or none. Wings pointed, longer than the narrow, nearly even tail (except in *Geothlypis*).

This group is specially characteristic of North America ; all the genera and the great majority of the species occurring within our limits in summer, though most of them winter in the West Indies, Mexico and Central America. *Dendrœca*, the largest and most beautiful genus, is particularly characteristic of the Eastern United States. All are strictly insectivorous, though not such expert flycatchers as the *Setophaginæ;* none rank high as songsters, though they have pleasing notes in springtime. With us, they are all migratory.

*** Genera 30, 31, are *creeping warblers,* having the hind toe longer than its claw, and the front toes more extensively soldered together at base than in any other forms. Gen. 36, 37, 38 are *ground warblers,* with the feet relatively stouter than in the rest. Gen. 32, 33, 34 are *worm-eating warblers;* these have no rictal bristles at all. Genus 35 comprehends the *wood warblers par excellence.*

30. Genus MNIOTILTA Vieillot.

Black and White Creeper. (Pl. II, figs. 12, 13, 14, 12a, 13a, 14a.) Entirely black and white, in streaks, except on the belly ; tail white-spotted, wings white-barred ; 5–5¼, wing 2¼–2¾, tail 2¼. Eastern North America ; a common bird, generally observed scrambling like a nuthatch about the trunk and larger branches of forest trees. WILS.,

iii, 22, pl. 19 ; NUTT., i, 384 ; AUD., ii, 105 ; pl. 114 ; BD., 236. . . . . VARIA.

FIG. 35. Black and White Creeper; natural size.

31. Genus PARULA Bonaparte.

Blue Yellow-backed Warbler. ♂ , in spring : blue, back with a golden-brown patch, throat and breast yellow with a rich brown or blackish patch, the former sometimes extending along the sides ; belly, eyelids, two wing-bars, and several tail-spots, white ; lores black ; upper mandible black, under flesh colored ; ♀ , in spring, with the blue less bright, the back and throat patches not so well defined; *young*, with the blue glossed with greenish, and these patches obscure or wanting; but always recognizable by the other marks and very small size; $4\frac{1}{2}$–$4\frac{3}{4}$; wing $2\frac{1}{4}$; tail $1\frac{3}{4}$. Eastern North America; an elegant, diminutive species, abundant in high open woods, where it is generally observed fluttering among the smallest twigs and terminal foliage. Wils., iv, 17, pl. 28, f. 3; Nutt., i, 397; Aud., ii, 57, pl. 91; Bd., 238. . . . . . americana.

32. Genus PROTONOTARIA Baird.

Prothonotary Warbler. Golden-yellow, paler on the belly, changing to olivaceous on the back, thence to bluish-ashy on the rump, wings and tail; most of the tail feathers largely white on the inner webs; no other special markings; bill entirely black, very large, at least $\frac{1}{2}$ long; $5\frac{1}{2}$, wing $2\frac{3}{4}$–3, tail $2\frac{1}{4}$. South Atlantic and Gulf States; straying, however, to Ohio, Missouri and even Maine; swamps and thickets; not common. Wils., iii, 72, pl. 24, f. 3; Nutt., i, 410; Aud., iii, 89, 106; Bd., 239. . . . . citræa.

Fig. 36. Prothono- tary Warbler.

33. Genus HELMITHERUS Rafinesque.

Worm-eating Warbler. Olive, below buffy, paler or whitish on the belly; head buff, with four sharp black stripes, two along sides of crown from bill to nape, one along each side of head through the eye; wings and tail olivaceous, unmarked: bill and feet pale; bill acute, unbristled, unnotched, at least $\frac{1}{2}$ long, stout at base; tail rounded; $5\frac{1}{2}$, wing $2\frac{3}{4}$, tail 2. The sexes are not particularly dissimilar. Eastern United States, rather southerly, but north to Maine; woods, shrubbery and swamps; rather common. Wils., iii, 74, pl. 24, f. 4; Nutt., i, 409; Aud., ii, 86, pl. 105; Bd., 252. . . . vermivorus.

Fig. 37. Worm-eating Warbler.

Swainson's Warbler. Somewhat similar; colors browner above, including the head, and more buffy below; a whitish superciliary line; no *decided* markings anywhere; bill still longer, shaped something like a meadowlark's; tail emarginate; nearly 6 long. A rare and curious species, confined to the South Atlantic States; said to have occurred in Massachusetts, but this is a mistake. Aud., ii, 83, pl. 104; Bd., 252. . . . swainsonii.

34. Genus HELMINTHOPHAGA Cabanis.

** The bill slender and exceedingly acute, unnotched, unbristled. The following analysis will determine the species in *adult* plumage — not otherwise : —

Tail feathers white-blotched — bluish, crown yellow, throat black, ... *chrysoptera.*
— greenish, crown and all under parts yellow, ... *pinus.*
— greenish, crown (partly) and throat black. . . . . *buchmanii.*
— upper tail coverts chestnut, crown patch chestnut, ... *lucia.*
Tail feathers all unmarked — upper tail coverts — yellow; crown patch chestnut, ... *virginiæ.*
— not yellow; crown patch — chestnut, ... *ruficapilla.*
— orange brown, . . . *celata.*
— wanting, ... *peregrina.*

Blue-winged Yellow Warbler. Crown and entire under parts rich yellow; upper parts yellow-olive, becoming slaty-blue on the wings and tail, former with two white or yellowish bars, latter with several large white blotches; bill and stripe through eye black; 5, wing 2½, tail 2¼. ♀ and young not very dissimilar. Eastern United States; common. The resemblance, in color, between this species and the prothonotary warbler, is striking. WILS., ii, 109, pl. 15; NUTT., i, 410; AUD., ii, 98, pl. 111; BD., 254. . PINUS.

Blue Golden-winged Warbler. ♂, in spring:—slaty-blue, paler or whitish below, where frequently tinged with yellowish; crown and two wing-bars rich yellow; broad stripe on side of head through eye, and large patch on throat, black, both these bordered with white; several tail feathers white-blotched; bill black. The back and wings are frequently glossed with yellowish-olive, especially in immature specimens, in which also the peculiar markings of the head and throat may be obscure. Size of

FIG. 38.   Blue Golden-winged Warbler.

pinus. Eastern United States; rather common, in woodland, like the preceding. WILS., ii, 113, pl. 15, f. 5; NUTT., i, 411; AUD., ii, 91, pl. 107; BD., 255. . . . . CHRYSOPTERA.

Bachman's Warbler. Greenish-olive, tinged with ashy on hind head; under parts, forehead, chin and lesser wing coverts, yellow; throat and band across crown, black; outer tail feathers white-blotched. Small; 4½; wing 2½, tail 2. An extremely rare species, confined to the South Atlantic States. AUD., ii, 93, pl. 108; BD., 255. . . . . . . BACHMANII.

Lucy's Warbler. Ashy-gray, below white, sometimes faintly buffy-tinted on the breast; upper tail coverts and crown patch chestnut, the latter often concealed, and wanting in the young; outer tail feathers obscurely white-blotched. Very small; 4½–4½, extent 7¼, wing 2¼–2½, tail 1¾–2, bill about ¼! A rare and curious species, lately discovered in Arizona; very unlike any other, and somewhat resembling a *Polioptila*. Colorado Valley. COOPER, Proc. Cala. Acad. 1861, 120, and B. Cal. 84; COUES, Proc. Acad. Philada. 1866, 35; BAIRD, Review, 178. . . . . . . LUCIÆ.

Virginia's Warbler. Plumbeous, washed with greenish-olive, especially in ♀ and autumnal specimens; below white, shaded on sides; throat with a yellow patch; upper and under tail coverts yellow (entirely yellow below when adult?); crown patch chestnut; a white ring around eye; 5; wing 2½, tail 2¼. Southern Rocky Mountain region. (Colorado, abundant, RIDGWAY; Arizona, rare, COUES.) Very near the next species! BAIRD, B. N. A. 1860, p. xi, pl. 79, f. 1, and Rev., 177; COOP., 85. . VIRGINIÆ.

Nashville Warbler. Olive-green, brighter on rump, changing to *pure ash* on head: below *bright yellow*, paler on belly, olive-shaded on sides; crown with a more or less concealed chestnut patch; lores and ring round eye pale; no superciliary stripe; ♀ and autumnal specimens have the head glossed with olive, and the crown patch may be wanting. 4½–4¾; wing 2¼–2½; tail 1¾–2. Eastern North America, common: also, California

(XANTUS, GRUBER). WILS., iii, 120, pl. 27, f. 3, and vi, 15; NUTT., i, 412; AUD., ii, 103, pl. 113; BD., 256; COOP., 82. . . . RUFICAPILLA.

Orange-crowned Warbler. Olive-green, nearly uniform, rather brightest on rump, *never* ashy on head : below, *greenish-yellow*, washed with olive on the sides; crown with more or less concealed *orange-brown* patch (sometimes wanting) ; eye-ring and obscure superciliary line yellowish. Size of the last, and often difficult to distinguish in immature plumage; but a general *oliveness* and *yellowness*, compared with the ashy of some parts of *ruficapilla*, and the different color of the crown-patch in the two species, will usually be diagnostic. North America ; common in the West, rare or irregular in the Eastern States. BONAP., Am. Orn., i, 45, pl. 5, f. 2 ; NUTT., i, 413 ; AUD., ii, 100, pl. 112 ; BD., 257 ; COOP., 83. . . . . CELATA.

Tennessee Warbler. Olive-green, brighter behind but never quite yellow on the tail coverts, more or less ashy towards and on head ; *no crown patch;* below, *white,* often glossed with yellowish but never quite yellow; a ring round eye, and superciliary line, whitish ; frequently an obscure whitish spot on outer tail feathers ; lores dusky ; in the ♀ and young the olivaceous glosses the whole upper parts. 4½–4¾, *wing about* 2¾, *tail* 2 *or less;* this comparative length of wing and tail, with other characters, probably always distinguishes the species from the foregoing. Eastern North America ; rare in New England. WILS., iii, 83, pl. 25, f. 2 ; NUTT., i, 412 ; AUD., ii, 91, pl. 110 ; BD., 258. . . . . PEREGRINA.

35. Genus DENDRŒCA Gray.

⁎⁎⁎ The coloration of the rectrices is a good clue to this genus; for all the species, excepting *æstiva* and its exotic conspecies or varieties, have the tail feathers at all ages blotched with white — a feature only shown, among North American allies, in gen. 30, 31, 32 and part of 34, 40. About thirty-five species pass current, but only twenty-seven of them are well established ; they all occur within our limits excepting these : —*pityophila* (Cuba), *adelaidæ* (Porto Rico), *pharetra* (Jamaica), *olivacea* (Mexico), and *petechia* with its several tropical forms, all like *æstiva,* and of which *eoa* (Jamaica) and *aureola* (Galapagos) seem most likely to prove genuine. Of the twenty-five species ascribed to North America, one, *olivacea,* has been admitted upon insufficient evidence ; of two others, "montana" and "carbonata," nothing is now known ; leaving twenty-two species to be here treated. *Kirtlandii* is exceedingly rare ; only two or three specimens have ever been discovered. *Tigrina* has been lately removed from the genus, as type of a new one (*Perissoglossa*), on account of a peculiar structure of the tongue, which resembles that of certain *Cœrebidæ;* but, as Sundevall remarks, we have yet to see whether other warblers do not possess the same character. This is an inviting problem ; the student may render good service to ornithology, and reflect credit on himself, by examining the tongues of some additional (see BAIRD, Rev., 164) species under a moderate magnifying power, and publishing his results. Baird's excellent analysis of the North American species known in 1858 was supplemented in 1865 by a more complete review of the whole genus, and in 1869 a monographic essay was given by Sundevall (Ofvers. Kongl. Vetensk. Akad. Forh., 615). The following artificial analysis will facilitate the determination of our twenty-two established

species; I believe it to be an infallible key to the perfect male plumages, and that it will probably hold good for spring specimens of both sexes of many species; but it will fail for nearly all autumnal and most female specimens of (b). It is difficult if not impossible to meet the varied requirements of these by rigid analysis; and recourse must be had to the detailed descriptions of the species arranged in what seems to be their natural sequence. The supplementary table of certain peculiarities may, however, prove of much assistance, though it is not a complete analysis.

ANALYSIS OF PERFECT SPRING MALES.

Tail feathers edged with yellow. . . . . *æstiva*.
Tail feathers blotched with white; a white spot at the base of primaries, ... *cærulescens*.
  — no white spot at base of primaries. (a)
(a) Wing-bars not white. Below, white, sides chestnut-streaked. crown yellow. . . . . . *pennsylvanica*.
   — yellow; sides reddish-streaked, crown reddish, ... *palmarum*.
    — black-streaked; above, ashy, ... *kirtlandii*.
     — olive, reddish-streaked, . *discolor*.
(a) Wing-bars white (sometimes fused into one large white patch). (b)
(b) Crown blue, like the back; below, white, sides and breast streaked, ... *cærulea*.
  — chestnut, like the throat; below, and sides of neck, buffy tinged, ... *castanea*.
  — clear ash; rump and under parts yellow, breast and sides black-streaked, ... *maculosa*.
  — blackish, with median line orange-brown, like the auriculars; rump yellow, ... *tigrina*.
  — perfectly black; throat black: a small yellow loral spot. . . . . *nigrescens*.
   — not black; no yellow; feet flesh-color, ... *striata*.
  — with yellow spot; throat flame-color; rump not yellow, ... *blackburniæ*.
   — white; rump and sides of breast yellow, ... *coronata*.
   — yellow; rump and sides of breast yellow, ... *audubonii*.
(b) Crown otherwise; throat black; back ashy, streaked, rump ash, crown yellow, ... *occidentalis*.
  — blackish, rump black, crown blackish, ... *chrysopareia*.
  — olive; crown like back, ... *virens*.
   — not like back, ... *townsendii*.
  — yellow; back olive; no black or ashy on head. . . . . *pinus*.
   — ashy-blue; cheeks the same; eyelids yellow. . . . . . *graciæ*.
    — black; eyelids white. . . . . . . *dominica*.

Diagnostic marks of certain Warblers in any plumage.

A white spot at base of primaries — *cærulescens*.
A yellow spot in front of the eye and nowhere else — *nigrescens*.
Wings and tail dusky, edged with yellow — *æstiva*.
Wing-bars and belly yellow — *discolor*.
Wing-bars yellow, and belly pure white — *pennsylvanica*.
Wing-bars white, tail-spots oblique, at end of two outer feathers only — *pinus*.
Wing-bars brownish, tail-spots square, at end of two outer feathers only — *palmarum*.
Wing-bars not evident (?), whole under parts yellow, back with no greenish — *kirtlandii*. .
Tail-spots at end of nearly all the feathers, and no definite yellow anywhere — *cærulea*.
Tail-spots at middle of nearly all the feathers, rump and belly yellow — *maculosa*.
Rump, sides of breast, and crown more or less yellow; throat white — *coronata*.
Rump, sides of breast, crown and throat, more or less yellow — *audubonii*. ·
Throat definitely yellow, belly white, back with no greenish — *dominica* or *graciæ*.
Throat yellow or orange, crown with at least a trace of a central yellow or orange spot, and outer tail feather white-edged externally — *blackburniæ*.
Throat, breast and sides black or with black traces, sides of head with diffuse yellow, outer tail feather white-edged externally — *virens* and its western allies.
Bill ordinary; and with none of the foregoing special marks — *striata* or *castanea*.
Bill extremely acute, perceptibly curved: rump (generally) yellow — *tigrina*.

Blue-eyed Yellow Warbler.   *Golden Warbler.*   *Summer Yellowbird.*
Golden-yellow; back olive-yellow, frequently with obsolete brownish
streaks; breast and sides streaked with orange-brown; wings and tail dusky,
yellow-edged; bill dark horn blue; ♀ and young paler, less or not streaked
below.   North America, everywhere a familiar and abundant bird.   *Sylvia
citrinella* WILS., ii, 111, pl. 15, f. 5; *S. childreni* AUD., Orn. Biog. i, 180,
pl. 35; *S. rathbonia* AUD., ii, 53, pl. 89; 50, pl. 88; NUTT., i, 364, 370;
BD., 282.   .   .   .   .   ÆSTIVA.

FIG. 39.   Black-throated Green Warbler.

Black-throated Green Warbler.   ♂, in spring: back and crown clear
yellow-olive, forehead, superciliary line and whole sides of head rich yellow
(in very high plumage, middle of back with dusky marks, and dusky or
dark olive lines through eyes and auriculars, and even bordering the crown);
chin, throat and breast jet black, prolonged behind as streaks on the sides;
other under parts white, usually yellow-tinged; wings and tail dusky,
former with two white bars and much whitish edging, latter with outer
feathers nearly all white; bill and feet blackish; ♂ in the fall and ♀ in
spring, similar, but the black restricted, interrupted or veiled with yellow;
young similar to the ♀, but black still more restricted or wanting altogether, except a few streaks along sides.   Small: about 5; wing 2½.
(Compare Blue Mountain warbler, beyond.)   Eastern United States, abundant in forests; breeds in New England in pine woods.   WILS., ii, 127, pl.
27, f. 3; NUTT., i, 376; AUD., ii, 42, pl. 84; BD., 267.   .   .   .   VIRENS.

Western Warbler.   Somewhat similar to the last; crown and back not
continuously olive; back olivaceous-ash, with blackish streaks; crown and
sides of head clear yellow, former with the feathers black-tipped or dusky-
clouded; no black stripe through eye; chin, throat and fore-breast pure
black, *ending behind with a sharp convex outline;* sides faintly or not
streaked with black; belly, wings and tail as in *virens.*   Rocky Mountains
to the Pacific, U. S.   The seasonal and sexual changes are not well made

out, but are doubtless parallel with those of *virens*. AUD., ii, 60, pl. 93; BD., 268. . . . . OCCIDENTALIS.

Townsend's Warbler. Somewhat similar to *virens;* upper parts olive-green, much black-streaked, crown mostly black with olive edgings of the feathers, chin and throat not perfectly black? Perfect plumage probably not known, and changes not well understood. Rocky Mountains to the Pacific; said to have once occurred near Philadelphia. NUTT., 2d ed., i, 446; AUD., 59, 92; BD., 269, and Rev., 185; COOP., 91. . TOWNSENDII.

Golden - cheeked Warbler. Prevailing color of the upper parts black, pure on the rump, elsewhere mixed with olive-green; sides of the head yellow, with narrow black stripe through the eye; below, with the wings and tail, as in *virens;* size of this species. Guatemala (*Salvin*) to Texas (San Antonio, *Heermann*). A species I have never seen; the description is abridged from BAIRD, Rev., 183, 267, who took it from the type of the species. SCL. and SALV., Proc. Zool. Soc. London, 1860, 298, and Ibis, 1860, 273. **** This and the two preceding species require further investigation to place their relationship to each other and to *virens* upon firm footing. . . . . CHRYSOPAREIA.

Black-throated Gray Warbler. ♂, in spring: back bluish-ash, with black streaks; head and neck all round pure black, with a white stripe over and behind eye, another, broader and longer, from the corner of the bill on each side of the chin and throat, and a little *yellow spot* just before and above the eye (no other yellow anywhere); below from the throat white, the sides with numerous black streaks; wings and tail blackish, former with two white bars and much whitish edging, latter with outer feathers almost entirely white; bill and feet black. Young, and ♀, differ chiefly in having the black of the head and throat clouded with ashy, and the black streaks of the back obsolete: the curious yellow loral spot seems to be persistent and diagnostic of the species. Size of *virens*, and much the same pattern of coloration, bluish-ash replacing the olive; stands between *virens* and *cærulescens;* the western analogue of the latter. Rocky Mountains to the Pacific. NUTT., i, 2d ed. 471; AUD., ii, 62, pl. 94; BD., 270; COOP., 90. . . . . NIGRESCENS.

FIG. 40. Black-throated Gray Warbler.

Black-throated Blue Warbler. ♂ in spring: above, uniform slaty-blue, the perfect continuity of which is only interrupted, in very high plumages, by a few black dorsal streaks; below, pure white; the sides of the head to above the eyes, the chin, throat, and whole sides of the body continuously jet-black; *wing-bars wanting* (the coverts being black, edged with blue), *but a large white spot at base of primaries;* quill feathers blackish, outwardly edged with bluish, the inner ones mostly white on their inner webs; tail with the ordinary white blotches, the central feathers edged with bluish; bill black; feet dark. Young ♂, similar, but the blue glossed with olivaceous, and the black interrupted and restricted. ♀ entirely different:

dull olive-greenish, with faint bluish shade, below pale soiled yellowish ; but recognizable by the white spot at base of primaries, which, though it may be reduced to a mere speck, is always evident, at least on pushing aside the primary coverts ; no other wing markings ; tail-blotches small or obscure ; feet rather pale.   Size of *virens*.   Eastern United States, abundant, in woodland.   *S. pusilla*, WILS., v, 100, pl. 43, f. 4 ; *S. sphagnosa*, NUTT., i, 406 ; AUD., Orn. Biog. ii, 279, are ♀ or young.   *S. canadensis*, WILS., ii, 115, pl. 15, f. 7 ; NUTT., i, 398 ; AUD., ii, 63, 95 ; BD., 271.   *S. cæru- lescens*, BD. Rev. 186. . . . . CÆRULESCENS.

OBS.   The only other warbler with a white spot at base of primaries is the *D. olivacea* of Mexico, and ascribed also to Texas ; it is olivaceous, the head, neck and breast orange-brown, with a black bar through the eye. CASS., Ill. 283, pl. 48 ; BD., Rev. 205.

Cærulean Warbler.   ♂ in spring : azure blue, with black streaks ; below, pure white, breast and sides with blue or blue-black streaks ; two white wing-bars ; tail-blotches small, but occupying every feather, except, perhaps, the central pair ; bill black, feet dark.   ♀ and young with the blue impure, strongly glossed with greenish, and the white similarly soiled with yellow- ish ; a yellowish eye-ring and superciliary line.   Eastern United States, not common in most places ; north to Connecticut Valley ; "Nova Scotia."   A small and very beautiful species ; 4–4¼.   *Sylvia rara*, WILS., iii, 119, pl. 27, f. 2 ; NUTT., i, 393.   *S. azurea*, NUTT., i, 407 ; *S. cærulea*, WILS., ii, 141, pl. 17, f. 5 ; AUD., ii, 45, pl. 86 ; BD., 280. . . . . . CÆRULEA.

Yellow-rumped Warbler.   Yellow-crowned Warbler.   Myrtle Bird.   ♂ , in spring : slaty-blue, streaked with black ; below, white, breast and sides mostly black, belly, and especially the throat, pure white, immaculate ; *rump, central crown patch, and sides of breast sharply yellow*, there being thus *four* definite yellow places ; sides of head black ; eyelids and super- ciliary line white ; ordinary white wing-bars and tail-blotches ; bill and feet black ; ♂ in winter, and ♀ in summer, similar, but slate color less pure, or quite brownish ; *young* birds are quite *brown* above, with a few obscure streaks in the whitish of the under parts.   It is im- possible to specify the endless intermediate styles ; but I never saw a specimen without the yellow rump, and at least a trace of the other yellow marks ; these points therefore are diagnostic.   The only other obscure-looking brownish warblers with

FIG. 41.   Yellow-rumped Warbler.

yellow rump are *maculosa* and *tigrina*, when young.   One of the larger species ; 5¼–5¾ ; wing 3, tail 2½.   North America, but chiefly eastern ; Alaska (*Dall*) ; Washington Territory (*Suckley*) ; California (*Cooper*, 89).   United States rarely in summer, but during the migrations the most abun- dant of all the warblers ; winters as far north at least as Washington, D.C. ; occurs, however, in Mexico and Central America ; seen everywhere, but is particularly numerous in shrubbery, along hedge-rows, in flocks, associating with troops of sparrows. WILS., ii, 138, pl. 17, f. 4; pl. 45, f. 3; NUTT., i, 361; AUD., ii, 23, pl.76; BD., 272; Rev., 187. CORONATA. *Audubon's Warbler.* With a close general resemblance to the last, but throat *yellow*, not white; eyelids white, but *no* white superciliary line; cheeks *not* definitely black; wing-bars generally fused into one large white patch, and tail-blotches larger; otherwise like *coronata*, of which it is the western representative; and with which its changes of plumage are entirely correspondent. North America, from Rocky Mountains to Pacific; very abundant. AUD., ii, 26, pl. 77; BD., 273; Coop., 88. . . AUDUBONII. *Blackburnian Warbler. Hemlock Warbler.* ♂ in spring: back black, more or less interrupted with yellowish; crown black, with a central orange spot; a broad black stripe through eye, enclosing the orange under eyelid; rest of head, with whole throat, most brilliant orange, or flame color; other under parts whitish, more or less tinged with yellow, and sides streaked with black; wing-bars fused into a large white patch; tail-blotches occupying nearly all the outer feathers; bill and feet dark. ♀ and young ♂: upper parts and crown olive and black, streaked (much like adult ♀ and young *striata*, but is smaller, with more black, and usually a yellow trace on the crown); superciliary line and throat clear yellow (pale for this species, but as rich as is usual for adults of the various yellow-throated species), fading insensibly on the breast; lower eyelid yellow, confined in the dusky ear-patch; sides streaked much as in the adult; wing-patch resolved into two bars; tail-blotches nearly as extensive as in the adult, the outer feathers showing white on the *outer* webs at base (*this is a strong feature*). Eastern United States, abundant in woodland; the loveliest of the warblers; none can compare with the exquisite hue of the throat. *S. parus*, WILS., v, 114, pl. 44, f. 3; NUTT., i, 392; AUD., ii, 40, pl. 83 (young). WILS., iii, 64, pl. 23, f. 3; NUTT., i, 379; AUD., ii, 48, pl. 87; BD., 274. . . . . BLACKBURNIÆ.

Black-poll Warbler. (PLATE II, figs. 15, 16, 15*a*, 16*a*.) ♂ in spring: upper parts thickly streaked with black and olivaceous-ash; *whole crown pure black;* head below the level of the eyes, and whole under parts, white, the sides thickly marked with black streaks crowding forward on the sides of the neck to form two stripes that converge to meet at base of the bill, cutting off the white of the cheeks from that of

FIG. 42. Black-poll Warbler.

the throat; wing-bars and tail-blotches ordinary; inner secondaries white-edged; primaries usually edged externally with olive; feet and under mandible flesh color, or pale yellowish; upper mandible black. ♀ in spring: upper parts, including the crown, greenish-olive, both thickly and rather sharply black-streaked; white of under parts soiled anteriorly with very pale olivaceous-yellow, the streaks smaller and not so crowded as in the ♂, but still plain enough. Young: closely resembling the adult ♀, but a brighter and more greenish olive above, with fewer streaks,

often obsolete on the crown; below more or less completely tinged with pale greenish-yellow, the streaks very obscure and sometimes altogether wanting; under tail coverts usually pure white; a yellowish superciliary line; wing-bars tinged with the same color. When the streaks on the sides are obsolete, the species bears an extraordinary resemblance to young *castanea*, which see. One of the larger species; 5¼–5¾, wing 2¾–3, tail 2–2¼. Eastern North America, very abundant; a late migrant; when the black-polls appear in force, the collecting season is about over! Wils., iv, 40, pl. 30, f. 3; vi, 101, pl. 54, f. 3; Nutt., i, 383; Aud., ii, 28, pl. 78; Bd., 280. . . . . striata.

Bay-breasted Warbler. *Autumnal Warbler.*   ♂ in spring: back thickly streaked with black and grayish-olive; *forehead and sides of head black enclosing a large deep chestnut patch;* a duller chestnut (exactly like a blue-bird's breast) occupies the whole chin and throat and thence extends, more or less interrupted, along the entire sides of the body; rest of under parts ochrey or buffy whitish; a similar buffy area behind the ears; wing-bars and tail-spots ordinary; bill and feet blackish. The ♀ in spring is more oliva-ceous than the male, with the markings less pronounced; but always shows evident *chestnut* coloration; and probably traces of it persist in all *adult* birds in the fall. The *young,* however, so closely resemble young *striata,* that it is sometimes impossible to distinguish them with certainty. The upper parts, in fact, are of precisely the same greenish-olive, with black streaks; but there is *generally* a difference below — *castanea* being there tinged with buffy or ochrey, instead of the clearer pale yellowish of *striata;* this shade is particularly observable on the belly, flanks and under tail coverts, just where *striata* is whitest; and moreover, *castanea* is usually not streaked on the sides at all. Mature spring birds vary interminably in the extent and intensity of the chestnut. Size of *striata.* Eastern United States, abundant. *Sylvia autumnalis.* Wils., iii, 65, pl. 23; Nutt., i, 390; Aud., Orn. Biog., i, 447, pl. 83 (young). Wils., ii, 97, pl. 14, f. 4; Nutt., i, 382; Aud., ii, pl. 80; Bd., 276. . . . . . castanea.

Chestnut-sided Warbler.   ♂ in spring: back streaked with black and pale yellow (sometimes ashy or whitish); *whole crown pure yellow* immedi-ately bordered with white, then enclosed with black; sides of head and neck and whole under parts *pure white,* former with an irregular black crescent before the eye, one horn extending back-ward over the eye to border the yellow crown and be dissipated on the sides of the nape, the other reaching downward and backward to connect with a chain of pure chestnut streaks that run the whole length of the body, the under eyelid and auriculars being left white; wing-bands generally fused into one large patch, and, like the edging of the inner secondaries, much tinged with yellow; tail-spots white, as usual; bill blackish, feet brown.   ♀ in spring, quite similar; colors less pure;

Fig. 13. Chestnut-sided Warbler.

black loral crescent obscure or wanting; chestnut streaks thinner. *Young:* above, including the crown, clear yellowish-green, perfectly uniform, or back with slight dusky touches; no distinct head-markings; below, *entirely white* from bill to tail, unmarked, or else showing a trace of chestnut streaks on the sides; *wing-bands* clear *yellow* as in the adult; this is a diagnostic feature, shared by no other species, taken in connection with the continuously white under parts; bill light colored below. 5–5¼, wing 2¼, tail 2. Eastern United States; abundant in woodland. WILS., i, 99, pl. 14. f. 5; NUTT., 1, 380; AUD., ii, 35, pl. 81; BD., 279. PENNSYLVANICA.

Black and Yellow Warbler. Magnolia Warbler. ♂ in spring: back black, the feathers more or less skirted with olive; *rump yellow; crown clear ash,* bordered by black in front to the eyes, behind the eyes by a white stripe; forehead and sides of head black, continuous with that of the back, enclosing the white under eyelid; *entire under parts* (except *white* under tail coverts) *rich yellow,* thickly streaked across the breast and along the sides with black, the pectoral streaks crowded and cutting off

FIG. 41. Black and Yellow Warbler.

the definitely bounded immaculate yellow throat from the yellow of the other under parts; wing-bars white, generally fused into one patch; tail spots *small, rectangular, at the middle of the tail and on all the feathers* excepting the central pair; bill black, feet brown. ♀, in spring, quite similar; black of back reduced to spots in the grayish-olive; ash of head washed with olive; other head-markings obscure, black streaks below smaller and fewer. *Young,* quite different; upper parts ashy-olive, still grayer on the head; no head-markings whatever, and streaks below wanting, or confined to a few small ones along the sides; but always known by the *yellow rump* in connection with extensively or completely *yellow under parts* (except white under tail coverts) and *small* tail spots near the *middle* of all the feathers except the central. Small; 5 or less, wing 2½, tail 2. Eastern United States; a dainty little species, abundant in woodland. *S. magnolia* WILS., iii, 63, pl. 23. *S. maculosa,* NUTT., i, 370; AUD., ii, 65, pl. 96; BD., 284. . . . . MACULOSA.

Cape May Warbler. ♂ in spring: back yellowish-olive, with dark spots; crown blackish, more or less interrupted with brownish; *ear-patch orange-brown;* chin, throat, and posterior portion of a yellowish superciliary line tinged with the same; a black loral line; *rump and under parts rich yellow,* paler on belly and crissum, the breast and sides streaked with black; wing-bars fused into a large whitish patch; tail-blotches large, on three pairs of rectrices; bill and feet black. ♀ in spring is somewhat similar, but lacks the distinctive head-markings; the under parts are paler and less streaked; the tail-spots small or obscure; the white on the wing less. *Young:* an insignificant-looking bird, resembling an overgrown ruby-crowned kinglet, without its crest; obscure greenish-olive above, rump olive-yellow, under parts yellowish white; breast and sides with the streaks

obscure or obsolete; little or no white on wings, which are edged with
yellowish; tail-spots very small. 5–5¼, wing 2¾, tail 2¼. Eastern North
America to Hudson's Bay; West Indies (where it also breeds). A species
not very common with us, remarkable for the very acute and somewhat
decurved bill, and the anatomical peculiarities of the tongue. *S. maritima*
WILS. vi, 99, pl. 54, f. 3; NUTT., i, 371; AUD., 44, pl. 85; *D. tigrina*
BD., 280; *Perissoglossa tigrina* BD., Rev. 181. . . . . TIGRINA.

Prairie Warbler. Yellow-olive: back with a patch of *brick-red spots;*
forehead, superciliary line, two wing-bars and entire under parts, rich
yellow; a V-shaped black mark on side of head, its upper arm running
through eye, its lower arm connecting with a series of black streaks along
the whole sides of the neck and body; tail-blotches very large, occupying
most of the inner web of the outer feathers. The sexes are almost exactly
alike, and the young only differ in not being so bright, and in having the
dorsal patch and head-markings obscure. Small; 4¾–5; wing 2¼; tail 2.
Eastern United States, to Massachusetts; an abundant little bird of the
Middle and Southern States, in sparse low woodland, cedar thickets and old
fields grown up to scrub-pines; remarkable for its quaint and curious song;
an expert flycatcher, constantly darting into the air in pursuit of winged
insects, like the redstart and the species of *Myiodioctes*. *S. minuta* WILS.,
iii, 87, pl. 25, f. 4. *S. discolor* NUTT., i, 394 ("294" by error of paging);
AUD., ii, 68, pl. 97; BD., 290. . . . . DISCOLOR.

Grace's Warbler. ♂ in spring: bluish-ash, back with black streaks,
crown with still more black streaks, so crowded anteriorly and on the sides
as to become continuous; *chin, throat and breast rich yellow,* ending abruptly
against the white of the other under parts; sides of neck and body with
numerous black streaks; a broad yellow superciliary line, changing to
white behind the eye; no white patch below auriculars; *lower eyelid yellow;*
a black line from bill to eye, with which the streaks of the side of the neck
connect; two white wing-bars, the anterior one much the stronger; tail
blotches large, the outer one occupying nearly all the feather; bill and feet
black. ♀ not particularly different. *Young:* dull brownish (like young
coronata) with few or no black streaks on back, crown or along sides; *throat,
eyelid and superciliary line rich yellow,* as in the adult; other under parts
soiled whitish. 5–5¼, wing 2¾, tail 2¼, bill under ½. New Mexico, Arizona
and southward; abundant, and breeding, at Fort Whipple (*Coues*). An
interesting lately discovered species, closely resembling the next. COUES,
Proc. Acad. Nat. Sci. Philada., 1866, p. 67. BAIRD, Rev. 210; COOPER,
p. — (appendix). . . . . GRACIÆ.

Yellow-throated Warbler. Much like the last species, with which its
changes of plumage are entirely correspondent; *no yellow* in the black
under the eye; a white patch separating the black of the cheeks from the
bluish ash of the neck; superciliary line usually yellow from bill to eye,
thence white to nape, sometimes entirely white; bill very long (at least ½),
extremely compressed, almost a little decurved. South Atlantic and Gulf

States, rather common ; north to Maryland and Ohio, but rare ; West Indies
(where it breeds), Mexico and Central America. *S. flavicollis* WILS., ii,
64, pl. 12, f. 6 ; *S. pensilis* NUTT., i, 374 ; AUD., ii, 32, 79 ; *D. super-
ciliosa* BD., 289 ; *D. dominica*, BD., Rev. 209. . . . . . DOMINICA.

Kirtland's Warbler. "Above slate-blue, the feathers of the crown with
a narrow, those of the back with a broader, streak of black ; a narrow
frontlet involving the lores, the anterior end of the eye and space beneath
it, black ; the rest of the eyelids white ; under parts clear yellow, almost
white on the under tail coverts, the breast with small spots and the sides
with short streaks of black ; greater and middle wing coverts, the quills and
tail feathers, edged with dull whitish ; two outer tail feathers with a dull
white spot on the inner web ; $5\frac{1}{2}$; wing 2^4, tail $2\frac{3}{4}$ " (*Baird*). Very rare ;
only two or three specimens known, from Ohio and the Bahamas. A species
I have never seen ; but I suspect that its relationships are with *dominica* and
graciæ, and that they may prove still closer with the Portorican species
of the same group (*adelaidæ*). BAIRD, Ann. Lyc. Nat. Hist. N. Y. v,
1852, 217, pl. 6 ; CASS., Ill. i, 278, pl. 47 ; BD., Rev. 206. . KIRTLANDII.

Yellow Red-poll Warbler. *Palm Warbler.* In spring : brownish-olive,
rump and upper tail coverts brighter yellowish-olive, back obsoletely
streaked with dusky, *crown chestnut ;* superciliary line and entire under
parts rich yellow, breast and sides with reddish-brown streaks, somewhat as
in the summer warbler ; a dusky loral line running through eye ; *no white
wing-bars*, the wing coverts and inner quills being edged with yellowish-
brown ; tail spots *at very end* of inner webs of two outer pairs of tail
feathers only, and *cut squarely off*—a peculiarity distinguishing the species
in any plumage. ♀ not particularly different from the ♂ : *young*, an obscure-looking species, brownish above like a young yellow-rump, but upper
tail coverts yellowish-olive, and under tail coverts apt to show quite bright
yellow in contrast with the dingy yellowish white or brownish white of other
under parts ; pectoral and lateral streaks obscure ; crown generally showing
chestnut traces ; but in any plumage, known by absence of white wing-bars
and peculiarity of the tail spots, as just said. $5\frac{1}{4}$, wing $2\frac{1}{2}$, tail $2\frac{1}{4}$. Eastern North America, abundant : usually found in fields, along hedgerows and
roadsides, with yellow-rumps and sparrows ; the most terrestrial species of
the genus, often recalling a titlark ; remains in the fall latest of any, except
the yellow-rump. Winters in Florida and the West Indies (*Allen*). *S.
petechia*, WILS., vi, 19, pl. 28, f. 4 ; NUTT., i, 364 ; AUD., ii, 55, pl. 90 ;
BD., 288. PALMARUM.

Pine Warbler. *Pine-creeping Warbler.* Uniform yellowish-olive above,
yellow below, paler or white on belly and under tail coverts, shaded and
sometimes obsoletely streaked with darker on the sides ; superciliary line
yellow ; wing-bars *white ;* tail-blotches *confined to two outer pairs of feathers,
large, oblique.* ♀ and *young*, similar, duller ; sometimes merely olive-gray
above and sordid whitish below. The variations in precise shade are inter-
minable ; but the species may always be known by the lack of any special sharp

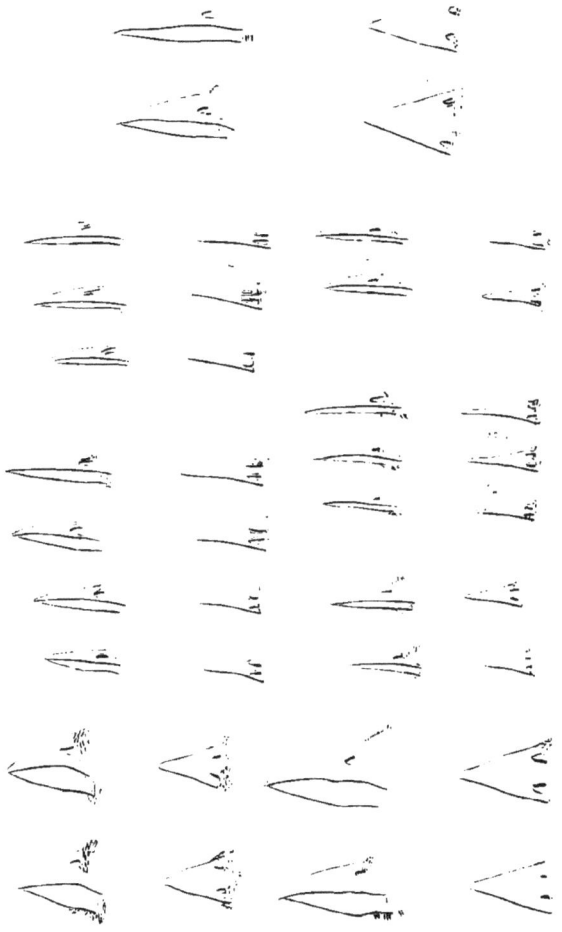

markings whatever, except the superciliary line; and by the combination of white wing-bars with large oblique tail-spots confined to the two outer pairs of feathers. One of the largest species: $5\frac{1}{2}$ to nearly 6. Eastern United States, very abundant in pine woods and cedar thickets; has an extensive breeding range, and is apparently resident in southern portions. *Vireo vigorsii* Nutt., i, 318; *S. pinus* Wils., iii, 25, pl. 19, f. 4; Nutt., i, 387; Aud., ii, 371, pl. 82; Bd., 277. . . . . pinus.

Obs. The two following species, ascribed to North America, are not now known:—

Blue Mountain Warbler. Sylvia montana Wils., v, 113, pl. 44, f. 2 (Blue Mountains of Virginia). Aud., ii, 69, pl. 98 ("California"). Bd., 278. Professor Baird suggests that some plumage of *D. pinus* or *striata* may furnish the clue to this lost species; but these are among the largest warblers, whilst Wilson says "length four inches and three-quarters." Mr. Turnbull (Birds of New Jersey, p. 18) says, without qualification, it is the young of *D. cærulea.* I think myself that it is simply the young of *D. virens!* of which, it seems, Wilson never recognized an autumnal example. A September specimen of *virens*, before me as I write, agrees almost precisely with Wilson's description — rich yellow olive; front, cheeks, chin and sides of neck, yellow; * * two exterior tail feathers white on the inner vanes from the middle to the tip, *and edged on the outer side with white*, etc. Now *D. virens* is the only Eastern species, showing this latter feature, that agrees with the other assigned characters at all. It is curious additional evidence that I am right in this surmise, that the original of Audubon's figure, in the British Museum, came from "California;" for I suppose that this specimen was the young of *occidentalis* or *townsendii*, some of the plumages of which, as well as can be made out, are with difficulty distinguishable from immature *virens*.

Carbonated Warbler. Sylvia carbonata Aud. Orn. Biog. i, 308, pl. 60; Nutt., i, 405; Aud., ii, 95, pl. 109; Bd., 287. Only known by the figure and description of a pair killed in Kentucky. I have no idea what this is; it may not be a *Dendrœca* at all. Audubon himself put it among the worm-eating warblers.

36. Genus SEIURUS Swainson.

. The birds of this genus have been classed with the thrushes, and also with the titlarks (which they somewhat resemble in habits, being walking birds), but they have no special affinity with either. They are simply terrestrial warblers, closely related to gen. 37, 38. Five species are enumerated, but the exotic representatives of *noveboracensis* and *ludovicianus* seem to be mere varieties.

* Crown orange-brown, with two black stripes; no superciliary line.

Golden-crowned Thrush. Oven Bird. Bright olive green; below pure white, thickly spotted with dusky on breast and along sides; a narrow maxillary line of blackish; under wing coverts tinged with yellow; a white eye-ring; legs flesh color; wings and tail unmarked. Sexes alike; young similar. Length $5\frac{1}{2}$–$6\frac{1}{2}$; wing 3; tail $2\frac{3}{4}$.

Fig. 45. Golden-crowned Thrush.

Eastern North America, West Indies, Mexico, Alaska (*Dall*). A very common bird in open woodland, spending much of its time on the ground rustling among the leaves; noted for its loud monotonous song, and its curious nest, which is placed on the ground and roofed over; whence the name "ovenbird." WILS., ii, 88, pl. 14, f. 2; NUTT., i, 355; AUD., Orn. Biog. ii, 253; v, 447; pl. 143; BD., 260. . . . . . . AUROCAPILLUS.

** Crown plain, like the back; a conspicuous superciliary line.

Water Thrush.   Water Wagtail. (PLATE II; figs. 9, 10, 11; 9*a*, 10*a*, 11*a*.) Deep olivaceous-brown; below, white, more or less tinged with *pale yellowish*, thickly and *sharply* spotted with the color of the back, except on lower belly and crissum: superciliary line yellowish; feet dark. Length 5½–6; wing 2¾; tail 2¼; *bill about ½*. North America, everywhere; a common bird of low watery thickets, in the habit of constantly vibrating the tail as it moves about in the underbrush. WILS., iii, 66, pl. 22, f. 5; NUTT., i, 353; AUD., Orn. Biog. v, 284, pl. 433; BD., 261. NOVEBORACENSIS.

Large-billed Water Thrush. (PLATE II, figs. 8, 8*a*.) Very similar to the last; rather larger, averaging about 6, with the wing 3; bill especially longer and stouter, over ½, and tarsus nearly 1. Under parts white, only faintly tinged, and chiefly on the flanks and crissum, with buffy (not sulphury yellow); the streaks sparse, pale, and not very sharp: throat, as well as belly and crissum, unmarked; legs pale. It may prove only a variety, but I have yet to see a specimen I cannot distinguish on sight; the size of the bill is not by any means the only character, as some seem to suppose, though it is the principal one. Eastern United States, rather southern, and not very common; north to Massachusetts (*Allen*). AUD. Orn. Biog. i, 99, pl. 19; BD., 262. . . . . LUDOVICIANUS.

37. Genus OPORORNIS Baird.

Connecticut Warbler. Olive-green, becoming ashy on the head; below, from the breast, yellow, olive-shaded on the sides; chin, throat and breast brownish-ash; a whitish ring round eye; wings and tail unmarked, glossed with olive; under mandible and feet pale; no decided markings anywhere; 5½; wing 2¾; tail 2. In spring birds the ash of the head, throat and breast is quite pure, and then the resemblance to *Geothlypis philadelphia* is close; but in the latter the wings are little if any longer than the tail. In the fall the upper parts from bill to tail are nearly uniform olive. Eastern United States, not common, and very rarely observed in the spring; a quiet, shy inhabitant of brushwood and thickets. Of late very abundant in the fall about Cambridge, Mass., where in two seasons over a hundred specimens have been taken (*Allen*). WILS., v, 64, pl. 24, f. 4; NUTT., 2d ed. i, 403; AUD., ii, 71, pl. 99; BD., 246. . . . . AGILIS.

Kentucky Warbler. Clear olive-green; entire under parts bright yellow, olive-shaded along sides; crown black, separated by a rich yellow superciliary line (which curls around the eye behind) from a broad black bar running from bill below eye and thence down the side of the neck; wings and tail unmarked, glossed with olive; feet flesh color; 5⅜; wing 2⅜–3; tail 2–2¼. Young birds have the black obscure if not wanting; in the fall, the black feathers of the crown of the adult are skirted with ash. Eastern United States, north to the Connecticut Valley; not abundant, but common in certain sections, as in southern Illinois (*Ridgway*) and Kansas (*Coues*). WILS., iii, 85, pl. 25, f. 2; NUTT., i, 399; AUD., ii, 19, pl. 74; BD., 247.　.　.　.　.　FORMOSUS.

38. Genus GEOTHLYPIS Cabanis.

Maryland Yellow-throat. ♂ in spring: olive-green, rather grayer anteriorly, forehead and a

FIG. 46.　Kentucky Warbler.

broad band through the eye to the neck pure black, bordered above with hoary ash; chin, throat, breast, under tail coverts and edge of wing rich yellow, fading into whitish on the belly; wings and tail unmarked, glossed with olive; bill black, feet flesh colored. ♀ in spring, without the definite black and ash on the head, the crown generally brownish, the yellow pale and restricted. The young, in general, resembles the ♀, at any rate lacking the head markings of the ♂; but it is sometimes buffy brownish below, sometimes almost entirely clear yellow. In any plumage, the bird is distinguished from warblers of any other genus, by having the wings shorter, or at most not longer, than the tail; and from the two following

FIG. 47.　Maryland Yellow-throat.

species by having no clear ash on the throat. Length 4⅜–5; wing and tail 1⅞–2¼. United States, from Atlantic to Pacific; Mexico, West Indies and Central America. An abundant and familiar inhabitant of shrubbery and underbrush, the sameness of which is enlivened by its sprightly presence and hearty song, throughout the summer months. WILS., i, 88, pl. 6, f. 1; NUTT., i, 401; AUD., ii, 78, pl. 102; BD., 241.　.　.　.　.　.　.　.　.　.　TRICHAS.

Mourning Warbler. Bright olive, below clear yellow; on the head the olive passes insensibly into ash; in high plumage the throat and breast are black, but are generally ash, showing black traces, the feathers being black skirted with ash, producing a peculiar appearance suggestive of the bird's wearing crape; wings and tail unmarked, glossed with olive; under mandible and feet flesh color; *no white about eyes.* Young birds have little or no ashy on the head, and no black on the throat, thus closely resembling *Oporornis agilis,* but are of course distinguishable by their generic characters; 5¼–5½; wing and tail, each, about 2¼. Eastern United States, rare; Minnesota, "abundant" (*Trippe*); a shy, retiring inhabitant of dense shrubbery. WILS., i, 101, pl. 14; NUTT., i, 404; AUD., ii, 76, pl. 101; BD., 243.　.　.　.　.　.　.　.　.　.　.　.　.　.　PHILADELPHIA.

Macgillivray's Warbler. Precisely like the last species, excepting that it has *white eyelids*. Rocky Mountains to the Pacific, U. S., and southward. One of the most abundant warblers in the mountains of Colorado, and common elsewhere in the West. *Trichas tolmiei* NUTT., 2d ed. i, 460; AUD., ii, 74, pl. 100; BD., 248; COOP., 96.   .   .   .   .   MACGILLIVRAYI.

Subfamily *ICTERIINÆ*. *Chats*.

A small group, recently framed to accommodate the following genus and its two tropical allies; it is perhaps questionable whether they are most naturally classed with the Warblers. *Icteria* shows the following points:— Larger than any other *Sylvicolidæ;* bill short, stout, compressed, culmen and commissure both curved, tip unnotched, rictus unbristled; wings much rounded, shorter or at most not longer than the tail. Sexes alike. Probably contains but one species.

39. Genus ICTERIA Vieillot.

Yellow-breasted Chat.   Bright olive green, below golden yellow, belly abruptly white; lore black, isolating the white under eyelid from a white superciliary line above and a short white maxillary line below; wings and

tail unmarked, glossed with olive; bill and feet blue-black; 7-7½; wing about 3; tail about 3¼.   Eastern United States, north to Massachusetts, abundant; an exclusive inhabitant of low tangled undergrowth, and oftener heard than seen, except during the mating season, when it performs the extravagant aerial evolutions for which, as well as for the variety and

FIG. 48. Yellow-breasted Chat.

volubility of its song, it is noted. Nest in a crotch of a bush near the ground: eggs 4-5, white, speckled with reddish brown. *Pipra polyglotta* WILS., i, 90 pl. 6, f. 2; *Icteria viridis* NUTT., i, 299: AUD. Orn. Biog. ii, 223, v, 433, pl. 137; BD., 248; *Icteria virens* BD., Rev. 228.   VIRENS.

Var. LONGICAUDA. *Long-tailed Chat.* Very similar; the olive duller and grayer, sometimes quite ashy on the head: tail usually but not always longer, averaging perhaps 3¼. Replaces *virens* from the Plains to the Pacific, U. S., and southward. BD., 249; COOPER, 98.

Subfamily *SETOPHAGINÆ*. *Flycatching Warblers.*

These have the bill depressed, considerably broader than high at base, notched and usually hooked at tip, and furnished with long stiff bristles that reach halfway or more from the nostrils to the end of the bill. In other respects they are not distinguished from the rest of the family. While many or most other *Sylvicolidæ* are expert in taking insects on the wing, these capture their prey in the air with special address, representing, in this respect, the true clamatorial flycatchers, with which some species of *Setophaginæ* used to be classed, in the extensive old genus "Muscicapa." As I have said, the *Sylvicolinæ* are peculiarly North American; while the *Setophaginæ* are most developed in Central and South America, where they are represented by three or four genera, and upwards of forty species. It is hardly necessary to add that, however closely some of them may resemble the

Tyrannidæ, they are at once distinguished from these clamatorial birds by the oscine character of the tarsi, and the presence of only nine primaries.

40. Genus MYIODIOCTES Audubon.

Hooded Flycatcher. Clear yellow-olive, below rich yellow shaded along the sides, whole head and neck pure black, enclosing a broad golden mask across forehead and through eyes; wings un-marked, glossed with olive; tail with large white blotches on the two outer pairs of feathers, as in *Dendrœca;* bill black, feet flesh color. ♀ with no black on the head; that of the crown replaced by olive, that of the throat by yellow; *young* ♂ with the black much restricted or interrupted, if not wholly wanting as in the ♀ (*Muscicapa selbyi* AUD. Orn. Biog. i, 46, pl. 9). Length 5-5¼;

FIG. 49. Hooded Flycatcher.

wing about 2¾, tail about 2¼. Eastern United States, apparently not very common. *Muscicapa cucullata* WILS., iii. 101, pl. 26, f. 3; NUTT., i, 373; AUD., ii, 12, pl. 71; BD., 292. . . . . MITRATUS.

Green Black-capped Flycatcher. Clear yellow-olive; crown glossy blue-black; forehead, sides of head and entire under parts bright yellow; wings and tail plain, glossed with olive; upper mandible dark, under pale; feet brown; ♀ and young similar, colors not so bright, the black cap. obscure. Small; 4¾-5; wing about 2¼; tail about 2. North America, at large; common. WILS., iii, 103, pl. 26, f. 4; NUTT., i, 408; AUD., ii, 21, pl. 75; BD., 293. . . . . PUSILLUS.

FIG. 50. Green Black-capped Flycatcher.

Canadian Flycatcher. Bluish-ash; crown speckled with lanceolate black marks, crowded and generally continuous on the forehead; the latter divided lengthwise by a slight yellow line; short superciliary line and edges of eyelids, yellow; lores black, continuous with black under the eye, and this passing as a chain of black streaks down the side of the neck and prettily encircling the throat like a necklace; excepting these streaks and the white under tail coverts, the entire under parts are clear yellow; wings and tail un-marked; feet flesh color. In the ♀ and young the black is obscure or much restricted, and the back may be *slightly* glossed with olive: but they cannot be mistaken. In this plumage the bird is *Myiodioctes*

FIG. 51. Canadian Flycatcher.

bonapartii AUD., ii, 17, pl. 73; NUTT., i, 2d ed. 330; BD., 295. Length about 5½; wing 2½; tail 2¼. Eastern United States, an abundant and beautiful woodland species. WILS., ii, 100, pl. 26, f. 2; NUTT., i, 372; AUD., ii, 14; pl. 72, BD., 294. . . . . CANADENSIS.

OBS. The *Small-headed Flycatcher,* MUSCICAPA MINUTA WILS., vi, 62, pl. 50, f. 2; NUTT., i, 2d ed. 334; AUD., i, 238, pl. 67; BD., 293, now unknown, is conjectured to belong to this genus; but this can hardly be, for Wilson says it has two white wing bands, a character not shown in *Myiodioctes*. There is no reasonable probability that any species of the family, inhabiting the Middle States in June, remains to be detected. I have no doubt that the bird is a *Dendrœca*, and nothing in the description forbids its reference to one of the endless plumages of *D. pinus!*

41. Genus SETOPHAGA Swainson.

Redstart. ♂ lustrous blue-black, belly and crissum white, sides of the breast, large spot at bases of the remiges, and basal half of the tail feathers (except the middle pair) *fiery-orange;* belly often tinged with the same; bill and feet black. ♀ olivaceous, ashier on the head, entirely white below, wings and tail blackish, with the flame color of the ♂ represented by yellow; young ♂ like the ♀ but browner, the yellow of an orange hue. From the circumstance that many spring males are shot in the general plumage of the female, but showing irregular isolated black patches, it is probable that the species requires at least two years to gain its perfect plumage. Length 5⅓; wing and tail about 2¼. Eastern North America, very abundant, in woodland. WILS., i, 103, pl. 6, f. 6; NUTT., i, 291; AUD., i, 240, pl. 68; BD., 297. . . . . RUTICILLA.

Painted Flycatcher. ♂ lustrous black, middle of breast and belly carmine red; eyelids, wing coverts and crissum white, inner quills edged with white, outer tail feathers mostly white. ♀ not particularly different. 5; wing and tail, each 2⅔; tarsus ⅔. A Mexican species, recently found in Arizona. (Tucson, *Bendire.*) BD., 298; Rev., 256. . . . . PICTA.

Obs. One other Mexican species of this genus (*S. miniata*), and two species of closely allied genera, *Basileuterus rufifrons* and *Cardellina rubra* (both Mexican), have been admitted to our fauna, though they have not, to my knowledge, been actually taken within our limits.

Family CŒREBIDÆ. Honey Creepers.

Primaries nine, and other external characters very nearly as in the last family; but the bill is generally slenderer and sharper, and often a little decurved. The line between the two families has never been drawn with precision, and has become the more difficult of expression since some of the *Sylvicolidæ* have proven possessed of a peculiarity of the *Cœrebidæ*—deeply bifid, penicillate tongue. A small group, containing perhaps forty species, of pretty little birds, confined to tropical America. Our species is merely a stray visitor to Florida.

42. Genus CERTHIOLA Sundevall.

Honey Creeper. Dark olivaceous ash; superciliary line and under parts dull white; belly, edge of wing, and rump, bright yellow; wings dusky, with a white spot at base of primaries; tail dusky, tipped with white; bill and feet black; "eyes blue." Length 4½; wing 2⅓; tail 1¾. Indian Key (*Wurdemann*). BD., 924. . . . . FLAVEOLA.

Family TANAGRIDÆ. Tanagers.

An extensive, brilliant family, confined to America, abounding in species between the tropics. Its position is a point at issue with ornithologists; it may however, not unnaturally follow the *Cœrebidæ* and *Sylvicolidæ*, though certainly no families should stand between it and *Fringillidæ*. In fact certain tropical forms might be assigned to either indifferently. The best definition of the tanagers I have seen is that given by the distinguished ornithologist who called them "dentirostral finches;" but this important generalization, like other happy epigrams, is insusceptible of application in detail, and the tanagers remain to be precisely characterized. As a consequence, the number of species can hardly be approximately estimated; but upwards of three hundred are usually enumerated.

The single well established North American genus may be recognized, among all the birds of our country, by the combination of nine primaries and scutellate tarsi with a turgid bill, notched at the tip and toothed or lobed near the middle of the superior maxillary tomia; though this last character is sometimes so obscure that it might be looked at without being seen. The species of *Pyranga* are birds of brilliant colors, with great seasonal and sexual differences of plumage. They are frugivorous and insectivorous, and consequently migratory in the United States. They inhabit woodland, lay 4–5 dark colored, speckled eggs, nest in trees, and are fair songsters. In distribution they are rather southerly, not passing northward beyond the United States. One species of another genus, *Euphonia elegantissima*, has been admitted to our fauna, but apparently upon insufficient evidence.

43. Genus PYRANGA Vieillot.

Scarlet Tanager. ♂ scarlet, with black wings and tail; bill and feet dark; ♀ *clear olive green*, below *clear greenish yellow*, wings and tail dusky, edged with olive; *no white wing-bars.* Young ♂, at first, like the ♀; afterward variegated with red, green and black. Length 7–7½; wing· 4; tail 3. Eastern United States, abundant. Wils., ii, 42, pl. 11, f. 3, 4; Nutt., i, 465; Aud., iii, 226, pl. 209; Bd., 300. . . . . . rubra.

Summer Red-bird. ♂ rich rose-red, or vermilion, including wings and tail; the wings, however, dusky on the inner webs; bill rather pale; feet darker; ♀ *dull brownish-olive*, below *dull brownish-yellow;* no white wing-bars; young ♂ like the ♀; the ♂ changing plumage shows red and green confused in irregular patches, but no black. The ♀, with a general resemblance to ♀ *rubra*, is distinguished by the dull brownish, ochre or buffy tinge, the greenish and yellowish of *rubra* being much purer; the bill and feet, also, are

FIG. 52. Summer Red-bird; *b*, Cooper's Tanager.

generally much paler in *æstiva*. Size of *rubra*, or rather larger. Eastern, Southern and South-western United States, hardly north to New England; abundant. Wils., i, 95, pl. 6, f. 3; Nutt., i, 469; Aud., iii, 222, pl. 208; Bd., 301. . . . . æstiva.

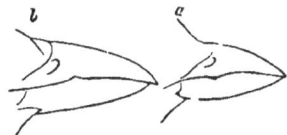

Obs. The *Pyranga cooperi*, lately based by Mr. Ridgway (Proc. Acad. Phila. 1869, 130) upon New Mexican specimens, seems scarcely tenable. The characters are not very tangible, and there is little probability of their proving constant. Though the difference in the shape of the bill of the type specimens is evident (fig.

Fig. 53. Hepatic Tanager.

52, *a* and *b*), yet this is no more than that occurring in Eastern specimens of unquestionable *æstiva*. (See PLATE II, figs. 19, 20, *a*, *b*.) It may, however, take rank as a geographical variety.

Hepatic Tanager. Ashy-red, or liver-brown, brighter red on the head and under parts; sides ashy-shaded; bill plumbeous black, conspicuously toothed; ♀ like that of the foregoing, but ashier on the back. Size of the last. New Mexico, Arizona, and southward. Bd., 302; RIDGWAY, Proc. Acad. Phila. 1869, 132; COOP., 144. . . . . HEPATICA.

Louisiana Tanager. ♂ bright yellow, middle of back, wings, and tail, black; head crimson; wings with two yellow bars. ♀ most nearly resembling that of *rubra*, but distinguished from this or any of the foregoing by presence of two whitish or greenish-yellow wing-bars, and much edging of the same color on the inner quills. Immature ♂ shows the black of the back mixed with olive, and the head only tinged with red; at first it is like the ♀. Size of the first species. U. S., Rocky Mountains to the Pacific (not in Louisiana!). WILS., iii, 27, pl. 20, f. 1; NUTT., i, 471; AUD., iii, 231, pl. 210; Bd., 303; COOP., 145. . . . . LUDOVICIANA.

Family HIRUNDINIDÆ. Swallows.

Fissirostral Oscines. Bill short, broad, flat, deeply cleft, the gape wide and about twice as long as the culmen — it generally reaches to about opposite the eyes. Nasal fossæ short, broad, the nostrils directed more or less upward, sometimes circular and completely open, sometimes overhung by a straight flat scale. Rictus with a few inconspicuous bristles or none. Wings extremely long, of nine primaries, of which the first equals or exceeds the second, the rest being rapidly graduated, the ninth hardly or not half as long as the first; secondaries and their coverts extremely short. Tail of 12 (rarely 10?) rectrices, usually forked, sometimes forficate with filamentous outer feathers. Feet short and weak; tarsi scutellate (occasionally feathered), commonly shorter than even the lateral toes; basal joint of middle toe adherent to one or both lateral toes; toes with the normal number of phalanges.

This is a perfectly natural group, well distinguished by the foregoing characters. The swallows alone represent, among Oscines, the fissirostral type of structure; they have a close superficial resemblance to the swifts and goat-suckers of another order, but the relation is one of analogy, not of affinity, though all these birds were formerly classed together in the highly unnatural " order" *Fissirostres*. (See beyond, under *Cypselidæ* and *Caprimulgidæ*.)

A hundred species of swallows are recorded; probably about three-fourths of them are genuine. They are distributed all over the world; the most generalized types, like *Hirundo* itself, are more or less cosmopolitan, but each of the great divisions of the globe has its peculiar subgenera or particular sets of species. Thus, all the American groups except *Hirundo* and *Cotyle* are peculiar to this continent.

Swallows are insectivorous, and therefore migratory in cold and temperate latitudes; unsurpassed in powers of flight, they are enabled to pass with ease and
swiftness from one country to another, as the state of the weather may require.
With us a few warm days in February and March often allure them northward, only
to be driven back again by the cold, giving rise to the well-known adage. No birds
are better known to all classes than these, and none so welcome to man's abode—
cherished witnesses of peace and plenty in the homestead, dashing ornaments of
the busy thoroughfare.

The habits of swallows best illustrate the modifying influences of civilization on
indigenous birds. Formerly, they all bred on cliffs, in banks, in hollows of trees,
and similar places, and many do so still. But most of our species have forsaken
these primitive haunts to avail themselves of the convenient artificial nesting places
that man, intentionally or otherwise, provides. Some are just now in a transition
state; thus the purple martin, in settled parts of the country, chooses the boxes
everywhere provided for its accommodation, while in the West it retains its old
custom of breeding in hollow trees.

44. Genus HIRUNDO Linnæus.

Barn Swallow. Lustrous steel blue; below, rufous or pale chestnut
of varying shade; forehead, chin and throat deep chestnut; breast with an
imperfect steel-blue collar; tail forficate,
its outer feathers attenuate, all but the
middle pair with white spots on the inner
web; bill and feet black. Sexes alike;
young less lustrous, much paler below,
tail simply forked. Wing $4\frac{1}{2}$-$4\frac{3}{4}$; tail $2\frac{1}{4}$
to 5 inches. North America, abundant
in the United States in the summer,
breeding in colonies in barns and outhouses; eggs white, speckled. WILS., v,
34, pl. 38; NUTT., i, 601; AUD., i, 181,
pl. 48; BD., 308. . . . . HORREORUM.

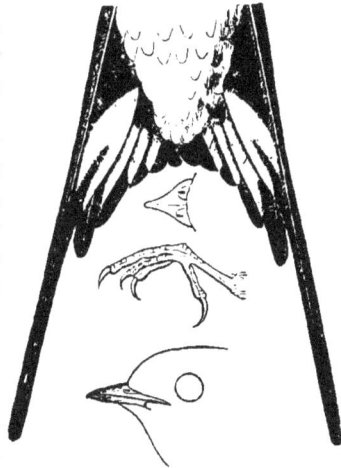

45. Genus TACHYCINETA Cabanis.

White-bellied Swallow. Lustrous
green, below pure white; tail simply
emarginate. Young similar, not so
glossy. 6-$6\frac{1}{2}$; wing 5; tail $2\frac{1}{4}$. North

Fig. 51. Barn Swallow.

America, abundant in the United States in summer. WILS., v, 49, pl. 38;
NUTT., i, 605; AUD., 1, 175, pl. 46; BD., 310. . . . . . BICOLOR.

Violet-green Swallow. Opaque velvety green, purple and violet; spot over
the eye, sides of rump, and whole under parts pure white. Young similar,
duller. $4\frac{3}{4}$-$5\frac{1}{4}$; wing $4\frac{1}{2}$; tail 2, emarginate. Rocky Mountains to the
Pacific, U. S.; an exquisite species, breeding in knotholes and woodpeckers'
holes, in pine woods and in weather-worn holes in cliffs. (ALLEN, Am.
Nat. 1872, 274.) AUD., i, 186, pl. 49; BD., 309; COOP., 107. THALASSINA.

46. Genus PETROCHELIDON Cabanis.

Cliff Swallow. Eave Swallow. Lustrous steel-blue; forehead whitish (or brown), rump rufous, chin, throat and sides of head chestnut; a steel-blue spot on the throat; breast, sides and generally a cervical collar rusty-gray, whitening on the belly. *Young* sufficiently similar. 5; wing 4½; tail 2¼, nearly square. North America, in all suitable places. Naturally this species builds on cliffs; but throughout the settled portions of the country it now places its curious bottle-shaped nests of mud under the eaves of barns and outhouses. NUTT. i,603; AUD., i, 177, pl.47; BD., 309. LUNIFRONS.

47. Genus COTYLE Boie.

Bank Swallow. Sand Martin. Lustreless gray, with a pectoral band of the same; other under parts white. A curious little tuft of feathers at the bottom of the tarsus. Sexes exactly alike; young similar, the feathers often skirted with rusty or whitish. 4½–4⅞; wing 3⅞–4; tail 2, simply emarginate. North America, very abundant; breeds in immense troops in holes excavated in banks of soft earth. WILS., v, 46, pl. 38; NUTT., i, 607; AUD., i, 187, pl. 50; BD., 313. . . . . RIPARIA.

48. Genus STELGIDOPTERYX Baird.

Rough-winged Swallow. Lustreless brownish-gray, paler below, whitening on the belly. Rather larger than the last; no feathery tuft on tarsus; outer web of outer primary, in the ♂, converted into a series of recurved hooklets, which are wanting, or much weaker, in the ♀. United States; rare or wanting in New England. AUD., i, 193, pl. 51; BD., 373; COUES, Proc. Phila. Acad., 1866, 37. .   . . . SERRIPENNIS.

49. Genus PROGNE Boie.

Purple Martin. Lustrous blue-black; no purple anywhere. The ♀ and young are much duller above, and more or less white below, streaked with gray. Bill very stout for this family, curved at the end; nostrils circular, opening upward, not roofed over. Length 7 or more; wing nearly 6; tail 3½, simply forked. United States, very abundant. WILS., v, 58, pl. 39, f. 2, 3; NUTT., i, 598; AUD., i, 170, pl. 45; BD., 314. . . PURPUREA.

Obs. Other species or varieties of *Progne*, requiring confirmation, are attributed to North America. See CASS., Ill., 246 (California); BD., 923, and Rev., 277 (Florida).

Family AMPELIDÆ.

This appears to be an arbitrary and unnatural association of a few genera that agree in some particulars, but are widely different in others. The composition and position of the group differ with almost every writer; some place it in *Clamatores*, next to the *Tyrannidæ*. I think that the family should be dismembered; Baird has already shown how near the *Myiadestinæ* are to the true Thrushes, and doubtless the other two subfamilies here presented may be properly dissociated.

Birds of the three following genera agree in this character :— Bill short, broad, flattened, plainly notched at tip, with wide rictus, and culmen or gonys hardly if at all exceeding half the length of the commissure; basal phalanx of middle toe joined with outer toe for about two-thirds its length, and to inner toe for about half its length. The three genera, or subfamilies, that follow, may be readily and precisely defined.

Subfamily AMPELINÆ.   Waxwings.

Bill as just described ; nasal fossæ broad, nasal opening exposed, but overarched by a broad scale more or less completely covered with close-set velvety antrorse feathers. Wings with ten primaries, but the first spurious, very short and displaced (on the outer side of the second) so as to be readily overlooked ; point of the wing formed by the third primary, closely supported by the second and fourth, the fifth being abruptly shorter. Inner quills, as a rule, and sometimes the tail feathers, tipped with horny appendages like red sealing-wax. Tail short, square, $\frac{1}{2}$ or $\frac{2}{3}$ as long as the wings, the under coverts highly developed, reaching nearly to its end. Feet weak ; the tarsus shorter than the middle toe and claw, its podotheca somewhat receding from strict oscine character.

Of this subfamily as here restricted there is only one genus with three species— one of Europe and America, one of Asia and Japan, and one confined to this country. They are songless, in this differing altogether from the <i>Myiadestinæ</i> and <i>Ptilogonytinæ;</i> and I should not be surprised if their relationships proved to be entirely with a certain exotic clamatorial family. Although by a strange misnomer sometimes called "chatterers" they are among the most silent of all birds, their only voice being a weak wheezy kind of whistle. They feed chiefly on berries and other soft fruits, but also on insects, and are gregarious and migratory. The sexes are alike ; the head is adorned with a beautiful crest ; the wings have unique ornaments, the use of which is unknown ; the tail is tipped with yellow (red in the Japanese species, <i>phœnicopterum</i>) ; the plumage is extremely smooth, and of a nameless color. Young birds lack the curious horny appendages, and have the general plumage streaked.

50. Genus AMPELIS Linnæus.

Bohemian Waxwing. Under tail coverts chestnut ; front and sides of the head tinged with a richer, more orange-brown shade ; primary wing coverts tipped with white ; each quill with a sharp white (or yellowish) stripe at the end of the outer web ; chin velvety black, in a large well defined area ; narrow line across forehead, along sides of head through eyes, meeting its fellow on the occiput behind the crest, also velvety black ; no white on under eyelid nor across forehead ; no yellowish on belly ; bill and feet black. 7 or 8 inches long, wing about 4½. Northern North America ; U. S. casually in winter, but sometimes appearing in immense roving flocks ; S. sometimes to 35°. Aud., iv, 269, pl. 246 ; Nutt., i, 246 ; Bd., 317. . . . . Garrulus.

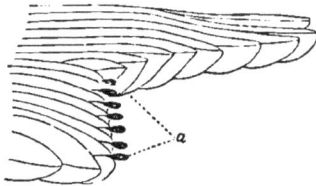

FIG. 55. Bohemian Waxwing.
a. appendages of the inner quills.

Carolina Waxwing. Cedar Bird. Cherry Bird.   Under tail coverts whitish; little or no orange-brown about head; no white on wings; chin black, shading gradually into the color of the throat; a black frontal, loral and transocular stripe, as in *garrulus*, but this bordered on the forehead with whitish; a white touch on lower eyelid, feathers on side of under jaw white; abdomen soiled yellowish. 6 or 7 long; wing about 3¾. Eastern North America to Hudson's Bay; an abundant bird, irregularly migratory, going in flocks nearly the whole year; breeds late (in June) in orchards and thickets; the nest is placed in the crotch of a tree; the eggs are 3–4, dull pale bluish, speckled with purplish and blackish. Wils., i, 107, pl. 7; Nutt., i, 248; Aud., iv, 165, pl. 245; Bd., 318. . . . . . . Cedrorum.

Subfamily PTILOGONYDINÆ. Ptilogonys.

Bill much as in the last subfamily, but slenderer for its length; nasal scale naked; a few short bristles about the base of the bill. Tarsus scutellate anteriorly, and sometimes also on the sides; about as long as the middle toe and claw; hind toe remarkably short. Wings not longer than the tail, much rounded, of ten primaries; the 1st spurious, less than half as long as the 2nd, which is only about as long as the 8th; point of the wing formed by the 4th, 5th and 6th or 3rd quills. Tail long, nearly even, with broad plane feathers (*Phænopepla*); or much graduated, with tapering central feathers (*Ptilogonys*). Head conspicuously crested; sexes (in our genus) dissimilar; young not streaked or spotted.   There are only two genera of the subfamily as thus restricted — *Phænopepla* and *Ptilogonys*, the latter with two strongly marked species of Mexico and Central America.

51. Genus PHÆNOPEPLA Sclater.

Black Ptilogonys.   ♂ uniform lustrous black; wings with a large white area, most of the inner web of each primary, except the first, being white; ♀ brown, the white on the wings restricted or obsolete; young ♂ gradating between the coloration of both sexes. 7½; wing and tail 3½–4. Valley of the Colorado and southward; *a delightful songster*, though the fact seems to have been ignored. Cass., Ill., 169, pl. 29; Bd., 320, and Rev., 416; Coop., 131. . . . . Nitens.

Subfamily MYIADESTINÆ. Flycatching Thrushes.

Bill as in the last subfamily. Tarsus *booted*, and toes deeply cleft, as in *Turdidæ*. Lateral toes very unequal in length, the tip of the inner claw falling short of the base of the middle. Wings of ten primaries, the 1st spurious, the 2nd about as long as the 6th, the point of the wing formed by the 3rd, 4th and 5th. Tail long, about equalling the wing, *double-rounded*, being forked centrally, graduated externally; all the feathers narrowing somewhat towards the end. Head subcrested;

plumage sombre, variegated on the wings; sexes alike; young spotted, like thrushes.

The birds of the group thus defined are, as Baird has pointed out, more closely related to the *Turdidæ* than to the family with which they are usually associated. They consist of about a dozen species, mostly of the genus *Myiadestes*, though there are others called *Cichlopsis* and *Platycichla*. With one exception, they are birds of Central and South America, and the West Indies. Our species, formerly called "*Ptilogonys*," simply for want of an English name, which I here supply, is not to be confounded with the foregoing. It is an exquisite songster.

52. Genus MYIADESTES Swainson.

Townsend's Flycatching Thrush. Nearly uniform ashy-gray, sometimes paler or mixed with whitish on throat, belly, crissum and under wing coverts; a whitish ring round the eye; quills variegated with pale cinnamon or buffy, showing as two oblique bands in the closed wing: tail blackish, central feathers like the back, the outermost pair edged and tipped, the two next pair tipped, with white. The *young* are speckled with round fulvous spots. Length about 8; wing and tail about 4½. Rocky Mountains to the Pacific, United States.

FIG. 57. Townsend's Flycatching Thrush. Bill and feet of natural size; wings and tail ⅜.

NUTT., i, 2d ed., 361; AUD., i, 243, pl. 69; BD., 321, and Rev., 429; COOP., 134. . . . . TOWNSENDII.

Family VIREONIDÆ. Vireos, or Greenlets. ·

Bill shorter than the head, stout, compressed, distinctly notched and hooked at tip; rictus with conspicuous bristles; nostrils exposed, overhung with a scale, but reached by the small bristly erect frontal feathers. Toes soldered at base for the whole length of the basal joint of the middle one, which is united with the basal joint of the inner and the two basal joints of the outer, all these coherent phalanges very short. (Lateral toes unequal in the genus *Vireo*.) Tarsus equal to or longer than the middle toe and claw, scutellate in front, laterally undivided, except at extreme base. Wings moderate, of ten primaries, of which the first is short (one-half to one-fourth the second), or spurious, or *apparently* wanting (being rudimentary and displaced).

This family was formerly united with the next (*Laniidæ*), chiefly on account of the resemblance in the shape of the bill of certain species to that of the shrikes; but the likeness is never perfect, and there are other more important characters, especially in the structure of the feet, by which the two groups may be discriminated. The *Vireonidæ* are peculiar to America; they are a small family of five or six genera and nearly seventy recorded species, of which about five-sixths appear to be genuine. The typical and principal genus, *Vireo*, containing nearly thirty species, is especially characteristic of North America, though several species occur in the West Indies and Central America; one genus and species, *Laletes osburni*, is exclusively West Indian; the rest — *Cyclarhis, Hylophilus, Vireolanius,* and *Neochloe* — are, with one exception, South and Central American. In further illustration of the characters of the group, I offer some remarks under the head of the only genus with which we have to do in the present connection.

53. Genus VIREO Vieillot.

The numerous species of this genus have been divided into several groups, but no violence will be done by considering them all as *Vireo* — in fact, it is difficult to do otherwise. For even the seemingly substantial division into two genera, according as there is an evident spurious first primary or apparently none, separates species, like *gilvus* and *philadelphicus*, hardly otherwise specifically distinguishable; while another division into two genera, according to shape of the wings and length of the spurious first primary or its absence, is subject to some uncertainty of determination, and unites species, like *olivaceus* and *flavifrons*, most dissimilar in other respects. The fact is, that almost every single species of *Vireo* has its own peculiar form, in shape of bill, proportions of primaries, etc., and these details cannot well be considered as of more than specific value. These slight differences are perfectly

FIG. 58. Warbling Vireo.

tangible and surprisingly constant, and render the determination of the species comparatively easy, though these birds bear to each other a close general resemblance in size and color. They are all more or less *olivaceous* above, sometimes inclining to gray or plumbeous, with the crown either like the back, or else ashy — in one species, however, brown, and in another black; and white or whitish below, usually more or less tinged with yellow. The coloration is very constant, the sexes being indistinguishable, and the young differing little, if at all, from the adults. All are small birds — about 5 or 6 inches long. As a group the student will probably have no difficulty in recognizing them by the foregoing diagnosis, as the character of the feet seems to be peculiar, among North American birds, and is at any rate diagnostic when taken in connection with the character of the bill — all those Oscines, as wrens, creepers, or titmice, that show much cohesion of the toes, having an entirely different bill. The bill of *Vireo* may be described as resembling that of a shrike in miniature — it is hooked and notched distinctly at the end, and there is sometimes a trace of a tooth behind the notch, and of a nick in the under mandible too. Some of the weaker-billed species might be carelessly mistaken for warblers — but there is no excuse for this, nor for confounding them with any of the little clamatorial flycatchers.

The Vireos were long supposed to possess either nine or ten primaries. But that the important character of number of primaries — one marking whole families as we have seen — should here subside to specific value only, seemed suspicious; and the fact is, as announced by Baird (Review, pp. 160, 325) that all the species really have ten, only that, in some instances, the first primary is rudimentary and displaced, lying concealed outside the base of the second quill.

The North American species are distributed over the temperate portions of this continent, and several of them are abundant birds of the Atlantic States, inhabiting woodland and shrubbery. They are exclusively insectivorous, and are therefore necessarily migratory in our latitudes. They build a neat pensile nest in the fork of a branchlet, and commonly lay four or five white speckled eggs. Next after the warblers, the greenlets are the most delightful of our forest birds, though their charms address the ear and not the eye. Clad in simple tints that harmonize with the verdure, these gentle songsters warble their lays unseen, while the foliage itself seems stirred to music. In the quaint and curious ditty of the white-eye — in the earnest, voluble strains of the red-eye — in the tender secret that the warbling vireo confides in whispers to the passing breeze — he is insensible, who does not hear the echo of thoughts he never clothes in words.

ANALYSIS OF SPECIES.

Primaries apparently 9 (the 1st rudimentary and displaced). (a)
Primaries evidently 10 (the 1st short or spurious). (b)
(a) Throat yellow, ... *flavifrons.*
   — white; crown ashy, not black-edged, hardly contrasting with back, ... *philadelphicus.*
      — black-edged, back olive; no maxillary streaks, ... *olivaceus.*
          — maxillary streaks, ... *barbatulus.*
(b) Crown black, ... *atricapillus.*
   — not black; spurious quill at least ½ as long as 2nd and wing 2½ long, ... *vicinior.*
      — not ½ as long as 2nd, or wing not 2¼ long (c)
(c) Wing-bands wanting: coloration as in *philadelphicus*, ... *gilvus.*
   — present; length over 5 in.; back olive, contrasting with ashy blue crown, ... *solitarius.*
      — plumbeous, crown scarcely different, ... *plumbeus.*
      — 5 in. or less; wing = tail, both about 2¼; 1st quill = ½ 2nd., ... *pusillus.*
         — > tail; crown ashy, chin and superc. line white, . . *bellii.*
          — olive, chin wht., superc. line yell., . *novebor.*
            — and under parts yell'sh, . *huttonii.*

Obs. The Bartramian Vireo of Aud., Orn. Biog. v, 296, pl. 434, f. 4; B. Am. iv, 153, pl. 242, and of Nutt., i, 2d ed. 358, has not been identified by later ornithologists; but there is little chance of its being a good species. The descriptions indicate a bird much like *V. olivaceus.* The original *Vireo bartramii* of Swainson, Fauna Bor.-Am. ii, 235, is a Brazilian species of the *olivaceus* group, wrongly ascribed to North America. The name *Vireo virescens* that Baird applied to the Bartramian Vireo, in B. N. A. p. 333, is doubtless an erroneous identification, as he has since shown, Vieillot's *virescens* being based on a Pennsylvania specimen, almost certainly *olivaceus.*— For the discussion of these questions, and a masterly review of the whole genus, see Baird, Review, pp. 322-370.

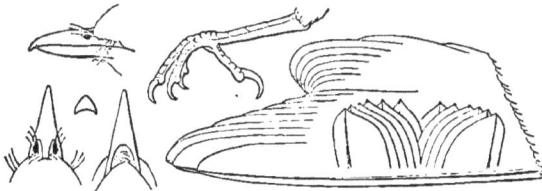

FIG. 59. Red-eyed Vireo. (This, and subsequent figs. of this family, of nat. size.)

Red-eyed Vireo. Above, olive-green; crown ash, edged on each side with a blackish line, below this a white superciliary line, below this again a dusky stripe through eye; under parts white, faintly shaded with olive along sides, and tinged with olive on under wing and tail-coverts; wings and tail dusky, edged with olive outside, with whitish inside: bill dusky, pale below; feet leaden-blue; eyes red; no dusky maxillary streaks; no spurious quill. Large; $5\frac{3}{4}$-$6\frac{1}{4}$; wing $3\frac{1}{4}$-$3\frac{1}{3}$; tail $2\frac{1}{3}$-$2\frac{1}{2}$; bill about $\frac{2}{3}$; tarsus $\frac{3}{4}$. Eastern North America; in most places the most abundant species of the genus, in woodland; a voluble, tireless songster. WILS., ii, 53, pl. 12, f. 3; NUTT., i, 312; AUD., iv, 155, pl. 243: BD., 331, and Rev. 333. OLIVACEUS.

Black-whiskered Vireo. Whip-tom-kelly. Very similar to the last; distinguished by a narrow dusky maxillary line, or line of spots, on each side of the chin; bill longer, $\frac{3}{4}$-$\frac{4}{5}$; proportions of quills slightly different (see the figs.). Cuba, Bahamas, and casually in Florida. *V. longirostris*, NUTT., i, 2d ed., 359. *V. altiloquus*, GAMBEL, Proc. Acad. Phila., 1848, 127; CASS., *ibid.*, 1851, 152, and Ill. pp. 8, 221, pl. 37; BD., 354. *V. barbatula*, BD., Rev. 331. . . . . ALTILOQUUS var. BARBATULUS.

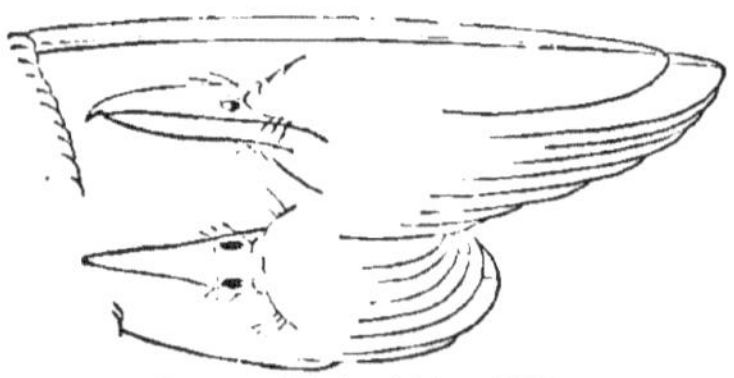

FIG. 60. Black-whiskered Vireo.

FIG. 61. Vireo flavoviridis.

OBS. Another species or variety of this long-billed, 9-primaried group, *V. flavoviridis* (CASS., Proc. Acad. Phila. 1851, 152: BD., 332 and Rev. 336), occurs in Mexico and may be expected over our border, though no specimens appear to have been taken within our limits; it has been admitted into late systematic works. It closely resembles *olivaceus*, but the under parts are yellow, brighter perhaps, at least on the axillars and crissum, than *olivaceus* ever becomes, even in the fall.

Brotherly-love Vireo. Above dull olive-green, brightening on the rump, fading insensibly into ashy on the crown, which is not bordered with blackish; a dull white superciliary line; below, palest possible yellowish, whitening on throat and belly, slightly olive-shaded on sides; sometimes a slight creamy or buffy shade throughout the under parts; no obvious wing-bars; no spurious quill. About 5 long; wing $2\frac{2}{3}$; tail $2\frac{1}{4}$; bill hardly or about $\frac{1}{2}$; tarsus $\frac{2}{3}$. Eastern North America; a small, plainly colored species, almost indistinguishable from *gilvus* except by absence of spurious quill; not very common. CASS., Proc. Acad. Phila. 1851, 153; BD., 335; Rev. 340. PHILADELPHICUS.

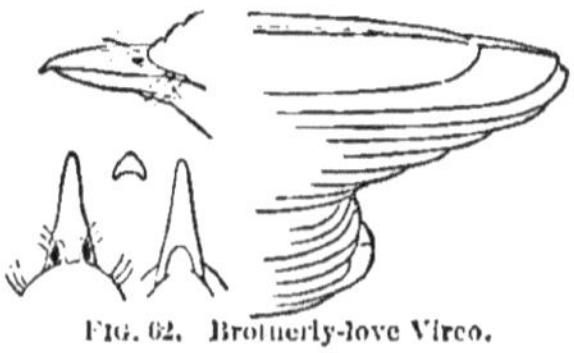

FIG. 62. Brotherly-love Vireo.

Warbling Vireo. Colors precisely as in the last species; spurious quill present, $\frac{1}{4}$-$\frac{1}{3}$ as long as the second primary. Eastern North America, an abundant little bird and an exquisite songster. Its voice is not strong, and many birds excel it in brilliancy of execution:

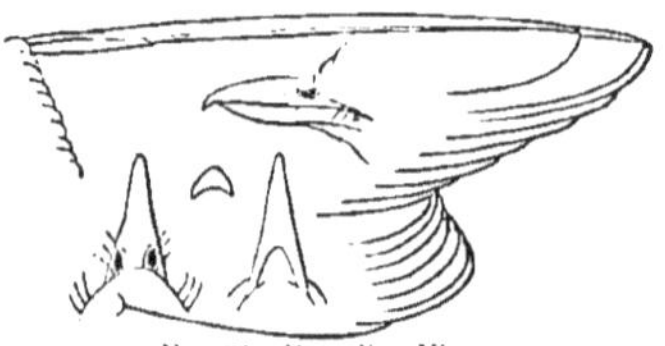

FIG. 63. Warbling Vireo.

but not one of them all can rival the tenderness and softness of the liquid strains of this modest vocalist. Not born to "waste its sweetness on the desert air," the warbling vireo forsakes the depths of the woodland for the park and orchard and shady street, where it glides through the foliage of the tallest trees, the unseen messenger of rest and peace to the busy, dusty haunts of men.— Wils., v, 85, pl. 42, f. 2; Nutt., i, 309; Aud., iv, 149, pl. 241; Bd., 335, and Rev. 342.   GILVUS.

Var. SWAINSONII. "Similar to *V. gilvus*, but smaller; colors paler; bill more depressed; upper mandible almost black; 2d quill much shorter than 6th." Baird, Rev. 343; Coop., 116; Elliot, pl. 7. Rocky Mountains to the

Fig. 64. Western Warbling Vireo.

Pacific, U. S. The Western form has been described as distinct, but I scarcely think the characters assigned will be found constant. In one of my Arizona skins the second quill is *longer* than it is in an Eastern specimen.

Yellow-throated Vireo. Above, rich olive-green, crown the same or even brighter, rump insensibly shading into bluish-ash; below, bright yellow, belly and crissum abruptly white, sides anteriorly shaded with olive, posteriorly with plumbeous; extreme forehead, superciliary line and ring round eye, yellow; lores dusky; wings dusky, with the inner secondaries broadly w h i t e-edged, and two broad white bars across tips of greater and median coverts; tail

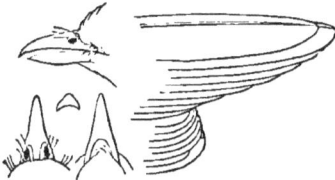

Fig. 65. Yellow-throated Vireo.

dusky, nearly all the feathers completely encircled with white edging; bill and feet dark leaden blue; no spurious quill; 5¾ - 6; wing about 3; tail only about 2¼. A large, stout, highly-colored species, common in the woods of the Eastern United States. Wils., i, 117, pl. 7, f. 3; Nutt., i, 302; Aud., iv, 141, pl. 238; Bd., 341, and Rev. 346. . . . . . . FLAVIFRONS.

Blue-headed, or *Solitary Vireo.* Above, olive-green, crown and sides of head bluish-ash in marked contrast, with a broad white line from nostrils to and around eye, and a dusky loral line; below, white, flanks washed with olivaceous, and axillars and crissum pale yellow; wings and tail dusky, most of the feathers edged with white or whitish, and two conspicuous bars of the same across tips of middle and greater coverts; bill and feet blackish horn-color. 5¼ - 5¾; wing 2¾ - 3; tail

Fig. 66. Blue-headed, or Solitary Vireo.

2¼-2⅜; spurious quill ½-⅔ long, about one-fourth as long as 2d. United States from Atlantic to Pacific, except perhaps Southern Rocky Mountains, where replaced by the next species; not rare, but not so common as *olivaceus, flavifrons* and *noveboracensis;* inhabits woodland. Wils., ii, 143, pl.

17, f. 6; NUTT., i, 305; AUD., iv, 144, pl. 239; BD., 340, and Rev. 347.
(*V. cassinii* XANTUS, Proc. Phila. Acad. 1858, 117; BD., 340, pl. 78, f. 1,
is not different.) . . . . SOLITARIUS.

Plumbeous Vireo. Leaden-gray, rather brighter and more ashy on the
crown, but without marked contrast, faintly glossed with olive on rump;
a conspicuous white line from nostril to and around eye, and below this a
dusky loral stripe; below, pure white, sides of neck and breast shaded with
color of the back, flanks, axillars
and crissum with a mere trace of
olivaceous, or none; wing and tail
dusky, with conspicuous pure white
edgings and cross-bars. Size of
the last or rather larger; bill nearly
½; tarsus ⅔; middle toe the same;

FIG. 67. Plumbeous Vireo.

spurious quill about ¾, one-third as long as the second quill. Central Plains
to the Pacific, U. S., and especially Southern Rocky Mountains, where it is
abundant. A large stout species, a near ally of *solitarius*, but nearly all
the olivaceous of that species replaced by plumbeous, and the yellowish by
white, so that it is a very different looking bird. It may prove only a
variety, but I have seen no intermediate specimens, and cannot reconcile
the obvious discrepancies, upon this supposition. COUES, Pr. Ac. Phila.,
1866, 74; BD., Rev. 349; COOP., 119; ELLIOT, pl. 7. . . PLUMBEUS.

Gray Vireo. With the general appearance of a small faded specimen of
plumbeus: leaden-gray, faintly olivaceous on the rump, below white, with
hardly a trace of yellowish on the sides; wings and tail hardly edged with
white; no markings about head except a whitish eye-ring. 5¾; extent 8¾;
wing and tail, each, 2½; tarsus nearly ¾; middle toe and claw hardly over ½;
tip of inner claw falling short of base of middle claw; tail decidedly
rounded; spurious quill ¾, half as long as the second primary, which latter
is not longer than the eighth. Arizona. If these peculiar proportions of the
single known specimen are constant, the species is distinct from any other.
It is our plainest colored species, resembling *plumbeus*, but apparently
more closely allied to the smaller rounder-winged species like *noveboracensis* and especially *pusillus;* the toes are almost abnormally short, and
the tail as long as
the wing. COUES,
Proc. Phila. Acad.
Sci. 1866, p. 75:
BD., Rev., 361;
COOP. 125; ELLIOT,
pl. 7. . VICINIOR.

White-eyed Vireo.
Above bright olive-

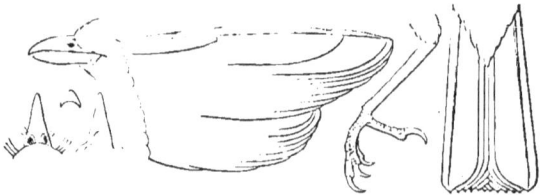

FIG. 68. White-eyed Vireo.

green, including crown; a slight ashy gloss on the cervix, and the rump
showing yellowish when the feathers are disturbed; below white, the sides .

of the breast and belly, the axillars and crissum, bright yellow ; a bright yellow line from nostrils to and around eye ; lores dusky ; two broad yellowish wing-bars ; inner secondaries widely edged with the same ; bill and feet blackish-plumbeous ; eyes white. About 5 inches long ; wing $2\frac{1}{8}$-$2\frac{1}{2}$; tail $2\frac{1}{4}$; spurious quill $\frac{3}{4}$, half as long as the second, which about equals the eighth ; tarsus about $\frac{3}{4}$; middle toe and claw $\frac{1}{2}$; bill nearly $\frac{1}{2}$. A small, compact, brightly-colored species, abundant in shrubbery and tangled undergrowth of the Eastern United States ; noted for its sprightly manners and emphatic voice ; eggs 4-5, white, speckled at large end. WILS., ii, 266, pl. 18 ; NUTT., i, 306 ; AUD., iv, 146, pl. 240 ; BD., 338, and Rev. 354. NOVEBORACENSIS.

FIG. 69. Hutton's Vireo.

Hutton's Vireo. A species or variety similar to the last, but differing much as *flavoviridis* does from *olivaceus*, in having the under parts almost entirely yellowish ; second quill about equal to the tenth. Lower California and southward. An accredited species, but one I have not tested, and cannot endorse. CASS., Proc. Acad. Phila. 1851, 150, 1852 ; pl. 1, f. 1 ; BD., 339, pl. 78, f. 2 ; Rev. 357. . . HUTTONII.

Bell's Vireo. Olive-green, brighter on rump, ashier on head, but without decided contrast ; head-markings almost exactly as in *gilvus ;* below, sulphury yellowish, only whitish on chin and middle of belly ; inner quills edged with whitish ; two whitish wing-bands, but one more conspicuous than the other. Hardly or not 5 long ; wing little over 2 ; tail under 2 ; spurious quill about $\frac{2}{3}$ the second, which equals or exceeds the seventh. A pretty little species, like a miniature *gilvus*, but readily distinguished from that species by its small size, presence of decided wing-bars, more yellowish under parts, and

FIG. 70. Bell's Vireo.

different wing-formula. Middle region, U. S., west to the Rocky Mountains, east to Kansas (*Coues*) and Illinois (*Ridgway*) ; an abundant species, inhabiting copses and shrubbery in open country, with much the same sprightly ways and loud song of *noveboracensis.* AUD., vii, 333, pl. 485 ; BD., 337 ; Rev. 358. BELLII.

Least Vireo. Olivaceous-gray, below white, merely tinged with yellowish on the sides ; head-markings obscure ; wing-bands and edg-

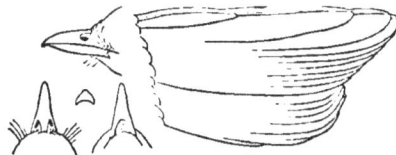

FIG. 71. Least Vireo.

ings, though evident, narrow and whitish ; no decided olive or yellow anywhere. Size of *bellii ;* wing and tail of equal lengths, little over 2 inches ; bill $\frac{1}{4}$; tarsus $\frac{3}{4}$; middle toe and claw $\frac{1}{2}$; spurious quill about $\frac{1}{2}$ as long as the second, which is intermediate between the seventh and eighth. A small obscure-looking species near *bellii*, which it replaces in Southwestern U. S ; possibly a grayer, longer-tailed, geographical race, but more specimens will be required to prove this.   Its habits are the same as those of Bell's vireo. COUES, Proc. Acad. Phila., 1866, 76 ; BD., Rev. 360 ; COOP. 124.   (*V. bellii* COOP., Proc. Cala. Acad. 1861, 122.). . . . . PUSILLUS.

Black-headed Vireo.   Olive-green, the crown and sides of head *black;* below white, olive-shaded on sides ; $4\frac{3}{4}$; wing $2\frac{1}{4}$; tail 2.   Southwestern Texas, extremely rare ;   only three specimens known.   WOODHOUSE, Proc. Phila. Acad. 1852, 60 ; Rep. Expl. Zuñi River, 75, pl. 1 ; CASS., Ill., 153, pl. 24 ; BD., 337, and Rev. 353. . ATRICAPILLUS.

Family LANIIDÆ.   Shrikes.

Essentially characterized by the combination of comparatively weak, strictly passerine feet with a notched, toothed and hooked bill, the size, shape and strength of which recalls that of a bird of prey.   The family comprises about two hundred recorded species, referable to numerous genera, and divisible into three groups, of which the following is the only one occurring in America.

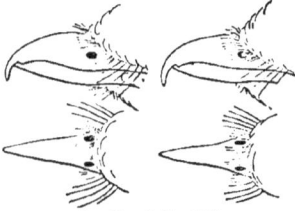

FIG 72. Shrikes' bills.

Subfamily LANIINÆ.   *True Shrikes.*

The genus *Collurio* is the only representative of this group in North America. In this genus the wing has ten primaries and the tail twelve rectrices ; both are much rounded and of nearly equal lengths.   The rictus is furnished with strong bristles.   The circular nostrils are more or less perfectly covered and concealed by dense tufts of antrorse bristly feathers.   The tarsi are scutellate in front and on the outside — in the latter respect deviating from a usual Oscine character. Our shrikes will thus be easily distinguished ; additional features are, the point of the wing formed by the 3d, 4th and 5th quills, the 2d not longer than the 6th, the 1st about half the 3d ; the tarsus equalling or slightly exceeding the middle toe and claw ; the lateral toes of about equal lengths, their claws reaching the base of the middle claw.   In coloration our species are much alike, and curiously similar to the mockingbird, being bluish-, grayish- or brownish-ash above, white more or less evidently vermiculated with black below ; wings and tail black variegated with white, rump and scapulars more or less whitish, and a black bar through the eye.

These shrikes are bold and spirited birds, quarrelsome among themselves, and tyrannical toward weaker species ; in fact, their nature seems as highly rapacious as that of the true birds of prey.   They are carnivorous, feeding on insects and such small birds and quadrupeds as they can capture and overpower ; many instances have been noted of their dashing attacks upon cage-birds, and their reckless pursuit of other species under circumstances that cost them their own lives.   But the most remarkable fact in the natural history of the shrikes is their singular and inexplicable habit of impaling their prey on thorns or sharp twigs, and leaving it sticking there.   This has occasioned many ingenious surmises, none of which, however, are entirely satisfactory.   They build a rather rude and bulky nest of twigs, and lay 4–6 speckled eggs.   They are not strictly migratory, although our northernmost species usually retires southward in the fall.   The sexes are alike, and the young differ but little. There are only two well determined American species, of nine that compose the genus. .

54. Genus COLLURIO Vigors.

Great Northern Shrike, or *Butcherbird.* Clear bluish-ash blanching on the rump and scapulars, below white always vermiculated with fine wavy blackish lines ; a black bar along side of head *not* meeting its fellow across forehead, interrupted by a white crescent on under eyelid, and bordered above by hoary white that also occu-pies the extreme forehead ; wings and tail black, the former with a large spot near base of the prima-ries, and the tips of most of the quills, white, the latter with nearly all the feathers broadly tipped with white, and with concealed w h i t e bases ; bill and feet black ; 9 - 10

FIG. 73. Butcherbird.

long ; wing 4½ ; tail rather more. The young is similar, but none of the colors are so pure or so intense ; the entire plumage has a brownish suffu-sion, and the bill is flesh colored at base. North America, northerly ; breeds, however, in mountainous parts of the United States (Alleghanies, *Turnbull*) ; in winter, usually extends southward about to 35° (*Coues*). WILS., i, 74, pl. 5, f. 1 ; NUTT., i, 258 ; AUD., iv, 130, pl. 236 ; BD., 324, and Rev. 440. BOREALIS.

Loggerhead Shrike. Slate-colored, slightly whitish on the rump and scapulars, below white, with a few obscure wavy black lines, or none ; black bar on side of head meeting its fellow across the forehead, *not* interrupted by white on under eyelid, and scarcely or not bordered above by hoary white ; otherwise like *borealis* in color, but smaller ; 8-8½ ; wing about 4 ; tail rather more. Young birds differ much as described under *borealis,* and are decid-edly waved below as in that species ; but the other characters readily distin-guish them. South Atlantic States. WILS., iii, 57, pl. 22, f. 5 ; NUTT., i, 561 ; AUD., iv, 135, pl. 237 ; BD., 325, and Rev. 443. . LUDOVICIANUS.

Var. EXCUBITOROIDES. *White-rumped Shrike.* With the size, and the essential characters of the head-stripe, of *ludovicianus,* and the under parts, as in that species not, or not obviously, waved, but with the clear light ash upper parts, and hoary whitish superciliary line, scapulars and rump of *borealis.* Middle and West-ern N. Am. ; N. to the Saskatchewan, E. to Illinois, S. into Mexico. BD., 327, 328, and Rev., 344, 345 ; Coor., 138.

OBS. Extreme examples of *ludovicianus* and *excubitoroides* look very different, but they are observed to melt into each other when many specimens are compared, so that no specific character can be assigned. To this species I must also refer the *C. elegans* of Baird, considering that the single specimen upon which it was based, represents an individual peculiarity in the size of the bill. This specimen is sup-posed to be from California, but some of Dr. Gambel's to which the same locality is assigned, were certainly procured elsewhere, and it may not be a North American bird at all. The highest authority on this genus, Messrs. Dresser and Sharpe, have shown from examination of Swainson's type specimen, that *his elegans* is the *C. lahtora*, a widely-spread Asiatic species probably erroneously attributed to North America.

Family FRINGILLIDÆ. Finches, etc.

The largest North American family, comprising between one-seventh and one-eighth of all our birds, and the most extensive group of its grade in ornithology. As ordinarily constituted, it represents, in round numbers, five hundred current species and one hundred genera, of nearly all parts of the world, except Australia, but more particularly of the northern hemisphere and throughout America, where the group attains its maximum development.

Any one United States locality of average attractiveness to birds, has a bird-fauna of over two hundred species; and if it be away from the sea-coast, and consequently uninhabited by marine birds, about one-fourth of its species are *Sylvicolidæ* and *Fringillidæ* together — the latter somewhat in excess of the former. It is not easy, therefore, to give undue prominence to these two families.

The *Fringillidæ* are more particularly what used to be called "conirostral" birds, in distinction from "fissirostres," as the swallows, swifts and goatsuckers, "tenuirostres," as humming birds and creepers, and "dentirostres," as warblers, vireos and most of the preceding families. The bill approaches nearest the ideal cone, combining strength to crush seeds, with delicacy of touch to secure minute objects. The cone is sometimes nearly expressed, but is more frequently turgid or conoidal, convex in most directions, and sometimes so contracted that some of its outlines are concave. The nostrils are usually exposed, but in many, chiefly boreal, genera, the base of the bill is furnished with a ruff, or two tufts of antrorse feathers more or less completely covering the openings. The cutting edges may be slightly notched, but are usually plain; there are usually a few inconspicuous bristles about the rictus, sometimes wanting, sometimes highly developed, as in our grosbeaks. The wings are endlessly varied in shape, but agree in possessing only nine developed primaries; the tail is equally variable in form, but always has twelve rectrices. The feet show a strictly Oscine podotheca, scutellate in front, covered on the side with an undivided plate, producing a sharp ridge behind. None of these members offer extreme phases of development or arrestation, in any of our species.

But the most tangible characteristic of the family is *angulation of the commissure*. The commissure runs in a straight line, or with a slight curve, to or near to the base of the bill, and is then more or less abruptly bent down at a varying angle — the cutting edge of the upper mandible forming a reëntrance, that of the lower mandible a corresponding salience. In the great majority of cases the feature is unmistakable, and in the grosbeaks, for example, it is very strongly marked indeed; but in some of the smaller-billed forms, and especially those with slender bill, it is hardly perceptible. On the whole, however, it is a good character, and at any rate it is the most reliable external feature that can be found. It separates our fringilline birds pretty trenchantly from other Oscines except *Icteridæ*, and most of these may be distinguished by the characters given beyond.

When we come, however, to consider this great group of conirostral Oscines in its entirety, as compared with bordering families like the Old World *Ploceidæ*, or the *Icteridæ*, and especially the *Tanagridæ*, of the New, the difficulty if not the impossibility of framing a perfect diagnosis becomes apparent, and I am not aware that a rigid definition has been successfully attempted. Ornithologists are nearly agreed what birds to call fringilline, without being very well prepared to say what " fringilline " means. The division of the family into minor groups, as might be expected, is a conventional matter at present — the subfamilies vary with every leading writer. Our species might be thrown into several groups, but the distinctions would be more or less arbitrary, not readily perceived, and doubtless negatived upon consideration of exotic material. It becomes necessary, therefore, to waive this matter, and simply collocate the genera in orderly sequence.

The *Fringillidæ* are popularly known by several different names. Here belong all the *sparrows*, with the allied birds called finches, buntings, linnets, grosbeaks and crossbills. In the following pages I describe seventy-one species, well determined, and ascertained to occur within our limits, referring them to thirty-four genera, as the custom is, although I think this number of genera altogether too large. Species occur throughout our country, in every situation, and many of them are among our most abundant and familiar birds. They are all granivorous — seed-eaters, but many feed extensively on buds, fruits and other soft vegetable substances, as well as on insects. They are not so perfectly migratory as the exclusively insectivorous birds, the nature of whose food requires prompt removal at the approach of cold weather ; but, with some exceptions, they withdraw from their breeding places in the fall to spend the winter further south, and to return in the spring. With a few signal exceptions they are not truly gregarious birds, though they often associate in large companies, assembled in community of interest. The modes of nesting are too various to be here summarized. Nearly all the finches sing, with varying ability and effect ; some of them are among our most delightful vocalists. As a rule, they are plainly clad — even meanly, in comparison with some of our sylvan beauties ; but among them are birds of elegant and striking colors. Among the highly-colored ones, the sexes are more or less unlike, and other changes, with age and season, are strongly marked ; the reverse is the case with the rest.

55. Genus HESPERIPHONA Bonaparte.

Evening Grosbeak. Dusky olivaceous, brighter behind, forehead, line over eye and under tail coverts yellow ; crown, wings, tail and tibiæ black, the secondary quills mostly white ; bill greenish-yellow, of immense size, about $\frac{3}{4}$ of an inch long and nearly as deep ; $7\frac{1}{2}$-$8\frac{1}{2}$; wing 4-4$\frac{1}{2}$; tail 2$\frac{1}{2}$. The ♀ and young differ somewhat, but cannot be mistaken. Plains to the Pacific, U. S., and somewhat northward ; occasional eastward to Ohio and Illinois, and even straying to Canada (*McIlwraith*) and New York (*Lawrence*). AUD., iii, 217, pl. 207 ; BD., 409 ; COOP., 174. . VESPERTINA.

56. Genus PINICOLA Vieillot.

Pine Grosbeak. ♂ carmine red, paler or whitish on the belly, darker and streaked with dusky on the back ; wings and tail dusky, much edged with white, former with two white bars ; ♀ ashy-gray, paler below, marked with brownish-yellow on the head and rump. 8-9 long ; wing 4$\frac{1}{2}$; tail 4, emarginate ; bill short, stout, convex in all directions. Northern North America, appearing in the United States in winter, generally in flocks, in pine woods ; resident in the Sierra Nevada of California (*Cooper*). WILS., i, 80, pl. 5 ; AUD., iii, 179, pl. 199 ; BD., 410 ; COOP., 152. . ENUCLEATOR.

57. Genus PYRRHULA Auctorum.

Cassin's Bullfinch. ♂ above clear ashy gray, below cinnamon gray, rump and under wing and tail coverts white ; wings and tail, crown, chin and face black ; outer tail feathers with a white patch, greater wing coverts tipped and primaries edged with whitish : bill black, feet dusky ; ♀ unknown. Length 6½ ; wing 3½ ; tail 3¼. Nulato, Alaska (*Dall*), only one specimen known, originally described as a variety of *P. coccinea* of Europe, but later determined to be distinct. Bᴅ., Trans. Chicago Acad. 1869, 316, pl. 29, f. 1 ; Nᴇwᴛᴏɴ, Ibis, 1870, 251 ; Tʀɪsᴛʀᴀᴍ, Ibis, 1871, 231. cᴀssɪɴɪɪ.

58. Genus CARPODACUS Kaup.

* Adult ♂ with the red diffuse, belly unstreaked, and edging of wings reddish.

Fɪɢ. 74. Cassin's Bullfinch.

Purple Finch. ♂ crimson, rosy, or purplish-red, most intense on the crown, fading to white on the belly, mixed with dusky streaks on the back ; wings and tail dusky, with reddish edgings, and the wing coverts tipped with the same ; lores and feathers all around base of bill hoary. ♀ and *young* with no red—olivaceous-brown, brighter on the rump, the feathers above all with paler edges, producing a streaked appearance ; below white, thickly spotted and streaked with olive-brown, except on middle of belly and under tail coverts ; obscure whitish superciliary and maxillary lines. Young males show every gradation between these extremes, in gradually assuming the red plumage, and are frequently brownish-yellow or bronzy below. 5¾-6¼ ; wing 3-3¼ ; tail 2¼-2½, forked ; tarsus ⅝ ; middle toe and claw ⅞ ; bill under ½, turgid, with a little ruff of antrorse feathers. Not crested, but the coronal feathers erec-tile. The foregoing description should prevent con-founding young birds with any of the streaked and spotted sparrows. United States from Atlantic to Pacific, and somewhat northward in summer ; an abundant species, particularly in spring and fall, in

Fɪɢ. 75. Bill of Purple Finch.

woods and orchards, generally found in flocks except when breeding ; feeds on seeds, buds and blossoms ; a delightful songster. Wɪʟs., i, 119, pl. 7, f. 4 ; Auᴅ., iii, 170, pl. 196 ; Bᴅ., 412 ; also, *C. californicus* Bᴅ., 413, Cooᴘ., 154, which I cannot distinguish at all.   .   .   .   .   .   pᴜʀᴘᴜʀᴇᴜs.

Cassin's Purple Finch. Similar ; the red paler, more streaked with dusky on the upper parts, crown rich crimson in marked contrast ; larger ; 6¼-7 ; wing 3½ ; tail 2¾ ; bill about ½, comparatively less turgid ; tarsus ⅝. Southern Rocky Mountain Region. Bᴅ., 414 ; Cᴏᴜᴇs, Pr. Acad. Nat. Sci. Phila. 1766, 45 ; Cooᴘ., 155.   .   .   .   .   cᴀssɪɴɪɪ.

**Adult ♂ with the red partly in definite areas, the belly streaked, the edging of the wings whitish.

Crimson-fronted Finch. House Finch. Burion. ♂ with the forehead and a line over the eye, the rump, and the chin, throat and breast, crimson; other upper parts brown, streaked with darker, and marked with dull red, and other under parts white or whitish, streaked with dusky; wings and tail dusky with slight whitish edgings and cross bars. The changes of plumage are parallel with those of *C. purpureus*, but the species may easily be distinguished in any plumage by its smaller size, with relatively longer wings and tail, these members being absolutely as long or nearly as long as in *purpureus;* the tail barely or not forked; and especially by the much shorter and more inflated bill, which is almost exactly as represented in the foregoing figure of *Pyrrhula cassinii*. Rocky Mountains to the Pacific, U. S., a very abundant species in the towns and gardens of New Mexico, Arizona and California, where it is as familiar as the European Sparrow has become in many of our large eastern cities; nests about the houses; a pleasant songster. AUD., iii, 175, pl. 197; BD., 415; COOP., 156. . . . FRONTALIS.

59. Genus CURVIROSTRA Scopoli.

*** Distinguished from all other birds by the falcate mandibles with crossed points. Nasal ruff conspicuous; wings long, pointed; tail short, forked; feet strong. Sexes dissimilar; ♂ some shade of red, nearly uniform, with dusky wings and tail; ♀ brownish or olivaceous, more or less streaked, head and rump frequently washed with brownish-yellow; young like the ♀. Irregularly migratory, according to exigencies of the weather, eminently gregarious, and feed principally on pine seeds, which they skilfully husk out of the cones with their singular bill. Our two species inhabit the northern parts of America, coming southward in flocks in the fall; but they are also resident in northern and mountainous pine-clad parts of the United States, where they sometimes breed *in winter.*

FIG. 76. White-winged Crossbill.

White-winged Crossbill. Wings in both sexes with two conspicuous white bars; ♂ rosy red, ♀ brownish-olive, streaked and speckled with dusky, the rump saffron; about 6; wing 3½; tail 2¼. WILS. iv, 48, pl. 31, f. 3; AUD., iii, 190, pl. 201; BD., 427. . . . . LEUCOPTERA.

Red Crossbill. Common Crossbill. (PLATE III, figs. 13, 14, 15, 13*a*, 14*a*, 15*a*.) Wings blackish, unmarked; ♂ bricky red; ♀ as in *leucoptera*, but wings plain. WILS., iv, 44, pl. 31, f. 1, 2; AUD., iii, 186, pl. 200; BD. 426; COOP., 148. . . . . AMERICANA.

Var. *mexicana.* Similar to the last; bill large, about ¾ of an inch long. Moun-

tainous parts of New Mexico, and southward. Bd., 427 (in text), 924. My New
Mexican specimens show a bill almost matching that of *C. pytiopsittacus* of Europe.

60. Genus LEUCOSTICTE Swainson.

*** Sides of the under mandible with a small sharp oblique ridge ; nasal tufts
conspicuous.

. *Gray-crowned Finch.* Chocolate or liver-brown, the feathers posteriorly
skirted with rosy or lavender, wings and tail dusky, rosy-edged, chin dusky
with little or no ashy, crown alone clear ash, forehead alone black, bill and
nasal feathers whitish, feet black ; ♀ not particularly different ; about 7 ;
wing 4¼ ; tail 2¾ ; a little forked. In midsummer, the black frontlet extends
over the crown, the rosy heightens to crimson, and the bill blackens ; the
whole plumage is likewise darker. Rocky Mountain region, south to Colorado. Aud., iii, 176, pl. 198 ; Bd., 430 ; Coop., 164. . TEPHROCOTIS.

Var. CAMPESTRIS Bd., in Coop., 163. Colorado. In the specimen described, the
ash of the head extends a little below the eyes but not on the auriculars, and forms
a narrow border on the chin ; thus approximating to the
next.

FIG. 77. Gr. y-ared Finch (⅓).

Var. GRISEINUCHA. *Gray-eared Finch.* The ash of the
head extending over the whole cheeks and ears and
part of the chin ; the black frontlet extending over most
of the crown. Larger than average *tephrocotis.* Aleutian
Islands. Bd., 430 (footnote) ; Trans. Chicago Acad.
1869, pl. 28, f. 2 ; Coop., 161.

Var. LITTORALIS Bd., Trans. Chicago Acad. 1869, p. 317, pl. 28, f. 1 ; Coop., 163.
In the specimens described, from Sitka and British Columbia, the whole head
including the chin, except the black frontlet, is ashy. The gradations noted in the
foregoing paragraphs show that there is but a single species, although *griseinucha*
and *littoralis* look quite different from *tephrocotis* and *campestris.*

Siberian Finch. Dusky purplish ; neck above pale yellowish ; forehead
and nasal feathers blackish ; outer webs of quills and wing coverts, tail
coverts, rump and crissum silvery gray, rosy-margined. Kurile and Aleutian Islands ; Siberia. Bd., 430 (footnote) ; Coop., 165. . . ARCTOA.

61. Genus AEGIOTHUS Cabanis.

*** Small species (5¼–5¾ ; wing 2¾–3 ; tail 2¼–2½), with the bill extremely acute,
overlaid at the base with nasal plumules, the wings long, pointed, the tail short,
forked, the feet moderate. Conspicuously *streaked*, the crown with a crimson patch
in both sexes, the face and chin dusky, wings and tail dusky with whitish edgings ;
the males with the whole breast rosy and the rump tinged with the same. Boreal
birds, occurring in the United States in winter, in large flocks.

Red-poll Linnet. (PLATE III, figs. 1, 1*a*, 2, 2*a*, 4, 4*a*, 5, 5*a*.) Upper parts
streaked with dusky and flaxen in about equal amounts, rump white or rosy,
always streaked with dusky ; below, streaked on the sides, the belly dull
white ; bill mostly yellow ; feet blackish ; middle toe and claw as long
as the tarsus. WILS., iv, pl. 42 ; NUTT., i, 512 ; AUD., iii, 122 ; BD., 428 ;
COUES, Proc. Phila. Acad. 1861, 373. . . . . LINARIUS.

Var. FUSCESCENS. *Dusky Redpoll*. (PLATE III, figs. 3, 3a.) Upper parts nearly uniform dusky, rump scarcely lighter, sides heavily streaked; bill dusky. AUD., iii, pl. 179? COUES, Proc. Phila. Acad., 1861, 222, 390; 1869, 186; ELLIOT, pl. 10. I am not sure that this is anything more than a state of plumage of *linarius*, as the dusky appearance may be due to wearing away of the lighter edges of the feathers.

Var. EXILIPES. *American Mealy Redpoll*. (PLATE III, figs. 6, 6a.) Colors pale, the flaxen of *linarius* bleaching to whitish; rump white or rosy, entirely unstreaked in the adults; breast pale rosy, and streaks on the sides small and sparse; bill very small, with heavy plumules; feet small, the middle toe and claw hardly or not equal to the tarsus. AUD., iii, 120, pl. 178; COUES, Proc. Phila. Acad., 1861, 385, 1869, 187; ELLIOT, pl. 9. An Arctic race, not difficult to recognize, representing in America the true Mealy Redpoll, *A. canescens*, of Greenland.

OBS. In addition to the foregoing, a large variety with a very large yellow bill, apparently corresponding to what is called *holbœlli* in Europe, has been noted from Canada. COUES, Proc., Phila. Acad., 1862, 40; 1869, 184.

61 bis. Genus LINOTA Bonaparte.

Brewster's Linnet. With the general appearance of an immature *Aegiothus*, this bird will be recognized by absence of any crimson on the crown, a peculiar yellowish shade on the lower back, and somewhat different proportions. Massachusetts, one specimen (*Brewster*). *Aegiothus flavirostris*, var. *brewsterii* RIDGWAY, Am. Nat. vi, July, 1872. An interesting discovery, of which I learn as these pages go to press; may be same as the European. (Not in the Key.)   .   .   .   FLAVIROSTRIS var. BREWSTERII.

62. Genus CHRYSOMITRIS Boie.

* Sexes alike. Bill extremely acute; nasal plumules sometimes deficient. Everywhere thickly streaked; no definite black on head; no red anywhere.

Pine Linnet. (PLATE III, figs. 11, 11a, 12, 12a.) Continuously streaked above with dusky and olivaceous brown or flaxen, below with dusky and whitish, the whole plumage in the breeding season more or less suffused with yellowish, particularly bright on the rump; the bases of the quills and tail feathers extensively sulphury yellow, and all these feathers more or less edged externally with yellowish. $4\frac{3}{4}$; wing $2\frac{3}{4}$; tail $1\frac{3}{4}$; forked. The plumage is extremely variable; young birds frequently show a buffy or flaxen suffusion, and resemble a redpoll; but the yellowish coloration of the wings and tail is peculiar, and distinctive of the species. North America, breeding northerly, ranging throughout most of the United States, in flocks, in the winter; abundant. WILS., ii, 133, pl. 17, f. 1; NUTT., i, 511; AUD., iii, 125, pl. 180; BD., 425; COOP., 172.   .   .   .   .   .   . PINUS.

* * Sexes unlike. Bill moderately acute. Not noticeably streaked. The adult males with definite black on the crown, wings and tail.

American Goldfinch. Yellowbird. Thistlebird. (PLATE III, figs. 7, 8, 9, 10, 7a, 8a, 9a, 10a.) ♂ in summer, rich yellow, changing to whitish on the tail coverts; a black patch on the crown; wings black, more or less edged and barred with white; lesser wing coverts yellow; tail black, every feather with a white spot; bill and feet flesh-colored. In September, the black cap disappears, and the general plumage changes to a pale flaxen brown above, and whitey-brown below, with traces of the yellow, especially about the head: this continues until the following April or May.   ♀ olivaceous, including the crown; below soiled yellowish, wings and tail dusky, whitish-edged : young like the ♀ . About 4¾ long ; wing 2¾ ; tail 2, a little forked; ♀ rather smaller than the ♂ . North America, especially the Eastern United States ; an abundant and familiar species, conspicuous by its bright colors, and plaintive lisping notes ; in the fall, collects in large flocks, and so remains until the breeding season ; irregularly migratory ; feeds especially on the seeds of the thistle and buttonwood ; flies in an undulating course. Nest small, compact, built of downy and very soft pliant substances, with stucco-work of lichens, placed in a crotch ; eggs 4-5, white, speckled. WILS., i, 20, pl. 1, f. 2 ; NUTT., i, 507 ; AUD., iii, 129, pl. 181 ; BD., 421. . . . . TRISTIS.

Lawrence's Goldfinch.   ♂ gray, whitening on the belly and crissum : rump, a large breast patch, and often much of the back, rich yellow ; crown, face and chin black ; wings black, variegated with yellow, most of the coverts being of this color, and the same broadly edging the quills ; tail black, most of the feathers with large square white spots on the inner web ; bill and feet dark. The ♀ resembles the male, but there is no black on the head, and the yellow places are not so bright. Size of *tristis*, or rather less ; an elegant species.

FIG. 78. Lawrence's Goldfinch.

California, Arizona, and probably New Mexico. BD., 424 ; ELLIOT, pl. 8 ; COOP., 171. . . . . LAWRENCEI.

Arkansas Goldfinch.   ♂ olive-green, below yellow ; crown black, this *not* extending below eyes ; wings black, most of the quills and the greater coverts white-tipped, and the primaries white at base ; tail black, the outermost three pairs of feathers with a long rectangular white spot on the inner web.   ♀ and young similar, but not so bright, and no black on the head ; sometimes, also, no decided white spots on the tail. 4¼–4½ ; wing 2⅜ ; tail 2. Plains to the Pacific,

FIG. 79. Arkansas Goldfinch.
(Arizona variety.)

U. S., rather southerly. AUD., iii, 134, pl. 183 ; NUTT., i, 510 ; BD., 422 ; COOP., 168. . . . . PSALTRIA.

Var. ARIZONÆ COUES. Proc. Phila. Acad., 1866, p. 46 ; COOP., 170. The upper parts mixed olive and black in about equal amounts, thus leading directly into

Var. MEXICANA, with the upper parts continuously black. and the black of the crown extending below the eyes, enclosing the olive under eye-lid. Mexican border and southward. Bp., 423; Coop., 169. This bird looks quite unlike typical *psaltria*, but the gradation through var. *arizonœ* is perfect; and *mexicana*, moreover, leads directly into var. *columbiana*, a Central American form in which the tail-spots are very small or wanting. The females of these several varieties cannot be distinguished with certainty.

Obs. *Chrysomitris magellanica*, a South American species with the whole head black, is said by Audubon to have been taken in Kentucky, where probably it will not be found again. *Chrysomitris stanleyi* and *C. yarrellii*, of Audubon, were apparently cage-birds, improperly attributed to North America.

FIG. 80.   Mexican Goldfinch.

63. Genus PLECTROPHANES Meyer.

* Bill small, truly conic, ruffed at base ; hind claw decidedly curved.

Snow Bunting. Snowflake. In breeding plumage, pure white, the back, wings and tail variegated with black ; bill and feet black. As generally seen in the United States, the white is clouded with clear, warm brown, and the bill is brownish. Length about 7; wing $4\frac{1}{2}$; tail $2\frac{3}{4}$. Arctic America, irregularly southward, in flocks, in the winter, to about 35° ; but its movements depend much on the weather. WILS., iii, 86, pl. 21 ; NUTT., i, 458 ; AUD., iii, 55, pl. 155 ; Bp., 432. . . . . NIVALIS.

FIG. 81.   Foot in Centrophanes.

** Bill moderate, unruffed, but with a little tuft of feathers at the base of the rictus ; hind claw straightish, with its digit longer than the middle toe and claw. Sexes dissimilar ; ♂ with a cervical collar, and oblique white area on the outer tail feathers ; ♀ resembling some of the streaked sparrows. (*Centrophanes*.)

Lapland Longspur. Adult ♂ : whole head and throat jet black bordered with buffy or whitish which forms a postocular line separating the black of the crown from that of the sides of the head ; a broad chestnut cervical collar ; upper parts in general blackish streaked with buffy or whitish that edges all the feathers ; below, whitish, the breast and sides black-streaked ; wings dusky, the greater coverts and inner secondaries edged with dull bay ; tail dusky with white areas as above mentioned ; bill yellowish tipped with black, *legs and feet black*. $6-6\frac{1}{2}$; wing $3\frac{1}{4}-3\frac{1}{2}$; tail $2\frac{1}{2}-2\frac{3}{4}$. Winter males show less black on the head, and the cervical chestnut duller ; the ♀ and young have no continuous black on the head, and the crown is streaked like the back ; but there are traces of the cervical collar, whilst the generic characters will prevent confusion with any of the ordinary streaked sparrows. Arctic America, irregularly southward into the United States in winter, frequently in company with *nivalis*, but not so common. NUTT., i, 463 ; AUD., iii, 50, pl. 152 ; BD., 433. . . . . LAPPONICUS.

Painted Lark Bunting. Adult ♂ : cervical collar and entire under parts rich buffy brown or dark fawn ; crown and sides of head black, bounded below by a white line, and interrupted by white superciliary and auricular line and white occipital spot ; upper parts streaked with black and brownish yellow ; lesser and middle wing coverts black, tipped with white forming conspicuous patches ; one or two outer tail feathers mostly white ; no white on the rest ; *legs pale.* Size of *lapponicus:* seasonal and sexual changes of plumage correspondent. British America into United States in the *interior;* not common with us. NUTT., ii, 589 ; AUD., iii, 52, pl. 153 ; vii, 337, pl. 487 (*smithii*) ; BD., 484. . . . . PICTUS.

Chestnut-collared Lark Bunting. Adult ♂ : a chestnut cervical collar, as in *lapponicus,* and upper parts streaked much as in that species, but grayer ; nearly all the under parts continuously *black,* the throat yellowish ; lower belly and crissum only whitish ; in high plumage the black of the under parts is more or less mixed with intense ferrugineous, and sometimes this rich sienna color becomes continuous ; crown and sides of head black, interrupted with white auricular and postocular stripes, and in high plumage with a white occipital spot ; lesser wing coverts *black* or brownish-black ; outer tail feathers mostly or entirely white, and all the rest largely white from the base — a character that distinguishes the species in any plumage from the two preceding ; legs not black ; ♀ with or without traces of the cervical collar ; crown exactly like the back, generally no black on head or under parts ; below whitish, with slight dusky maxillary and pectoral streaks and sometimes the whole breast black, edged with grayish. Immature males have the lesser wing coverts like the back ; but they show the black of the breast, veiled with gray tips of the feathers, long before any black appears on the head. Size less than in the foregoing. 5½-6 ; wing 3-3½ ; tail 2-2½. Missouri Region, Kansas, and westward ; S. to the Table-lands of Mexico. AUD., iii, 53, pl. 154 ; NUTT., 2d ed. 1, 539 ; BD., 435. *P. melanomus* BD., 436, appears to be merely a high plumage, perhaps not always assumed by northerly birds. . . . . ORNATUS.

* * * Bill large, turgid, unruffed ; hind claw as before, but shorter ; sexes dissimilar ; no cervical collar ; outer tail feathers white, the rest, except the middle pair, white on the inner webs to near the tip, the line of demarcation running straight across. (*Rhynchophanes.*)

Maccown's Bunting. Adult ♂ : crown and a broad pectoral crescent black ; superciliary line and under parts white ; bend of wing chestnut ; above, streaked with blackish and yellowish-brown. Size of the last, or rather larger ; 6-6½ ; wing 3¾ ; tail 2¼-2½ ; bill nearly ½ inch long. The ♀ lacks the black and chestnut, but in any plumage the species may be known by the peculiar markings of the tail feathers, the white areas being cut squarely off, except in the outer pair, which are wholly white. Plains to the Rocky Mountains, U. S., rather northern ; breeds abundantly about Cheyenne, Wyoming. (*Allen.*) LAWR., Ann. Lyc. Nat. Hist. N. Y., 1851, v, 122; CASS. Ill., 228, pl. 39; BD., 437. . . . . MACCOWNII.

64. Genus CENTRONYX Baird.

Baird's Bunting. Hind claw rather longer than its digit; hind toe and claw not shorter than the middle one. Wings pointed, but inner secondaries not lengthened as in *Passerculus*. Tail emarginate. Thickly streaked everywhere above, on the sides, and across the breast; above, grayish streaked with dusky, below white, with blackish maxillary, pectoral and lateral streaks; crown divided by a brownish-yellow line; a faint superciliary whitish line; no yellowish on bend of wing; outer tail feathers whitish. A curious bird, apparently related to *Plectrophanes* in form, but with the general appearance of a savanna sparrow or bay-winged bunting. Only one specimen known. Yellowstone, AUD., vii, 359, pl. 500; BD., 441. "Massachusetts," MAYNARD, Am. Nat., 1869, 554, and Guide, 112, frontispiece; ALLEN, Am. Nat. 1869, 631; BREWSTER, Am. Nat. 1872, 307. I have seen the later supposed specimens, the fresh measurements of one of which (6¼; wing 3¼; tail 2⅔; bill .4; tarsus nearly an inch) are much larger than those recorded by Audubon, and there are many other discrepancies. The bird should be diligently sought for, as a full investigation will reveal something not now anticipated. . . . . BAIRDII.

65. Genus PASSERCULUS Bonaparte.

Savanna Sparrow. (PLATE III, figs. 16, 17, 18, 16*a*, 17*a*, 18*a*.) Thickly streaked everywhere above, on sides, and across breast; a superciliary line, and edge of the wing, *yellowish;* lesser wing coverts *not* chestnut; legs flesh-color; bill rather slender and acute; tail nearly even, its outer feathers not pure white; longest secondary nearly as long as the primaries in the closed wing. Above, brownish-gray, streaked with blackish, whitish-gray and pale bay, the streaks largest on interscapulars, smallest on cervix, the crown divided by an obscure whitish line; sometimes an obscure yellowish suffusion about head besides the streak over the eye.

FIG. 82. Savanna Sparrow.
(Bill too slender.)

Below, white, pure or with faint buffy shade, thickly streaked, as just stated, with dusky—the individual spots edged with brown, mostly arrow-shaped, running in chains along the sides, and often aggregated in an obscure blotch on the breast. Wings dusky, the coverts and inner secondaries black-edged and tipped with bright bay; tail feathers rather narrow and pointed, dusky, not noticeably marked. 5¼–5¾; wing 2½–2¾; tail 2–2¼; middle toe and claw together 1¼; bill under ½. North America; a terrestrial species, abundant everywhere in fields, on plains, by waysides, and along the seashore; migratory, gregarious. With a close general resemblance to several other species, it may be readily distinguished by the foregoing marks. It varies but little with sex

and age, though the colors may be darker and sharper, or brighter and more diffuse, according to season and wear of the feathers. WILS., iii, 55, pl. 22, f. 1 ; NUTT., i, 489 ; AUD., iii, 68, pl. 160 ; BD., 442. *P. alaudinus* BD., 446 ; COOP., 181, is indistinguishable. . . . . SAVANNA.

Var. ANTHINUS BD., 445 ; COOP., 183 ; ELL., pl. 13, may be recognized. Bill longer, slenderer (as in fig. 82) ; spots below very numerous, close, sharp, dark. California coast, abundant in the salt marshes.

Var. SANDVICENSIS BD., 444 ; COOP., 180.   A large northern race : 6 or more long ; bill ½, stout ; head more yellowish. Northwest coast.

Sea-shore Sparrow.   With the form of a savanna sparrow, but the bill elongated as in *Ammodromus*, yet very stout and turgid, with decidedly convex culmen, ½ an inch long.   No evident yellowish over eye or on edge of

FIG. 83.   Sea-shore Sparrow.

wing ; no evident median stripe on crown.   Brownish-gray, back and crown streaked with dusky, below dull white, confluently streaked with brown everywhere except on belly and crissum.   Wings and tail dusky gray, the rectrices with paler edges, the primaries with whitish edges, the wing coverts and secondaries broadly edged and tipped with grayish-bay ; an obscure whitish superciliary line ; under mandible yellowish, legs pale ; 5¼ ; wing 2⅔ ; tail 2.   Pacific coast, U. S. ; a curious species, common, maritime, representing, with var. *anthinus*, the *Ammodromi* in the marshes of the seashore.   CASS., Ill., 226, pl. 28 ; BD., 446 ; COUES, Ibis, 1866, 268 ; COOP., 184. . . . . ROSTRATUS.

St. Lucas Sparrow.   Similar to *rostratus;* same size ; bill not so heavy : "A stripe of pale yellow runs from the bill to the eye, a longer stripe of pale yellow extends from the under mandible down the side of the throat * * * differs from all its allies in the obscure grayish coloring of the upper parts, with no reddish-brown, and in having its under plumage more closely and fully spotted."   San José, L. California ; a variety of the last ?   LAWR., Ann. Lyc. Nat. Hist. N. Y. 1867, 473 ; COOP., 185.   . . . GUTTATUS. †

66. Genus POOECETES Baird.

Bay-winged Bunting.   Grass Finch.   Thickly streaked everywhere above, on sides and across breast ; *no* yellow anywhere ; lesser wing coverts *chestnut* and 1 – 3 pairs of outer tail feathers partly or wholly *white*.   Above grayish-brown, the streaking dusky and brown, with grayish-white ; below white, usually noticeably buffy-tinged, the streaks very numerous on the fore parts and sides ; wing coverts and inner quills much edged and tipped with bay ; crown like back, without median stripe, line over, and ring round, eye, whitish ; feet pale ; 5¾–6¼ ; wing 2⅞–3¼, with inner secondaries lengthened ; tail 2¼–2¾.   North America ; a rather large, stout species, known on sight by combination of chestnut lesser wing coverts and white outer tail feathers ; the sexes are alike, and the variations in color are only such as are indicated under *P. savanna;* western specimens average paler and grayer, representing var. *confinis* BD., 448.   A very abundant bird,

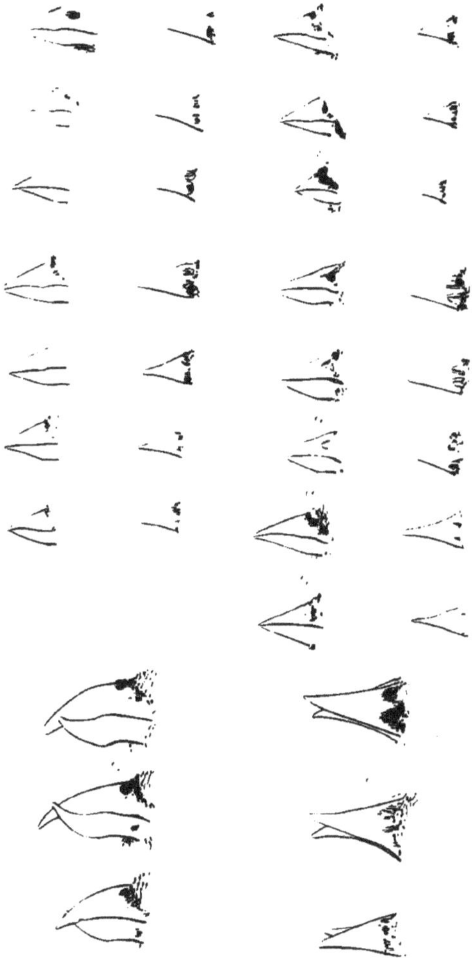

in fields, etc., terrestrial, migratory, gregarious in the fall. WILS., iv, 51, pl. 31, f. 5; NUTT., i, 482; AUD., iii, 65, pl. 159; BD., 447. GRAMINEUS.

67. Genus COTURNICULUS Bonaparte.

Yellow-winged Sparrow. Edge of wing conspicuously yellow; lesser wing coverts, and short line over eye, yellowish; below, not or not evidently streaked, but fore parts and sides, buff, fading to dull white on the belly. Above, singularly variegated with black, gray, yellowish-brown and a peculiar purplish-bay in short streaks and specks, the crown being nearly black with a sharp median brownish-yellow line, the middle of the back chiefly black with bay and brownish-yellow edgings of the feathers, the cervical region and rump chiefly gray mixed with bay; wing coverts and inner quills variegated like the back; feet pale. Small; only 4⅝–5¼ long; wing 2⅜, much rounded; tail 2 or less, with very narrow pointed feathers, the outstretched feet reaching to or beyond its end; bill short, turgid. Sexes alike; young similar, not so buffy below, and with pectoral and maxillary dusky spots; but in any plumage known from other sparrows (except the next species) by amount of yellow on wings, and peculiar proportions of parts. United States; abundant in tall grass and weeds of plains and fields; strictly terrestrial, migratory, with a peculiar chirring note, like a grasshopper's; nests on the ground, eggs 4–5, white, speckled. Specimens from dry western regions are paler and grayer (var. *perpallidus* RIDGWAY, Mss.). WILS., iii, 76, pl. 26, f. 5; AUD., iii, 73, pl. 162; NUTT., i, 494; BD., 450; COOP., 189. . . . . PASSERINUS. +

Henslow's Sparrow. Resembling the last; smaller; more yellowish above, and with sharp maxillary, pectoral and lateral black streaks below; tail longer, reaching beyond feet; bill stout. Eastern U. S., not very common. AUD., iii, 75, pl. 163; NUTT., i, 2d ed. 571; BD., 451. HENSLOWII.

Leconte's Sparrow. Like the last; bill *much* smaller; fore and under parts and sides of head buff, with black touches on sides; no yellow loral spot; median crown-stripe buff, white posteriorly; 4½; wing 2⅜; tail 1⅞. Missouri region; Texas. A long-lost species, rediscovered in No. 50, 222, Mus. S. I. (*Lincecum.*) AUD., vii, 338, pl. 488; BD., 452. . LECONTEI. +

68. Genus AMMODROMUS Swainson.

*** Small streaked sparrows, remarkable for the slender lengthened form of the bill, and the narrow, acute tail feathers. Wing short, much rounded, its edge yellow; tail short; feet very large, reaching nearly to end of tail. Confined to salt-marshes of the Atlantic and Gulf States; abundant, migratory.

Sea-side Finch. Olive-gray, obscurely streaked on the back and crown with darker and paler; below, whitish, often washed with brownish, and shaded on the sides with the color of the back, with ill-defined streaks on the breast and sides; wings and tail plain dusky, with slight olivaceous edgings, wing coverts and inner quills somewhat margined with brown; *a yellow spot over eye*, and often some vague brownish and dusky markings on side of head;

bill plumbeous, feet dark; 5¾–6¼; wing 2¼–2½; tail about 2. WILS., iv, 68, pl. 34, f. 2; NUTT., i, 2d ed. 592, 593; AUD., iii, 103, 106, pl. 172, 173 (*macgillivrayi*); BD., 454. . . . . MARITIMUS.

Sharp-tailed Finch. Olive-gray, sharply streaked on the back with blackish and whitish; crown darker than nape, with brownish-black streaks and obscure median line; no yellow loral spot, but long line over eye and sides of head rich buff or orange-brown enclosing olive-gray auriculars and a dark speck behind these; below, white, the fore parts and sides tinged with yellowish brown or buff of variable intensity, the breast and sides sharply streaked with dusky. Rather smaller than the last; bill still slenderer, and tail feathers still narrower and more acute. WILS., iv, 70, pl. 34, f. 3; NUTT., i, 504; AUD., iii, 108, pl. 174; BD., 453. . . . CAUDACUTUS.

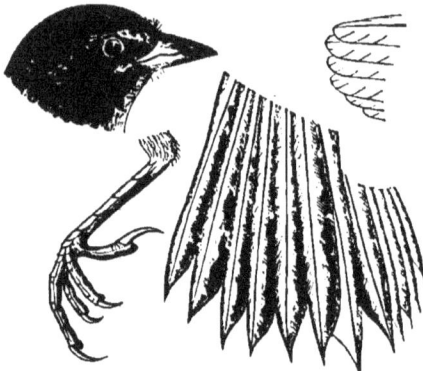

FIG. 81. Details of structure in Sharp-tailed Finch.

69. Genus MELOSPIZA Baird.

*Breast streaked, and with a transverse belt of brownish-yellow; tail nearly equal to wings.

Lincoln's Finch. Below, white, breast banded and sides often shaded with yellowish, everywhere except on the belly thickly and sharply streaked with dusky; above grayish-brown, crown and back with blackish, brownish and paler streaks; tail grayish-brown, the feathers usually showing blackish shaft lines; wings the same, the coverts and inner quills blackish with bay and whitish edgings; no yellow on wings or head; 5½; wing and tail about 2¼. North America; not common in the Eastern States. AUD., iii, 116, pl. 177; NUTT., i, 2d ed. 569; BD., 482; COOP., 216. . . LINCOLNII.

** Breast ashy, unbelted, with few streaks, or none; tail about equal to wings.

Swamp Sparrow. Crown bright bay, or chestnut, blackening on the forehead, often with obscure median ashy line, and usually streaked with black; cervix, entire sides of head and neck, and the breast, strongly ashy, with vague dark auricular and maxillary markings, the latter bounding the whitish chin, the ashy of the breast obsoletely streaky; belly whitish; sides, flanks and crissum strongly shaded with brown, and faintly streaked; back and rump brown, rather darker than the sides, boldly streaked with black and pale brown or grayish. Wings so strongly edged with bright bay as to appear almost uniformly of this color when viewed closed, but inner secondaries showing black with whitish edging; tail likewise strongly edged with bay, and usually showing sharp black shaft lines. No yellowish anywhere; no tail feathers white; further distinguished from its allies by the emphasis of its black, bay and ash; 5½-6; wing and tail 2¼-2¾. Eastern North America; a common inhabitant of low thickets, swamps and marshes. WILS., iii, 49, pl. 22, f. 1; AUD., iii, 110, pl. 175; NUTT., i, 502; BD., 483.   PALUSTRIS.

* * * Breast white, with numerous streaks aggregated into a central blotch; tail obviously longer than the wings, both rounded. Thickly streaked everywhere above, on sides and across breast. No yellowish anywhere.

Song Sparrow. Below, white, slightly shaded with brownish on the flanks and crissum, the numerous streaks just mentioned being dusky with brown edges, forming a pectoral blotch and also usually coalescing into maxillary stripes bounding the white throat; crown dull bay, with fine black streaks, divided and bounded on either side by ashy-whitish lines; vague brown or dusky and whitish markings on the sides of the head; the interscapular streaks black, with bay and ashy-white edgings; rump and cervix grayish-brown, with merely a few bay marks; wings with dull bay edgings, the coverts and inner quills marked like the interscapulars; tail plain brown, with darker shaft lines, on the middle feathers at least, and often with obsolete wavy markings. Very constant in plumage, the chief differences being in the sharpness and breadth of the markings, due in part to the wear of the feathers. 6-6½; wing about 2½; tail nearly or quite 3. Eastern United States; one of our most abundant birds everywhere, and a well known pleasing songster. WILS., ii, 125, pl. 16, f. 4; NUTT., i, 486; AUD., iii, 147, pl. 189; BD., 477. . . . . MELODIA.

Obs. The Eastern song sparrow is simply one variety of a bird distributed from Atlantic to Pacific, and which in the West is split into numerous geographical races, some of them looking so different from typical *melodia* that they have been considered as distinct species, and even placed in other genera. This differentiation affects not only the color, but the size, relative proportion of parts, and particularly the shape of the bill; and it is sometimes so great, as in case of *M. insignis*, that less dissimilar looking birds are commonly assigned to different genera. Nevertheless, the gradation is complete, and effected by imperceptible degrees. The following varieties have been described, and may usually be recognized.

Var. FALLAX BD., 481; COOP., 215. Extremely similar; wings and tail slightly longer; paler, grayer; the streaks not so obviously blackish in the centre. Whole of Rocky Mountains and Great Basin; scarcely distinguishable.

Var. GUTTATA NUTT., i, 2d ed. 581; *Fringilla cinerea*, AUD., iii, 145, pl. 187. Darker and more rufous, the colors more blended, from indistinctness of the streaks; below, quite brownish, except on middle of belly. Pacific coast, U. S., and British Columbia.

Var. RUFINA BD., 480; COOP., 214. Similar, but larger; color more fuliginous. Sitka, Alaska.

Var. HEERMANNI BD., 478; COOP., 212. Grayish, the streaks numerous, broad, distinct, mostly lacking pale edging. California.

Var. GOULDII BD., 479. Similar to the last, in distinctness of the black streaks, but very small, under 5; wing little over 2; tail 2¼. California. *Ammodromus samuelis* BD., 455, pl. 71, f. 1; COOP., 191, is the same bird.

Var. INSIGNIS BD., Trans. Chicago Acad. 1869, 319, pl. 29, f. 1.   Plumbeous brown, not rufous, the dorsal streaks fine; beneath, plumbeous white, with almost confluent ashy-brown streaks.   Large; 7; wing and tail 3¼; bill slender and very long, .60.   Kadiak, Alaska.

70. Genus PEUCÆA Audubon.

* Edge and bend of wing yellowish, as in *Coturniculus*, which the species much resembles in the color of the upper parts; but it has no obvious yellowish about the head; the wings are not longer than the tail, and the tail feathers, though very narrow and lanceolate, are not acute at tip.

Bachman's Finch.   Upper parts, including crown, continuously streaked with blackish, dull chestnut and ashy-gray; wing coverts and inner second-aries marked like the back; below, dull brownish-ash, or brownish-gray, whitening on the belly, deepest on sides and across breast, nowhere obviously streaked in adult plumage; some obscure dusky maxillary streaks, some vague dusky markings on auriculars, a slight ashy superciliary line and very obscure median ashy line on crown; bill dark above, pale below; legs very pale; lateral claws falling far short of base of middle claw; hind claw much shorter than its digit; tarsus not longer than middle toe and claw; tail much rounded, with obscure grayish-white area on the lateral feathers. *Young* have the breast and sides evidently streaked.   5½–6; wing 2¼; tail 2⅝.   Southern States.   NUTT., i, 568; AUD., iii, 113, pl. 176. BD., 484. . . . . ÆSTIVALIS.

Var. CASSINII.   Similar; paler and grayer; wings and tail rather longer. Texas, New Mexico and Arizona, and southward.   BD., 485; COOP., 219.

* * Edge and bend of wing without yellow.

Rufous-crowned Finch.   Similar; rather smaller; crown uniform chest-nut, and maxillary streaks sharp, distinct.   California.   CASS., Ill. 135, pl. 20; BD., 486; COOP., 218. . . . . RUFICEPS.

71. Genus POOSPIZA Cabanis.

⁎ Southwestern species, with rounded blackish tail not shorter than the wings, plumbeous-black bill and feet, and few decided streaks, or none.

Black-throated Finch.   Face, chin and throat sharply jet-black; a strong pure white superciliary line, and another bounding the black of the throat; under eyelid also pure white; auriculars dark slate; no yellow anywhere.   Below, pure white; the sides, flanks and crissum shaded with ashy or fulvous-brownish, but no streaks.   Above, uniform grayish-brown; wings dusky, coverts and inner quills edged with the colors of the back; tail black, with narrow grayish edgings, the outer feather sharply edged and tipped with white and several others similarly tipped.

FIG. 85.   Black-throated Finch.

Small, 5–5½; wing about 2½; tail 2⅞.   In the young the head-markings are obscure, there is little or no black on the throat, and a few pectoral streaks. Texas, New Mexico, Arizona and California. Cass., Ill. 150, pl. 23; Bd., 470; Coop., 203. . . . . BILINEATA.

Bell's Finch. No definite black about head, and edge of wing slightly yellowish. Forehead, line over eye and edges of eyelids, inconspicuously white. Below white, more or less tinged with pale brownish, the sides with slight sparse streaks that anteriorly become aggregated into rather vague maxillary stripes cutting off from the white throat a whitish line that runs from the corner of the bill; lores and circumocular region dusky. Above grayish-brown, ashier on head, the crown and back with small sparse sharp black streaks; wing coverts and inner quills with much fulvous edging; tail black with slight pale edgings, the outer web of the outer feather simply whitish. About 6 long; wing and tail 3. Utah, New Mexico, Arizona and California. Bd., 470; Elliot, pl. 14; Coop., 204. . . . . . BELLII.

72. Genus JUNCO Wagler.

*** Unspotted, unstreaked, the colors in large definite areas; 2–3 outer tail feathers white; bill flesh-colored. Length, 6–7; wing and tail about 3.

Snowbird. Blackish-ash, below abruptly pure white from the breast. In the ♀, and in fact in most fall and winter specimens, the upper parts have a more grayish, or even a decidedly brownish cast, and the inner quills are edged with pale bay. One of our most abundant and familiar sparrows, in flocks, from October to April; retires to high latitudes and mountains, to breed. Chiefly Eastern; but also found in Alaska (*Dall*), Washington Territory (*Suckley*) and Arizona (*Coues*). A western variety has the middle and greater wing coverts tipped with white, forming two conspicuous crossbars. Wils., ii, 129, pl. 16, f. 6; Nutt., i, 491; Aud., iii, 88, pl. 187; Bd., 468. . . . . HYEMALIS.

Oregon Snowbird. Head and neck all round, and breast, black; middle of back dull reddish-brown, and wings much edged with the same; below from the breast abruptly white, tinged on the sides with pale reddish-brown. In the ♀ and young the black is obscured by brownish, but the species may always be distinguished by an evident contrast in color between the interscapulars and head, and the fulvous wash on the sides. Rocky Mountains to the Pacific; as abundant there as *hyemalis* is with us. Aud., iii. 91, pl. 168; Bd., 466; Coop., 199. . . . . OREGONUS.

Cinereous Snowbird. Clear grayish-ash, fading rather gradually into white on belly; interscapulars abruptly, definitely, chestnut or rusty-brown; lores blackish; no fulvous wash on sides; no chestnut on wings. Rocky Mountains, U. S., and southward. Bd., 467, 468; Coop., 201; Coues, Proc. Phila. Acad. 1866, 50. . . . . . . CINEREUS var. CANICEPS.

Obs. The true *cinereus*, a Mexican bird, has the wing coverts edged with rusty like the back, the bill black and yellow. *Junco dorsalis* Henry, Proc. Phila. Acad. 1858, 117, is *caniceps* approaching *cinereus*—perhaps a hybrid. *J. annectens* Bd., in Coop., 564, based on specimens I procured in Arizona, is intermediate between *oregonus* and *caniceps*—in all probability a hybrid. See Coues, Proc. Acad.

Phila. 1866, p. 50. Specimens intermediate between *oregonus* and *hyemalis* have been instanced (RIDGWAY, Proc. Phila. Acad. 1869, 126), and all these forms of the genus, in fact, seem to be nascent species, still unstable in character; but the modification of the *Junco* stock has passed the merely varietal stage.

73. Genus SPIZELLA Bonaparte.

*** Small species, 5–6 inches long, with the long, broad-feathered, forked tail about equalling the rather pointed wings, with no yellowish anywhere, and no streaks on the under parts *when adult.*

* Species with the crown *of the adult* chestnut.

Tree Sparrow. Bill black above, yellow below; legs brown, toes black; no black on forehead; crown chestnut (in winter specimens the feathers usually skirted with gray), bordered by a grayish-white superciliary and loral line, and some vague chestnut marks on sides of head; below impurely whitish, tinged with ashy anteriorly, washed with pale brownish posteriorly, the middle of the breast with an obscure dusky blotch; middle of back boldly streaked with black, bay and flaxen; middle and lesser wing coverts black, edged with bay and tipped with white, forming two conspicuous cross-bars; inner secondaries similarly variegated, other quills and tail feathers dusky, with pale edges. A handsome sparrow, the largest of the genus, at least 6 inches long; the wing and tail almost 3; abundant in the United States in winter, flocking in shrubbery; breeds in mountainous and boreal regions. WILS., ii, 12, pl. 12, f. 3; NUTT., i, 2d ed. 572; AUD., iii, 83, pl. 166; BD., 472; COOP., 206. . . . . MONTICOLA.

Chipping Sparrow. Chipbird or Chippy. Hairbird. Adult : bill black; feet pale; crown chestnut, extreme forehead black, a grayish-white super-

FIG. 86. Chipping Sparrow.

ciliary line, below this a blackish stripe through eye and over auriculars. Below, a variable shade of pale ash, nearly uniform and entirely unmarked; back streaked with black, dull bay and grayish-brown, inner secondaries and wing coverts similarly variegated, the tips of the greater and lesser coverts forming whitish bars; rump ashy, with slight blackish streaks; primaries and tail dusky with paler edges. Smaller; 5–5½; wing about 2⅔; tail rather less. Sexes alike, but very young birds are quite different, the crown being streaked like the back, the breast and sides thickly streaked with dusky, the bill pale brown and the head lacking definite black. In this stage, which, however, is of brief duration, it resembles some other species, but may be known by a certain ashiness the others lack, and from the small sparrows that are streaked below when adult, by its generic characters. North America, extremely abundant, and the most familiar species about houses, in gardens, and elsewhere, nesting in shrubbery; nest of fine dried grass, lined with hair; eggs 4–5, bluish, speckled. WILS., ii, 127, pl. 16, f. 5; NUTT., i, 497; AUD., iii, 80, pl. 165; BD., 473. . SOCIALIS.

Var. ARIZONÆ COUES, *n. v.* Like an immature *S. socialis.* Paler than this species, the ashiness in great measure brown ; crown grayish-brown streaked with dusky like the back, and showing evident traces of rich chestnut, but never becoming wholly chestnut ; black frontlet lacking, and no definite ashy superciliary line, the sides of the crown merely lighter brown ; bill brown above, pale below. Arizona, and probably other portions of the same region. A curious form, as it were an arrested stage of *socialis.* Some specimens, with the least chestnut on the head, look remarkably like *pallida* var. *brewerii,* but this last is evidently smaller, without chestnut on the head, and otherwise different.

Field Sparrow. Bill pale reddish ; feet very pale ; crown dull chestnut ; no decided black or whitish about head. Below white, unmarked, but much washed with pale brown on breast and sides ; sides of the head and neck with some vague brown markings ; all the ashy parts of *socialis* replaced by pale brownish. Back bright bay, with black streaks and some pale flaxen edgings ; inner secondaries similarly variegated ; tips of median and greater coverts forming decided whitish cross-bars. Size of *socialis,* but more nearly the colors of *monticola;* sexes alike ; young for a short time streaked below, as in *socialis.* Eastern United States, very abundant in fields, copses and hedges, in flocks when not breeding. WILS., ii, 121, pl. 16, f. 2 ; AUD., iii, 77, pl. 164 ; NUTT., i, 499 ; BD., 473.   . . . . .   PUSILLA.

** Western species, with the crown not chestnut, and streaked like the back.

Clay-colored Sparrow. Crown and back clay-colored or flaxen, distinctly streaked with black, without evident bay, the dorsal streaks noticeably separated from those of the crown, by an ashier, less streaked, cervical interval ; rump brownish-gray. Crown divided by a pale median stripe ; a distinct whitish superciliary line ;' loral and auricular regions decidedly brown ; wing coverts and inner secondaries variegated like the back. Below, white, soiled with clay-color. Bill and feet pale. Small ; $4\frac{3}{4}$–$5\frac{1}{4}$; wing and tail, each, $2\frac{1}{4}$. Central region of the United States into British America. *Emberiza pallida,* SWAINSON, Fauna Bor.-Am. ii, 251 ; *E. shattuckii* AUD., vii, 347, pl. 493. BD., 474.   . . . .   PALLIDA.

Var. BREWERII. Similar ; paler and duller, all the markings indistinct ; streaks of crown and back small, numerous, not separated by a cervical interval ; no definite markings on sides of head. Size of the last, but tail relatively longer, rather exceeding the wings — about $2\frac{3}{4}$ long, thus equalling, if it does not somewhat exceed, that of *socialis,* although the latter is a larger bird. It bears an extraordinary resemblance to the curious western variety of *socialis,* above described ; but in this, as in typical *socialis,* the tail is appreciably shorter than the wings. Southwestern U. S. *Emberiza pallida* AUD., iii, 71, pl. 161. *S. breweri* CASS., Proc. Phila. Acad. 1856, 40 ; BD., 475 ; COOP., 209.

*** Western species, with the crown of the adult dark ash.

Black-chinned Sparrow. Dark ash, fading insensibly into whitish on the belly, deepening to black on the face and throat ; interscapulars bright bay, streaked with black ; wing coverts and inner secondaries variegated with the same colors ; tail blackish, with pale edgings ; bill reddish, feet dark. A small species, but measuring full 6 long, on account of the great length of the tail (fully 3), which greatly exceeds the wings ($2\frac{1}{2}$) ; the young lack the black on the face, and have the crown washed with ashy-brown, and the bill dusky above ; but may be known by the length of the tail.   Mexico, north to Colorado Valley ; Cape St. Lucas.   (*S. atrimentalis* COUCH.   *S. cana* BD.   *S. evura* COUES, Ibis, 1865, 118, 164.) BD., 476; COUES, Proc. Phila. Acad., 1866, 87; COOP., 210. . . . . ATRIGULARIS.

74. Genus ZONOTRICHIA Swainson.

FIG. 87. Black-chinned Sparrow.

** Embracing our largest and handsomest sparrows, $6\frac{1}{4}$ to $7\frac{1}{4}$ inches long, the rounded wings and tail each 3 or more ; the under parts with very few streaks, or none, the middle of the back streaked, the rump plain, the wings with two white cross-bars, and the head of the adults with black.

White-throated Sparrow.   Peabody-bird.   Adult ♂ with the crown black, divided by a median white stripe, bounded by a white superciliary line and *yellow spot* from nostrils to eye ; below this a black stripe through the eye ; below this a maxillary black stripe bounding the definitely pure white throat, sharply contrasted with the dark ash of the breast and sides of the neck and head. *Edge of wing yellow.*   Back continuously streaked with black, chestnut and fulvous-white ; rump ashy, unmarked.   Wings much edged with bay, the white tips of the median and greater coverts forming two conspicuous bars ; quills and tail feathers dusky, with pale edges.   Below, white, shaded with ashy-brown on sides, the ash deeper and purer on the breast ; bill dark, feet pale.   ♀ , and immature birds, with the black of the head replaced

FIG. 88. White-throated Sparrow.

by brown, the white of the throat less conspicuously contrasted with the duller ash of surrounding parts, and frequently with obscure dusky streaks on the breast and sides ; but the species may always be known by the yellow over the eye and on the edge of the wing (these never being imperceptible), coupled with the large size and the general characters above given.   A fine sparrow, abundant throughout Eastern North America in all situations, generally in flocks, except when breeding ; a pleasing if not brilliant songster.   WILS., iii, 51, pl. 22, f. 2 ; NUTT., i, 481 ; AUD., iii, 153, pl. 191 ; BD., 463. . . . . ALBICOLLIS.

White-crowned Sparrow.   Adults of both sexes with the crown pure white, enclosing on either side a broad black stripe that meets its fellow on the forehead and descends the lores to the level of the eyes, and bounded by another narrow black stripe that starts behind the eye and curves around the side of the hind head, nearly meeting its fellow on the nape ; edge of under eyelid white.   Or, we may say, crown black, enclosing a median white stripe and two lateral white stripes, all confluent on the hind head.

General color a fine dark ash, paler below, whitening insensibly on chin and belly, more brownish on the rump, changing to dull brownish on the flanks and crissum, the middle of the back streaked with dark purplish-bay and ashy-white. No bright bay, like that of *albicollis*, anywhere, except some edging on the wing coverts and inner secondaries; middle and greater coverts tipped with white, forming two bars; no yellow anywhere; bill and feet reddish. Young birds have the black of the head replaced by very rich warm brown, the white of the head by pale brownish, and the general ash has a brownish suffusion, and the back is more like that of *albicollis;* but the two species can hardly be confounded. North America, especially eastern and rather northerly, not nearly so abundant in the United States as *albicollis*, but common in many sections in winter and during the migrations. WILS., iv, 49, pl. 31, f. 4; NUTT., i, 479; AUD., iii, 157, pl. 192; BD., 458; COOP., 196. . . . . LEUCOPHRYS.

FIG. 80. White-crowned Sparrow.

Var. GAMBELII. Exactly like the last, but the lores gray or ashy, continuous with the white stripe over the eye, *i. e.*, the black of the forehead does not descend to the eye. Perhaps averaging a trifle smaller, and duller colored. Mr. Allen tells me he has seen specimens that resembled *leucophrys* on one side of the head, and *gambelii* on the other! Rocky Mountains to the Pacific, there mostly replacing true *leucophrys*. NUTT., 2d ed. i, 556; BD., 460; COOP., 195.

Golden-crowned Sparrow. Adults of both sexes with the forehead and sides of the crown black, enclosing a dull yellow coronal patch; edge of the wing yellow. Above, much like *albicollis*, but with less bay; below, much like *leucophrys*, but the ashy not so pure; larger than either. Young have the black of the crown replaced by brown, but there are always traces of the yellow on crown and wings. Pacific coast (to the Rocky Mountains?), abundant. *Fringilla atricapilla* AUD., iii, 162, pl. 193; *F. aurocapilla* NUTT., 2d ed. i, 555. BD., 461; COOP., 197. . . . . . CORONATA.

Harris's Sparrow. Adult ♂ with whole crown, face and throat jet-black; sides of the head pale ash, the auriculars darker ash, bounded by a black line starting behind the eye and curving around them. Under parts nearly pure white, but slightly ashy before and faintly brownish-washed behind, the sides with a few dusky streaks, the breast with a few black ones continued from the black throat-patch; back nearly as in *coronata;* bill and toes dark, tarsi pale; no yellow anywhere; very large, tail about 3½; ♀ similar, but with much less black on head and throat. This and *coronata* represent the maximum dimensions above given, while the other two species are at or near the minimum. Missouri region; a bird of imposing appearance — for a sparrow. *F. querula* NUTT., i, 2d ed. 555; *F. harrisii* AUD., vii, 331, pl. 484; BD., 462. . . . . QUERULA.

OBS. Morton's finch, *F. mortonii* AUD., iii, 151, is a South American species of this genus (*Z. matutina*), erroneously attributed to California.

75. Genus CHONDESTES Swainson.

Lark Finch. Head curiously variegated with chestnut, black and white; crown chestnut blackening on forehead, divided by a median stripe, and bounded by superciliary stripes, of white; a black line through eye, and another below eye, enclosing a white streak under the eye and the chestnut auriculars; next, a sharp black maxillary stripe not quite reaching the bill, cutting off a white stripe from the white chin and throat. A black blotch on middle of breast. Under parts white, faintly shaded with grayish-brown; upper parts grayish-brown, the middle of the back with fine black streaks. Tail very long, its central feathers like the back, the rest jet-black, broadly tipped with pure white in diminishing amount from the lateral pair inward, and the outer web of the outer pair entirely white; 6½–7; wing 3½, pointed; tail 3, rounded. A beautiful species, abundant from the eastern edge of the prairies to the Pacific; the young differ somewhat, particularly about the head, but the bird is unmistakable in any plumage; the coloration of the tail alone is diagnostic. A sweet songster; nest on the ground, of dried grass; eggs 4–5, white, with straggling zigzag dark lines, as in many *Icteridæ.* AUD., iii, 63, pl. 158; NUTT., i, 480; BD., 456; COOP., 193. . . . . GRAMMACA.

FIG. 90. Lark Finch.

76. Genus PASSER Auctorum.

English Sparrow. Bill shaped much as in the purple finch, with a slight basal ruff; tarsus as long as the middle toe; wings pointed; tail forked a little, ⅔ as long as the wing. ♂ , above, reddish-brown, the back black-streaked, the crown and under parts brownish-ash, the chin and throat black; ♀ lacking the latter marks. A species lately imported from Europe, now thoroughly naturalized, and already abundant in many towns and cities of the Eastern and Middle States, though not yet generally dispersed over the country. It has also been recently introduced into Salt Lake City, where it seems to thrive equally well. It has proved highly beneficial by destroying canker-worms, the pest of our shade trees, and our dusty streets are enlivened with its presence; but if it continues to multiply at the present rate, it must soon overflow municipal limits, and then the results of the contact of this hardy foreigner with our native birds may cause us to regret its introduction, unless it finds natural enemies to check its increase. LAWR., Ann. Lyc. Nat. Hist. N. Y. viii, 1866, 287; Proc. Bost. Soc. 1867, 157; 1868, 389; COUES, Proc. Essex Inst. 1868, 283; ALLEN, Am. Nat. iii, 635; ELLIOT, pl. 12. . . . . DOMESTICUS.

OBS. Two other European finches, the Goldfinch, *Carduelis elegans*, and the Serin finch. *Serinus meridionalis*, are reported from Massachusetts, but believed to have been escaped cage-birds. ALLEN, Am. Nat. iii, 635.

77. Genus PASSERELLA Swainson.

Fox Sparrow. General color ferrugineous or rusty red, purest and brightest on the rump, tail and wings, on the other upper parts appearing as streaks laid on an ashy ground; below, white, variously but thickly marked except on the belly and crissum with rusty red—the markings anteriorly in the form of diffuse confluent blotches, on the breast and sides consisting chiefly of sharp sagittate spots and pointed streaks; tips of middle and greater coverts forming two whitish wing-bars; upper mandible dark, lower mostly yellow; feet pale—the lateral toes so lengthened that the tips of their claws fall far beyond the base of the middle claw; this is a diagnostic feature, not shared by any other streaked sparrow. A large handsome species. 6¾-7¼ long; wing and tail, each, 3 or more; sexes alike, and young not particularly different. Eastern North America, abundant. WILS., iii, 53, pl. 24, f. 4; NUTT., i, 514; AUD., iii, 139, pl. 186; BD., 488. *P. obscura* VERRILL, Proc. Bost. Soc. Nat. Hist. ix, 1862, p. 143. . . . . ILIACA.

FIG. 91. Bill of Fox Sparrow.

Var. TOWNSENDII. With the same size and pattern of coloration, but darker; above, continuous olive-brown, with a rusty shade; rump, tail and wings rather brighter; no whitish wing-bars; below, the markings of the color of the back, close, and illy defined. Pacific coast. AUD., iii, 143, pl. 187; NUTT., i, 2d ed. 533; BD., 489; COOP., 221.

Var. SCHISTACEA. Similar to the last; above continuous slate-gray, with little rufous on wings and tail, the spots below slate-colored, sparse, small, sharp. Rocky Mountain region, U. S. BD., 490, 925, pl. 69, f. 3, 4 (*megarhynchus*—large-billed form from Cala.); COOP., 222.

OBS. *P. schistacea* and *townsendii* are certainly not distinct specifically from each other, but it may be a question whether they do not form two races of a species different from *iliaca.* In all three cases, however, the difference is solely in the relative intensity and predominance of certain common colors; and although the Western and Eastern forms may not have been shown to intergrade, they differ from each other less than some of the recognized varieties of *Melospiza* do from the Eastern song sparrow, and in a parallel manner.

78. Genus CALAMOSPIZA Bonaparte.

Lark Bunting. White-winged Blackbird. ♂ entirely black, with a large white patch on the wings, and the quills and tail feathers frequently marked with white; bill dark horn above, paler below; feet brown; 6-6½; wing 3½; tail 2¾. Sexes unlike: ♀ resembling one of the sparrows, brown above, streaked, white below, somewhat streaked, but always known by the whitish wing-patch; ♂ said to wear the black plumage only during the breeding season, like the bobolink (*Allen*). In the form of the bill, this interesting species is closely allied to the grosbeaks (*Goniaphea*); and this, with the singularly enlarged tertiaries, as long as the primaries in the closed wing, renders it unmistakable in any plumage. A prairie bird, abundant on the western plains, to the Rocky Mountains; Cape St. Lucas. Aud., iii, 195, pl. 201; Nutt., 2d ed. 1, 303; Bd., 492; Coop., 225.   . bicolor.

79. Genus EUSPIZA Bonaparte.

Black-throated Bunting. ♂ above grayish-brown, the middle of the back streaked with black, the hind neck ashy, becoming on the crown yellowish-olive with black touches ; a yellow superciliary line, and maxillary touch of the same ; eyelid white ; ear coverts ashy like the cervix ; chin white ; throat with a large jet-black patch ; under parts in general white, shaded on the sides, extensively tinged with yellow on the breast and belly ; edge of wing yellow ; lesser and middle coverts rich chestnut, other coverts and inner secondaries edged with paler ; bill dark horn blue, feet brown ; 6½–7 ; wing 3½, sharp-pointed ; tail 2¾, emarginate. ♀ smaller ; above, like the ♂ , but head and neck plainer ; below, less tinged with yellow, the black throat patch wanting and replaced by sparse sharp maxillary and pectoral streaks, wing coverts not chestnut. An elegant species, of trim form, tasteful colors and very smooth plumage, abundant in the fertile portions of the Eastern U. S. ; N. to Connecticut Valley ; W. to Kansas and Nebraska. Wils., iii, 86, pl. 3, f. 2 ; Nutt., i, 461 ; Aud., iii, 58, pl. 156 ; Bd., 494. americana.

Townsend's Bunting. "Upper parts, head and neck all round, sides of body and forepart of breast, slate-blue ; back and upper surface of wings tinged with yellowish-brown ; interscapulars streaked with black ; super-ciliary and maxillary line, chin and throat and central line of under parts from breast to crissum, white ; edge of wing, and gloss on breast and middle of belly, yellow ; a black spotted line from lower corner of lower mandible down the side of the throat, connecting with a crescent of streaks in the the upper edge of the slate portion of the breast." Bd., 495. Aud., iii, 62, pl. 157 ; Nutt., i, 2d. ed. 528. Pennsylvania ; one specimen known, a standing puzzle to ornithologists, in the uncertainty whether it is a good species or merely an abnormal plumage of the last. . . . townsendii.

80. Genus GONIAPHEA Bowditch.

⁎⁎⁎ Embracing large species, of beautiful and striking colors, the sexes dissimilar. Bill extremely heavy, with the lower mandible as deep as the upper or deeper, the commissural angle strong, far in advance of the feathered base of the bill, the rictus overhung with a few long stiff bristles. Brilliant songsters.

Rose-breasted Grosbeak. Adult ♂ with the head and neck all round and most of the upper parts black, the rump, upper tail coverts and under parts white, the breast and under wing coverts exquisite carmine or rose-red ; wings and tail black, variegated with white ; bill pale, feet

Fig. 92. Rose-breasted Grosbeak.

dark. ♀ above, streaked with blackish and olive-brown or flaxen-brown, with median white coronal and superciliary line ; below, white, more or less tinged with fulvous and streaked with dusky; *under wing coverts saffron-yellow*; upper coverts and inner quills with a white spot at end; bill brown. Young ♂ at first resembling the ♀. 7½–8½; wing about 4; tail about 3½. Eastern United States, common. Wils., ii, 135, pl. 17, f. 2; Nutt., i, 527; Aud., iii, 209, pl. 205; Bd., 497. . . . . LUDOVICIANA. ╪

Black-headed Grosbeak. Adult ♂ with the crown and sides of head, back, wings and tail black; the two latter variegated with white blotches; neck all around and under parts rich orange-brown, changing to bright yellow on the belly and under wing coverts; bill and feet dark horn. Size of the last; the ♀ and young differ much as in the last species, but may be recognized by the *rich sulphur-yellow* under wing coverts; the bill is shorter and more tumid. Plains to the Pacific, United States; common. Aud., iii, 211, pl. 206; Bd., 498; Coop., 228. . . . . . MELANOCEPHALA. ╪

Blue Grosbeak. Adult ♂ rich dark blue, uniform; feathers around base of bill, wings and tail, black; middle and greater wing coverts tipped with chestnut; bill dark horn, feet blackish; 6½–7; wing 3½; tail 3. ♀ smaller, plain warm brown, paler and rather flaxen below, wings with whitey-brown cross-bars, bill and feet

FIG. 93. Blue Grosbeak.

brown. Young ♂ at first like ♀; when changing, shows confused brown and blue, afterward blue interrupted with white below. United States, ratherly southerly, but N. to Massachusetts, and even Maine (*Boardman*). Wils., iii, 78, pl. 21, f. 6; Nutt., i, 529; Aud., iii, 204, pl. 204; Bd., 499; Coop., 230. . . . . CÆRULEA.

81. Genus CYANOSPIZA Baird.

Painted Finch. Nonpareil. Adult ♂ with the head and neck rich blue, the rump, eyelids and under parts intense red, the lores, back and wings glossed with golden-green, the tail purplish-blue. ♀ above plain greenish, below yellow; young ♂ at first like the ♀. 5½; wing 2⅞; tail 2½. South Atlantic and Gulf States, common; an exquisite little creature of matchless hues. Wils., iii, 68, pl. 24, f. 1, 2; Aud., iii, 93, pl. 169; Nutt., i, 477; Bd., 563. . . . . CIRIS.

Western Nonpareil. Adult ♂ with the forehead, cervix, bend of wing and rump purplish-blue, throat and hind head dusky red, belly reddish-purple, wings and tail dusky glossed with blue; ♀ "similar to that of *C. amœna*, but distinguished by the absence of the two white bands on wings, and by the legs being black." Size of the last. Mexico; Cape St. Lucas. Bd., 503; Coop., 234. . . . . VERSICOLOR.

Lazuli Finch. Adult ♂ lazuli-blue, obscured on the back, the lores black, the breast definitely brown, the rest of the under parts, and the wing-bands, white; tibiæ blue; bill and feet dark. Size of the first. ♀ plain brown above, whitish below, the breast browner, the wings with whitish bars. Plains to the Pacific, replacing *cyanea*; common. NUTT., i, 478 ; AUD., iii, 100, pl. 171 ; BD., 504 ; COOP., 233. . . . . . AMŒNA.

Indigo bird. Adult ♂ indigo-blue, intense and constant on the head, glancing greenish with different lights on other parts ; wings and tail blackish glossed with greenish-blue ; feathers around base of bill black ; bill dark above, rather paler below, with a curious black stripe along the gonys. ♀ above plain warm brown, below whitey-brown, obsoletely streaky on the breast and sides, wing coverts and inner quills pale-edged, but not whitish ; upper mandible blackish, lower pale, with the black stripe just mentioned — this is a pretty constant feature, and will distinguish the species from any of our little brown birds. Young ♂ is like the ♀, but soon shows blue traces, and afterward is blue with white variegation below. Size of the foregoing. Eastern United States, abundant, in fields and open woodland, in summer ; a well meaning but rather weak vocalist. WILS., i, 100, pl. 6, f. 5 ; NUTT., i, 473 ; AUD., iii, 96, pl. 170 ; BD., 505. . . . . CYANEA.

82. Genus SPERMOPHILA Swainson.

Morelet's Finch. Top and sides of head, back of neck, broad band across upper part of breast, middle of back, wings and tail, black ; chin, upper throat and neck all round, except behind, rump, and remaining under parts, white, the latter tinged with brownish-yellow ; two wing-bands, and concealed bases of all the quills, also white. ♀ olivaceous above, brownish-yellow below, wings and tail somewhat as in the ♂.

FIG. 94. Morelet's Finch.

Length about 4 inches ; wing 2 ; tail less. Mexico to Texas. BD., 507. *S. albigularis* LAWR., Ann. Lyc. Nat. Hist., v, 1851, 124. . . . . MORELETII.

82bis. Genus PHONIPARA Bonaparte.

Black-faced Finch. ♂ dark olive green, paler below, grayish-white on the belly ; head, throat and breast more or less blackish ; wings and tail dusky, unmarked, with olivaceous gloss ; upper mandible blackish, lower pale. The ♀ lacks the black of the ♂, but is otherwise similar. About 4 inches long ; wing 2 ; tail 1¾. A West Indian bird, the occurrence of which, in Florida, I learn from advance sheets of Mr. C. J. Maynard's work on the Birds of Florida, now publishing. (Not in the Key.) . . . . BICOLOR.

83. Genus PYRRHULOXIA Bonaparte.

Texas Cardinal. Conspicuously crested, and otherwise like the common cardinal in form, but the bill extremely short and swollen. ♂ ashy-brown, paler below ; the crest, face, throat, breast and middle line of the belly, with the wings and tail, more or less perfectly crimson or carmine red ; bill whitish. ♀ similar, rather brownish-yellow below, with traces of the red on the breast and belly.

FIG. 95. Texas Cardinal.

Length about 8½; wing 3¾; tail 4½. Mexico, Texas, Arizona, Cape St. Lucas. Cass., Ill. 204, pl. 33; Bd., 508; Coop., 236. . . . SINUATA.

84. Genus CARDINALIS Bonaparte.

Cardinal Red-bird. Virginia Nightingale. Conspicuously crested; tail longer than the wings, both rounded. ♂ rich vermilion or rosy red, obscured with ashy on the back, face black, bill reddish, feet brown. ♀ ashy-brown, paler below, with evident traces of the red on the crest, wings, tail and under parts. Length 8–9; wing about 3¾; tail 4; ♀ rather less than the ♂. Eastern United States, somewhat southern, seldom north to the Connecticut Valley; a bird of striking appearance and brilliant vocal powers, resident in thickets and undergrowth, abundant. Its rolling notes recall those of the Carolina wren, but are stronger. Wils., ii, 38, pl.

FIG. 96. Cardinal Red-bird.

6, f. 1, 2; Nutt., i, 519; Aud., iii, 198, pl. 203; Bd., 509. VIRGINIANUS.

Var. IGNEUS. Like the last, but paler, with the black frontlet interrupted at the base of the culmen, where the red comes down to the bill. Cape St. Lucas; Colorado Valley. Bd., Proc. Phila. Acad. 1859, 305; Elliot, pl. 16; Coop., 238.

85. Genus PIPILO Vieillot.

* Colors of the male black, white and chestnut in definite areas.

† No white on the scapulars or wing coverts. Sexes very unlike.

Towhee Bunting. Marsh Robin. Chewink. (PLATE II, figs. 17, 18, 17a, 18a.) Adult male black, belly white, sides chestnut, crissum fulvous brown; primaries and inner secondaries with white touches on the outer webs; outer tail feather with the outer web and nearly the terminal half of the inner web, white, the next two or three with white spots decreasing in size; bill blackish, feet pale brown, iris red in the adult, white or creamy in the young, and generally in winter specimens; ♀ rich warm brown where the ♂ is black, otherwise similar. Very young birds are streaked brown and dusky above, below whitish tinged with brown and streaked with dusky; but this plumage, corresponding to the very early speckled condition of thrushes and warblers, is of brief duration; sexual distinctions may be noted in birds just from the nest, and they rapidly become much like the adults. ♂ 8½; wing 3¼, much rounded; tail 4; ♀ rather less. Eastern United States, an abundant and familiar inhabitant of thickets, undergrowth and briery tracts, spending much of its time on the ground, scratching among fallen leaves; migratory. Nest on the ground, bulky, of leaves, grasses and other fibrous material; eggs 4–5, white, thickly speckled with reddish. Wils., vi, 90, pl. 53; Nutt., i, 515; Aud., iii, 167, pl. 195; Bd., 512. . . . . ERYTHROPHTHALMUS.

Var. ALLENII COUES, Am. Nat. 1871, 366. Similar; smaller; less white on the wings and tail; claws longer; iris white. Florida. *P. leucopis* MAYNARD, Birds of Florida (in press).

† † Scapulars and wing coverts with white spots; sexes more alike.

Spotted Towhee. A Mexican species. . . . . MACULATUS.

Represented in the United States by the following varieties : —

Var. OREGONUS. *Oregon Towhee.* Very similar to *erythrophthalmus;* wing coverts with small rounded, and scapulars with larger oval, white spots on the outer web of the feathers near the end ; white marks on the quills very small or wanting ; white spots on tail feathers very small, the outer web of the outer rectrix not white except just at the end. Excepting these particulars, this variety looks more like *erythrophthalmus*, than like the typical *maculatus*, in which the body colors are *olivaceous;* nevertheless, it shades into the latter. On the other hand, *erythrophthalmus*, which might seem to be merely the extreme link in the chain, may be fairly considered a different bird ; its sexes are very unlike, whereas in the western black Pipilos the ♀ is blackish-brown, more like the ♂ ; its note is entirely different, the words "towhee" and "chewink" being an attempt to imitate the sound, while the cry of the western varieties of *maculatus* is exactly like the scolding mew of a catbird.—Pacific coast. *Pipilo oregonus* BELL, Ann. Lyc. Nat. Hist. N. Y. v, 1852, 6 ; BD., 513 ; COOP., 241. *P. arcticus* AUD., iii, 164, pl. 194.

Var. ARCTICUS. *Arctic Towhee.* Similar to the foregoing; the white spots of the wing coverts larger, those of the scapulars still larger and lengthening into streaks, the interscapulars also spotted with white ; the white on the quills and tail feathers at a maximum, as in *erythrophthalmus;* there are usually, also, concealed white specks in the black of the throat. ♀ comparatively dark. Central region of N. A. *P. arcticus* SWAINSON, Fauna Bor.-Am. 1831, 11, 260. NUTT., i, 589 ; 2d ed. i, 610 ; BD., 514.

Var. MEGALONYX BD., 515, pl. 73 ; COOP., 242, is the prevailing form in the Southern Rocky Mountain region, New Mexico, Arizona, and California. It is precisely like *arcticus*, but the feet are larger, with highly developed claws; the hind claw is decidedly longer than its digit, while the lateral claws reach to or beyond the middle of the middle claw. In this form at any rate, the ♀ is hardly distinguishable in color from the ♂, being blackish with an appreciable olivaceous shade, thus exhibiting an approach to the typical Mexican stock. (See COUES, Proc. Phila. Acad. 1866, 89 ; ALLEN, Bull. Mus. Comp. Zool., iii.)

* * Colors not definitely black, white and chestnut ; no greenish ; sexes alike.

Brown Towhee. *Cañon Towhee.* Above, uniform grayish-brown with a slight olivaceous shade, the crown brown in appreciable contrast ; wings and tail like the back, unmarked ; below, a paler shade of the color of the back, whitening on the belly, tinged with fulvous and streaked with dusky on the throat and breast, washed with rusty brown on the flanks and crissum. 8½ ; wing 3¾ ; tail 4¼. New Mexico, Arizona, and southward. This is the *P. mesoleucus* BD., 518 ; COUES, Proc. Phila. Acad. 1866, 90 ; COOP., 247, which is *P. fuscus* SWAINSON, Philos. Mag. 1827, 434, of Mexico. FUSCUS.

Var. ALBIGULA. Exactly like the last, but the white of the under parts extending further up the breast, the gular spots more restricted, sparser, and better defined. Cape St. Lucas. BD., Proc. Phila. Acad. 1859, 305 ; ELLIOT, pl. 1 ; COOP., 248.

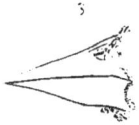

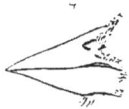

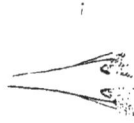

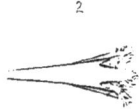

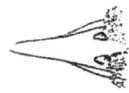

Var. CRISSALIS. Similar to the first; crown like the back; rather darker above, decidedly so below, the middle of the belly scarcely or not whitening, the gular fulvous strong and, with its dusky streaks, definitely restricted to the throat; the flanks and crissum chestnut or deep cinnamon brown. Upwards of 9 inches long; wing 4; tail 5; ♀ rather less. Coast region of California (and northward?), abundant. This is the dark coast form, bearing the same relation to *fuscus* (*mesoleucus*) that the coast *Harporhynchus redivivus* bears to the paler *H. lecontii* of the interior. It is the *P. fuscus* of CASS., Ill. 124, pl. 17; BD., 517; COOP., 215, but not the true *fuscus* of SWAINSON; and its earliest name appears to be *Fringilla crissalis* VIGORS, Zool. Voy. Blossom, 1839, 19.

Abert's Towhee. Somewhat similar to the foregoing species of this section; no decided markings anywhere. Dull brown, paler and more fulvous below, *the face dusky;* otherwise nearly uniform. Very large, 9; wing 4 or less; tail 5 or more. New Mexico and Arizona; abundant in the Colorado Valley; COUES, Pr. Phila. Acad. 1866, 90. BD., 516; COOP., 244. ABERTII.

*** Colors greenish; sexes alike.

Green-tailed, or Blanding's Finch. Above dull olive-green, brighter on the wings and tail, crown chestnut, forehead blackish, edge of wing yellow; chin and throat pure white, bounded by dusky maxillary stripes, and contrasting with the dark ash of the breast and sides of head and neck (very much as in the white-throated sparrow); this ash fades to white on the belly; the flanks and crissum are washed with dull brownish; bill dark horn, feet brown; about 7¼ long; wing 3¼; tail rather more. In the young the markings, especially of the head and throat, may be obscure, but the species is unmistakable. Rocky Mountain region, U. S. and southward, northeast to Kansas; abundant. AUD., Orn. Biog. v, 339; CASS., Ill., 70, pl. 12; BD., 519; COOP., 250. . . . . CHLORURUS.

86. Genus EMBERNAGRA Lesson.

Green Finch. "Above uniform olivaceous green; sides of the hood and a stripe behind eye, dull brownish rufous, not very conspicuous; an ashy superciliary stripe, rather yellowish anteriorly; under parts brownish-white, tinged with yellowish anteriorly, and with olivaceous on the sides, white in the middle of the belly; edge of wing, under coverts and axillaries, bright yellow. Length 5½; wing 2⅔; tail 2¾. Valley of the Rio Grande, and probably of the Gila, and southward." BD., 487. . . . RUFIVIRGATA.

Family ICTERIDÆ. American Starlings.

A family of moderate extent, confined to America, where it represents the *Sturnidæ*, or Starlings, of the Old World. It is nominally composed of a hundred and fifty species, half of which may prove valid, distributed among fifty genera or subgenera, of which one-fourth may be considered worthy of retention. The relationships are very close with the *Fringillidæ* on the one hand; on the other, they grade toward the crows (*Corvidæ*). They share with the fringilline birds the characters of angulated commissure and nine developed primaries, and this distinguishes them from all our other families whatsoever; but the distinctions from the *Fringillidæ* are not easily expressed. In fact, I know of no character that, for example, will relegate the bobolink and cowbird to the *Icteridæ* rather than to the *Fringillidæ*, in the current acceptation of these terms. In general, however, the *Icteridæ* are distinguished by the length, acuteness and not strictly conical shape of the unnotched, unbristled bill, that shows a peculiar extension of the culmen on the forehead, dividing the prominent antiæ (p. 29, § 52) of close-set, velvety feathers that reach to or on the nasal scale.

Among our comparatively few species are representatives of each of the three subfamilies into which the group is conveniently and probably naturally divisible. In most of them, black is predominant, either uniform and of intense metallic lustre, or contrasted with masses of red or yellow. In nearly all, the sexes are conspicuously dissimilar, the female being smaller, and plain brownish or streaky in the iridescent black species, olivaceous or yellowish in the brilliantly colored ones. All are migratory in this country.

Subfamily *AGELÆINÆ*. Marsh Blackbirds.

Gregarious, granivorous species, more or less completely terrestrial, and chiefly palustrine, not ordinarily conspicuous vocalists, building rather rude, not pensile, nests, laying 4 – 6 spotted or curiously limned eggs. With the feet strong, fitted both for walking and for grasping swaying reeds, the wings more or less pointed, equalling or exceeding the tail in length, the bill conic-acute, shorter or little longer than the head, its cutting edges more or less inflected. *₄* In gen. 87, 91, the tail feathers are acute; in 91, the wing is unusually rounded for this family; in 87, 88, the bill has an ordinary fringilline character.

87. Genus DOLICHONYX Swainson.

Bobolink; Northern States. *Reedbird;* Middle States. *Ricebird;* Southern States.   ♂, in spring: black; cervix buff; scapulars, rump and upper tail coverts ashy white; interscapulars streaked with black, buff and ashy;

Fig. 97. Bobolink.

outer quills edged with yellowish; bill blackish horn; feet brown. ♂ in fall, ♀ and young, entirely different in color; yellowish-brown above, brownish-yellow below, crown and back conspicuously, nape, rump and sides less broadly, streaked with black; crown with a median and lateral light stripe; wings and tail blackish, pale-edged; bill brown. The ♂ changing shows confused characters of both sexes; but in any plumage the species may be recognized by the stiflish, extremely acute tail feathers, in connection with these dimensions; length 6½–7½; wing 3½–4; tail 2½–3; *tarsus about* 1; *middle toe and claw about* 1¼. Eastern United States, very abundant. In its black livery, only worn for a short time, the "bobolink" is dispersed over the meadows of the Northern States to breed, and is a voluble, spirited songster. After the midsummer change, the "reedbird" or "ricebird" throngs the marshes in immense flocks, with the blackbirds; has simply a chirping note, feeds on the wild oats, becomes extremely fat and is accounted a great delicacy. The name "ortolan," applied by some to this bird, by others to the Carolina rail, is a strange misnomer, the ortolan being a fringilline bird of Europe. In the West Indies, where the present species retires in winter, it is called "butterbird." WILS., ii, 48, pl. 12, f. 1, 2; NUTT., i, 185; AUD., iv, 10, pl. 211; BD., 522. . . . . ORYZIVORUS.

88. Genus MOLOTHRUS Swainson.

Cowbird. ♂ iridescent black, head and neck purplish-brown; 7½–8; wing over 4; tail over 3. ♀ 7–7½; wing 3¾; tail 2¾; an obscure looking bird, nearly uniform dusky grayish-brown, but rather paler below, and appearing somewhat streaky, owing to darker shaft lines on nearly all the feathers; bill and feet black in both sexes. The young ♂ at first resembles the ♀, but is decidedly streaked below. North America, abundant; gregarious, polygamous, parasitic. The singular habits of this bird, probably shared by others of the genus, form one of the most interesting chapters in ornithology. Like the European cuckoo, it builds no nest, laying its eggs by stealth in the nests of various other birds, especially warblers, vireos and sparrows; and it appears to constitute, furthermore, a remarkable exception to the rule of conjugal affection and fidelity among birds. A wonderful provision for the perpetuation of the species is seen in its instinctive selection of smaller birds as the foster-parents of its offspring; for the larger egg receives the greater share of warmth during incubation, and the lustier young cowbird asserts its precedence in the nest; while the foster-birds, however reluctant to incubate the strange egg (their devices to avoid the duty are sometimes astonishing) become assiduous in their care of the foundling, even to the neglect of their own young. The cowbird's egg is said to hatch sooner than that of most birds; this would obviously confer additional advantage.—WILS., ii, 145, pl. 18, f. 1, 2, 3; NUTT., i, 178; AUD., iv, 16, pl. 212; BD., 524. . . . . PECORIS.

Var. OBSCURUS. *Dwarf Cowbird.* Similar; smaller; ♂ the size of ♀ *pecoris;* ♀ under 7; wing 3¼; tail 2¼. The difference is very strongly marked, and apparently constant. Arizona, Lower California, and southward. CASS. Proc. Phila. Acad. 1866, 18; COUES, *ibid.*, 90; COOP., 260.

89. Genus AGELÆUS Vieillot.

₊ The ♂ uniform lustrous black, with the bend of the wing red; 8–9 long; wing 4½–5; tail 3½–4. The ♀ everywhere streaked; above blackish-brown with pale streaks, inclining on the head to form median and superciliary stripes; below whitish with very many sharp dusky streaks, the sides of the head, throat, and the bend of the wing, tinged with reddish or fulvous; under 8; wing about 4; tail 3¼. The young ♂ at first like the ♀, but larger, apt to have a general buffy or fulvous suffusion, and bright bay edgings of the feathers of the back, wings and tail, and soon showing black patches.— Upon investigation of the variations in the character of the wing-patch, upon which our three accredited species rested, I am satisfied of the propriety of treating them as varieties of one. The ♀'s are indistinguishable.

Red-winged Blackbird. (PLATE IV, all the figs.) Lesser wing coverts scarlet, broadly bordered by brownish-yellow, or brownish-white, the middle row of coverts being entirely of this color; sometimes the greater row, likewise, are mostly similar, producing a patch on the wing nearly as large as the red one; occasionally, there are traces of red on the edge of the wing and below. In some Eastern specimens the bordering is almost pure white. United States from Atlantic to Pacific, very abundant.   WILS., iv, 30, pl. 30; NUTT., i, 167; AUD., iv, 31, 216; BD., 526. . .   PHŒNICEUS.

Var. GUBERNATOR. Lesser wing coverts scarlet, narrowly or not at all bordered, the next row having black tips for all or most of their exposed portion, so that the brownish-yellow of their bases does not show much, if any. Pacific Coast. NUTT., i, 2d ed., 187; AUD., iv, 29, pl. 215; BD., 529; Coor., 263.

Var. TRICOLOR. Lesser wing coverts dark red, bordered with pure white. California. NUTT., i, 2d ed., 186; AUD., iv, 27, pl. 214; BD., 530; COOP., 265.

90. Genus XANTHOCEPHALUS Baird.

Yellow-headed Blackbird. ♂ black, whole head (except lores), neck and upper breast yellow, and sometimes yellowish feathers on the belly and legs; a large white patch on the wing, formed by the primary, and a few of the outer secondary coverts; 10–11; wing 5½; tail 4½. ♀ and young brownish-

FIG. 98. Yellow-headed Blackbird.

black, with little or no white on the wing, the yellow restricted or obscured; ♀ *much* smaller than the ♂ — 9½, etc. A handsome bird, abundant on the prairies and marshes from Illinois and Wisconsin, westward; N. to 58° and even Greenland (*Reinhardt*). NUTT., i, 176; AUD., iv, 24, pl. 213; BD., 531; COOP., 267; COUES, Am. Nat., 1870, 195. . ICTEROCEPHALUS.

91. Genus STURNELLA Vieillot.

*** Plumage highly variegated ; each feather of the back blackish, with a terminal reddish-brown area, and sharp brownish-yellow borders ; neck similar, the pattern smaller ; crown streaked with black and brown, and with a pale median and superciliary stripe ; a blackish line behind eye ; several lateral tail feathers white, the others, with the inner quills and wing coverts, barred or scalloped with black, and brown or gray. Edge of wing, spot over eye, and under parts generally, bright yellow, the sides and crissum flaxen-brown, with numerous sharp blackish streaks, the breast with a large black crescent (obscure in the young) ; bill horn color, of peculiar shape ; feet light brown, very large and strong, reaching beyond the very short tail. Length 10–11 ; wing 5 ; tail 3½ ; bill ¼ ; ♀ smaller (9½ ; wing 4½ ; tail 3), similar in color ; young not particularly different.

Fieldlark. (PLATE vi, figs. 1, 2, 3, 4, 1*a*, 2*a*, 3*a*, 4*a*.) The colors, as above described, rich and pure, the prevailing aspect brown ; yellow of chin usually confined between rami of under mandible ; black bars on wings and tail usually confluent along the shaft of the feathers, leaving the gray in scallops. Eastern United States, everywhere abundant in open country ; imperfectly migratory ; partially gregarious when not breeding ;

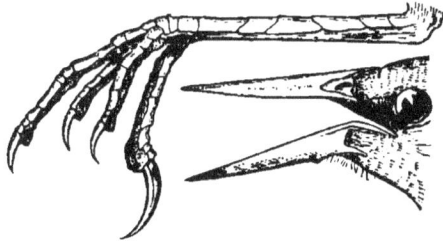

FIG. 99. Fieldlark.

strictly terrestrial ; an agreeable vocalist ; nest of dried grass, on the ground, eggs 4–6, white, speckled with reddish. WILS., iii, 20, pl. 19 ; AUD., iv, 70, pl. 223 ; NUTT., i, 47 ; BD., 535. . . . . . . MAGNA.

Var. NEGLECTA. The colors duller and paler, the prevailing aspect gray ; yellow of chin usually mounting on sides of lower jaw ; black on wings and tail usually resolved into distinct bars alternating with gray bars. Western U. S. Song said to be different. AUD., vii, 339, pl. 487 ; BD., 537 ; COOP., 270.

OBS. It does not appear that the Red-breasted Lark, *Trupialis militaris*, was ever taken in this country. It is a South American species resembling ours, but with red in place of the yellow. BD., 533.

Subfamily ICTERINÆ. *Orioles.*

Non-gregarious, insectivorous and frugivorous species, strictly arboricole, of brilliant or strikingly contrasted colors, and pleasing song, distinguished as architects, building elaborately woven pensile nests. With the bill relatively longer, slenderer and more acute than in most of the last subfamily, and the feet weaker, exclusively fitted for perching. Three of our species are abundant migratory birds in summer ; the rest merely reach our southern border from tropical America.

92. Genus ICTERUS Brisson.

* The ♂ black and chestnut.

Orchard Oriole. ♂ black, lower back, rump, lesser wing coverts and all under parts from the throat, deep chestnut; a whitish bar across tips of greater wing coverts; bill and feet blue-black; about 7; wing 3¼; tail 3; ♀ smaller, plain yellowish-olive above, yellowish below; wings dusky; tips of the coverts and edges of the inner quills, whitish; known from the ♀ of the other species by its small size and very slender bill. Young ♂ at first like ♀, afterward showing confused characters of both sexes; in a particular stage, it has a black mask on the face and throat. Eastern U. S., very abundant in parks, orchards, and the skirts of woods. WILS., i, 64, pl. 4, f. 1, 2, 3, 4; AUD., iv, 46, pl. 219; NUTT., i, 165; BD., 547. . SPURIUS.

Var. AFFINIS. Much smaller; ♂ 6¼; wing under 3. Texas. LAWR. Ann. Lyc. Nat. Hist. N. Y. 1851, 113.

* * The ♂ black and orange.

Baltimore Oriole. Golden Robin. Firebird. Hangnest. ♂ with the head and neck all round, and the back, black; rump, upper tail coverts, lesser wing coverts, most of the tail feathers, and all the under parts from the throat, fiery orange, but of varying intensity according to age and season; middle tail feathers black; wings black, the middle and greater coverts, and inner quills, more or less edged and tipped with white, but the white on the coverts not forming a continuous patch; bill and feet blue-black; 7½-8; wing 3¾; tail 3. ♀ smaller, and much paler, the black obscured by olive, sometimes entirely wanting. Young ♂ entirely without black on throat and head, otherwise colored nearly like the ♀. Eastern United States, west to the mountains of Colorado (*Allen*); abundant, in orchards and streets, as well as in woodland, conspicuous by its brilliant colors and spirited song. WILS., i, 23, pl. 1, f. 3; vi, pl. 53; AUD., iv, 37, pl. 217; NUTT., i, 152; BD., 548. . . . . . BALTIMORE.

FIG. 100. Bullock's Oriole.

Bullock's Oriole. Similar; the orange invading the sides of the head and neck and the forehead, leaving only a narrow space on the throat, the lores, and a line through the eye, black; a large continuous white patch on the wing, formed by the middle and greater coverts. ♀ olive-gray, below whitish, all the fore parts of the body and head tinged with yellow, the wings dusky, with two white bars, but the tail and its under coverts quite yellowish. Young ♂ at first like the ♀, soon however showing black and orange. Rather larger than the last. Western United States, in woodland, abundant, replacing the Baltimore. AUD., iv, 43, pl. 218; BD., 549; COOP., 273; COUES, Am. Nat., v, 1870, 678. . . . . . . BULLOCKII.

Hooded Oriole.   ♂ orange ; wings, tail, a narrow dorsal area and a large mask on the face and throat, black ; tips of wing coverts, and edges of many quills, white ; size of the foregoing, but wings shorter and tail longer.   The ♀ lacks the black mask ; but the species may be distinguished in any plumage from either of the foregoing by not having the wings evidently longer than the tail.   Valley of the Rio Grande and Colorado, Lower California, and southward.   Cass., Ill. 42, pl. 8 ; Bd., 546 ; Coop., 275. cucullatus.

*** The ♂ black and clear yellow.

Scott's Oriole.   ♂ black ; below from the breast, rump and upper tail coverts, lesser, middle and under wing coverts, both above and below, and basal portions of all the tail feathers, except the central ones, clear yellow ; greater wing coverts tipped, inner quills edged, with white ; in the ♀ or young the black is replaced by brownish, and the yellow is not pure ; about 8 ; wing and tail about 4.   Valley of the Rio Grande and Colorado, Lower California, and southward.   Bd., 544 ; Coop., 276.   .   .   .   parisorum.

Audubon's Oriole.   ♂ yellow, somewhat olivaceous on the middle of the back ; head, neck, breast, wings and tail black ; wings with a white cross bar and white edging ; about 9 ; wing 4 ; tail 4½.   Texas, and southward. Cass., Ill., p. 137, pl. 21 ? (apparently represents the Southern smaller true *melanocephalus*) ; Bd., 542.   .   .   .   .   .   (melanocephalus var ?) audubonii.

Obs.   Several additional orioles have been ascribed to this country, but the foregoing are all that appear to have been actually taken within our limits ; others, however, may be confidently expected to occur on our Southern border.

Subfamily QUISCALINÆ.   *Crow Blackbirds.*

Closely resembling the *Agelæinæ* both in structure and in habits, these birds are distinguished by the length and attenuation of the bill, with decidedly curved culmen, especially towards the end, and strongly inflected tomia.   The typical *Quiscali* have a certain crow-like aspect, but they are readily distinguished by several features. The feet are large and strong, and the birds spend much of their time on the ground, where they walk or run instead of advancing by leaps.   They generally build rude, bulky nests, lay spotted or streaked eggs, and their best vocal efforts are hardly to be called musical.   The ♂ of most of the species is uniform lustrous black, the ♀ brown and much smaller.   There is only one genus (*Cassidix*) besides the two of this country ; in 93, the tail is slightly rounded and shorter than the wings ; in 94, the tail is graduated, and about equals or exceeds the wings.

93. Genus SCOLECOPHAGUS Swainson.

Rusty Grackle.   ♂ in summer lustrous black, the reflections greenish, and not noticeably different on the head ; but not ordinarily found in this condition in the U. S. ; in general simply glossy black, nearly all the feathers skirted with warm brown above, and brownish-yellow below, frequently continuous on the fore parts ; the ♂ of the first season, like the ♀, is entirely rusty brown above, the inner quills edged with the same ; a pale superciliary stripe ; below, mixed rusty and grayish-black, the primaries and tail alone black; bill and feet black at all times; ♂ about 9; wing 4½; tail 3½; bill ¾; very slender for the family, somewhat resembling a thrush's; ♀ smaller. Eastern North America, N.W. to Alaska (*Dall*), very common in the U. S. in the fall and winter, in flocks, in fields; breeds in Labrador and other Northern regions, laying speckled, not streaky, eggs. WILS., iii, 41, pl. 21, f. 3; NUTT., i, 199; AUD., iv, 65, pl. 222; BD., 551. . FERRUGINEU S.

Blue-headed Grackle. Brewer's Blackbird. Similar; the general iridescence green as before, changing abruptly on the head to purplish, violet or steel-blue, the difference obvious; larger; ♂ 9½-10½; wing 5-5½; tail 4-4½; bill much stouter, more like that of *Agelæus*, and altogether it seems to be quite another bird. The ♀ and young ♂ differ much as in the last species, but they are never so rusty. Plains to the Pacific, U. S. and southward, abundant. AUD., vii, 345, pl. 492; BD., 552; COOP., 278. . . . CYANOCEPHALUS.

94. Genus QUISCALUS Vieillot.

* The ♂ iridescent black throughout.

Great-tailed Grackle. ♂ about 18 inches long; wing 7½; tail 9, its lateral feathers about 3½ inches shorter than the central ones; bill about 1¾. Texas,

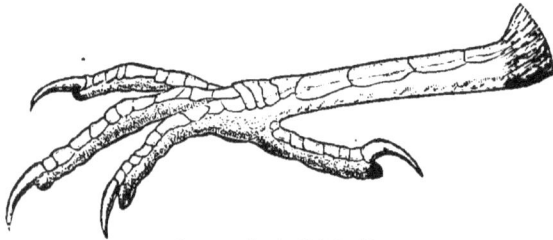

FIG. 101.   Great-tailed Grackle.

and southward. It may prove only an extreme form of the following species, but presents dimensions that the latter has not shown. BD., 254. MACROURUS.

Boat-tailed Grackle. Jackdaw. ♂ 15½-17 long; wing and tail 7-8; bill about 1½; graduation of the tail under 3 inches; tarsus nearly 2, middle toe and claw about the same; the general iridescence green, purple or violet mainly on the head. ♀ astonishingly smaller than the ♂, lacking entirely the great development of the tail, and easily to be mistaken for ♀ *purpureus*, but is rarely so glossy; 12-13½; wing 5½-6; tail 4¾-5½. ♀ and young apt to be quite brown, only blackish on the wings and tail, below grayish-brown, frequently whitening on the throat and breast. South Atlantic and Gulf States, on the coast; strictly maritime, abundant; N. regularly to the Carolinas, frequently to the Middle districts, but *not* to New England as currently reported. AUD., iv, 52, pl. 220; BD., 555. . . . . . . MAJOR.

Purple Grackle. (PLATE V, figs. 1, 3, 4, 5, 1*a*, 3*a*, 4*a*, 5*a*.) ♂ 12-13; wing averaging 5¾; tail 5¼; but either from 5 to 6; bill about 1¼; tarsus 1⅜; graduation of the tail 1½ or less; ♀ 11-12; wing about 5; tail about 4½. Iridescence of the male variable with age, season and other circumstances,

but in the adults always intense, inclining to bronzy, purplish or violet rather than the uniform green of the last species ; ♀ blackish-brown, sometimes quite lustrous. Eastern United States, abundant and generally distributed, migratory, gregarious. WILS., iii, 44, pl. 21, f. 4 ; NUTT., i, 194 ; AUD., iv. 58, pl. 221 ; BD., 555. . . . . . PURPUREUS.

Var. AGLÆUS. (PLATE V, figs. 2, 6, 2a, 6a.) Similar ; averaging smaller, but dimensions inosculating with those of the last ; bill relatively larger, or at least longer, with more attenuated and decurved tip. Florida. *Q. baritus* BD., 556 ; *Q. agleeus* BD., Am. Jour. Sc. 1866, 84 ; CASS., Proc. Phila. Acad., 1866, 404 ; RIDGWAY, *ibid.*, 1869, 135.

OBS. The *Quiscalus æneus*, lately described as a new species by Mr. Ridgway (*l. c.* 134), appears to be based upon a special plumage of *Q. purpureus;* and since it does not prove to be confined, as its describer believed, to any particular region, I should judge it not entitled to rank as a geographical variety. The brilliant coloration is that represented in Audubon's plate, above cited.

FIG. 102. Bills of *Quiscali.*

Family CORVIDÆ. Crows, Jays, etc.

A rather large and important family, comprising such familiar birds as ravens, crows, rooks, magpies, jays, with their allies, and a few diverging forms not so well known ; nearly related to the famous birds of paradise. There are 10 primaries, of which the 1st is short, generally about half as long as the 2d, and several outer ones are more or less sinuate-attenuate on the inner web toward the end. The tail has 12 rectrices, as usual among higher birds ; it varies much in shape, but is generally rounded — sometimes extremely graduated, as in the magpie, and is not forked in any of our forms. The tarsus has scutella in front, separated on one or both sides from the rest of the tarsal envelope by a groove, sometimes naked, sometimes filled in by small scales. The bill is stout, about as long as the head or shorter, tapering, rather acute, generally notched, with convex culmen ; it lacks the commissural angulation of the *Fringillidæ* and *Icteridæ*, the deep cleavage of the *Hirundinidæ*, the slenderness of the *Certhiidæ*, *Sittidæ*, and most small insectivorous birds. The rictus usually has a few stiffish bristles, and there are others about the base of the bill. An essential character is seen in the dense covering of the nostrils with large long tufts of close-pressed antrorse bristly feathers (excepting, among our forms, in gen. 97, 98). These last features distinguish the *Corvidæ* from all our other birds excepting *Paridæ;* the mutual resemblance is here so close, that I cannot point out any obvious technical character of external form to distinguish, for example, *Cyanurus* from *Lophophanes*, or *Perisoreus* from *Parus*. But as already remarked (p. 79), *size* is here perfectly distinctive, all the *Corvidæ* being much larger birds than the *Paridæ*.

Owing to the uniformity of color in the leading groups of the family, and an apparent plasticity of organization in many forms, the number of species is difficult to determine, and is very variously estimated by different writers. Mr. G. R. Gray admits upwards of two hundred species, which he distributes in fifty genera and subgenera ; but these figures are certainly excessive, probably requiring reduction by at least one-third, in both cases. They have been divided into five sub-families; three of these are small and apparently specialized groups confined to the Old World, where they are represented most largely in the Australian and Indian regions; the other two, constituting the great bulk of the family, are more nearly cosmopolitan. These are the *Corvinæ* and *Garrulinæ*, readily distinguishable, at least so far as our forms are concerned.

Subfamily CORVINÆ. Crows.

With the wings long and pointed, much exceeding the tail, the tip formed by the 3d, 4th and 5th quills; the legs stout, fitted for walking as well as perching. As a rule, the plumage is sombre or at least unvariegated — blue, the characteristic color of the jays, being here rare. The sexes are alike, and the changes of plumage slight. Although technically oscine, these birds are highly unmusical; the voice of the larger kinds is raucous, that of the smaller strident. They frequent all situations, and walk firmly and easily on the ground. They are among the most nearly omnivorous of birds, and as a consequence, in connection with their hardy nature, they are rarely if ever truly migratory. Their nesting is various, according to circumstances, but the fabric is usually rude and bulky; the eggs, of the average oscine number, are commonly bluish or greenish, speckled. Although not properly gregarious, as a rule, they often associate in large numbers, drawn together by community of interest. In illustration of this, may be instanced the extensive roosting-places in the Atlantic States, comparable to the rookeries of Europe, whither immense troops of crows resort nightly, often from great distances, recalling the fine line of the poet —

" The blackening trains of Crows to their repose."

95. Genus CORVUS Linnæus.

₊ The species throughout uniform lustrous black, including the bill and feet; nasal bristles about half as long as the bill.

* *Ravens*, with the throat-feathers acute, lengthened, disconnected.

Raven. About 2 feet long; wing 16–18 inches; tail about 10. North America; but now rare in the United States, east of the Mississippi, and altogether wanting in most of the States; Labrador, ranging southward, rarely, along the coast to the Middle districts; very abundant in the west, there generally supplanting the crow. WILS., ix, 136, pl. 75, f. 3; NUTT., i, 202; AUD., iv, 78, pl. 224; BD. 560. *C. cacolotl* BD., 563. CORAX (var?).

White-necked Raven. Smaller; concealed *bases* of cervical feathers pure white. Southwestern U. S. BD., 565; COOP., 284. CRYPTOLEUCUS.

** *Crows*, with the throat-feathers oval and blended.

Crow. Length 18–20; wing 13–14; tail about 8; bill 1¾–2, its height at base ¾; tarsus about equal to the middle toe and claw. Eastern North America, chiefly U. S., not ordinarily found westward in the interior,

FIG. 103. Bill of Crow.

where the raven abounds. WILS., iv, 79, pl. 25, f. 3; NUTT., i, 209; AUD., iv, 87, pl. 225; BD., 566. . . . . AMERICANUS.

Var. FLORIDANUS BD., 568, represents the greater relative size of the bill and feet shown by many birds of Florida and corresponding latitudes.

Var. CAURINUS BD., 569; COOP., 285, is a smaller race from the Pacific Coast; maritime; piscivorous; voice said to be different.

Fish Crow. Small; 14–16 inches long; wing 10–11; tail 6–7; tarsus about equal to middle toe alone; a bare space about the gape? South Atlantic and Gulf States, north to New England, common; maritime, piscivorous. Apparently a different bird, as it presents some tangible distinctions, although constantly associated with the last. WILS., v, 27, pl. 37, f. 2; NUTT., i, 216; AUD., iv, 94, pl. 226; BD., 571. . OSSIFRAGUS.

96. Genus PICICORVUS Bonaparte.

Clarke's Crow. Gray, often bleaching on the head; wings glossy black, most of the secondaries broadly tipped with white; tail white, the central feathers black; bill and feet black. About a foot long; wing 7½–8 inches; tail 4½–5; bill 1¾; nasal feathers very short for this family; claws very large and much curved. Coniferous belt of the West, N. to Sitka, S. to Mexico, E. to Nebraska, W. to the Coast Range; the American representa-

FIG. 104. Clarke's Crow.

tive of the European nutcracker, *Nucifraga caryocatactes;* abundant, imperfectly gregarious. WILS., iii, 29, pl. 20; NUTT., i, 2d ed. 251; AUD., iv, 127, pl. 235; BD., 573; COOP., 289. . . . . COLUMBIANUS.

97. Genus GYMNOKITTA Maximilian.

Blue Crow. Dull blue, very variable in intensity, nearly uniform, but brightest on the head, fading on the belly; the throat with whitish streaks; wings dusky on the inner webs; bill and feet black; ♂ 11–12; wing about 6; tail about 4½; bill 1⅓; ♀ smaller, duller. Rocky Mountain region; much the same elevated distribution as the last, but apparently rather more southerly; decidedly gregarious, and very abundant in some places, though still rare in collections.

FIG. 105. Blue Crow.

A remarkable bird, combining the form of a crow with the color and rather the habits of a jay, and a peculiarly shaped, slender, lengthened and acute bill; the antiæ are prominent and somewhat antrorse, but do not hide the nostrils. CASS., Ill. 165, pl. 28; BD., 574; COOP., 292. CYANOCEPHALUS.

Subfamily GARRULINÆ. Jays.

With the wings much shorter than or about equalling the tail, both rounded, the tip of the wing formed by the 4th–7th quills. The feet, as well as the bill, are usually weaker than in the true crows, and the birds are more strictly arboricole, usually advancing by leaps when on the ground, to which they do not habitually resort. In striking contrast to most *Corvinæ*, the jays are usually birds of bright and striking colors, among which blue is the most prominent, and the head is frequently crested. The sexes are nearly alike, and the changes of plumage do not appear to be as great as is usual among highly colored birds, although some differences are frequently observable. Our well known blue jay is a familiar illustration of the habits and traits of the species in general. They are found in most parts of the world, and reach their highest development in the warmer portions of America. With one boreal exception (*Perisoreus*), the genera of the Old and New World are entirely different.

It is proper to observe, that, while the American *Corvinæ* and *Garrulinæ*, upon which the foregoing paragraphs are mainly drawn up, are readily distinguishable, the characters given may require modification in their application to the whole family, the different divisions of which appear to intergrade closely.

98. Genus PSILORHINUS Rueppel.

Brown Jay. Smoky brown, darker on head, fading on belly; wings and tail with bluish gloss; bill and feet black, sometimes yellow; about 16 long; wing 7½; tail 8½, much graduated; bill 1¼, very stout; *nostrils naked;* head uncrested. Rio Grande Valley and southward. Bd., 592. . . . MORIO.

99. Genus PICA Brisson.

Magpie. Lustrous black, with green, purple and violet, and even golden iridescence, especially on the tail and wings; below from the breast, a scapular patch, and edging of the quills, white; some whitish touches on the throat; bill and feet black. Length 15 or 20 inches, according to the development of the tail, which is a foot or less long, extremely graduated; wing about 8, the outer primary short, slender, and falcate. Arctic America, and U. S. from Plains to Pacific, except California; common. Wils., iv, 75, pl. 35; Nutt., i, 219; Aud., iv, 99, pl. 227; Bd., 576. . . . . MELANOLEUCA var. HUDSONICA.

FIG. 106, Magpie.

Var. NUTTALLII. *Yellow-billed Magpie.* Bill yellow; otherwise precisely like the last, of which it is a perpetuated accident! The European Magpie sometimes shows the same thing, and in some other species, like *P. morio*, the bill is indifferently black or yellow. California. Aud., iv, 104, pl. 228; Nutt., i, 2d ed., 236; Bd., 578; Coop., 295.

Obs. The Columbian Magpie. *Pica bullockii* of Aud., iv, 105, pl. 229, and Nutt., i, 220, is the *Calocitta colliei*, a magnificent species of the West Coast of Mexico, erroneously attributed to California and Oregon.

100. Genus CYANURUS Swainson.

*** Conspicuously crested ; wings and tail blue, black-barred ; bill and feet black. Length 11–12 ; wing or tail 5–6.

Blue Jay. Purplish-blue, below pale purplish-gray, whitening on the throat, belly and crissum ; a black collar across the lower throat and up the sides of the neck and head behind the crest, and a black frontlet bordered with whitish ; wings and tail pure rich blue,- with black bars, the greater coverts, secondaries and tail feathers, except the central, broadly tipped with pure white ; tail much rounded, the graduation over an inch. Eastern North America, especially the United States, everywhere abundant. Wils., i, 2, pl. 1, f. 1 ; Aud., iv, 110, pl. 231 ; Nutt., i, 224 ; Bd., 580. CRISTATUS.

Steller's Jay. Sooty brown, darker on the head, passing insensibly into rich blue on the rump and below from the breast ; wings and tail deep prussian blue, with black bars (wanting in very young birds) ; crest faced with some blue touches, and throat with some whitish streaks ; no white on

Fig. 107. Steller's Jay; long-crested variety.

the eyelids ; tail moderately rounded ; crest about two inches long when full grown. Western North America ; the typical bird rather northerly. Nutt., ii, 229 ; Aud., iv, 107, pl. 230 ; Bd., 581 ; Coop., 298. . . STELLERII.

Var. MACROLOPHUS. *Long-crested Jay.* Similar ; head quite black ; crest longer? the facing bluish-white, and some white touches on the eyelids. Southern Rocky Mountain region. Bd., 582 ; Ell., pl. 17 ; Coop., 300 ; Coues, Am. Nat. v. 1871, 770.

101. Genus APHELOCOMA Cabanis.

*** Not crested ; wings and tail blue, not barred.

Florida Jay. Blue ; back with a large well-defined gray patch, belly and sides pale grayish, under tail coverts and tibiæ blue in marked contrast ; much hoary whitish on forehead and sides of crown ; chin, throat and middle of breast vague streaky whitish ; ear-coverts dusky ; the blue that seems to encircle the head and neck well defined against the gray of back and breast ; bill comparatively short, very stout at the base. About 12 ; wing 5 or less ;

tail about 6, much rounded; bill about 1.   Florida (and Gulf States?), abundant.   NUTT., i, 230; AUD., iv, 118, pl. 233; BD., 586.   FLORIDANA.

Var. WOODHOUSEI.   The dorsal patch dark, somewhat glossed with blue, shading into the blue of surrounding parts; under parts rather darker, somewhat bluish-gray; the tail coverts pale bluish but not contrasted; on the breast the blue and gray shading into each other, the gular and pectoral streaks whitish and well defined, the superciliary line definite white, but no hoary on forehead; bill slenderer. Southern Rocky Mountain region.   BD., 585, pl. 59; COOP., 304.

Var. CALIFORNICA.   *California Jay.*   The dorsal patch light and distinct as in true *floridana*, but the under parts, including tail coverts and tibiæ, nearly white; gular streaks very large, aggregated, and white, causing this part to be nearly uniform; a white superciliary line, but no hoary on forehead, as in *woodhousei;* bill slender.   Thus it is seen that each of the three forms presents a varying emphasis of common characters.   Pacific Coast, U. S.   AUD., iv, 115, pl. 232; BD., 584; COOP., 302.

FIG. 108. Florida Jay.

Sieber's Jay.   Bright blue, scarcely duller on the middle of the back, below white, the throat and breast tinged with blue.   Length 13; wing 6⅔; tail about the same, rounded, the graduation nearly an inch; tarsus 1⅔; bill 1, its height at base nearly ⅓.   BD., 587; COOP., 305. . . SORDIDA.

OBS.   Not having seen this species, I take the name and description from the works cited, without raising the question of its relationships to its allies, especially *C. ultramarinus.*

102. Genus XANTHOURA Bonaparte.

Rio Grande Jay.   Green, below greenish-yellow, inside of the wings and all the tail feathers except the central pair, clear yellow; crown, nape and stripe from bill to eye, rich blue; forehead hoary white; rest of the head and whole throat velvety black; central tail feathers greenish-blue; bill black; feet dark.   About 11 long; wing 4½, rounded, with elongated inner quills; tail 5, graduated an inch or more: bill very short and stout.   Southern Texas and southward.   CASS., Ill. i, pl. 1; BD., 589.   YNCAS var. LUXUOSA.

FIG. 109. Rio Grande Jay.

103. Genus PERISOREUS Bonaparte.

Canada Jay.   *Whiskey Jack.*   Gray, whitening anteriorly, with a darker nuchal area; wings and tail plumbeous, the feathers obscurely tipped with whitish; bill and feet black.   *Young* much darker, sooty or smoky brown; the bleaching progresses indefinitely with age.   10–11 long; wing 5½–6; tail rather more, graduated; tarsus 1⅓; bill under 1, shaped like a titmouse's. Arctic America, into the N. States, S. along the Rocky Mountains to 40° and perhaps further; breeds in Maine in winter.   WILS., iii, 33, pl. 21; NUTT., i, 232; AUD., iv, 121, pl. 234; BD., 590; COOP., 307. . CANADENSIS.

Obs. Several additional species of jays have been ascribed to our country, but apparently upon insufficient evidence or erroneous information.

Suborder CLAMATORES. Non-melodious Passeres.

As already intimated (p. 70), the essential character of this group, as distinguished from Oscines, is an anatomical one, consisting in the non-development of a singing apparatus; the vocal muscles of the lower larynx (syrinx) being small and weak, or else forming simply a large fleshy mass, not separated into particular muscles. This character, however, appears subject to some uncertainty of determination, and probably does not always correspond with the only external character assignable to the group, namely, a certain condition of the tarsal envelope rarely if ever seen in the higher Passeres. If the leg of a kingbird, for example, be closely examined, it will be seen covered with a row of scutella forming cylindrical plates continuously enveloping the tarsus like a segmented scroll, and showing on its postero-internal face a deep groove where the edges of the envelope come together; this groove widening into a naked space above, partially filled in behind with a row of small plates. With some minor modifications, this condition marks the clamatorial birds, and is something tangibly different from the ordinary oscine character of the tarsus, which consists in the presence on the sides of entire corneous laminæ meeting behind in a sharp ridge; and even when, as in the case of Eremophila and Ampelis, there is extensive subdivision of the laminæ on the sides or behind, the arrangement does not exactly answer to the above description. The Clamatores represent the lowest Passeres, approaching the large order Picariæ (see beyond) in the steps by which they recede from Oscines, yet well separated from the Picarian birds. The families composing the suborder, as commonly received, are few in number; only one of them is represented in North America, north of Mexico.

Family TYRANNIDÆ. Flycatchers.

While having a close general resemblance to some of the foregoing insectivorous Passeres, the North American representatives of this family will be instantly distinguished by the above-described condition of the tarsus; and from the birds of the following order by the Passerine characters of twelve rectrices, greater wing coverts not more than half as long as the secondaries, and hind claw not smaller than the middle claw.

This family is peculiar to America; it is one of the most extensive and characteristic groups of its grade in the New World, the Tanagridæ and Trochilidæ alone approaching it in these respects. There are over four hundred current species, distributed among about a hundred genera and subgenera. As well as I can judge at present, at least two-thirds of the species are valid, or very strongly marked geographical races, the remainder being about equally divided between slight varieties and mere synonyms. Only a small fragment of the family is represented within our limits, giving but a vague idea of the numerous and singularly diversified forms abounding in tropical America. Some of these grade so closely toward other families, that a strict definition of the Tyrannidæ becomes extremely difficult; and I am not prepared to offer a satisfactory diagnosis of the whole group. Our species, however, are closely related to each other, and may readily be defined in a manner answering the requirements of the present volume. With a possible exception, not necessary to insist upon in this connection, they belong to the

Subfamily *TYRANNINÆ*, *True Tyrants*,

presenting the following characters :—Wings of 10 primaries, the 1st never spurious nor very short, one or more frequently emarginate or attenuate on the inner web near the end. Tail of 12 rectrices, usually nearly even, sometimes deeply forficate. Feet small, weak, exclusively fitted for perching ; the tarsus little if any longer than the middle toe and claw, the anterior toes, especially the outer, extensively coherent at base. Bill very broad and more or less depressed at the base, and tapering to a fine point, thus presenting a more or less perfectly triangular outline when viewed from above ; the tip abruptly deflected and usually plainly notched just behind the bend ; the culmen smooth and rounded transversely, straight or nearly so lengthwise, except towards the end ; the commissure straight (or slightly curved) except at the end ; the gonys long, flat, not keeled. Nostrils small, circular, strictly basal, overhung but not concealed by bristles. Mouth capacious, its roof somewhat excavated, the rictus ample and deeply cleft, the commissural point almost beneath the anterior border of the eye. Rictus beset with a number of long stiff bristles, sometimes reaching nearly to the end of the bill, but generally shorter, and flaring outward on each side ; there are other bristles or bristle-tipped feathers about the base of the bill. The bill is very light, gives a resonant sound, in dried specimens, when tapped, and on being broken open, the upper mandible will be found extensively hollow.

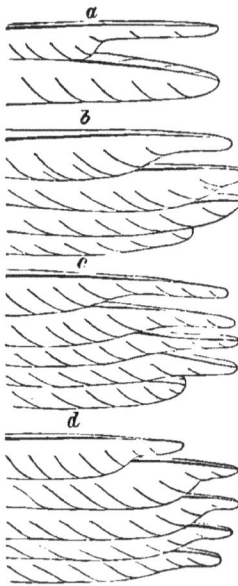

FIG. 110. Emargination of primaries in *Tyrannidæ.*

These several peculiarities of the bill are the more obvious and important features of the group ; and will prevent our small olivaceous flycatchers from being confounded with insectivorous Oscines, as the warblers and vireos.

The structure of the bill is admirably adapted for the capture of winged insects ; the broad and deeply fissured mandibles form a capacious mouth, while the long bristles are of service in entangling the creatures in the trap and restraining their struggles to escape. The shape of the wings and tail confers the power of rapid and varied aërial evolutions necessary for the successful pursuit of active flying insects. A little practice in field ornithology will enable one to recognize the flycatchers from their habit of perching in wait for their prey upon some prominent outpost, in a peculiar attitude, with the wings and tail drooped and vibrating in readiness for instant action ; and of dashing into the air, seizing the passing insect with a quick movement and a click of the bill, and then returning to their stand. Although some Oscines have somewhat the same habits, these pursue insects from place to place, instead of perching in wait at a particular spot, and their forays are not made with such admirable *élan*. Dependent entirely upon insect food, the species are necessarily migratory in our latitudes ; they appear with great regularity in spring, and depart on the first approach of cold weather in the fall. They are distributed over temperate North America ; many of them are common birds of the Eastern States. The voice, susceptible of little modulation, is usually harsh and strident, though some species have a not unpleasant whistle or twitter. The sexes are not ordinarily distinguishable (remarkable exception in gen. 111), and the changes of plumage with age and season are not very great. The larger kinds are unmistakable, but several of the smaller species (of gen. 107, 108, 109) look very much alike, and their discrimination becomes a matter of much tact and diligence.

104. Genus MILVULUS Swainson.

*_** Adults with the tail much longer than the body, deeply forficate, one or more outer primaries strongly emarginate, and a brightly colored crown-patch.

Fork-tailed Flycatcher. Three or four primaries emarginate; crown-patch yellow. Clear ash, below white; top and sides of head, and tail, black; the outer tail feather white on the outer web for about half its length; wings dusky, unmarked. Sexes alike; young similar, but primaries not emarginate, nor tail lengthened. Wing about 4; tail from 3 inches to a foot long. A beautiful bird of tropical America, accidental in the U. S. (Louisiana, New Jersey, *Audubon*). Aud., i, 196, pl. 52; Nutt., i, 274, 2d ed. 307; Bd., 168. . . . . Tyrannus.

Swallow-tailed Flycatcher. Scissor-tail. First primary alone emarginate (fig. 110a); crown-patch orange or scarlet. Hoary ash, paler or white below, sides at insertion of the wings scarlet or bloody-red, and other parts of the body variously tinged with the same, or a paler shade; wings blackish, generally with whitish edgings; tail black, several outer feathers extensively white or rosy; wing about 4½; tail upwards of a foot long. Young similar, lacking the crown-patch, less tinged with red, tail not elongated, primary not emarginate. Lower Mississippi Valley, Texas and southward; accidental in New Jersey (Abbot, Am. Nat., vi, 367). A most elegant and graceful bird. Nutt., i, 275; Aud., i, 197, pl. 53; Bd., 169. . . Forficatus.

105. Genus TYRANNUS Cuvier.

*_**Adults with the tail not forficate, shorter than the lengthened wings, of which several outer primaries are emarginate or gradually attenuate, and crown with a yellow or flame-colored patch. Young with the crown plain and primaries not emarginate. Sexes alike in color; primaries said to be less emarginate in the ♀.

* No olive nor decided yellow.

Kingbird. Bee-martin. (Plate ii, figs. 1, 2, 1a, 2a.) Only two outer primaries obviously emarginate (fig. 110b). Blackish-ash, still darker on head, below pure white, the breast shaded with plumbeous; wings dusky, with much whitish edging; tail black, broadly and sharply tipped with white, the outer feather sometimes edged

Fig. 111. Kingbird. Bee-martin.

with the same; bill and feet black; very young birds show rufous edging of the wings and tail. Length about 8 inches; wing 4½; tail 3½, even or slightly rounded; bill small, under an inch long. Temperate North America, but chiefly Eastern United States to Rocky Mountains; rare or casual on the Pacific slope; abundant in summer. Destroys a thousand noxious insects for every bee it eats! WILS., i, 66, pl. 13; AUD., i, 204, pl. 56; NUTT., i, 265; BD., 171; COOP., 311. . . . . CAROLINENSIS.

Gray Kingbird. Five or six outer primaries usually emarginate. Grayish-plumbeous, rather darker on the head, the auriculars dusky; below white, shaded with ashy on breast and sides, the under wing and tail coverts faintly yellowish; wings and tail dusky, edged with whitish or yellowish; the tail feathers merely indistinctly lighter at the extreme tip. Larger than the last; about 9; wing 5½; tail nearly 5, more or less emarginate; bill very turgid, an inch long. West Indies; Florida regularly; N. to. Carolina rarely (*Audubon*), to Massachusetts accidentally (*Allen*). AUD., i, 201, pl. 55; BD., 172. . . . . DOMINICENSIS.

* * Olivaceous and yellow; belly and under tail coverts clear yellow, back ashy olive, changing to clear ash on the head, throat and breast, the chin whitening, the lores and auriculars usually dusky, wings dark brown with whitish edging, tail black or blackish, bill and feet black. Very young birds paler below, with rufous traces above. 8-9 long; wing nearly 5; tail about 4; bill $\frac{2}{3}-\frac{3}{4}$.

Arkansas Flycatcher. Several outer primaries gradually attenuated for a long distance (fig. 110c). Outer web of outer tail feather entirely white. Ash of the fore parts pale, contrasting with dusky lores and auriculars, fading insensibly into white on the chin, and changing gradually to yellow on the belly; olive predominating over ashy on the back. Western U. S., abundant; accidental in Louisiana, New Jersey. NUTT., i, 273; AUD., i, 199, pl. 54; BD., 173; COOP., 312. . . VERTICALIS.

FIG. 112. Arkansas Fly-catcher.

Cassin's Flycatcher. Several outer primaries abruptly emarginate for a short distance (fig. 110d). Outer web of outer tail feather barely or not edged with whitish. Ash of fore parts dark, little different on the lores and auriculars, changing rather abruptly to white on the chin and to yellow on the belly; ashy predominating over olive on the back. Southwestern U. S., and southward, common. *Tyrannus cassinii* LAWR., Ann. Lyc. Nat. Hist. N. Y. v, 39, pl. 3, f. 2; *T. vociferans* BD., 174; COOP., 314. . VOCIFERANS.

Couch's Flycatcher. Very similar to the last; tail dark brown, like the wings, and obviously forked (about ½ an inch; in *cassinii* the tail is quite black and slightly emarginate or nearly even), all its feathers with slight pale edges, and their shafts pale on the under surface; yellow of under parts very bright, reaching high up the breast; throat as well as chin extensively white. A universally distributed S. and Cent. Am. species, of which a slight northern variety (*T. couchii* BD., 175) reaches our Mexican border. S. Arizona (*Bendire*); COUES, Am. Nat. vi, Aug. 1872. . MELANCHOLICUS.

106. Genus MYIARCHUS Cabanis.

⁎ No colored patch on the crown, but head slightly crested; primaries not emarginate. Olivaceous; more or less yellow below, the throat ash, the primaries

margined with chestnut; the tail feathers the same or mostly chestnut; wings rounded, about as long as the nearly even tail; feet black, bill blackish, usually pale at the base below. Fig. 113a.

Great Crested Flycatcher. Decidedly olivaceous above, a little browner on the head, where the feathers have dark centres; throat and fore breast pure dark ash, rest of under parts bright yellow, the two colors meeting abruptly; primaries margined on both edges with chestnut: secondaries and coverts edged and tipped with yellowish-white, tail with all the feathers but the central pair chestnut on the whole of the inner web excepting perhaps a very narrow space next the shaft; outer web of outer feathers edged with yellowish; the middle feathers, outer webs of the rest, and wings except as stated, dusky brown. Very young birds have rufous skirting of many feathers, in addition to the chestnut above described, but this soon disappears. Large; 8½–9½; wing and tail about 4; bill ¾; tarsus ¾. Eastern United States, west to Missouri, Kansas, Arkansas and Texas, north to Massachusetts; Mexico and Central America in winter. An abundant bird, in woodland, of loud harsh voice and quarrelsome disposition, noted for its habitual use of cast off snake skins in the structure of its nest. WILS. ii, 75, pl. 13; NUTT., i, 271; AUD., i, 209, pl. 57; BD., 178. . CRINITUS.

Ash-throated Flycatcher. Rather olivaceous-brown above, quite brown on the head, the throat very pale ash, sometimes almost whitish, changing gradually to very pale yellow or yellowish-white on the rest of the under parts; primaries edged as before, but secondaries and coverts edged with grayish-white; tail feathers as in the last, but the chestnut of the inner webs hardly or not reaching the tip, being cut off from the end by invasion of the dusky. In young birds, in which the quills and tail feathers are more extensively rufous-edged, this last distinction does not hold. Southwestern U. S., Mexico, common; very near the last species, but apparently a different bird. It is rather smaller, but with longer (⅘) tarsi; the bill obviously narrower, only about as wide as high at the base; but in Cape St. Lucas specimens (*M. pertinax* BD., Proc. Phila. Acad. 1859, 303; COOP., 318), again, the bill is shaped as in *crinitus*, although smaller. *Tyrannula cinerascens* LAWR., Ann. Lyc. N. Y. 1851, 109; *M. mexicanus* BD., 179; COOP., 316 (*not of* KAUP, Proc. Zoöl. Soc. 1851, 51). . . . . CINERASCENS.

Lawrence's Flycatcher. Very similar in color to *crinitus*, but *much* smaller; about 7 long, wing and tail about 3½; wing coverts and inner quills as well as the primaries edged with rufous (rarely yellowish on the inner secondaries); *no* chestnut on tail feathers except a narrow bordering on the *outer* webs, and, in the *young*, an inner *margining* also; bill broad, flattened. Texas (?), Mexico and Central America. BD., 181. LAWRENCEI.

107. Genus SAYORNIS Bonaparte.

₊ The three following species do not particularly resemble each other; most authors place them in separate genera, and even under different subfamilies. The discrepancies of form, however, are not startling, and for the purposes of this work the species may be properly put together, as they agree in presenting a certain aspect not shown by the other North American groups. Fig. 113*b*.

Say's Flycatcher. Grayish-brown, paler below and changing to cinnamon on the belly and crissum; wings dusky with paler edgings on the inner quills and coverts; tail perfectly black; bill and feet black. Younger birds are much more extensively fulvous or paler cinnamon than the old, this color extending far up the breast, skirting the feathers of the back and rump, forming conspicuous crossbars and edging on the wings, and sometimes tipping the tail. 7 or 8 inches long; wing 4; tail 3½, emarginate. Western America, in open country, common. NUTT., i, 277; AUD., i, 217, pl. 59; BD., 185; COOP., 320. . . . . SAYUS.

Black Flycatcher. Sooty-brown, deepest on head and breast, belly and crissum abruptly pure white; lining of wings and edging of outer tail feather and inner quills, whitish; bill and feet black; "iris red;" about 7; wing 3¾; tail 3⅓. Southwestern United States and southward, in unwooded country, cañons, and along rocky streams. NUTT., i, 2d ed. 311; AUD., i, 217; pl. 59; BD., 185; COOP., 320. . . . . NIGRICANS.

Pewee. Pewit. Phœbe. Dull olivaceous-brown, the head much darker fuscous-brown, almost blackish, usually in marked contrast with the back; below soiled whitish, or palest possible yellow, particularly on the belly; the sides, and the breast nearly or quite across, shaded with grayish-brown; wings and tail dusky, the outer tail feather, inner secondaries, and usually the wing coverts, edged with whitish; a whitish ring round the eye; bill and feet black. Varies greatly in shade; the foregoing is the average spring condition. As summer passes, the plumage becomes much duller and darker brown, from wearing of the feathers, and then, after the moult, fall specimens are much brighter than in spring, the under parts being frequently decidedly yellow, at least on the belly. Very young birds have some feathers skirted with rusty, particularly on the edges of the wing and tail feathers. The species requires

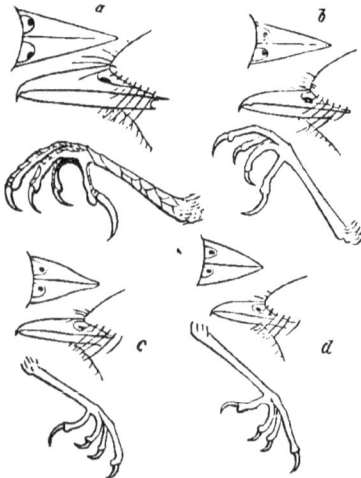

FIG. 113. Generic details in the smaller Flycatchers.

careful discrimination, in the hands of a novice, from any of the little olivaceous species of the next two genera. It is larger; 6¾–7; wing 3–3½; tail about the same, slightly emarginate; bill ½ or slightly more, little depressed, not so broad for its length as is usual in *Contopus* and *Empidonax*, its lateral outlines straight; tarsus equalling or slightly exceeding the middle toe and claw, these together about 1⅓ long; point of the wing formed by the 2d to

5th quill; 1st shorter than 6th; 3d and 4th generally rather the longest.
Eastern North America, very abundant, in open places, fields, along streams,
etc.; one of the very earliest arrivals in spring, a late loiterer in the fall;
winters in the Southern States. Voice short, abrupt, unlike the drawling
note of the wood pewee. WILS., ii, 78, pl. 13; NUTT., i, 278; AUD., i,
223, pl. 63; BD., 184. . . . . FUSCUS.

108. Genus CONTOPUS Cabanis.

*** With the feet extremely small, the tarsus shorter than the middle toe and
claw; the tarsus, middle toe and claw together, barely or not one-third as long as
the wing; the bill flattened, very broad at base; the pointed wings much longer
than the emarginate tail. Medium sized and rather small species, brownish-
olivaceous, without any bright colors, or very decided markings; the coronal
feathers lengthened and erectile, but hardly forming a true crest. Fig. 113c.

* Species 7–8 long, with a tuft of white fluffy feathers on the flank.

Olive-sided Flycatcher. Dusky olivaceous-brown, usually darker on the
crown, where the feathers have blackish centres, and paler on the sides;
chin, throat, belly, crissum and middle line of breast, white, more or less
tinged with yellowish; wings and tail blackish, unmarked, excepting inconspicuous grayish-brown tips of the wing coverts, and some whitish edging
on the inner quills; feet and upper mandible black, lower mandible mostly
yellowish. The olive-brown below has a peculiar *streaky* appearance hardly
seen in other species, and extends almost entirely across the breast. *Young*
may have the feathers, especially of the wings and tail, skirted with rufous.
Wing 3⅗–4⅘, remarkably pointed; second quill longest, supported nearly to
the end by the first and third, the fourth abruptly shorter; tail about 3;
tarsus, middle toe and claw together only about 1¼; bill ⅔–¾. North America, apparently nowhere very abundant. NUTT., i, 282; 2d ed. 298; AUD.,
i, 212, pl. 58; BD., 188; COOP., 323. . . . . BOREALIS.

Coues' Flycatcher. Somewhat similar; colors more uniform and more
clearly olive; below, fading insensibly on the throat and belly into yellowish
white, and lacking the peculiar streaky appearance; cottony tufts on the
flanks less conspicuous; wing-formula entirely different; second, third and
and fourth quills nearly equal and longest, first abruptly shorter; tail longer,
about 3¾. Mexico; north to Arizona. CAB., Mus. Hein. ii, 72; COUES,
Proc. Phila. Acad. 1866, 60; ELL., pl. 18; COOP., 324. . . PERTINAX.

** Species under 7 long, without an evident cottony white tuft on the flank.

Wood Pewee. Olivaceous-brown, rather darker on the head, below with
the sides washed with a paler shade of the same reaching nearly or quite
across the breast; the throat and belly whitish, more or less tinged with dull
yellowish; under tail coverts the same, usually streaked with dusky; tail
and wings blackish, the former unmarked, the inner quills edged, and the
greater and middle coverts tipped, with whitish; feet and upper mandible
black, under mandible usually yellow, sometimes dusky. Spring specimens
are purer olivaceous; early fall birds are brighter yellow below; in

summer, before the now worn features are renewed, the plumage is quite brown, and dingy whitish; very young birds have the wing-bars and edging of quills tinged with rusty, the feathers of the upper parts skirted, and the lower plumage tinged, with the same; but in any plumage the species may be known from all the birds of the following genus, by these dimensions: Length 6–6½; wing 3¼–3½; tail 2¾–3; tarsus, middle toe and claw together hardly one inch, or evidently less; tarsus alone about ½, not longer than the *bill*. North America, in woodland: extremely abundant in most United States localities, May—Sept. *Muscicapa rapax* WILS., ii, 81, pl. 13, f. 5; *M. virens* AUD., i, 231, pl. 64; NUTT., i, 285; *C. virens* BD., 190. VIRENS.

Var. RICHARDSONII. *Western Wood Pewee.* Similar; darker, more fuscous olive above, the shading of the sides reaching almost uninterruptedly across the breast; belly rather whitish than yellowish; outer primary usually not obviously white-edged; bill below oftener dusky than yellow, sometimes quite black. I fail to appreciate any reliable differences in size or shape. Note not exactly like that of *virens*; nesting said to be different (*Audubon*, *Allen*). Rocky Mountains to the Pacific; "Labrador" (*Audubon*). *Tyrannula Richardsonii* SW., Fn. Bor.-Am. ii, 146? *Contopus richardsonii* BD., 189; COOP., 325. *Muscicapa phœbe* AUD., i, 219, pl. 61; NUTT., i, 2d ed. 319.

109. Genus EMPIDONAX Cabanis.

*** Species 5–6 (rarely 6¼) long; wing 3⅜ *or less;* tail 2¾ *or less;* whole foot at least ⅓ as long as wing; tarsus more or less obviously longer than middle toe and claw, much longer than bill; 2d, 3d and 4th quills entering into point of wing, 1st shorter or not obviously longer than 5th; tail not over ½ an inch shorter than wings; breast not buffy. (Compare 107, 108, 110.) As in allied genera, several outer primaries are slightly emarginate on the inner web, but this character is obscure, and often inappreciable. Fig. 113*d*.

Small Green-crested or *Acadian Flycatcher*. Above, olive-green, clear, continuous and uniform (though the crown may show rather darker, owing to dusky centres of the slightly lengthened, erectile feathers); below, whitish, olive-shaded on sides and nearly across breast, yellowish-washed on belly, flanks, crissum and axillars; wings dusky, inner quills edged, and coverts tipped, with tawny yellow; all the quills whitish-edged internally; tail dusky, olive-glossed, unmarked; a yellowish eye-ring; feet and upper mandible brown, under mandible pale. In midsummer, rather darker; in early fall, brighter and especially more yellowish below; when very young, the wing-markings more fulvous, the general plumage slightly buffy-suffused. Largest; 5¾–6¼; wing 2¾–3 (rarely 3⅛); tail 2½–2¾; bill nearly or quite ½, about ¼ wide at nostrils; tarsus ⅗; middle toe and claw ½; point of wing reaching nearly an inch beyond the secondaries; 2d, 3d and 4th quills nearly equal and *much* (¼ inch or more) longer than 1st and 5th, which about equal each other; 1st *much* longer than 6th. Eastern United States, abundant, in woodland; readily diagnoscible by the points of size and shape, without regarding coloration. *Muscicapa querula* WILS., ii, 77, pl. 13, f. 3; *M. acadica* NUTT., i, 208; AUD., i, 221, pl. 62; BD., 197. . . ACADICUS.

Traill's Flycatcher. Above, olive-brown, lighter and duller brownish posteriorly, darker anteriorly, owing to obviously dusky centres of the coronal feathers; below, nearly as in *acadicus*, but darker, the olive-gray shading quite across the breast; wing-markings grayish-white with slight yellowish or tawny shade; under mandible pale; upper mandible and feet black. Averaging a little less than *acadicus;* 5½–6; wing 2⅔–2¾, more rounded, its tip only reaching about ⅔ of an inch beyond the secondaries, formed by 2d, 3d and 4th quills, as before, but 5th not so much shorter, (hardly or not ¼ of an inch), the 1st ranging between 5th and 6th; tail 2½; tarsus ⅔, as before, but middle toe and claw ⅗, the feet thus differently proportioned, owing to length of toes. Eastern North America to the Plains, common; an entirely different bird from *acadicus*, but difficult to distinguish from the following species. AUD., i, 234, pl. 65; NUTT., i, 2d ed. 323; BD., 192, 193. . . . . TRAILLII.

Var. PUSILLUS of BD., 194, which replaces true *traillii* from the Plains to the Pacific, may usually be recognized by its more fuscous coloration, the olivaceous and yellowish shades of *traillii* being subdued; by its larger bill, and the feet nearly as in *acadicus*. The original *Tyrannula pusilla* of SWAINSON, Fn. Bor.-Am. ii, 144; AUD., ii, 236, pl. 66, is uncertain, just as likely have been *minimus* as this bird. I therefore pass over the name, which, if belonging here, antedates *traillii*, and adopt *traillii* for the eastern form (although AUDUBON says "Arkansas to the Columbia"), taking *pusillus* of BAIRD for the western variety.

Least Flycatcher. Colors almost exactly as in *traillii*; usually however olive-gray rather than olive-brown; the wing-markings, eye-ring and loral feathers plain grayish-white; the whole anterior parts often with a slight ashy cast; under mandible ordinarily dusky; feet perfectly black. It is a smaller bird than *traillii*, and not so stoutly built; the wing-tip projects only about ½ an inch beyond the secondaries; the 5th quill is but very little shorter than the 4th, the 1st apt to be nearer 6th than 5th; the feet are differently proportioned, being much as in *acadicus;* the bill is obviously under ½ an inch long. Length 5–5½; wing 2⅗ *or less;* tail about 2¼. Although it grades up to *traillii* in size, and has no obviously different coloration, yet I am satisfied that it is a different bird. Eastern North America to the Plains, very abundant in the U. S. during the migrations, in orchards, coppices, hedgerows and the skirts of woods rather than in heavy forests. AUD., vii, 343, pl. 491; BD., 195. . . . . MINIMUS.

Yellow-bellied Flycatcher. Above, olive-green, clear, continuous and uniform as in *acadicus*, or even brighter; below, not merely yellow*ish*, as in the foregoing, but emphatically *yellow*, bright and pure on the belly, shaded on the sides and anteriorly with a paler tint of the color of the back; eye-ring and wing-markings yellow; under mandible yellow; feet black. In respect of color, this species differs materially from all the rest; none of them, even at their autumnal yellowest, quite match it. Size of *traillii*, or rather less; feet proportioned as in *acadicus;* bill nearly as in *minimus*, but rather larger; 1st quill usually equal to 6th. Eastern United States, common. Aud., vii, 341, pl. 490; Bd., 198. Var. DIFFICILIS Bd., 198 (in text), Coop., 328, is the paler western form. . . . FLAVIVENTRIS. *Hammond's Flycatcher.* Above, olive-gray, decidedly grayer or even ashy on the fore parts, the whole throat and breast almost continuously olive-gray but little paler than the back, the belly alone more or less decidedly yellowish; wing-markings and eye-ring dull soiled whitish; bill very small, and extremely narrow, being hardly or not ¼ wide at the nostrils; this distinguishes the bird from all but *minimus* and *obscurus;* under mandible usually blackish; tail usually decidedly forked, more so than in other species, though in all of them it varies from slightly rounded to slightly emarginate; outer tail feather usually whitish-edged externally (a character often shown by *traillii* and *minimus*), but not decidedly white. About the size of *traillii*, but not so stoutly built; wings and tail relatively longer; feet as in *minimus*. Western United States. Bd., 199; Coop., 330. . . . . . HAMMONDII.

Wright's Flycatcher. Colors not tangibly different from those of *traillii* or *minimus*, but outer web of outer tail feather abruptly white in decided contrast. General dimensions approaching those of *acadicus*, owing to length of wings and tail; wing 2⅔ to nearly 3; tail 2¼–2¾; tarsi about ¾; bill about ½, extremely narrow (much as in *Sayornis fuscus*), its width at the nostrils only about ½ its length. Southwestern U. S. Bd., 200, 922; Coop., 329. . . . . OBSCURUS.

Obs. The foregoing account, carefully prepared after examination of a great amount of material from all parts of the country, will probably suffice to determine ninety out of a hundred specimens; but I confess it does not entirely satisfy me, and, as it does not fully answer all the requirements of the case, it must be regarded as provisional. At the same time I must say, that the only alternative seems to be, to consider all the foregoing (excepting *acadicus* and *flaviventris*, perhaps) as varieties of one species; but for this I am not prepared.

110. Genus MITREPHORUS Sclater.

Buff-breasted Flycatcher. Coronal feathers and rictal bristles longer than in *Empidonax*, and general cast of the plumage buffy. Above, dull grayish-brown tinged with olive, particularly on the back; below, pale fulvous, strongest across the breast, whitening on the belly; no fulvous on the fore-head; sides of head light brownish-olive; wings and tail dusky, outer web of outer tail feathers, edges of inner primaries except at the base, and tips of wing coverts, whitish; iris brown; bill yellow below, black above; feet black; 4¾ long; extent 7½; wing 2¼; tail 2; tarsus .55; middle toe and claw .45; bill .40. Fort Whipple, Arizona. *Empidonax pygmæus* Coues, Ibis, 1865, 537; *M. pallescens* Coues, Proc. Phila. Acad. 1866, 63; Coop., 334; Elliot, pl. 19. My original specimens, affording the descriptions quoted, and the first known to have been taken in the United States, do not appear to be specifically distinct from *fulvifrons* of Giraud (B. of Tex. pl. 2, f. 2), which may itself be the same as a Mexican species of prior name. . . . . FULVIFRONS var. PALLESCENS.

111. Genus PYROCEPHALUS Gould.

Vermilion Flycatcher.   ♂ pure dark brown; wings and tail blackish with slight pale edgings; the full globular crest, and all the under parts, scarlet; bill and feet black.   ♀ dull brown, including the little crested crown; below, white, tinged with red or reddish in some places, the breast with slight dusky streaks.   Immature ♂ shows gradation between the characters of both sexes; the red is sometimes rather orange.   5½–6; wing 3¼; tail 2¼.   Valleys of the Rio Grande and Colorado, and southward.   CASS., Ill. 127, pl. 17; BD., 201; COOP., 333.   .   .   RUBINEUS var. MEXICANUS.

FIG. 114.   Vermilion Flycatcher.

Order PICARIÆ. Picarian Birds.

This is a miscellaneous assortment (in scientific language, "a polymorphic group,") of birds of highly diversified forms, grouped together more because they differ from other birds in one way or another, than on account of their resemblance to each other.   As commonly received, this order includes all the non-passerine *Insessores* down to those with a cered bill (parrots and birds of prey).   Excluding the parrots, which constitute a strongly marked natural group, of equal value with those called orders in this work, the *Picariœ* correspond to the *Strisores* and *Scansores* of authors, including, however, some that are often referred to *Clamatores*.   This "order" *Scansores*, or *Zygodactyli*, containing all the birds that have the toes arranged in pairs, two in front and two behind (and some that have not), is one of the most unmitigated inflictions that ornithology has suffered; it is as thoroughly unnatural as the divisions of my artificial key to our genera.

As at present constituted, the *Picariœ* are insusceptible of satisfactory definition; but we may indicate some leading features, mostly of a negative character, that they possess in common.   The sternum rarely if ever conforms to the particular Passerine model, its posterior border usually being either entire or else doubly notched.   The vocal apparatus is not highly developed, having not more than three pairs of separate intrinsic muscles; the birds, consequently, are never highly musical.   There are some modifications of the cranial bones not observed in *Passeres*.   According to Sundevall, they, like lower birds, lack a certain specialization of the flexor muscles of the toes seen in *Passeres*.   The feet are very variously modified; one or another of all the toes, except the middle one, is susceptible of being turned, in this or that case, in an opposite from the customary direction; the fourth one being frequently capable of turning either way; while in two genera the first, and in two others the second, toe is deficient; and, moreover, the tarsal envelope is never entire behind as in the higher *Passeres*.   Another curious peculiarity of the feet is, that the claw of the hind toe is smaller, or at most not larger, than that of the third toe.   The wings, endlessly varied in shape, agree in possessing ten developed primaries, of which the first is rarely spurious or very short.   A notable exception to this occurs in the *Pici*.   A very general and useful wing-character is, that the greater coverts are at least half as long as the secondary quills they cover, and they sometimes reach nearly to the ends of these quills.   This is the common case among lower birds, but it distinguishes most of the *Picariœ* from *Passeres;* it is not shown, however, in the *Picidæ* and some others. The tail is indefinitely varied in shape, but the number of its feathers is a good clue to the order. There are not ordinarily more than *ten* perfect rectrices, and occasionally there are only eight; the woodpeckers have twelve, but one pair is abortive; there are twelve, however, in the kingfishers, and some others.

With this slight sketch of some leading features of the group (it will enable the student to recognize any Picarian bird of this country at least), I pass to the consideration of its subdivision, with the remark, that a precedent may be found for any conceivable grouping of the families that is not simply preposterous, and for some arrangements that are nearly so. As well as I can judge from the material at my command, and relying upon excellent authority for data that I lack, the *Picariæ* fall naturally into THREE divisions. These I shall call suborders, not however insisting in the least upon the question of taxonomic rank, but simply employing the terms conformably with my usage in other cases. The three groups may be here tabulated, with remarks calculated to give an idea of their composition : —

I. CYPSELI — including only the three families *Cypselidæ*, *Caprimulgidæ*, and *Trochilidæ* — the swifts, goatsuckers, and hummingbirds. They are birds of remarkable volitorial powers ; the wing is pointed, and very long, in its feathers and terminal portions, though the upper arm is very short. The feet are extremely small and weak, and are scarcely if at all serviceable for progression. The hind toe is sometimes versatile (among the swifts) or somewhat elevated (in the goatsuckers and some swifts) ; the front toes are frequently connected at base by movable webbing (goatsuckers), and sometimes lack the normal number of phalanges (among swifts and goatsuckers). The variously shaped tail has ten rectrices. One family (hummingbirds) shows the tenuirostral type of bill ; the other two, the fissirostral, on which account they used to be classed with the swallows. The sternum is broad, with a deep keel, entire or doubly notched (rarely singly notched) behind ; the syrinx has not more than one pair of intrinsic muscles.

II. CUCULI — comprehending the great bulk of the order ; in all, about fifteen families, rather more than less. They are only readily limited by exclusion of the characters of the preceding and following groups. The sternum is usually notched behind ; the syringeal muscles are two pairs at most. The feet are *generally* short ; the disposition of the toes varies remarkably. In the *Coliidæ*, or colies, of Africa, all the toes are turned forward. In the *Trogonidæ*, the second toe is turned backward, so that the birds are zygodactyle, but in a different way from all others. Families with the feet permanently zygodactyle in the ordinary way by reversion of the fourth, or partially so, the outer toe being versatile, are — the *Cuculidæ*, or cuckoos, with their near relatives the *Indicatoridæ*, or guide-birds of Africa ; the *Rhamphastidæ*, or toucans, confined to tropical America and distinguished by their enormous vaulted bill ; the *Musophagidæ*, plantain-eaters or touracos, of Africa ; the *Bucconidæ* and *Capitonidæ*, or barbets of the New and Old World respectively ; and the *Galbulidæ*, or jacamars, of America. In the remaining groups, the toes have the ordinary position, but sometimes offer unusual characters in other respects. Thus in the *Alcedinidæ* (kingfishers), and *Momotidæ* (motmots or sawbills), the middle and outer toes are perfectly coherent for a great distance, constituting the *syndactyle* or *anisodactyle* foot. The *Bucerotidæ*, or hornbills, of the Old World, characterized by an immense corneous process on the bill, are near relatives of the kingfishers ; so are the *Todidæ*, a group of small brightly colored birds of Mexico and the West Indies. Other forms, all Old World, are the *Meropidæ* or bee-eaters, the *Upupidæ* or hoopoes, the *Coraciidæ* or rollers, with their allies the *Leptoso-matidæ*, of Madagascar.

III. PICI—comprising only three families, the *Iyngidæ*, or wrynecks, with one genus and four species, of Europe, Asia and Africa; the *Picumnidæ*, with one or two genera and nearly thirty species, chiefly American; and the *Picidæ* or true woodpeckers. The digits are permanently paired by reversion of the fourth, except in two tridactyle genera; there is a modification of the lower end of the metatarsus, corresponding to the reverse position of the fourth toe, and the upper part of the same bone is perforated by canals for flexor tendons. The basal phalanges of the toes are short. The wing has ten primaries, with short coverts, contrary to the rule in this order; the tail ten rectrices, soft and rounded in *Iyngidæ* and *Picumnidæ*, rigid and acuminate in *Picidæ*, where also a supplementary pair of spurious feathers is developed. The nostrils vary: they are large and of peculiar structure in *Iyngidæ*, usually covered with antrorse plumules in the rest. The bill is straight or nearly so, hard and strong, acute or truncate, the mandibles equal; the tongue is lumbriciform, and very generally extensile to a remarkable degree, by a singular elongation of the bones and muscles. The salivary glands have an unusual development, in the typical species at any rate. The sternum is doubly notched behind. A very strongly marked group; in some respects it approaches the *Passerine* birds more nearly than other *Picariæ* do.

Suborder *CYPSELI. Cypseliform Birds.*

See p. 178, where some leading characters of the group are indicated.

Family CAPRIMULGIDÆ. Goatsuckers,

So called from a traditional superstition. *Fissirostral Picariæ:* head broad, flattened; eyes and ears large; bill extremely small, depressed, triangular when viewed from above, with enormous gape reaching below the eye, and generally with bristles that frequently attain an extraordinary development; nostrils basal, exposed, roundish, with a raised border, sometimes prolonged into a tube. Wings more or less lengthened and pointed, of ten primaries and more than nine secondaries; tail variable in shape, of ten rectrices. Feet extremely small; tarsus usually short, and partly feathered; hind toe commonly elevated and turned sideways; front toes connected at base by movable webbing, and frequently showing abnormal ratio of phalanges; middle toe lengthened beyond the short lateral ones, its claw frequently pectinate. A definitely circumscribed, easily recognized group of about fourteen genera and rather more than a hundred species, of temperate and tropical parts of both hemispheres. It is divisible, according to the structure of the feet, into two subfamilies, *Podarginæ*, chiefly Old World, with the normal ratio of phalanges, and *Caprimulginæ*, as below. Considering, however, other points, particularly the shape of the sternum, a more elaborate division is into *Podarginæ*, phalanges normal, but tarsus naked and lengthened, and sternum doubly notched, with three genera of the Old World — *Nyctibiinæ*, phalanges normal, tarsus short, feathered, sternum doubly notched, upper mandible toothed, containing one genus of tropical America — *Steatornithinæ*, phalanges normal, sternum singly notched, with one genus of tropical America — and finally *Caprimulginæ*, comprising the rest

[NOTE. An erroneous sequence of two genera having been discovered since the key was printed, and therefore too late to rectify the numbering, Gen. 112 and Gen. 113, will be found next after Gen. 125.]

of the family. The eggs are colorless in the first and third of these, colored in the second and fourth.

Subfamily CAPRIMULGINÆ. True Goatsuckers.

Sternum singly notched on each side behind, its body not square. *Outer toe 4-jointed; middle claw pectinate;* hind toe very short, elevated, semi-lateral; anterior toes movably webbed at base; tarsus very short, commonly much feathered. Besides the semipalmation of the feet, there is another curious analogy to wading birds; for the young are downy at birth, as in *Præcoces,* instead of naked, as is the rule among *Altrices.* The plumage is soft and lax, much as in the owls; the birds have the same noiseless flight, as well as, in many cases, nocturnal or crepuscular habits; and they sometimes bear an odd resemblance to owls in their general appearance. An evident design of the capacious mouth, is the capture of insects; the active birds quarter the air with wide open mouth, and their minute prey is readily taken in. But they also secure larger insects in other ways; and to this end the rictus is frequently strongly bristled, as in the *Tyrannidæ.* Our two genera are readily discriminated by the enormous rictal bristles, rounded tail and comparatively short wings of *Antrostomus,* the slight bristles, forked tail and long pointed wings of *Chordeiles;* they each represent one of the two sections of the subfamily. In both, the feet are so extremely short that the birds cannot perch in the usual way, but sit lengthwise on a large branch, or crouch on the ground. They lay two lengthened, dark colored, thickly spotted eggs, on or near the ground, in stumps, etc.; the sexes are distinguishable but nearly alike; the colors are subdued, blended and variegated; the voice is peculiar. Migratory.

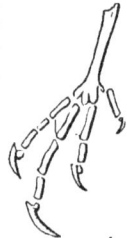

FIG. 115. Bones of Caprimulgine foot.

114. Genus ANTROSTOMUS Gould.

* The rictal bristles with lateral filaments.

Chuck-will's-widow. Singularly variegated with black, white, brown, tawny and rufous, the prevailing tone fulvous; a whitish throat-bar; several lateral tail feathers tipped with white in the ♂, with rufous in the ♀. Large; a foot long; wing 8–9; tail 5½–6½, slightly rounded. South Atlantic and Gulf States, strictly; resident in Florida. WILS., vi, 95, pl. 54, f. 2; NUTT., i, 612; AUD., i, 151, pl. 41; BD., 147. CAROLINENSIS.

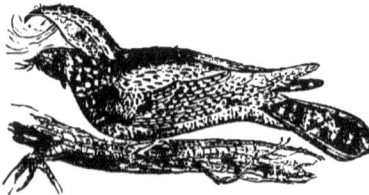

* * The rictal bristles simple.

Whippoorwill. Night-jar. Upper parts variegated with gray, black, whitish and tawny; black streaks sharp on the head and back, the colors elsewhere delicately marbled, including the four median tail feathers; prevailing tone gray; wings and their coverts with bars of rufous spots; lateral tail feathers black, with tawny marbling in distant broken bars, and tipped with white (♂) or

FIG. 116. Whippoorwill. .

tawny (♀) ; a bar across the throat white (♂) or tawny (♀) ; below mottled with dusky and whitish; 9–10 long; wing 5–6; tail 4–5, much rounded. Eastern United States, abundant; a nocturnal bird, rarely seen, but well known for its loud strange cry, whence its name is taken. Eggs 2, elliptical, 1¼ by ⅞, white, speckled and blotched. WILS., v, 72, pl. 41, f. 1, 2, 3; NUTT., i, 614; AUD., i, 155, pl. 42; BD., 148. . . . . . VOCIFERUS.

Nuttall's Whippoorwill. Somewhat similar; small; about 8; tail under 4; much paler in tone; crown barred transversely; throat patch very large; tawny prevailing over black on the wings; terminal white tail spots short. Note different, the first syllable being omitted. Plains to the Pacific, U. S. AUD., vii, 350, pl. 495; CASS., Ill. 237; BD., 149; COOP., 340. NUTTALLII.

115. Genus CHORDEILES Swainson.

Night-hawk. Bull-bat. Above, mottled with black, brown, gray and tawny, the former in excess; below from the breast transversely barred with blackish and white or pale fulvous; throat with a large white (♂) or tawny (♀) cross-bar; tail blackish, with distant pale marbled cross-bars and a large white spot (wanting in the ♀) on one or both webs of nearly all the feathers toward the end; quills dusky, unmarked except by one large white spot on *five* outer primaries about midway between their base and tip; in the ♀ this area restricted or not pure white. Length about 9; wing about 8; tail 5. Temperate North America, abundant. This species flies abroad at all times, though it is perhaps most active towards evening and in dull weather; and is generally seen in companies, busily foraging for insects with rapid, easy and protracted flight; in the breeding season it performs curious evolutions, falling through the air with a loud booming cry.

FIG. 117. Night-hawk.

Eggs 2, elliptical, 1¼ by ⅞, finely variegated. WILS., v, 65, pl. 40; f. 1, 2; NUTT., i, 619; AUD., i, 159, pl. 43; BD., 151. . . . VIRGINIANUS.

⊦ Var. HENRYI is the lighter colored form prevailing in the dryer or unwooded portions of western United States; the gray and fulvous in excess of the darker hues, the white patches on the wing, tail and throat usually larger. CASS., Ill. 233; BD., 153, 922, pl. 17; COOP., 344.

Texas Night-hawk. Similar to the first; smaller; wing 7; tail 4; fine gray mottling much predominant above; below rufous prevailing over the dark bars; many broad fulvous bars on the tail, besides the white spots (♂) wanting in the ♀ ; primaries all sprinkled toward the base with numerous fulvous spots; the large white (♂) or tawny (♀) area nearer the tip than the bend of the wing, and on only *four* primaries. Southwestern U. S. and southward; unquestionably different from the common bird of this country, but in adopting the name *texensis*, I must say that I have not investigated its relationships to the South American form. LAWR., Ann. Lyc. v. 1851, 114; CASS., Ill. 238; BD., 154; COOP., 344. . . . . . TEXENSIS.

Family CYPSELIDÆ. Swifts.

Fissirostral Picariæ: bill very small, flattened, triangular when viewed from above, with great gape reaching below the eyes. Wings extremely long, thin and pointed (frequently as long as the whole bird); the secondaries extremely short (nine?). Tail of ten rectrices, variable in shape. Feet small, weak; tarsi naked or feathered; hind toe frequently elevated, or versatile, or permanently turned sideways or even forward; anterior toes completely cleft, the basal phalanges extremely short, the penultimate very long, the number of phalanges frequently abnormal; claws sharp, curved, never pectinate. Sternum deep-keeled, widening behind, its posterior margin entire. Eggs narrowly oval, white. For pterylosis see PLATE I.

"One of the most remarkable points in the structure of the *Cypselidæ* is the great development of the salivary glands. In all the species of which the nidification is known, the secretion thus produced is used more or less in the construction of the nest. In most cases it forms a glue by which the other materials are joined together, and the whole nest is affixed to a rock, wall, or other object against which it is placed. In some species of *Collocalia*, however, the whole nest is made up of inspissated saliva, and becomes the 'edible bird's-nest' so well known in the East." (SCLATER.)

A well defined family of six or eight genera and about fifty species, inhabiting temperate and warm parts of the globe. They are rather small birds of plain plumage, closely resembling swallows in superficial respects, but with no real affinity to these *Oscines*. The family is divisible into two subfamilies, according to the structure of the feet.

FIG. 118. Bones of Cypseline foot.

Subfamily CYPSELINÆ. Typical Swifts.

Ratio of the phalanges abnormal, the 3d and 4th toes having each 3 joints like the 2d; hind toe reversed (in *Cypselus*, where nearly all the species belong) or lateral (in *Panyptila*); tarsi feathered (in *Cypselus*); toes also feathered (in *Panyptila*). Contains only these two genera and nearly half the species of the family. Of *Panyptila* there are only three well determined species, all American; while *Cypselus* has upward of twenty, mostly of the Old World; the three or four American ones are sometimes detached under the name of *Tachornis*.

116. Genus PANYPTILA Cabanis.

White-throated Swift. Black or blackish; chin, throat, breast and middle line of belly, tips of secondaries, edge of outer primary, bases of tail feathers and a flank patch, white. Length 5½–6; wing the same; tail about 2¾, forked, soft. Southwestern U. S. and southward, breeding in colonies on cliffs. *Acanthylis saxatilis* WOODHOUSE, Expl. Zuñi River, 1853, 64; *Cypselus melanoleucus* BD., Proc. Phila. Acad. 1854, 118. COUES, *ibid.* 1866, 57; BD., 141; COOP., 347. . . . . SAXATILIS.

Subfamily CHÆTURINÆ. *Spine-tailed Swifts.*

Toes with the normal number of phalanges; hind toe not reversed, but some-times versatile; our species have it obviously elevated, and should have come in the Key under A, like gen. 114, 115; but it has not been technically so considered (compare § 87, p. 49). Tarsi never feathered. In the principal genus, *Chætura*, containing about half the species of the subfamily, of various parts of the world, the tail feathers are stiffened and *mucronate* by the projecting rhachis. The other genera are *Collocalia* and *Dendrochelidon* of the Old World; *Cypseloides*, and the scarcely different *Nephœcetes*, of the New.

117. Genus NEPHŒCETES Baird.

Black Swift. Blackish, nearly uniform. Length nearly 7; wing as much; tail about 3, forked, stiffish, but not mucronate. Western America. BD., 142; ELLIOT, pl. 20; COOP., 349. . . . . . . NIGER var. BOREALIS.

118. Genus CHÆTURA Stephens.

Chimney Swift. *Chimney "Swallow."* Sooty brown with a faint green-ish gloss above, below paler, becoming gray on the throat; wings black. Length about 5; wing the same; tail 2 or less, even or a little rounded, spiny. Eastern United States, migratory, very abundant in summer. Like the swallows, which this bird so curiously resembles, not only in its form, but in its mode of flight, its food, and twitter-ing notes, it has mostly forsaken the ways of its ancestors, who bred in hollow trees, and now places its curious open-work nest, of bits of twig glued together, inside disused chimneys. WILS., v, 48, pl. 39, f. 1; NUTT., i, 609; AUD., i, 164, pl. 44; BD., 144. . . . . PELASGIA.

FIG. 119. Chimney Swift, with mucronate rectrix.

Vaux's Swift. Similar; paler; the throat whitish; smaller; length 4½; wing the same. Pacific Coast, U. S. Seems to be different from *pelasgia*, but perhaps the same as a S. Am. species. BD., 145; COOP., 351. VAUXII.

Family TROCHILIDÆ. Hummingbirds.

Tenuirostral Picariæ. These beautiful little creatures will be known on sight; and as the limits of this work preclude any adequate presentation of the subject, I prefer merely to touch upon it. The hummers are peculiar to America. Species occur from Alaska to Patagonia, but we have a mere sprinkling in this country; the centre of abundance is in tropical South America, particularly New Granada. Nearly five hundred species are current; the number of positively specific forms may be estimated at about three hundred. The genera or subgenera vary with authors from fifty to a hundred and fifty; perhaps half the latter number of generic names may be eligible. The birds appear to fall naturally into two groups; one of these, *Phæthornithinæ*, representing about one-tenth of the whole, is composed of duller colored species especially inhabiting the dense forests of the Amazon; the other is the

Subfamily TROCHILINÆ.

119. Genus HELIOPÆDICA Gould.

- *Xantus Hummingbird.* Tarsi feathered; tail nearly even; first primary not attenuate; frontal feathers ending abruptly at base of bill; ♂ above, and the throat, metallic grass-green; below, cinnamon-rufous; face blue-black; a white stripe through the eye; wings purplish-dusky; tail purplish-chestnut, the central feathers glossed with golden green; bill flesh-colored, black-tipped. ♀ shining green above, including central tail feathers; below, and the face, pale rufous, whitening about the vent, and the sides greenish; head-stripes rufous, whitening on the auriculars; tail feathers, except the central, chestnut, with a dark terminal spot. 3½; wing 2½; tail 1½; bill ⁊. Cape St. Lucas. *Amazilia xantusii* and *Heliopædica castaneicauda* LAWR., Ann. Lyc. N. Y. 1860, pp. 105, 109; ELL., pl. 22; COOP., 365. XANTUSII. -

120. Genus LAMPORNIS Swainson.

Black-throated Hummingbird. Tomia serrate near the end; bill depressed, not quite straight; no metallic scales on throat; ♂ golden-green above and on the sides; below, opaque black, with white flank-tufts; wings and tail dusky-purplish. ♀ white below, with median black stripe. 4½; wing 2½; tail 1¾; bill nearly 1. Straggler to Florida. BD., 130, 922. . MANGO? ⊣

121. Genus TROCHILUS Linnæus.

Ruby-throated Hummingbird. ♂ with the tail forked, its feathers all narrow and pointed; no scales on crown; metallic gorget reflecting ruby-red, etc.; golden-green, below white, the sides green; wings and tail dusky-purplish. ♀ lacking the gorget; the throat white; the tail somewhat double-rounded, with black bars, and the outer feathers white-tipped. 3½; wing 1⅝; tail 1¼; bill ⅜. Eastern North America, abundant in summer, generally seen hovering about flowers, sometimes *in flocks.* Feeds on insects, and the sweets of flowers. Nest a beautiful structure, of downy substances, stuccoed with lichens outside; eggs two, white. WILS., ii, 26, pl. 10; NUTT., i, 588; AUD., iv, 190, pl. 253; BD., 131. . . . . . COLUBRIS.

Black-chinned Hummingbird. Similar; tail merely emarginate; gorget opaque black, reflecting steel blue, etc., posteriorly. ♀ with the tail simply rounded. California, Arizona and southward. CASS., Ill. 141, pl. 22; BD., 133; COOP., 353. . . . . ALEXANDRI. ⊦

122. Genus SELASPHORUS Swainson.

* No metallic scales on crown; ♂ throat scales not much prolonged into a ruff; outer primary attenuate; tail graduated, the middle feathers broader than the lateral.

Rufous-backed Hummingbird. ♂ chiefly cinnamon-rufous above, below and on the tail; traces of green above, especially on crown; gorget red, etc.; a white collar behind it. ♀ with a trace of the gorget; upper parts more or less green; tail barred with black and tipped with white. 3½; wing 1⅞; tail 1⅜. Rocky Mountains to the Pacific, from Mexico to Alaska; abundant; the sole boreal representative of the family. AUD., iv, 200, pl. 254; BD., 134; COOP., 355. . . . . RUFUS. †

Broad-tailed Hummingbird. Outer primaries much attenuated, outer tail feather linear, very narrow; others broad; ♂ glittering green, with much white below, the gorget purplish-red, etc.; wings and tail dusky-purplish, most of the tail feathers with rufous edging basally. ♀ with no gorget; no green, but much rufous, below; 4; wing 2. Rocky Mts. to lat. 42°; W. to Sierra Nevada; S. into Mex. BD., 135; COOP., 357. PLATYCERCUS. +

** Crown of ♂ with metallic scales like the gorget, which is prolonged into a ruff; outer primary not attenuate; tail of ♂ forked, the outer feather abruptly narrow and linear. (*Calypte.*)

Anna Hummingbird. ♂ above, and the breast, green; crown and gorget ruby-red, etc.; ♀ lacking the scales; the tail slightly rounded, black-barred, white-tipped. Size of the last. California; resident, abundant. NUTT., i, 2d. ed. 712; AUD., iv, 188, pl. 252; BD., 137; COOP., 358. . ANNA.

Costa Hummingbird. ♂ above, and on the sides, green; mostly white below; crown and gorget purplish, steel-blue, etc.; the latter much prolonged into a ruff; tail lightly forked; ♀ like that of *anna;* smaller; about 3½; wing under 2; tail 1⅛; bill ⅔. Valley of the Colorado; S. and Lower California. BD.,138; COOP., 360. . . . . COSTÆ. †

*** Crown of ♂ not metallic like the gorget, which is prolonged into a ruff; outer primary of ♂ attenuate; tail graduated, the feathers rounded at the end, the lateral black-barred and white-tipped (in both sexes). (*Atthis.*)

Heloise Hummingbird. ♂ golden-green above, including crown; gorget lilac-red, bordered with white; below white; sides with green and rufous; tail feathers cinnamon-rufous at base, the central otherwise like the back, the others black-barred and white-tipped; ♀ similar, lacking the gorget; outer primary not attenuate. Very small; 2¾; wing 1⅜; tail 1; bill ½. Texas and southward. ELLIOT, pl. 21; COOP., 361. . . . . . HELOISÆ.

123. Genus STELLULA Gould.

Calliope Hummingbird. ♂ golden-green; below white, with green and rufous on sides; gorget violet or lilac, the bases of the scales, and sides of the neck, pure white; tail feathers brown, including the central pair, with pale tips and slight rufous edgings; under mandible light; ♀ with dusky specks in place of the gorget; throat feathers not ruffed; no green on sides; tail feathers variegated with green, rufous, black and white. 2¾; wing 1⅞; tail 1; bill ½. Mountains of Washington, Oregon and California, to Mexico. ELLIOT, pl. 23; COOP., 363. . . . . CALLIOPE. †

124. Genus AGYRTRIA Cabanis.

Linné Hummingbird. ♂ bronzy-green, including middle tail feathers; throat and breast grass-green, paler on sides; middle of belly, and crissum, white; wings purplish-brown; lateral tail feathers black with paler tips; ♀ duller, more white below, no green on throat; wing 2; tail 1¼; bill ⅝. South America; accidental in Massachusetts, one instance (Aug. 1865, *Brewster*); but I am advised that the occurrence is open to suspicion. ALLEN, Am. Nat. iii, 1869, p. 645; MAYNARD, Guide, 128. . LINN.EI. ⊥

Suborder *CUCULI. Cuculiform Birds.*

The nature of this large group has been indicated on a preceding page (178).

Family TROGONIDÆ. Trogons.

Feet zygodactyle by reversion of the second toe. The base of the short, broad, dentate bill is hidden by appressed antrorse feathers; the wings are short and rounded, with falcate quills; the tail is long, of twelve broad feathers; the feet are very small and weak. The general plumage is soft and lax, the skin tender, the eyelids lashed. A well-marked family of about fifty species and perhaps a dozen genera, chiefly inhabiting tropical America. They are of gorgeous colors, and among them are found the most magnificent birds of this continent.

125. Genus TROGON Auctorum.

Mexican Trogon. Metallic golden-green; face and sides of head black; below from the breast carmine; a white collar on the throat; middle tail feathers coppery-green, the outer white, barred with black;

FIG. 120. Mexican Trogon. quills edged with white; about 11; wing 5⅓; tail 6¾. Valley of the Rio Grande, southward. BD., 69, pl. 40. . . MEXICANUS?

Family MOMOTIDÆ. Sawbills.

Feet syndactyle by cohesion of third and fourth toes; tomia serrate. A very small family of tropical American birds, comprising about fifteen species. Neither this nor the foregoing has really rightful place here, but they come on our border, and are included to illustrate the suborder. In the following species, the central tail feathers are long-exserted, and spatulate by absence of webs along a part of the shaft—a mutilation effected, it is said, by the birds themselves; the bill is about as long as the head, gently curved;

FIG. 121. Blue-headed Sawbill.

the nostrils are rounded, basal, exposed; the wings are short and rounded; the tarsi are scutellate anteriorly.

112. Genus MOMOTUS Lesson.

Blue-headed Sawbill. Greenish, rather paler below ; purer on wings and tail ; the crested crown blue, encircled with black ; face mostly black. 15 ; wing 5½. Mexico. BD., 161, pl. 46. . . . . CÆRULEICEPS.

Family ALCEDINIDÆ. Kingfishers.

Feet syndactyle by cohesion of third and fourth toes; tomia simple. Bill long, large, straight, acute (rarely hooked), "fissirostral," the gape being deep and wide ; tongue rudimentary or very small ; nostrils basal, reached by the frontal feathers. Feet very small and weak, scarcely or not ambulatorial ; tarsi extremely short, reticulate in front ; hallux short, flattened underneath, its sole more or less continuous with the sole of the inner toe ; soles of outer and middle toe in common for at least half their length ; inner toe always short, in one genus rudimentary. in another wanting (an abnormal modification, overlooked in penning § 86. p. 49 ; but see § 81) ; wings long, of 10 primaries ; tail of 12 rectrices, variable in shape.

FIG. 122.   Syndactyle foot.

"The kingfishers form a very natural family of the great Picarian order, and are alike remarkable for their brilliant coloration and for the variety of curious and aberrant forms which are included within their number. . . 'Their characteristic habit is to sit motionless watching for their prey, to dart after it and seize it on the wing, and to return to their original position to swallow it.' . . The *Alcedinidæ* nest in holes and lay white eggs. It is, however, to be remarked that, in accordance with a modification of the habits of the various genera, a corresponding modification has taken place in the mode of nidification, the piscivorous section of the family nesting for the most part in holes in the banks of streams, while the insectivorous section of the family generally nest in the holes of trees, not necessarily in the vicinity of water." (SHARPE.)

The nearest allies of the kingfishers are considered to be the hornbills and bee-eaters of the Old World, and the sawbills and todies of the New. One would gain an imperfect or erroneous idea of the family to judge of it by the American fragment, of one genus and six or eight species. According to the author of the splendid monograph just cited, there are in all 125 species, belonging to 19 genera ; the latter appear to be very judiciously handled, but a moderate reduction of the former will be required. They are very unequally distributed ; *Ceryle* alone is nearly cosmopolitan, absent only from the Australian region ; the northern portion of the Old World has only 2 peculiar species ; 3 genera and 24 species are characteristic of the Ethiopian region, one genus and 25 species are confined to the Indian, while no less than 10 genera and 59 species are peculiar to the Australian. Mr. Sharpe recognizes two subfamilies ; in the *Daceloninæ* (with 14 genera, and 84 species) the bill is more or less depressed with smooth, rounded or sulcate, culmen. In the

Subfamily ALCEDININÆ,

the bill is compressed, with carinate culmen. The American species all belong here. It is the more particularly piscivorous section ; the *Daceloninæ* feed for the most part upon insects, reptiles, and land mollusks.

113. Genus CERYLE Boie.

Belted Kingfisher. Upper parts, broad pectoral bar, and sides under the wings, dull blue with fine black shaft lines; lower eyelid, spot before eye, a cervical collar and under parts except as said, pure white; the ♀ with a chestnut belly-band, and the sides of the same color; quills and tail feathers black, speckled, blotched or barred on the inner webs with white; outer webs of the secondaries and tail feathers like the back; wing coverts frequently sprinkled with white; bill black, pale at base below; feet dark, tibiæ naked below; a long, thin, pointed occipital crest; plumage compact and oily to resist water, into which the birds constantly plunge after their finny prey. Length a foot or more; wing about 6; tail 3½; whole foot 1⅓; culmen about 2¼. North America, common everywhere, resident or only forced

FIG. 123. Belted Kingfisher. ♀.

southward by freezing of the waters. WILS., iii, 59, pl. 23, f. 1; NUTT., i, 594; AUD., iv, 205, pl. 255; BD., 158. . . . . ALCYON.

Cabanis' Kingfisher. Glossy green; a cervical collar and the under parts white; ♂ with a rufous, ♀ with an imperfect, greenish, pectoral bar; quills and tail feathers black, partly like the back, with numerous white spots, mostly paired. Small; about 8; wing 3½; tail 2¼. Valleys of the Rio Grande and Colorado, and

FIG. 124. Cabanis' Kingfisher.

southward. CASS., Ill. 255; BD., 159, and Mex. Bound. Surv. ii, pl. 7; COOP., 339. . . . . AMERICANA var. CABANISII.

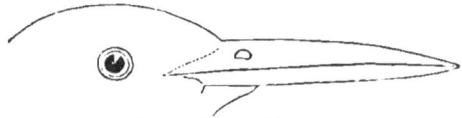

Family CUCULIDÆ. Cuckoos.

Feet zygodactyle by reversion of the fourth toe. This character, in connection with those given below, will answer present purposes; and in my ignorance of some of the exotic forms, I cannot attempt to give a full diagnosis. The family is a large and important one. It comprehends quite a number of leading forms showing peculiar minor modifications; these correspond in great measure with certain geographical areas of faunal distribution, and are generally held to constitute subfamilies. Three or four such are confined to America; about twice as many belong exclusively to the Old World; among them are the *Cuculinæ*, or typical cuckoos allied to the European *C. canorus*, famous, like our cowbird, for its parasitism. This section comprehends the great majority of the Old World species; the *Couinæ* are a peculiar Madagascan type; others rest upon a special condition of the claws or plumage. There are about two hundred current species of the family.

Subfamily CROTOPHAGINÆ. Anis.

Tail of *eight* feathers, graduated, longer than the rounded wings. Bill exceedingly compressed, the upper mandible rising into a thin vertical crest, the sides usually sulcate, the tip deflected. Plumage uniform (black), lustrous, the feathers of the head and neck lengthened, lanceolate, distinct, with scale-like margins; face naked. Terrestrial. Nest in bushes. One genus, and two or three species, of the warmer parts of America.

126. Genus CROTOPHAGA Linnæus.

Ani. About a foot long; wing 6; tail 8. Florida (and Gulf?) coast, southward; accidental north to Philadelphia. *C. ani* and *C. rugirostris* Bd., 71, 72. . . . . ANI.

Subfamily SAUROTHERINÆ. Ground Cuckoos.

Tail of *ten* feathers, graduated, longer than the short, rounded, concave wings. Bill about as long as the head, compressed, straight at base, tapering, with deflected tip, gently curved culmen, and ample rictus. Feet large and strong, in adaptation to terrestrial life; tarsus longer than the toes, scutellate before and behind. One West Indian genus, *Saurophaga*, with three or four species, and the following, with one or two : —

127. Genus GEOCOCCYX Wagler.

Ground Cuckoo. Chaparral Cock. Road Runner. Snake Killer. Paisano. Most of the feathers of the head and neck bristle-tipped; a naked area around eye; crown crested; plumage coarse. Above, lustrous bronzy green, the crest dark blue, everywhere sharply streaked with whitish or tawny brown; sides and front of the neck tawny, with sharp black streaks; other under parts dirty white; quills and tail feathers much edged with white; central rectrices like the back, others darker green, violet, etc., with broad white tips. Nearly 2 feet long; tail a foot or more; wing 6–7 inches; tarsus 2; bill 1¾. Sexes nearly alike. Texas, New Mexico, Arizona, California and southward. A bird of remarkable aspect, noted for its swiftness of foot; aided by its wings held as outriggers, it taxes the horse in a race; feeds on reptiles, insects and land mollusks. Cass., Ill. 213, pl. 36; Bd., 73; Coop., 363. . . . . . CALIFORNIANUS.

FIG. 125. Ground Cuckoo.

Subfamily COCCYZINÆ. American Cuckoos.

Tail of *ten* soft feathers, much graduated, little longer than the wings, which are somewhat pointed, although the first and second quills are shortened. Bill about equalling or rather shorter than the head, stout at base, then much compressed, curved throughout, tapering to a rather acute tip; nostrils basal, inferior, exposed, elliptical; feet comparatively small, the tarsus naked, not longer than the toes.

Four or five genera, and perhaps twenty species; none parasitic. Ours are strictly arboricole birds of lithe form, blended plumage and subdued colors; the head is not crested; the tibial feathers are full, as in a hawk; the sexes are alike, and the young scarcely different. In the following, the upper parts are uniform satiny olive-gray, or "quaker color," with bronzy reflections. Migratory, insectivorous; lay plain greenish eggs, in a rude nest of twigs saddled on a branch or in a fork. They are well known inhabitants of our streets and parks as well as of woodland, noted for their loud jerky cries, which they are supposed to utter most frequently in falling weather, whence their popular name, "rain crow."

128. Genus COCCYZUS Vieillot.

Black-billed Cuckoo. Bill blackish except occasionally a trace of yellowish below. Below, pure white, sometimes with a faint tawny tinge on the fore parts. Wings with little or no rufous. Lateral tail feathers not contrasting with the central, their tips for a short distance blackish, then obscurely white. Bare circumocular space red. Length 11–12; wing 5–5½; tail 6–6½; bill under an inch. Eastern U. S. and Canada. WILS., iv, 16, pl. 28; NUTT., i, 556; AUD., iv, 300, pl. 276; BD., 77. ERYTHROPHTHALMUS.

Yellow-billed Cuckoo. Bill extensively yellow below and on the sides. Below, pure white. Wings extensively cinnamon-rufous on inner webs of the quills. Central tail feathers

Fig. 126. Yellow-billed Cuckoo.

like the back, the rest black with large white tips, the outermost usually also edged with white. Size of the last. United States, rather more southerly than the last species, and chiefly Eastern; also, Pacific Coast (*Cooper, Nuttall*). WILS., iv, 13, pl. 28; NUTT., i, 551; AUD., iv, 293, pl. 275; BD., 76; COOP., 371. . . . . AMERICANUS.

Mangrove Cuckoo. Bill much as in the last. Below, pale orange-brown. Auriculars dark, in contrast. Tail as in the last, but outer feathers not white-edged. Size of the others, or rather less. West Indies and Florida. NUTT., i, 558; AUD., iv, 303; pl. 277; BD., 78. . SENICULUS.

Suborder PICI. Piciform Birds.

See p. 179 for characters of this suborder.

Family PICIDÆ. Woodpeckers.

These birds have been specially studied, with more or less gratifying success, by Malherbe, Sundevall and Cassin. There are nearly two hundred and fifty well determined species, of all parts of the world except Madagascar, Australia and Polynesia. Their separation into minor groups has not been agreed upon; our species are commonly thrown into three divisions, which, however, I shall not present. The ivory-bill and the flicker stand nearly at extremes of the family, the little diversity of which is thereby evident. One of our genera, without very obvious external peculiarities, stands apart from the rest in the character of the tongue—a fact that seems to have escaped general attention. In ordinary *Pici* the " horns" of the tongue are extraordinarily produced backward, as slender jointed bony rods curling up over the skull behind, between the skin and the bone, to the eyes or even further; these rods are enwrapped in highly developed, specialized muscles, by means of which the birds thrust out the tongue sometimes several inches beyond the bill. This is not the case in *Sphyrapicus*, where the hyoid cornua do not extend beyond the base of the skull, and the tongue, consequently, is but little more extensible than in ordinary birds. I have determined this by examination of all our species but one, in the flesh. The tongue of *Sphyrapicus* is beset at the end by numerous brushy filaments, instead of the few acute barbs commonly observed in the family. (See also under gen. 133.) In most of our species the bill is perfectly straight, wide and stout at the base, tapering regularly to a compressed and vertically truncate tip, chisel-like, and strengthened by sharp ridges on the side of the upper mandible — an admirable tool for cutting into trees; and in all such, the nostrils are hidden by dense tufts of antrorse feathers. In others, like the flicker, the bill is smooth, barely curved, the tip acute and the nostrils exposed. The claws are always large, strong, sharp and much curved; the feet do not present striking modifications, except in the three-toed genus *Picoides*. The wings offer nothing specially noteworthy, unless it be the shortness of the coverts, in exception to the Picarian rule; and the shortness of the first primary, which may fairly be called spurious. The remarkable character of the tail has been already mentioned. This member offers indispensable assistance in climbing, when the stiff strong quills are pressed against the tree, and form a secure support To this end, the muscles are highly developed, and the last bone (*vomer* or *pygostyle*) is large and peculiar in shape. Woodpeckers rarely if ever hang head downward, like Nuthatches, nor are the tarsi applied to their support.

Species are abundant in all the wooded portion of this country, and wherever found are nearly resident. For, although insectivorous, they feed principally upon dormant or at least stationary insects, and therefore need not migrate; they are, moreover, hardy birds. They dig insects and their larvæ out of trees, and are eminently beneficial to the agriculturist and fruit grower. Contrary to a prevalent impression, their boring does not seem to injure fruit trees, which may be riddled with holes without harmful result. The number of noxious insects these birds destroy is simply incalculable; what little fruit some of them steal is not to be mentioned in the same connection, and they deserve the good will of all. The birds of the genus *Sphyrapicus* are probably an exception to most of these statements. Woodpeckers nest in holes in trees, which they excavate for themselves, sometimes to a great depth, and lay numerous rounded pure white eggs, of which the shell has a crystalline texture, on the chips and dust at the bottom of the hole. The voice is loud and harsh, susceptible of little inflection. The plumage as a rule presents bright colors in large areas or in striking contrasts, and is sometimes highly lustrous. The sexes are ordinarily distinguishable by color-markings.

Obs. *Campephilus imperialis*, the largest and most magnificent bird of the family, inhabiting Central America, has been attributed to the United States, but upon unsatisfactory evidence. (AUD., iv, 213; CASS., 285, pl. 49; BD., 82.) *Dryocopus lineatus*, likewise, was improperly introduced by Audubon (iv, 233).

129. Genus CAMPEPHILUS Gray.

Ivory-billed Woodpecker. Black; a stripe down the side of the neck, one at base of bill, the scapulars, under wing coverts, and ends of secondaries, white; bill and nasal feathers white; ♂ scarlet-crested; ♀ black-crested. A large, powerful bird of the South Atlantic and Gulf States; about 21 long; wing 10–11; tail 7–8. WILS., iv, 20, pl. 39, f. 6; NUTT., i, 564; AUD., iv, 214, pl. 256; BD., 81.   . . .   PRINCIPALIS.

130. Genus HYLOTOMUS Baird.

Pileated Woodpecker. Black; the head, neck and wings much varied with white or pale yellowish; bill dark;

FIG. 127. Ivory-billed Woodpecker.

♂ scarlet-crested, scarlet-moustached; ♀ with the crest half black, half scarlet, and no maxillary patches. . Only yielding to the ivory-bill in size; length 15–19; wing 8½–10; tail 6–7. North America, anywhere, in heavy timber. WILS., iv, 27, pl. 29, f. 2; NUTT., i, 567; AUD., iv, 226, pl. 257; BD., 107. . . PILEATUS.

131. Genus PICUS Linnæus.

⁎⁎⁎ All the following species are black-and-white, the ♂ with red on the head; and all but the first have numerous, small, round, white spots on the quills.

FIG. 128. Pileated Woodpecker.

* Body not banded, streaked, nor spotted.

White-headed Woodpecker. Uniform black; whole head white, in the ♂ with a scarlet nuchal band; a large patch of white on the wing, commonly resolved into a number of blotches; about 9; wing 5; tail 3½. Mountains of California, Oregon and Washington. CASS., Journ. Phila. Acad. 1853, 257, pl. 22; BD., 96; ELL., pl. 24; COOP., 382. . . . . . ALBOLARVATUS.

** Spotted and crosswise banded, but not streaked.

Red-cockaded Woodpecker. Head black on top, with a large silky white auricular patch embracing the eye and extending on the side of the neck, bordered above in the ♂ by a scarlet stripe *not* meeting its fellow on the nape; nasal feathers and those on the side of the under jaw white;

FIG. 129. Red-cockaded Woodpecker.

black of the crown connected across the lores with a black stripe running from the corner of the bill down the side of the throat and neck to be dissipated on the side of the breast in black spots continued less thickly along the whole side and on the crissum; under parts otherwise soiled white; central tail feathers black, others white, black-barred; back and wings barred with black and white, the larger quills and many coverts with the white bars resolved into paired spots; 8–8½; extent 14–15; wing 4½; tail 3½. Pine swamps and barrens of the South Atlantic and Gulf States; North to Pennsylvania. WILS., ii, 103, pl. 15; NUTT., i, 577; AUD., iv, 254, pl. 264; BD., 96. . . . . BOREALIS.

Texan Woodpecker. Crown black, frequently speckled with white, in the ♂ the hind head and nape extensively crimson; sides of the head white, with a long black stripe from the bill under the eye, widening behind, there joining a black postocular stripe and spreading over the side of the neck; nasal feathers usually brown; under parts ranging from soiled white to smoky gray, with numerous black spots on the sides, flanks and crissum; lateral tail feathers perfectly barred with black and white in equal amounts, the central ones black; back and wings as in the last species. Small; about 7; wing 3½–4; tail under 3; bill ⅔–¾. Southwestern U. S. and southward. BD., 94; COOP., 379. . . . . SCALARIS.

Var. NUTTALLII. Rather larger; more white, this rather prevailing on the back over the black bars, the hind neck chiefly white, the nasal tufts white, the lateral tail feathers, especially, sparsely or imperfectly barred. The Californian coast race; BD., 93; COOP., 378. *Picus lucasanus*, from Cape St. Lucas, is a local form like *nuttallii*, with rather larger bill and feet; bill 1 inch. XANTUS, Proc. Phila. Acad. 1859, 298, 302; CASS., *ibid.* 1863, 195; COOP., 381. *P. parvus* CABOT; *P. bairdii* SCLATER; *PP. vagatus* and *orizabæ* CASSIN, all belong to *scalaris*.

*** Spotted and lengthwise streaked, but not banded.

† Usually 9–10 long; outer tail feathers wholly white.

Hairy Woodpecker. Back black, with a long white stripe; quills *and wing coverts* with a profusion of white spots; four middle tail feathers black, next pair black and white, next two pair white, as stated; under parts white; crown and sides of head black, with a white stripe over and behind the eye, another from the nasal feathers running below the eye to spread on the side of the neck, and a scarlet nuchal band in the ♂, wanting in the ♀; young with the crown mostly red or bronzy, or even yellowish. Eastern North America, abundant. Wing nearly 5; tail 3½; bill 1⅛; whole foot 1⅜. Varies greatly in size, mainly according to latitude. Large whiter northern birds are—*P. leucomelas* BODD., Pl. Enlum. 345, f. 1; *P. canadensis* GM., i. 437; *P. phillipsii* AUD., iv, 238, pl. 259 (young with crown yellowish); *P. septentrionalis* NUTT., i, 2d ed. 684 (same); var. *major* BD., 84. Ordinary birds are—*P. villosus* WILS., i, 150, pl. 9; NUTT., i, 575; AUD., iv, 244, pl. 262; *P. martinæ* AUD., iv, 240, pl. 260 (young with crown reddish); *P. rubricapillus* NUTT., i, 2d ed. 685 (same); var. *medius* BD., 84. Small southern birds are—*P. auduboni* SWAINSON, Fn. Bor.-Am. ii, 306; *P. auduboni* TRUDEAU, Journ. Phila. Acad. 1837, 404 (young with crown yellowish); AUD., iv, 259, pl. 265; NUTT., i, 2d ed. 684; var. *minor* BD., 85. . . . . VILLOSUS.

Var. HARRISII.   Exactly like *villosus*, excepting fewer wing-spots ; generally *none* on the coverts and inner quills ; with specimens enough we can see the spots disappear one by one.   Generally white below, but in some regions smoky-gray (a thing not observed in Eastern birds, but apparently due, sometimes at least, to soiling with carbonaceous matter). Rocky Mountains to the Pacific. AUD., iv, 242, pl. 261 (dark-bellied) ; NUTT., i, 2d ed. 627 ; BD., 87 ; COOP., 375.   *P. hyloscopus* CABANIS.

†† Usually 6–7 long ; outer tail feathers barred with black and white.

Downy Woodpecker.   Exactly like *P. villosus*, except in the above respects ; wing under 4 ; tail under 3 ; bill about ⅔ ; whole foot 1¼.   Eastern North America, abundant in orchards, and all wooded places.   WILS., i, 153, pl. 9 ; NUTT., i, 576 ; AUD., iv, 249, pl. 263 ; BD., 89. *P. meridionalis* SWAINS., F. B.-A. ii, 308 (small southern race) ; *P. medianus* ID., *ibid.* 308. . . . . PUBESCENS.

FIG. 130.   Downy Woodpecker.

Var. GAIRDNERII. Bearing the same relation to *P. pubescens*, that *harrisii* does to *P. villosus*, and inhabiting the same regions ; the wing spots few or wanting on the inner quills and the coverts, the belly smoky-gray in some localities. AUD., iv, 252 ; BD., 91, pl. 85 ; COOP., 377. · *P. meridionalis* NUTT., i, 2d ed. 690.

132. Genus PICOIDES Lacepede.

⁎ Three-toed ; the hallux absent.   Crown with a yellow patch in the ♂ ; sides of head striped, of body barred, with black and white ; under parts otherwise white ; quills with white spots ; tail feathers unbarred, the outer white, the central black. Length 8–9 ; wing 4½–5 ; tail 3½–4.

Black-backed Woodpecker.   Back uniform black.   Arctic America to the Northern States. AUD., iv, 266, pl. 268 ; NUTT., i, 578 ; BD., 98 ; COOP., 384.   *P. tridactylus* BONAP., Am. Orn. ii, 14, pl. 14, f. 2.   . ARCTICUS.

Banded Woodpecker.   Back with a white lengthwise stripe, banded with black tips of the feathers.   Arctic America into Northern States. *P. hirsutus* AUD., iv, 268, pl. 269 ; NUTT., i, 2d ed. 622 ; BD., 98 ; *P. tridactylus* SWAINS., F. B.-A. ii, 311, pl. 56 ; *P. americanus* COOP., 385. AMERICANUS.

Var. DORSALIS. Back with an uninterrupted white stripe ; BD., 100, pl. 85, f. 1. Rocky Mountain region. *⁎* All the species of this genus are unquestionably modified derivatives of one circumpolar stock ; the American seem to have become completely differentiated from the Asiatic and European, and further divergence seems to have perfectly separated *arcticus* from *americanus;* but *dorsalis* and *americanus* are still linked together.

133. Genus SPHYRAPICUS Baird.

⁎ Tongue not extensible ; the tip brushy ; hyoid bones short. Birds of this genus feed much upon fruits, as well as insects, and also, it would seem, upon soft inner bark (cambium) ; they injure fruit trees by stripping off the bark, sometimes in large areas, instead of simply boring holes. Of the several small species commonly called "sapsuckers," they alone deserve the name. In declaring war against

woodpeckers, the agriculturist will do well to discriminate between the somewhat injurious and the highly beneficial species.

Yellow-bellied Woodpecker.   ♂ with the crown crimson, bordered all around with black; chin, throat and breast black, enclosing a large crimson patch on the former (in the ♂; in the ♀ this patch white); sides of head with a white line starting from the nasal feathers and dividing the black of the throat from a trans-ocular black stripe, this separated from the black of the crown by a white postocular stripe; all these stripes frequently y e l l o w i s h; under parts dingy yellow, brown-ish and with sagittate dusky marks on the sides; back variegated with black and yellowish-brown; wings black with a large oblique white bar on the coverts, the quills with numerous paired white spots on

FIG. 131.   Yellow-bellied Woodpecker.

the edge of both webs; tail black, most of the feathers white-edged, the inner webs of the middle pair, and the upper coverts, mostly white.   Young birds lack the definite black areas of the head and breast, and the crimson throat-patch, these parts being mottled gray; but in any plumage the bird is recognized by its *yellowness*, different from what is seen in any other Eastern species, and the broad white wing-bar, to say nothing of the generic characters.   About 8½; wing 4½–5; tail 3½.   Eastern North America, abundant.   WILS., i, 147, pl. 9, f. 2; NUTT., i, 574; AUD., iv, 263, pl. 267; BD., 103. . . . .   VARIUS.

Var. NUCHALIS.   With an additional band of scarlet on the nape, and the throat-patch more extensive; it is often seen in the ♀.   Rocky Mountains to the Pacific, but apparently not exclusively western; I am informed that birds of this descrip-tion are found in New England.   BD., 103, 897; COOP., 390.

Red-breasted Woodpecker.   Exactly like the last, but the whole head, neck and breast carmine red, in both sexes; gray in the young.   Size of the last, with which it is said to intergrade, and of which it is apparently only a variety.   Pacific Coast, U. S.   AUD., iv, 261, pl. 266; BD., 104; COOP., 392. . . . .   RUBER.

Brown-headed Woodpecker.   General plumage closely banded with black and grayish-white; rump white; middle of belly yellow, of breast black; whole head nearly uniform brown; quills sprinkled with white along the edges; tail black with the middle feathers white-barred; 9–9½; wing 5 or more; tail 4 or less.   Wooded mountainous regions, Pacific slope.   CASS., Ill. 200, pl. 32; BD., 106; ELLIOT, pl. 25; COOP., 393. .   THYROIDEUS.

Williamson's Woodpecker.   Glossy black; sides and crissum mixed black and white; belly yellow; rump white; tail black, unmarked; wings black, with a large oblique white bar on the coverts, and a few white spots on the edges of the quills; throat with a narrow crimson patch (white in the ♀ ?) ; head with a white postocular stripe meeting its fellow on the nape, and another from the nasal feathers to below the auriculars; size of the last. Same habitat. A beautiful species; this, and *thyroideus*, resemble no others. Bd., 105; Coop., 393. . . . . WILLIAMSONII.

134. Genus CENTURUS Swainson.

*** Back and wings, except larger quills, closely banded with black and white ; primaries with large white blotches near the base, and usually a few smaller spots ; below, immaculate, except sagittate black marks on the flanks and crissum ; the belly tinged with red or yellow ; 9–10 long ; wing about 5 ; tail about 3½.

Red-bellied Woodpecker. Whole crown and nape scarlet in the ♂ , partly so in the ♀ ; sides of head and under parts grayish-white, usually with a yellow shade, *reddening* on the belly ; tail black, one or two outer feathers white-barred ; inner web of central feathers white with black spots, outer web of the same black with a white space next the shaft for most of its length; white predominating on the rump. Eastern United States, somewhat southerly, rarely N. to New England ; common. WILS., i, 113, pl. 7, f. 2 ; NUTT., i, 572 ; AUD., iv, 270, pl. 270 ; BD., 109. . . CAROLINUS.

Yellow-faced Woodpecker. Extreme forehead and a nuchal band yellow ; crown with a central square crimson patch, wanting in the ♀ ; rump and upper tail coverts entirely white ; tail feathers entirely black, except white touches on the outer pair ; lower parts sordid whitish, becoming yellow on the belly. Texas, southward. *C. flaviventris* BD., 110, pl. 42 ; *P. aurifrons*, WAGLER, Isis, 1829, 512 ; COOP., 399. . . . . AURIFRONS.

Gila Woodpecker. No yellow about the head ; crown with a square crimson patch, wanting in the ♀ ; rump and upper tail coverts barred with black ; tail feathers marked as in *carolinus*; head and under parts dull brown, becoming yellow on the belly. Valley of the Colorado and Gila. BD., 111 ; Coop., 399. . . . . UROPYGIALIS.

135. Genus MELANERPES Swainson.

Red-headed Woodpecker. Glossy blue-black ; rump, secondaries and under parts from the breast, pure white ; primaries and tail feathers black ; whole head, neck and breast crimson, in both sexes, grayish-brown in the young ; about 9 ; wing 5½ ; tail 3½. Eastern U. S. to the Rocky Mountains ; California? A very abundant and familiar bird, in orchards and gardens as well as in the woods, conspicuous by its gay tricolor plumage ; migratory in northerly sections. WILS., i, 142, pl. 9, f. 1 ; NUTT., i, p. —; AUD., iv, 274, pl. 271 ; BD., 113 ; Coop., 402. . . . . . ERYTHROCEPHALUS.

Californian Woodpecker. Glossy blue-black ; rump, bases of all the quills, edge of the wing, and under parts from the breast, white, the sides with sparse black streaks ; forehead squarely white, continuous with a stripe down in front of the eyes and thence broadly encircling the throat, there

2
2a

3

4

5

6
6a

becoming yellowish; this cuts off the black around base of bill and on the chin completely; crown in the ♂ crimson from the white front, in the ♀ separated from the white by a black interval; frequently a few red feathers in the black breast-patch, which is not sharply defined behind, but changes by streaks into the white of the belly. Bill black; eyes white, brown in the young, which are not particularly different, but have the head markings less defined. Size of the last. Rocky Mountains to the Pacific, U. S., abundant; noted for its habit of sticking acorns in little holes that it digs in the bark for the purpose; whole branches are frequently studded in this manner. CASS., Ill. ii, pl. 2; BD., 114; COOP., 403. . . . . . . FORMICIVORUS.

Var. ANGUSTIFRONS is said to have the white frontal bar narrower and the bill somewhat differently shaped. Cape St. Lucas. COOP., 405.

136. Genus ASYNDESMUS Coues.

Lewis' Woodpecker. Black, with bronzy-green iridescence; wings and tail the same, unmarked; face and sides of head dusky crimson; cervical collar and under parts hoary-ash, becoming crimson or bloody-red on the belly; the feathers of these parts of a peculiar loose bristly texture; sexes alike; young plainer black above, with little or no crimson on face or below. About 11 long; wing 6½; tail 4½. Wooded and especially mountainous parts of Western America; a remarkable looking bird. WILS., iii, 31, pl. 20; NUTT., i, 577; AUD., iv, 280, pl. 272; BD., 115; COUES, Proc. Phila. Acad. 1866, 56; COOP., 406. . . . . TORQUATUS.

137. Genus COLAPTES Swainson.

⁎⁎* Under parts with numerous circular black spots on a pale ground. A large black pectoral crescent. Rump snowy white. Back, wing coverts and innermost quills brown with an olive or lilac shade, and thickly barred with black; quills and tail black, excepting as below stated. About a foot long; wing about 6; tail 4½.

Golden-winged Woodpecker. Flicker. Wings and tail showing golden-yellow underneath, and the shafts of this color; a scarlet nuchal crescent in both sexes; ♂ with black maxillary patches, wanting in the ♀; crown and nape ash; chin, throat and breast lilac-brown; sides tinged with creamy brown, and belly with yellowish; shade of the back rather olivaceous. Eastern North America; Alaska (*Dall*). A very abundant and well known bird. WILS., i, 45, pl. 3, f. 1; NUTT., i, 561; AUD., iv, 282, pl. 273; BD., 118. . . . . AURATUS.

FIG. 132. Golden-winged Woodpecker.

Gilded Woodpecker. Wings and tail showing golden yellow underneath, and the shafts of this color; no nuchal crescent in either sex; ♂ with scarlet maxillary patches, wanting in the ♀ ; crown lilac-brown; chin, throat and breast ash ; sides tinged with creamy-brown, and belly with yellowish. Colorado Valley, Lower California, and southward. BD., 125, and Proc. Phila. Acad. 1859, 302; ELLIOT, pl. 26 ; COOP., 410. . . CHRYSOIDES.

Red-shafted or *Mexican Woodpecker.* Wings and tail showing orange-red underneath, and the shafts of this color; no nuchal crescent in either sex; ♂ with scarlet maxillary patches, wanting in the ♀ ; crown lilac-brown; chin, throat and breast ash ; under parts shaded with lilac-brown; no yellowish on the belly. Western North America, Sitka to Mexico. AUD., iv, 290, pl. 274 ; NUTT., ii, 603; BD., 120; COOP., 408. . MEXICANUS.

OBS. It will be noted, how curiously these species are distinguished mainly by a different combination of common characters.— *Colaptes ayresii* of AUD., vii, 348, pl. 494 ; *C. hybridus* of BAIRD, 122, is a form from the Missouri region in which the characters of *mexicanus* and *auratus* are blended in every conceivable degree in different specimens. Perhaps it is a hybrid, and perhaps it is a transitional form. According to Mr. Allen, Florida specimens of *auratus* sometimes show red touches in the black maxillary patch, as is frequently the case with Kansas examples.

Order PSITTACI. Parrots.

Feet permanently zygodactyle by reversion of the fourth toe ; bill short, extremely stout, strongly epignathous, and furnished with a (frequently feathered) cere, as in the birds of prey ; wings and tail variable. The parrots, including the macaws, cockatoos, lories, etc., form one of the most strongly marked groups of birds, as easily recognizable by their peculiar external appearance as defined by the technical points of structure. They were formerly included in an order *Scansores* on account of the paired toes, but this is a comparatively trivial circumstance ; they have no special affinity with other zygodactyle birds, and their peculiarities entitle them to rank with groups called orders in the present volume. They might not inaptly be styled *frugivorous Raptores;* and in some respects they exhibit a vague analogy to the quadrumana (monkeys) among mammals. The upper mandible is much more freely movable than is usual in birds, being articulated instead of suturally joined with the forehead ; and the bill is commonly used in climbing. The bony orbits of the eyes are frequently completed by union of the lachrymal bones with postorbital processes. The symphysis of the lower jaw is short and obtuse. The sternum is entire or simply fenestrated posteriorly ; the furculum is weak, sometimes defective, or wanting. The principal metatarsal bone is short and broad, and its lower extremity is modified to suit the position of the fourth toe. The lower larynx is peculiarly constructed, with three pairs of muscles. The plumage shows aftershafts ; the oil gland is often wanting.

" Parrots abound in all tropical countries, but, except in Australia and New Zealand, rarely extend into the temperate zone. The Indian and Æthiopian regions are poor in parrots, while the Australian is the richest, containing many genera and even whole families peculiar to it" (NEWTON). The highest authority, FINSCH, recognizes 354 species as well-determined, distributing them in 26 genera ; 142 are American, 23 African, and 18 Asiatic ; the Moluccas and New Guinea have 83, Australia 59, and Polynesia 29. Ornithologists are now nearly agreed to divide them into 5 families. The curious flightless ground-parrot of New Zealand (*Strigops habroptilus*) forms one of these, *Strigopidæ*. " The most highly organized group is the *Trichoglossidæ*, in which the whole structure is adapted to flower-feeding habits" (WALLACE) ; it belongs to the Australian region. The cockatoos are familiar examples of a third family, *Plictolophidæ*, of Australia and the East Indies. The great bulk of the order, however, is made up of the other two less specialized and more generally distributed groups, the *Psittacidæ* proper, and the

Family ARIDÆ,

of which the macaws (*.Ira*), and the following species, are characteristic examples.

138. Gen. CONURUS Kuhl.

Carolina   Parroquet.
Green ; head yellow ; face
red ; bill white ; feet flesh
color ; wings more or less
variegated with blue and
yellow. Sexes alike.
Young simply green. 13 ;
wing 7½ ; tail 6. Southern
States ; up the Mississippi
Valley to the Missouri
region ; formerly strayed
to Pennsylvania and New
York, but of late has
receded even from the
Carolinas ; still abundant
in Florida. Gregarious,

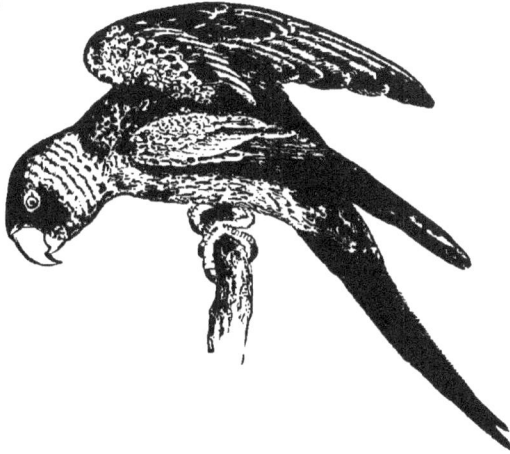

FIG. 133. Carolina Parroquet.

frugivorous and granivorous; not regularly migratory. WILS., iii. 89, pl.
26, f. 1 ; NUTT., i, 545 ; AUD., iv, 306, pl. 278 ; BD., 67.   CAROLINENSIS.

Order RAPTORES. Birds of Prey.

Bill epignathous, cered ; and feet not zygodactyle. The rapacious birds form a
perfectly natural assemblage, to which this expression furnishes a clue. The
parrots, probably the only other birds with strongly hooked and *truly* cered bill,
are yoke-toed. The *Raptores* present several osteological and other anatomical
peculiarities. There are two carotids ; the syrinx, when developed, has but one
pair of intrinsic muscles. The alimentary canal varies with the families, but
differs from that of vegetarian birds, in adaptation to an exclusively animal diet.
In the higher types, the whole structure betokens strength, activity and ferocity,
carnivorous propensities and predaceous nature. Most of the smaller, or weaker,
species feed much upon insects ; others more particularly upon reptiles, and fish ;
others upon carrion ; but the majority prey upon other birds, and small mammals,
captured in open warfare. Representatives of the order are found in every part of
the world. They are divisible into *four* families. One of these, *Gypogeranidæ*,
consists of the single remarkable species *Gypogeranus serpentarius*, the secretary-
bird or serpent-eater of Africa ; this shows a curious grallatorial analogy, being

mounted on long legs, like a crane, and has several other more important structural modifications. The other three families occur in this country; and the following accounts are sufficiently explicit to illustrate the order, without further remark in this connection.

Family STRIGIDÆ. Owls.

Head very large, and especially broad from side to side, but shortened length-wise, the "face" thus formed further defined by a more or less complete "ruff," or circlet of radiating feathers of peculiar texture, on each side. Eyes very large, looking more or less directly forward, set in a circlet of radiating bristly feathers, and overarched by a superciliary shield. External ears extremely large, often pro-vided with an operculum or movable flap, presenting the nearest approach, among birds, to the ear-conch of mammals. Bill shaped much as in other ordinary rapa-cious birds, but thickly beset at base with close-pressed antrorse bristly feathers. Nostrils large, commonly opening at the edge of the cere rather than entirely in its substance. Hallux of average length, not obviously elevated in any case; outer toe more or less perfectly versatile (but never permanently reversed), and shorter than the inner toe. Claws all very long, much curved and extremely sharp, that of the middle toe pectinate in some species. As a rule, the tarsi are more or less completely feathered, and the whole foot is often thus covered. Among numerous osteological characters may be mentioned the wide separation of the inner and outer tablets of the brain case by intervention of light spongy diploë; the commonly 4-notched sternum, and a peculiar structure of the tarso-metatarsus. The gullet is capacious but not dilated into a special crop; the gizzard is only moderately muscular; the intestines are short and wide; the cœca are extremely long and club-shaped. The syrinx has one pair of intrinsic muscles. The feathers have no aftershaft, and the general plumage is very soft and blended.

The Nocturnal Birds of Prey will be immediately recognized by their peculiar physiognomy, independently of the technical characters that mark them as a natural, sharply defined family. They are a highly monomorphic group, without extremes of aberrant form; but the ease with which they are collectively defined is a measure of the difficulty of their rigid subdivision, and the subfamilies are not yet satis-factorily determined. Too much stress appears to have been laid upon the trivial, although evident, circumstance of presence or absence of the peculiar ear-tufts that many species possess : more reliable characters may probably be drawn from the structure of the external ear, and facial disk, the modifications of which appear to bear directly upon mode of life, these parts being as a rule most highly developed in the more nocturnal species; while some points of internal structure may yet be found correspondent. One group, of which the barn owl, *Strix flammea*, is the type, seems very distinct in the angular contour and high development of the facial disk, pectination of the middle claw, and other characters; and probably the rest of the family fall in two other groups; but I do not deem it expedient to present subfamilies on this occasion.

As is well known, owls are eminently nocturnal birds; but to this rule there are numerous striking exceptions. This general habit is correspondent to the modifi-cation of the eyes, the size and structure of which enable the birds to see by night, and cause them to suffer from the glare of the sunlight. Most species pass the daytime secreted in hollow trees, or dense foliage and other dusky retreats, resuming their wonted activity after nightfall. Owing to the peculiar texture of the plumage their flight is perfectly noiseless, like the mincing steps of a cat; and no entirely fanciful analogy has been drawn between these birds and the feline carnivora that chiefly prey stealthily in the dark. ·Owls feed entirely upon animal substances, and capture their prey alive—small quadrupeds and birds, reptiles and insects, and even fish. Like most other Raptores, they eject from the mouth, after a meal, the bones, hair, feathers and other indigestible substances, made up into a round pellet. They are noted for their loud outcries, so strange and often so lugubrious, that it is no wonder that traditional superstition places these dismal night birds in the category of things ill-omened. The nest is commonly a rude affair of sticks gathered in the various places of diurnal resort; the eggs are several (commonly 3–6), white, sub-spherical. The female, as a rule, is larger than the male, but the sexes are alike in color; the coloration is commonly blended and diffuse, difficult of concise description.

Owls are among the most completely cosmopolitan of birds; with minor modifications according to circumstances, their general habits are much the same the world over. A difficulty of correctly estimating the number of species arises from the fact that many, especially of the more generalized types, have a wide geographical distribution, and, as in nearly all such cases, they split into more or less easily recognized races, the interpretation of which is at present a matter of opinion rather than a settled issue. About 200 species pass current; this number must be reduced by one-third; out of about 50 generic names now in vogue, probably less than one-half represent some structural peculiarity. Notable exotic genera are the Japanese *Phodilus* (*P. badius*), an ally of *Strix* proper; the Asiatic *Ketupa;* and the extensively distributed Old World *Athene*, in its broad acceptation.

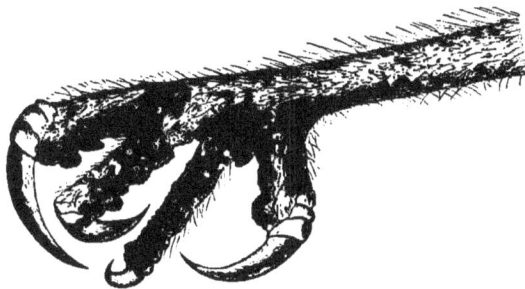

FIG. 134. Foot of Barn Owl.

139. Genus STRIX Linnæus.

Barn Owl. Tawny, or fulvous-brown, delicately clouded or marbled with ashy and white, and speckled with brownish-black; below, a varying shade from nearly pure white to fulvous, with sparse sharp blackish speckling; face white to purplish-brown, darker or black about the eyes, the disk bordered with dark brown; wings and tail barred with brown, and finely mottled like the back; bill whitish; toes yellowish. Facial disk highly developed, not circular; no tufts; ears very large, operculate; tarsi long, scant-feathered, below bristly, like the nearly naked toes; middle claw usually found serrate or at least jagged; plumage very downy. ♀ 17 long; wing 13; tail 5½; ♂ rather less. U. S., Atlantic to Pacific, southerly; rare in the interior, rarely N. to New England. WILS., vi, 57, pl. 50, f. 2 ; NUTT., i, 139 ; AUD., i, 127, pl. 34 ; CASS. in BD., 47 ; COOP., 415.   FLAMMEA var. AMERICANA.

140. Genus BUBO Cuvier.

Great Horned Owl. Distinguished by its large size, in connection with the conspicuous ear-tufts : the other species of similar dimensions are tuft-

less. The plumage varies interminably, and no concise description will meet all its phases ; it is a variegation of blackish, with dark and light brown, and fulvous. A white collar is the most constant color-mark. Var. *arcticus* is the northern bird, very light colored, and frequently nearly white, like the snowy owl, in arctic speci-mens. Var. *pacificus* is a littoral form, very dark colored, with little fulvous, "extending from Oregon northward, coastwise, to Labrador." (*Ridgway.*) Facial disks complete ; ear non-opercu-late ; feet entirely feathered. Length about 2 feet, rather less than more ; wing 14–16 inches ; tail 9–10. This powerful bird, only yielding to the great gray owl in size, and to none in spirit, is a common inhabitant of North America at large ; not

FIG. 135.   Great Horned Owl.

migratory ; breeds in late winter and early spring months, building a large nest of sticks, on the branches or in the hollows of trees : eggs white, nearly spherical, 2¼ by 1⅞. WILS., vii, 52, pl. 50, f. 1 ; NUTT., i, 124 ; AUD., i, 143, pl. 39 ; CASS. in BD., 49 ; COOP., 418.   .   .   VIRGINIANUS.

141. Genus SCOPS Savigny.

* *Toes bristly.*

Screech Owl.   Red Owl.   Mottled Owl. Like a miniature *Bubo* in form : 8 or 10 inches long ; wing 6–7 ; tail 3–3½. *One plumage :* — General aspect gray, paler or whitish below, above speckled with blackish, below patched with the same ; wings and tail dark-barred ; usually a lightish scapular area. *Another :* — General aspect brownish-red, with sharp black streaks ; below, rufous-white, variegated ; quills and tail with rufous and dark bars. These plumages shade insensibly into each other, and it has been determined that they bear no definite relations to age, sex, or season. Parallel varia-tions occur in some other species. North America at large ; one of the most abundant species. WILS., iii, 16, pl. 19, f. 1 ; v, 83, pl. 42, f. 1 ; NUTT., i, 120 ; AUD., i, 147, pl. 40 ; CASS. in BD., 51 ; COOP., 420. ·ASIO.

Var. KENNICOTTII. Large dark north-western form; general color sepia-brown, mottled and blotched with black; 11; wing 7¼; tail 4. Alaska to Washington and Idaho; three specimens known. ELLIOT, Proc. Phila. Acad. 1867, 69; ID., pl. 27; BD., Trans. Chicago Acad. 1869, 311: COOP., 423.

+ Var. MACCALLII. Small, pale. southern form; size at the minimum above given. Southwestern United States. CASS.. Ill. 180, and in BD., 52.

FIG. 136. Screech Owl.

** *Toes perfectly naked.*

Flammulated Owl. Above, grayish-brown, obscurely streaked with black, and finely speckled with white; below, grayish-white with some rufous mottling, each feather with a shaft streak, and several cross-lines, of black; .face and ruff varied with rufous; edges of the scapulars the same, forming a noticeable oblique bar; wing coverts tipped, and outer webs of the quills squarely spotted, with white, or rufous-white, and tail feathers imperfectly barred with the same. 6½–7; wing 5¼–5½; tail 2½. A small owl with the form and much the general aspect of an ungrown *S. asio*, but the feathering of the feet stops abruptly at the toes. Mexico; North to Fort Crook, California, where found breeding (*Feilner*). SCL., Proc. Zool. Soc. 1858, 96; SCL. and SALV., *ibid.* 1868, 57, and Exotic Ornithology, vii, 68, 99, pl. 50; COOP., 422. . . . . FLAMMEOLA.

142. Genus OTUS Cuvier.

Long-eared Owl. General plumage above, a variegation of dark brown, fulvous and whitish, in a small pattern: breast more fulvous, belly whiter. former sharply striped, and latter striped and elaborately barred, with blackish; quills and tail mottled and closely barred with fulvous and dark brown; face pale, with black touches and eye patches; bill and claws blackish.

Tufts long and conspicuous, of 8–12 feathers; ear parts immense, with a semicircular flap; facial disk complete; tarsi and toes feathered. 14–15 long; wing 11–12; tail 5–6. Temperate North America, common. Wils., vi, 73, pl. 51, f. 3; Nutt., i, 130; Aud., i, 136, pl. 37; Cass. in Bd., 53; Coop., 426.. . . . . vulgaris var. wilsonianus.

143. Genus BRACHYOTUS Gould.

Short-eared Owl. Fulvous or buffy-brown, paler or whitey-brown below; breast and upper parts broadly and thickly streaked with dark brown, belly usually sparsely streaked with the same, but not barred crosswise; quills and tail buff, with few dark bands, and mottling; facial area, legs and crissum pale, unmarked; eye patch blackish. With the size and form of the last species, but readily seen to be different; ear tufts small and inconspicuous, few-feathered. Temperate North America, abundant; not appreciably different from the European. Wils., iv, 64, pl. 33, f. 3; Nutt., i, 132; Aud., i, 140, pl. 38; Cass. in Bd., 54; Coop., 427. . . . palustris.

144. Genus SYRNIUM Savigny.

** Large owls, without ear-tufts, the facial disks complete and of great extent, the eyes comparatively small, the ear parts moderate, operculate, the tarsi and toes fully feathered.

Great Gray Owl. Above, cinereous-brown, mottled in waves with cinereous-white; below, these colors rather paler, disposed in *streaks* on the breast, in *bars* elsewhere; quills and tail with five or six darker and lighter bars; the great disk similarly marked in regular concentric rings. An immense owl, one of the largest of all, much exceeding any other of this country; about 2½ feet long, the wing 1½, the tail a foot or more. Arctic Am., irregularly S. into the northern U. S. in winter. Bonap., Am. Orn. pl. 23, f. 2; Sw. and Rich., F. B.-A. ii, 77, pl. 31; Aud., i, 130, pl. 35; Nutt., i, 128; Cass. in Bd., 56; Coop., 433. lapponicum var. cinereum.

Barred Owl. Above, cinereous-brown, barred with white, often tinged with fulvous; below, similar, paler, the markings in *bars* on the breast, in *streaks* elsewhere; quills and tail feathers barred with brown and white with an ashy or fulvous tinge. Length about 18; wing 13–14; tail 9. Eastern North America, common. Wils., iv, 61, pl. 33, f. 2; Nutt., i, 133; Aud., i, 132, pl. 36; Cass. in Bd., 56; Coop., 431. . . nebulosum.

Western Barred Owl. Resembling the last, but easily distinguished: general color warm brown; the white bars above broken into spots particularly towards and on the head; below, the markings in *bars everywhere;* wings and tail closely barred. Fort Tejon, Cala.; one specimen known. Xantus, Proc. Phila. Acad. 1859, 193; Bd., B. N. A. 1860, p. v, pl. 66 (not in the Government edition); Coop., 430. . occidentale.

FIG. 137. Barred Owl.

145. Genus NYCTEA Stephens.

Snowy Owl. Pure white, with more or fewer blackish markings. Nearly 2 feet long; wing 17 inches; tail 10. Head smooth; facial disks incomplete; eyes and ear parts moderate; feet densely clothed. This remarkable owl, conspicuous both in size and color, inhabits the boreal regions of both continents, coming southward in winter; it ordinarily enters the United States, and in extreme cases ranges irregularly through most of the States. It is not by any means exclusively nocturnal. WILS., iv, 53, pl. 32, f. 1; NUTT., i, 116; AUD., i, 113, pl. 28; CASS. in BD., 63; COOP., 447. . . . . . . NIVEA.

FIG. 138. Snowy Owl.

146. Genus SURNIA Dumeril.

Hawk Owl. *Day Owl.* Dark brown above, more or less thickly speckled with white; below, closely barred with brown and whitish, the throat alone streaked; quills and tail with numerous white bars; face ashy, margined with black. Length about 16 inches; wing 9; tail 7, graduated, the lateral feathers 2 inches shorter than the central. Except in the length of its tail, which produces linear measurements unusual for a bird of its bulk in this family, its general form is that of the snowy owl. Like that species, it is a bird of Arctic regions, coming southward in winter, but its range is more restricted, rarely extending to the Middle States. It is the most diurnal bird of the family, ranging abroad at all times, and approaches a hawk more nearly than any other. WILS., vi, 64, pl. 50, f. 6; NUTT., i, 115; AUD., i, 112, pl. 27; CASS. in BD., 64; COOP., 448. ULULA var. HUDSONICA.

147. Genus NYCTALE Brehm.

*** Small owls with the head untufted, the facial disks complete, the ears operculate, the tarsus longer than the middle toe, the tail nearly even, the 3d quill longest, the first 5 emarginate; color above chocolate-brown, spotted with white, the tail with transverse white bars; the *adult* with the facial area and forehead variegated with white, the face and superciliary line grayish-white, the lower parts white with spots or streaks of the color of the back; the *young* with the facial area and forehead dark brown, the face dusky, the eyebrows pure white, the lower parts brown, paler on the belly, unmarked. (See RIDGWAY, Am. Nat. vi, 284.)

Tengmalm's Owl. Large; wing 7¼; tail 4½, thus more than half the wing. Bill yellow, the cere not tumid, the nostrils presenting laterally, and obliquely oval. Arctic America, south to the borders of the United States.

Strix tengmalmi AUD., Orn. Biog. iv, 559, pl. 380; B. Am. i, 122, pl. 32.
Nyctale richardsoni CASS. in BD., 57. According to Mr. Ridgway, the
American bird is a distinguishable variety, being darker, the dark areas
larger, legs speckled instead of plain, etc. TENGMALMII var. RICHARDSONII.

Acadian Owl. Saw-whet Owl. Small; wing 5½; tail 2¾, thus not more
than half the wing. Bill black, the cere tumid, the circular nostrils presenting anteriorly.   United States and somewhat northward; México.
Common. *Nyctale albifrons* CASS. in BD., 57, and Ill. 187; COOP., 435;
N. kirtlandii HOY, Proc. Phila. Acad. 1852, 210; CASS., Ill. 63, pl. 11;
Strix frontalis LICHT.; these are the *young. Strix passerina* WILS., iv,
66, pl. 34, f. 1; *Strix acadica* NUTT., i, 137; *Ulula acadica* AUD., i, 123,
pl. 33; *N. acadica* CASS. in BD., 58; COOP., 436. .   .   .   .   .   ACADICA.

148. Genus GLAUCIDIUM Wagler.

** Very small; head untufted : facial disk nearly obsolete; ear parts moderate;
tarsus fully feathered, toes thickly bristled; wings short and much rounded, 4th
quill longest, the 3 outer ones emarginate; tail rather long, even; claws strong,
sharp, much curved.

Pygmy Owl. Above, uniform brown, everywhere dotted with small
round white spots, and with a collar of mixed white and blackish around the
back of the neck : breast with a mottled brown band separating the white
throat from the rest of the white under parts, which all have lengthwise
reddish-brown streaks; wings and tail dusky brown with round white spots
on both webs, largest on the inner; under wing coverts white with black
marks disposed in an oblique bar. ♂ 7, or a little less; extent 14½; wing
3¾; tail 3; ♀ larger; 7½, extent 15½, etc. Iris and soles yellow; toes
above, bill and cere, greenish-yellow. The shade of the upper parts ranges
from pure deep brown to pale grayish-brown, sometimes with a slight olivaceous shade. Rocky Mountains to the Pacific, U. S., common; a crepuscular and rather diurnal than strictly nocturnal species. *Surnia passerinoides*
AUD., i, 117, pl. 30. *G. infuscatum* CASS., Ill. 189; *G. gnoma* CASS. in
BD., 62; COOP., 444. .   .   .   .   .   .   PASSERINUM var. CALIFORNICUM.

Ferrugineous Owl. With the size, shape, and somewhat the coloration
of the foregoing, but readily distinguished : under parts and nuchal collar
much the same, but the former usually with a rusty tinge; upper parts
ranging from the color of *gnoma* to a rusty-red (the variation nearly as great
as in the two plumages of *Scops asio*), not continuously speckled, the
whitish or ochrey spots mostly confined to the wing coverts and scapulars,
those of the crown lengthened into sharp streaks; spots on the quills
enlarged into bars nearly confluent from one web to the other, rusty or
ashy next the shafts, white or tawny on the edges of the feathers, especially
the inner; tail in both plumages alike closely and continuously barred with
brown and rusty-red (same as the color of the upper parts in the red
plumage, conspicuously different in the gray plumage), the latter sometimes
fading on the inner webs. South and Central America and Mexico to the

U. S. border; Arizona (*Bendire*). COUES, Am. Nat. vi, 370. (Described from extra-limital specimens, No. 58,229, Mazatlan, and 43,055, Costa Rica, transmitted by the Smithsonian for the purpose.) . . . FERRUGINEUM.

149. Genus MICRATHENE Coues.

Whitney's Owl. Above light brown, thickly dotted with angular paler brown marks, the back also obsoletely marbled with darker; a concealed white cervical collar, forming a bar across the middle of the feathers, which are plumbeous at base and brown at tip; quills with 3–6 spots on each web, white on the inner webs of all and outer webs of several, brown on the rest; coverts with two rows of white spots, brown spots intervening; outer secondaries with a few white spots, and scapulars showing a white stripe; lower wing coverts tawny white, with a dark brown patch; other wing-feathers dark brown with pale ashy dots near the ends of the secondaries; tail feathers with light spots forming five broken bars, and a narrow terminal bar; feathers over eyes white, with black-spotted shaft; under eyes light brown obsoletely barred with darker; bristles about the bill black on their terminal half; chin and throat white, becoming light brown below, the white forming a broad crescent; sides of neck narrowly barred with ashy and brown, and breast imperfectly barred and blotched with the same, towards the abdomen forming large patches, margined with gray and white; tibiæ narrowly barred with light and dark brown; tarsal bristles whitish; bill pale greenish; iris and soles yellow. Length 6¼; extent 15¼; wing 4½; tail 2¼: gape of bill ½; bill ⅓ high, ⅔ wide at base. Facial disk imperfect; no ear tufts; wings very long, but rounded; 3d and 4th quills longest, 2d equal to 6th, 1st ⅔ the 3d; tail nearly even, with broad-tipped feathers; *tarsus nearly bare of feathers,* sparsely bristly, like the toes; middle toe and claw about as long as the tarsus; claws remarkably small, weak, and little curved. Colorado Valley and southward (Fort Mojave, *Cooper;* Southern Arizona, *Bendire;* Mazatlan and Socorro, *Grayson*). A diminutive owl of remarkable characters, only lately discovered. COOPER, Proc. Cala. Acad. 1861, 118, and B. Cal. 442; COUES, Proc. Phila. Acad. 1866, 51; LAWR., Proc. Bost. Soc. 1871, p.—; ELLIOT, pl. 29. . . . . . . WHITNEYI.

150. Genus SPHEOTYTO Gloger.

Burrowing Owl. Above, grayish-brown, with white, black-edged spots; below, tawny-whitish, variegated with reddish-brown, chiefly disposed in bars; face and throat whitish; crissum and legs mostly unmarked; quills with numerous paired tawny-white spots, and tail feathers barred with the same; bill grayish-yellow; claws black. 9–10 long; wing 6½–7½; tail 3½–4. No tufts; facial disk imperfect; tarsi very long, extensively denuded, bristly like the toes. Prairies and other open portions of the United States west of the Mississippi, abundant; lives in holes in the ground, in prairie-dog towns, and the settlements of other burrowing animals, using their deserted holes for its nesting place. There is certainly but one species in this country; it is merely a variety of the S. American bird. Bonap., Am. Orn. pl. 7, f. 2; Nutt., i, 118; Aud., i, 119, pl. 31; *Athene hypogæa* and *A. cunicularia* Cass. in Bd., 59, 60; Coop., 437, 440. CUNICULARIA var. HYPOGÆA.

Family FALCONIDÆ. Diurnal Birds of Prey.

Comprising the great bulk of the order, this large family may be best defined by exclusion of the special features marking the others. There is nothing of the grallatorial analogy exhibited by the singular *Gypogeranidæ;* the nostrils are not completely pervious, nor the hind toe obviously elevated. as in *Cathartidæ*, and other peculiarities of the American vultures are not shown. Comparing with the owls, we miss their peculiar physiognomy, the eyes looking laterally as in ordinary birds, the disk wanting (except in the *Circus* group, where it is imperfect), the aftershaft present (except in *Pandion*), the outer toe not versatile (except in *Pandion*), and not shorter than the inner. The external ears are moderate and non-operculate. The eyes, as a rule (but not always), are sunken beneath a projecting superciliary shelf, conferring a decided and threatening gaze. The bill shows the raptorial type in its perfection, and is always furnished with a cere in which the nostrils are pierced. The lores, with occasional exceptions owing to nakedness or dense soft feathering, are scantily clothed with radiating bristles, which however do not form, as in the owls, a dense appressed mass hiding the base of the bill. The feet are strong, with widely separable and highly contractile toes, and large sharp curved claws—efficient instruments of prehension, offence and defence. The toes are generally scabrous underneath, with wart-like pads at the joints, to prevent slipping, and commonly show a basal web. The podotheca is very variable; the whole tarsus is frequently feathered, and usually partially so; the horny covering takes the form of scutella, or reticulations, or rugous granulations, and is occasionally fused. The capacious gullet dilates into a crop; the gizzard is moderately muscular; the intestines vary; the cœca are extremely small. The syrinx has one pair of intrinsic muscles. There are several good osteological characters.

Birds of this family abound in all parts of the world, and hold the relation to the rest of their class that the carnivorous beasts do to other mammals. There are upwards of 300 good species or very strongly marked geographical races, justly referable to about 50 full genera. In round numbers, 1,000 specific and 200 generic names have been instituted for *Falconidæ*. No unexceptionable subdivision of the family has yet been proposed; and as this point is still at issue, I deem it best not to present subfamilies. Instead of an attempt in this direction, which would necessarily be premature, I will endeavor to give the student a general idea of the composition of the family.

1. The *Old World vultures* form a group standing somewhat apart from the others in many points of external structure and habits, although correspondent in more essential characters. Until Prof. Huxley's successful exhibition of this fact, they were usually united in a family, *Vulturidæ*, with the American vultures, from which, however, they differ decidedly, as stated beyond. It is a small group of six genera and about twelve species. The bearded griffin, *Gypaëtus barbatus*, is conspicuous for its raptorial nature. The other genera, more or less decidedly "vulturine," are *Vultur*, *Otogyps*. *Gyps*, *Neophron* and *Gypohierax;* the characteristic species are — *V. monachus*. *O. auricularis*, *G. fulvus*, *N. percnopterus* and *G. angolensis*.

2. The genus *Polyborus* (beyond), illustrates a small group of hawks partaking

somewhat of a vulturine nature; they feed much upon carrion, are rather sluggish in habit, and lack the spirit of the typical hawks. Details of form vary in the three genera *Polyborus*, *Ibycter*, and *Milvago*. There are less than twelve species, all confined to America.

3. The *harriers* are another small group, in which a ruff, forming an imperfect facial disk, as in the owls, is more or less developed. It consists of the genus *Circus* and its subdivisions (to which some add *Polyboroides*, of Africa), comprising about fifteen species of various parts of the world. Our species is a typical example.

4.. The *fish-hawks*, of the single genus *Pandion*, with four or five species or races of various parts of the world, are remarkably distinguished from other birds of the family by the lack of aftershafts, a special tract-formation, a peculiar conformation of the feet, and other characters as noticed further on.

5. The genus *Pernis* is distinguished from ordinary *Falconidæ*, in having the whole head softly and densely feathered. *P. apivorus*, the bee-eating hawk of Europe, is the type. It approaches the kites.

6. The *kites* form a rather extensive group of hawks averaging undersized and of no great strength, though very active, generally of lithe and graceful shape, with long pointed wings and often forked tail. They subsist on small game, especially insects, which they capture with great address. The eye is commonly unshielded. Besides the genera given beyond, there are several others: *Milvus*, near which our *Nauclerus* stands, of Europe and Asia; the Indian and East Indian *Baza*, the African *Aviceda*, and the remarkable *Machaerhamphus* of Africa and Malacca; with the American *Cymindis*, and *Gampsonyx*. There are some thirty species of the group as thus constituted; but some of the genera are questionably enumerated here. *Milvus*, *Nauclerus*, *Elanus* and *Ictinia* are true kites.

7. The *buzzards* form a large group, not easily defined, however, unless it be by exclusion of the peculiarities of the others. They are hawks of medium and rather large size, heavy-bodied, of strong but rather measured flight, inferior in spirit to the true hawks and falcons, and as a rule feed upon humble game, which they rather snatch stealthily than capture in open piracy. The extensive genus *Buteo* with its subdivisions, and its companion *Archibuteo*, typify the buzzards; they include, however, a variety of forms, shading into other groups. With them must be associated the *eagles;* for the popular estimate of these famous great birds as something remarkably different from ordinary hawks is not confirmed by examination of their structure, which is essentially the same as that of the buzzards, into which they grade. Although usually of large size, and powerful physique, they are far below the smallest falcons in raptorial character, prey like the buzzards, and often stoop to carrion. The genus *Aquila* may stand as the type of an eagle; its several species are confined to the Old World, with one exception. *Haliaëtus* represents a decided modification in adaptation to maritime and piscivorous habits. A celebrated bird of this group is the harpy eagle of South America, *Thrasiaëtus harpyia*, with immense bill and feet, and one of the most powerful birds of the whole family. There are several other genera in either hemisphere.

8. The *hawks proper* are another extensive group, of medium sized and small species, which, although less powerfully organized, are little, if any, inferior in spirit and relative strength to the true falcons. Their flight is swift, they capture their prey in active chase like hounds, and always kill for themselves. The wings are rather short, as a rule, with the tip formed by the 3d–5th quills, the 2d and 1st being shortened; the tail is generally lengthened. The eye is shaded by a bony brow.

The genera *Astur* and *Accipiter* are perfect illustrations of this group; the several other genera usually adopted are not very different. There appear to be about seventy-five species, of most parts of the world.

9. Lastly, the true *falcons* are prominently distinguished by the presence of a tooth behind a notch of the upper mandible, in the foregoing birds the tomia being simply lobed or festooned, or merely arched. The falcons are birds of medium and small size (one of them is not larger than some sparrows), but extremely compact and powerful organization, and bold ruthless disposition; they prey by sudden and violent assault, and exhibit the raptorial nature in its perfection. The wings are strong, long and pointed, the tip formed by the 2d and 3d quills supported nearly to the end by the 1st and 4th; the tail is generally short and stiff. The typical and principal genus is *Falco*, of which there are, however, several subdivisions corresponding to minor modifications. The Australian *Ieracidea*, the East Indian *Ierax*, and the Brazilian *Harpagus*, which is doubly-toothed, are the principal other forms. There are upwards of fifty species of true falcons.

With many exceptions, in this family the sexes are alike in color, but the female is almost invariably larger than the male. The changes of plumage with age are great, and render the determination of the species perplexing—the more so since purely individual, and somewhat climatic, color-variations, and such special conditions as melanism, are very frequent. The modes of nesting are various; the eggs as a rule are blotched, and not so nearly spherical as those of owls. The food is exclusively of an animal nature, though endlessly varied; the refuse of digestion is ejected in a ball by the mouth. The voice is loud and harsh. As a rule, the birds of prey are not strictly migratory, though many of them change their abode with much regularity. Their mode of life necessarily renders them non-gregarious.

In the following sequence of our genera, the student will observe an attempt to indicate affinities not only in the family itself, but with allied families, by the central position of the typical *Falco*, the series beginning with the most owl-like form, and ending with the vulturine buzzards. But it is hoped that he will detect the imperfection of the arrangement, and that his studies will soon convince him of the impossibility of expressing natural relationships in any linear series. With this hint, the inviting problem is left open to stimulate investigation.

151. Genus CIRCUS Lacepede.

Marsh Harrier. Adult ♂ pale bluish-ash, nearly unvaried, whitening below and on upper tail coverts; quills blackish toward the end; 16–18; wing 14–15; tail 8–9; ♀ larger, above dark brown streaked with reddish-brown, below the reverse of this; tail banded with these colors; the immature ♂ is like the ♀, though redder, but in any plumage the bird is known by its white upper tail coverts, and generic characters: face with ruffs; wings, tail and tarsi very long, the latter scutellate before and behind,

Fig. 139. Marsh Harrier.

and twice as long as the middle toe; nostrils oval, etc. North America, abundant. Nests on the ground. WILS., vi, 67, pl. 51, f. 2; NUTT., i, 109; AUD., i, 105, pl. 26; CASS. in BD, 38; COOP., 489. .   CYANEUS var. HUDSONIUS.

152. Genus ROSTRHAMUS Lesson.

Everglade Kite. Adult ♂ blackish; coverts and base of tail feathers white; cere and feet yellow; bill and claws black; iris red; 16–18; wing 13½–15½; tail 6½–7½, emarginate; bill about 1, *extremely slender* and with a long hook; tarsi scutellate in front, the bare part shorter than the middle toe; claws very long, gently curved. ♀ and young brown, more or less variegated with fulvous and whitish. Florida, and southward. CASS. in BD., 38; MAYNARD, Birds of Florida, pls. i, v (in press; best account of the bird extant). SOCIABILIS.

153. Genus ICTINIA Vieillot.

Mississippi Kite. Plumbeous, paler on the head and under parts, blackening on wings and tail; quills suffused with rich chestnut; sexes alike; young varied with rusty and whitish; 14–15; wing 11–12, pointed; tail 6–6½, nearly square. Bill very short and deep, the commissure with prominent festoon; nostrils small, circular; tarsus short, scutellate anteriorly; outer and middle toe webbed; claws short, stout, flattened beneath. S. Atlantic and Gulf States, N. to Illinois (*Ridgway*). WILS., iii, 80, pl. 25, f. 1; NUTT., i, 92; AUD., i, 73, pl. 17; CASS. in BD., 37. MISSISSIPPIENSIS.

154. Genus ELANUS Savigny.

White-tailed Kite. Black-shouldered Kite. Head, tail and under parts white; back cinereous; most of the wing coverts black; bill black; legs yellow; young variegated with brown above, the head and tail ashy. Rather larger than the last; nostrils nearly circular; tarsi reticulate, feathered above in front; outer toe scarcely webbed; claws rounded underneath; tail emarginate, but outer feather shorter than the next. South Atlantic and Gulf States, California, and southward, chiefly coastwise. NUTT., i, 93; AUD., i, 70, pl. 16; CASS. in BD., 37; COOP., 488. . . . . . LEUCURUS.

155. Genus NAUCLERUS Vigors.

Swallow-tailed Kite. Head, neck and under parts, white; back, wings and tail, lustrous black. Tail a foot or more long, deeply forficate; wing 15–18, pointed; feet small, greenish-blue; claws pale; tarsi reticulate and feathered half way down in front; toes hardly webbed; nostrils broadly oval. A beautiful bird, common in the South Atlantic and Gulf States, in its extensive wanderings sometimes reaching the Middle districts, and in the interior penetrating to Wisconsin (*Hoy*), Missouri (*Coues*) and even Minnesota (lat. 47°; *Trippe*). WILS., vi, 70, pl. 51, f. 3; NUTT., i, 95; AUD., i, 78, pl. 18; CASS. in BD., 36. . . . . FURCATUS.

156. Genus ACCIPITER Brisson.

** Tarsus feathered but little way down in front (in gen. 157 the feathering reaches half way to the toes); toes long, slender, much webbed at base and padded underneath ; height of bill at base greater than chord of culmen ; 4th quill longest, 2d shorter than 6th. 1st very short. The two following species are exactly alike in color ; one is a miniature of the other. The ordinary plumage is dark brown above (deepest on the head, the occipital feathers showing white when disturbed) with an ashy or plumbeous shade which increases with age, till the general cast is quite bluish-ash ; below, white or whitish, variously streaked with dark brown and rusty, finally changing to brownish-red (palest behind and slightly ashy across the breast) with the white then only showing in narrow cross-bars ; chin, throat and crissum mostly white with blackish pencilling ; wings and tail barred with ashy and brown or blackish, the quills white-barred basally, the tail whitish-tipped ; bill dark ; claws black ; cere and feet yellow.

Sharp-shinned Hawk. "*Pigeon Hawk.*" Feet extremely slender ; bare portion of tarsus longer than middle toe ; scutella frequently fused ; tail square. ♂ 10–12 ; wing 6–7 ; tail 5–6. ♀ 12–14 ; wing 7–8 ; tail 6–7. Whole foot 3½ or less. North America, abundant. *Falco velox* WILS., v, 116, pl. 45, f. 1 ; *F. pennsylvanicus* WILS., vi, 13, pl. 46, f. 1 ; Sw. and RICH., F. B.-A. ii, 74 ; NUTT., i, 87 ; AUD., i, 100, pl. 25 ; CASS. in BD.,

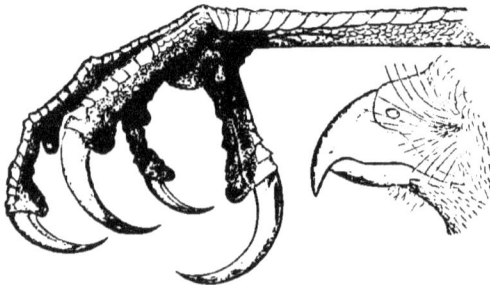
FIG. 110. Cooper's Hawk.

18 ; COOP., 466. . . . . FUSCUS.

Cooper's Hawk. *Chicken Hawk.* Feet moderately stout ; bare portion of tarsus shorter than middle toe ; scutella remaining distinct ; tail a little rounded. ♂ 16–18 ; wing 9–10 ; tail 7–8 ; ♀ 18–20 ; wing 10–11 ; tail 8–9. Whole foot 4 or more. N. Am., especially U. S. ; common. BONAP., Am. Orn. i, 1, pl. 1. f. 1 ; AUD., i, 98, pl. 24 ; CASS. in BD., 16 ; COOP., 464. *Falco cooperi* and *F. stanleii* NUTT., i, 90, 91. *A. mexicanus* CASS. in BD., 17 ; COOP., 465, is the same bird. . . . . COOPERII.

157. Genus ASTUR Lacepede.

Goshawk. Adult dark bluish-slate blackening on the head, with a white superciliary stripe ; tail with four broad dark bars ; below, closely barred with white and pale slate, and sharply streaked with blackish. Young dark brown above, the feathers with pale edges, streaked with tawny-brown on the head and cervix ; below fulvous-white with oblong brown markings. ♀ 2 feet long ; wing 14 inches ; tail 11 ; ♂ smaller. A large, powerful, and, in perfect plumage, a very handsome hawk, inhabiting northern North America ; the northern half of the United States chiefly in winter, but also breeding in mountainous parts. WILS., vi, 80, pl. 52, f. 3 ; NUTT., i, 85 ; AUD., i, 95, pl. 23 ; CASS. in BD., 15 ; COOP., 467. A variety of the European *Astur palumbarius?* . . . . ATRICAPILLUS.

158. Genus FALCO Linnæus.

* Tarsus more or less feathered above, elsewhere irregularly reticulate in small pattern ; 2d quill longest ; 1st alone decidedly emarginate on inner web.

Jerfalcon, or *Gyrfalcon.* Tarsus feathered fully half-way down in front, with only a narrow bare strip behind, and longer than middle toe ; 1st quill shorter than 3d. Upward of 2 feet long ; wing about 16 inches ; tail 10. *White,* with dark markings much as in the snowy owl ; or, ash-colored with numerous lighter bars ; young striped longitudinally beneath. An arctic falcon, of circumpolar distribution, in this country reaching the northern states in winter. It is split into several varieties which, however, do not seem to be strictly geographical, and concerning which ornithologists are singularly agreed to disagree. In var. *candicans,* the white predominates over the dark markings, and the bill and claws are white ; N. Greenland ; Iceland ; Arct. Am. and Eur. Aud., i, 81, pl. 19 ; Elliot, pl. 30 ; Cass. in Bd., 13. In var. *islandicus,* dark markings predominate, and the bill and claws are dark ; the crown is lighter than the back, and the dark maxillary patches are slight ; S. Greenland ; Iceland ; N. Eur. and Am. ; S. to U. S. in winter. Cass. in Bd., 13 ; Elliot, pl. 31. Var. *gyrfalco* is like the last, but with the crown darker than the back, and the moustaches heavy. Other strains are sometimes recognized by name. See Newton, Proc. Phila. Acad. 1871, 95 ; Ridgway, *ibid.* 1870, 140 ; Baird, Trans. Chicago Acad. i, 271. . . . . sacer (Forst. 1772).

Lanier Falcon. Tarsus feathered a third way down in front, broadly bare behind, and longer than middle toe ; 1st quill shorter than third. A foot and a half long ; wing 13–14 ; tail 7–8. Above, plain brown, the feathers bordered with rusty, the nape, forehead and superciliary line white ; below, white, with brown maxillary patches and other streaks on the breast and belly, the flanks barred ; tail barred and tipped with whitish ; adult with yellow iris and yellowish legs ; young with brown iris and bluish legs. Western United States and southward ; E. to Illinois (*Sargent, Ridgway*). *F. polyagrus* Cass., Ill. 88, pl. 16 ; Bd., 12 ; Coop., 458. . mexicanus.

Peregrine Falcon. Duck Hawk. Tarsus feathered but a very little way above in front, and not longer than the middle toe ; 1st quill not shorter than 3d. Size of the last, or rather less. Above, blackish-ash, with more or less evident paler waves ; below, and the forehead, white with more or less fulvous tinge, and transverse bars of blackish ; conspicuous black cheek-patches. Young with the colors not so intense, tending rather to brown ; the

Fig. 111. Peregrine Falcon.

tawny shade below stronger, the lower parts longitudinally striped. North America ; generally distributed, not abundant. *F. peregrinus* Wils., ix,

120, pl. 76; Sw. and Rich., F. B.-A. ii, 23; Nutt., i, 53; Aud., i, 84, pl. 20.   *F. anatum* and *F. nigriceps* Cass. in Bd., 7, 8.   .   .   communis.

Obs.   *F. rufigularis*, a bird of this section of the genus, admitted to our fauna under the name of *F. aurantius* (Cass. in Bd., 10 ; Elliot, pl. 32), does not appear to have been taken within our limits.

** Tarsus scarcely feathered above, with the plates in front enlarged, appearing like a double row of alternating scutella (and often with a few true scutella at base) ; 1st and 2nd quills emarginate on inner web.

Pigeon Falcon.   Pigeon Hawk.   Adult ♂ above ashy-blue, sometimes almost blackish, sometimes much paler ; below pale fulvous, or ochraceous, whitish on the throat, the breast and sides with large oblong dark brown spots with black shaft lines ; the tibiæ reddish, streaked with brown ; inner webs of primaries with about 8 transverse white or whitish spots ; tail tipped with white, and with the outer feather whitening ; with a broad subterminal black zone and 3–4 black bands alternating with whitish ; cere greenish-yellow, feet yellow.   ♀ with the upper parts ashy-brown ; the tail with 4–5 indistinct whitish bands ; about 13 ; wing 8 ; tail 5 ; ♂ smaller.   N. Am., generally distributed, common.   Observe that *Accipiter fuscus* is also called "pigeon hawk."   Wils., ii, 107, pl. 15, f. 3 ; Sw. and Rich., ii, 35 ; Nutt., i, 60 : Aud., i, 88, pl. 21 ; Cass. in Bd., 9.   .   .   .   .   .   columbarius.

Richardson's Falcon.   Similar ; sexes nearly alike, both lighter and more earthy-brown than the ♀ of the last ; head nearly white anteriorly ; streaks on the cheeks fine and sparse, those on the breast broad and sharp, light brown, with black shaft lines ; tail with 6 ashy-white bands ; ♀ above with pairs of ochraceous spots on the feathers, and secondaries with three ochraceous bands ; wing 9 ; tail 6 ; tarsus nearly 1½ ; ♂ smaller.   Interior N. Am., especially from the Mississippi to the Rocky Mountains.   Very near the last ; both are very closely related to *F. æsalon* of Europe, the fewer bars of the wings and tail being a principal character.   Ridgway, Proc. Phila. Acad. 1870, 145.   *F. æsalon* Rich. and Sw., Fn. Bor.-Am. ii, 37, pl. 25 ; Nutt., ii, 558 ; Coues, Proc. Phila. Acad. 1866, 42.   richardsonii.

Rusty-crowned Falcon.   Sparrow Hawk.   Crown ashy-blue, with a chestnut patch, sometimes small or altogether wanting, sometimes occupying nearly all the crown : conspicuous black maxillary and auricular patches, which with three others around the nape make seven black places in all, but a part of them often obscure or wanting ; back cinnamon brown, in the ♂ with a few black spots or none, in the ♀ with numerous black bars ; wing coverts in the ♂ ashy-blue, with or without black spots, in the ♀ like the back ; quills in both sexes blackish with numerous pale or white bars on inner webs ; tail chestnut, in the ♂ with one broad black subterminal bar, white tip, and outer feather mostly white with several black bars ; in the ♀ the whole tail with numerous imperfect black bars ; below white, variously tinged with buff, or tawny, in the ♂ with a few small black spots or none, in the ♀ with many brown streaks ; throat and vent nearly white and immaculate in both sexes ; bill dark horn, cere and feet yellow to bright orange ;

10–11 ; wing 7 ; tail 5, more or less. North America, everywhere, very abundant. This elegant little hawk will be immediately recognized by its small size, and entirely peculiar coloration, although the plumage varies almost interminably. However the case may be with the West Indian and other exotic forms, no races have been discovered in this country sufficiently marked to require designation by name. But we may, perhaps, with Mr. Ridgway (Proc. Phila. Acad. 1870, 149), recognize var. *isabellinus*, as a Middle American coast form occurring in the Gulf States, although of course it shades directly into the ordinary plumage (no rufous on crown ; several lateral tail feathers variegated, the black zone an

Fig. 112. Sparrow Hawk.

inch wide ; black spots on back and sides very sparse ; breast ochraceous ; ♀ with the black bars above unusually broad, upon a ferrugineous ground). WILS., ii, 117, pl. 16, f. 1 ; iv, 57, pl. 32, f. 2 ; NUTT., i, 58 ; AUD., i, 90, pl. 22 ; CASS. in BD., 13 ; COOP., 462. . . . . SPARVERIUS.

Femoral Falcon. Ashy-brown or pale slate, according to age ; forehead and superciliary line white, deepening to orange-brown on the auriculars ; two ashy stripes on side of head ; wings and tail with numerous white bars ; under wing coverts buffy with numerous black spots ; throat and breast white or tawny ; belly with a broad black zone ; tibiæ and crissum orange-brown. Length 15 or more ; wing 10½ ; tail 7½. A widely distributed South and Central American species, reaching just over our Mexican border ; it belongs to the same section of the genus as the sparrow hawk, but is not at all like this or any of the foregoing species. CASS. in BD., 11, pl. 1 ; DRESSER, Ibis, 1865, 333 ; COUES, Proc. Phila. Acad. 1866, 42 ; COOP., 461. FEMORALIS.

159. Genus BUTEO Cuvier.

* *Five* outer primaries emarginate on inner web ; bill high ; nostrils oval, horizontal, with eccentric tubercle ; feet robust. (Subgenus *Craxirex*.)

Harris' Buzzard. Dark chocolate-brown, nearly uniform ; wing coverts and tibiæ brownish-red ; upper tail coverts, base and tip of tail, white ; young duller brown, varied with fulvous ; ♀ nearly 24 ; wing 15 ; tail 10 ; ♂ smaller. A South and Central American species, reaching our Gulf border. Very different from any of the following species ; approaching the *Polybori* in habits. AUD., i, 25, pl. 5 ; BD., 46. UNICINCTUS var. HARRISII.

** *Four* outer primaries emarginate on inner web.

Cooper's Buzzard. Very pale ; below, pure white, the tibiæ tawny, the throat, breast and flanks with a few dark streaks ; a blackish patch on under wing coverts ; crown and hind neck with the feathers largely white at base, with dark tips and streaks ; upper tail coverts white, rufous-tinged, dark-barred ; tail mostly white, with ashy clouding, marked with rufous and darker in *lengthwise* pattern, and with dark subterminal zone ; back dark brown with an ashy shade ; 21½ ; wing 15 ; tail 9. Santa Clara Co., Cala. ;

one specimen known, which has not been referred to any described species, but which cannot be considered as establishing one. Cass., Proc. Phila. Acad. 1856, 253, and in Bd., 31 ; Coop., 472. . . . . . . cooperii.

Harlan's Buzzard. General color blackish, nearly uniform, the tail nearly concolor with the rest of the plumage, or mottled lengthwise with ashy, rufous and white, and having a dark subterminal bar (in the young brown banded with black) ; inner webs of quills extensively white. Of nearly the size and form of the following species ; tibial feathers remarkably long and flowing. "Louisiana ;" Aud., Orn. Biog. i, 441, v, 380, pl. 86, and B. Amer. i, 38, pl. 8 ; Nutt., i, 105. An obscure species, variously interpreted by different writers. See Lawr., Ann. Lyc. N. Y. v, 220 ; Cass., Ill. 101, and in Bd., 24 ; Bryant, Proc. Bost. Soc. N. H. viii, 109 ; Coues, Proc. Phila. Acad. 1866, 45 ; Ridgway, *ibid.* 1870, 142 ; Coop., 473. Different "black hawks" appear to have received this name, but Mr. Ridgway informs me that he believes he has the true *harlani,* and that it is a good species.   harlani.

Red-tailed Buzzard. Hen Hawk. Adult dark brown above, many feathers with pale or tawny margins, and upper tail coverts showing much whitish ; below white or reddish-white, with various spots and streaks of different shades of brown, generally forming an irregular zone on the abdomen ; *tail above bright chestnut red,* with subterminal black zone and narrow whitish tip, below pearly gray ; wing coverts dark ; young with the tail grayish-brown barred with darker, the upper parts with tawny streaking. A large stoutly-built hawk ; ♀ 23 ; wing 15½ ; tail 8½ ; ♂ 20 ; wing 14 ; tail 7. Wils., vi, 76, 78, pl. 52, f. 1, 2 (adult and young) ; Nutt., i, 102 ; Aud., i, 32, pl. 7 ; Cass. in Bd., 25. This is the ordinary bird, abundant in Eastern North America, where it is subject to comparatively little variation. In the West, a form with the throat dark colored, and the under parts extensively rufous, is *B. montanus* Cass., Proc. Phila. Acad. 1856, 39, and in Bd., 26 (but not of Nuttall). Coop., 469 ; *B. "swainsonii"* Cass., Ill. 98 (*not the true swainsonii;* see below). Another western, melanotic form, in which the whole plumage is dark chocolate-brown, with the tail red and sometimes a large red patch on the breast, is *B. calurus* Cass., Proc. Phila. Acad. vii, 1855, 281, and in Bd., 22 ; figured in Pacific R. R. Rep. x, pt. iii, pl. 14 ; Coues, Proc. Phila. Acad. 1866, 44. An unpublished variety from Cape St. Lucas is *B. lucasanus* Ridgway, Mss. . . borealis.

Red-Shouldered Buzzard. General plumage of the adult of a rich *fulvous* cast ; above, reddish-brown, the feathers with dark brown centres ; below a lighter shade of the same, with narrow dark streaks and white bars ; quills and tail blackish, conspicuously banded with pure white, *the bend of the wing orange-brown.* Young plain dark brown above, below white with dark streaks ; quills and tail barred with whitish ("winter falcon," *F. hyemalis* Wils., iv, 73, pl. 35, f. 1 ; Aud., Orn. Biog. i, 364, pl. 71 ; *F. buteoides* Nutt., i, 100). Nearly as *long* as *B. borealis,* but not nearly so heavy ; tarsi more naked ; ♀ 22 ; wing 14 ; tail 9 ; ♂ 19 ; wing 13 ; tail 8 (average). Eastern North America, very abundant. Wils., vi, 86, pl. 53, f. 3 ; Nutt.,

i, 106; AUD., i, 40, pl. 9; CASS. in BD., 28. In adult plumage, this handsome hawk is unmistakable; but the student may require to look closely
after the young. The western form, even darker "red" than the eastern, is
B. elegans CASS., Proc. Phila. Acad. 1855, 281, and in BD., 28, figured in
P. R R. Rep. x, Cala. Route, pl. 2; COOP., 477. . . . . . LINEATUS.

Band-tailed Buzzard. Black or blackish, upper parts with an indefinite
number of pure white spots; bases of primaries white with black bars; tail
of the adult with three broad white bars, of the young with several narrower
imperfect ones; young varied with rusty? Smaller than any of the foregoing, more slightly built, and otherwise obviously different; about 18;
wing 15; tail 9. California (*Cooper*), Arizona (*Coues*), and southward.
SCLATER, Trans. Zool. Soc. 1858, 263, pl. 59 (Mexico); COUES, Proc.
Phila. Acad. 1866, 46; COOP., 479. . . . . ZONOCERCUS.

*** *Three* outer quills emarginate on inner web.

Swainson's Buzzard. Extremely variable in color, but usually showing
a broad dark pectoral band contrasted with light surroundings, and numerous (8–12) narrow dark tail bars. A smaller bird than the foregoing
(except *zonocercus*); ♀ about 20; wing 16; tail 8½; ♂ less. Not so stoutly
built; wings and tail relatively longer. Chiefly Western North America;
also, Canada and Massachusetts. It comes nearest *B. vulgaris* of Europe.
B. vulgaris SW. and RICH., F. B.-A. ii, 47, pl. 27; NUTT., ii, 559; AUD.,
i, 30, pl. 6; *Falco buteo* AUD., Orn. Biog. iv, 208, pl. 372; *B. montanus*
NUTT., i, 2d ed. 112; *B. swainsoni* CASS. in BD., 19 (not of Ill. 98);
COOP., 476; BD., P. R R. Rep. x, pt. iii, pls. 12, 13. *B. bairdii* HOY,
Proc. Phila. Acad. 1853, 451 (Wisconsin); CASS., Ill. i, 99, 257, pl. 41,
and in BD., 21, is the *young*, differing materially in color. *B. insignatus*
CASS., Ill. 102, 198, pl. 31, and in BD., 23 (Canada; Nebraska; California);
COOP., 474, is a melanotic plumage. . . . . SWAINSONII.

Broad-winged Buzzard. Above, umber-brown, the feathers with paler,
or even with fulvous or ashy-white, edging, those of the hind head and nape
cottony-white at base; quills blackish, most of the inner webs white, barred
with dusky; tail with about three broad dark zones alternating with narrow
white ones, and white-tipped; *conspicuous dusky maxillary patches;* under
parts white, or tawny, variously streaked, spotted or barred with rusty or
rufous, this color usually predominating in adult birds, when the white
chiefly appears as oval or circular spots on each feather; throat generally
whiter than elsewhere, narrowly dark-lined. In the young, the upper parts
are duller brown, varied with white, the under parts tawny-whitish with
linear and oblong dark spots, the tail grayish-brown with numerous dark
bars. ♀ 18; wing 11; tail 7; ♂ less. Eastern North America, and
throughout Middle America to Panama; common. A rather small but stout
species, with short broad wings, very different from any of the foregoing,
and easily recognized; the maxillary patches are a strong feature. WILS.,
vi, 92, pl. 54, f. 1; NUTT., i, 105; AUD., i, 43, pl. 10; CASS. in BD., 29.
Falco latissimus WILS., *l. c.* (later copies). . . . . PENNSYLVANICUS.

OBS. I cannot admit *Buteo oxypterus* (CASS., Proc. Phila. Acad. 1855, 282, and in BD., 30; Fort Fillmore, N. M.) as a valid species, although I am not prepared to assign it as a synonym of any one of the foregoing. The type and only recognized specimen is apparently a young bird, very near *swainsonii*, if not the same. (Compare *B. fuliginosus* SCL., Proc. Zool. Soc. 1858, p. 356.)

160. Genus ARCHIBUTEO Brehm.

*** Large hawks with the tarsi feathered in front to the toes; upward of 2 feet long; wing 16–18; tail 8–10. Four outer primaries emarginate on inner web.

Rough-legged Buzzard. Below, white, variously dark-marked, and often with a broad black abdominal zone; but generally no ferrugineous. North America; abundant. The black hawk, *A. sancti-johannis*, is a melanotic state, in which the whole plumage is nearly uniform blackish. This does not appear to have been observed in the European bird, of which ours is a variety. The name adopted, it must be observed, is not intended to discriminate this black plumage, but to distinguish the American bird from the European *lagopus*, as a geographical race. *Falco lagopus* WILS., iv, 59, pl. 33, f. 1; v, 216, pl. 53, f. 1, 2; *F. niger* WILS., vi. 82; *Buteo lagopus* SW. and RICH., F. B.-A. ii, 52, pl. 28; NUTT., i, 97, 98; AUD., i, 46, pl. 11. CASS. in BD., 32, 33; COOP., 483. . LAGOPUS var. SANCTI-JOHANNIS.

Ferrugineous Buzzard. Below, pure white, scarcely or not marked, excepting that the legs are rich rufous with black bars, in marked contrast; above, varied with dark brown, rufous, and white; quills brown, with much white; tail silvery-ash, clouded with brown or rufous. Young duller above, more marked below, tibiæ not so strongly contrasted in color. Our handsomest and one of our largest hawks, inhabiting Western U. S., especially California, Arizona and New Mexico. CASS., Ill. 159, pl. 26, and in BD., 34; COUES, Proc. Phila. Acad. 1866, 46; COOP., 482. . FERRUGINEUS.

160bis. Genus ASTURINA Vieillot.

Gray Hawk. Nostrils horizontal, without tubercle, upper outline straight, lower semicircular; 4 outer primaries emarginate on inner web. Adult above cinereous, darkening on the rump; below closely barred with cinereous and white; tail blackish, with about three white bars, its upper and under coverts white; quills ashy-brown, with darker bars and much white edging on inner webs; crown with a lateral white stripe; cere and feet yellow; ♀ 18; wing 10; tail 7½; ♂ less. Young above umber-brown, below white with longitudinal brown stripes; tail light brown with numerous dark bands; tibiæ barred. A handsome species, resembling a goshawk, but belonging to the buteonine group; admitted to our fauna in 1858 (*A. nitida* CASS. in BD., 35; COOP., 486) upon the strength of its occurrence in Northern Mexico, but only lately detected in the United States. Illinois, RIDGWAY, Am. Nat. 1872, 130. S. Arizona, breeding (*Bendire, in epist.*). *A. plagiata* SCL., Mus. P.-B. *Asturinæ*, 1; SCL. and SALV., Proc. Zool. Soc. 1869, 130. (Not in the Key.) . . . . PLAGIATA.

161. Genus ONYCHOTES Ridgway.

Gruber's Buzzard. "Nostrils nearly circular, with a ·conspicuous (not central) tubercle; tarsus very long and slender; toes moderate; claws very long, strong and sharp, but only slightly curved; tibial feathers short, close, not reaching beyond the joint; wing very short, much rounded, and very concave beneath; 4th quill longest, 1st shorter than 9th; tail moderate, rounded; outstretched feet reaching beyond tail." No white about head or neck; general color dark bistre-brown, darkest on crown and back, below paler and more rusty; primaries uniform black above, below showing white basally; tail crossed by 7–8 obscure narrow dark bars; wing 10; tail 5⅓; tarsus 2¾. One specimen known, supposed to come from California. RIDGWAY, Proc. Phila. Acad. 1870, 149.   .   GRUBERII.

162. Genus PANDION Savigny.

Osprey. *Fish Hawk.* Plumage lacking aftershafts, compact, imbricated, oily, to resist water; that of the legs short and close, not forming the flowing tufts seen in most other genera, that of the head lengthened, acuminate: primary coverts stiff and acuminate. Feet immensely large and strong, the tarsus entirely naked, granular-reticulate, the toes all of the same length, unwebbed at base, very scabrous underneath, the outer versatile; claws very large, rounded underneath. Hook of the bill long; nostrils touching edge of the cere. Above, dark brown; most of the head and neck, and the under parts, white, latter sometimes with a tawny shade, and streaked with brown. 2 feet long; wing 18–20 inches; tail 8–10. Temperate North America, abundant; migratory, piscivorous. WILS., v, 13, pl. 37; NUTT., i, 18; AUD., i, 64, pl. 15; CASS. in BD., 44; COOP., 454.   .   HALIAETUS.

163. Genus AQUILA Auctorum.

Golden Eagle. Tarsus completely feathered. Dark brown with a purplish gloss; lanceolate feathers of head and neck, golden-brown; quills blackish; in the young, tail white, with a broad terminal black zone. About 3 feet long; wing upward of 2 feet; tail a foot or more. North America, rather northerly, in winter south ordinarily to about 35°. WILS., vii, 13, pl. 55, f. 1; NUTT., i, 62; AUD., i, 50, pl. 12. *A. canadensis* CASS. in BD., 41; COOP., 449. . . . . CHRYSAETUS.

164. Genus HALIAETUS Savigny.

Bald Eagle. Tarsus naked. Dark brown; head and tail white after the third year; before this, these parts like the rest of the plumage. About the size of the last species. Immature birds average larger than the adults; the famous "Bird of Washington" (AUD., Orn. Biog., i, 58, pl. 11, and B. Amer., i, pl. 13, Kentucky) is a case in point. North America, common; piscivorous; a piratical parasite of the osprey; otherwise notorious as the emblem of the Republic. WILS., iv, 89, pl. 36; vii, pl. 55; NUTT., i, 72; AUD., i, 57, pl. 14; CASS. in BD., 43; COOP., 451. . . .   LEUCOCEPHALUS.

OBS. The Greenland Sea Eagle, *H. albicilla;* and the Northern Sea Eagle, *H. pelagicus* (CASS., Ill. 31, pl. 6, and in BD., 42, 43 ; ELLIOT, pl. 34, 35), both usually attributed to our fauna, remain to be detected, the former in N. E., the latter in N. W., portions. *H. pelagicus* has 14 rectrices, and is otherwise distinct.

FIG. 143. Bald Eagle.

165. Genus POLYBORUS Vieillot.

Caracara Buzzard. Bill long, high, compressed, little hooked, commissure nearly straight to the deflected end ; nostrils linear, oblique, in the front upper corner of the cere, which is truncate and bristly ; sides of head extensively denuded ; occipital feathers lengthened ; 3d and 4th quill longest, 1st shorter than 7th ; outer 4–5 emarginate ; tarsus almost naked, longer than middle toe. Brownish-black, barred on the neck, breast and most of the upper parts, with yellowish-white ; auriculars whitish ; tail whitish, narrowly black-barred and with broad black terminal zone ; primaries likewise barred with whitish ; feet yellow ; bill greenish-white. Length 23 ; wing 15–17 ; tail about 10. Southern border, Florida to California ; a remarkable form, allied in some respects to the vultures. AUD., i, 21, pl. 4 ; NUTT., i, 52 ; CASS. in BD., 45 ; COOP., 492. THARUS var. AUDUBONII.

Family CATHARTIDÆ. American Vultures.

Head, and part of the neck, more or less completely bare of feathers ; eyes flush with the side of the head, not overshadowed by a superciliary shield ; ears small and simple. Bill lengthened, contracted toward the base, moderately hooked and comparatively weak. Nostrils very large, completely perforated, through lack of a bony septum. Wings very long, ample and strong ; tail moderate. Anterior toes long for the order, webbed at base ; hind toe elevated, very short ; claws comparatively lengthened, obtuse, little curved and weak. To these external characters, which distinguish our vultures, I may add, that there are numerous osteological peculiarities. A lower larynx is not developed. The capacious gullet dilates into an immense crop. The cœca are extremely small. The feathers lack an aftershaft.

The American vultures differ in so many essential respects from those of the Old World, that they should unquestionably rank as a separate family, whatever may be the propriety of uniting the others with the *Falconidæ*. In a certain sense, they represent the gallinaceous type of structure; our species of *Cathartes*, for instance, bear a curious superficial resemblance to a turkey. They lack the strength and spirit of typical *Raptores*, and rarely attack animals capable of offering resistance; they are voracious and indiscriminate gormandizers of carrion and animal refuse of all sorts—efficient and almost indispensable scavengers in the warm countries where they abound. They are uncleanly in their mode of feeding; the nature of their food renders them ill-scented, and when disturbed they eject the fœtid contents of the crop. Although not truly gregarious, they assemble in multitudes where food is plenty, and some species breed in communities. When gorged,

FIG. 114. Californian Vulture.

they appear heavy and indisposed to exertion, usually passing the period of digestion motionless, in a listless attitude, with the wings half-spread. But they spend most of the time on wing, circling high in the air: their flight is easy and graceful in the extreme, and capable of being indefinitely protracted. On the ground, they habitually walk instead of progressing by leaps. Possessing no vocal apparatus, the vultures are almost mute, emitting only a weak hissing sound. The plumage in *Cathartes* is sombre and unvaried; its changes are slight; the sexes are alike in color; the ♀ is not larger than the ♂. The famous condor of the Andes, *Sarcor-hamphus gryphus*, the king vulture, *S. papa*, and the following species of *Cathartes*, with their one or two South American analogues, compose the family.

166. Genus CATHARTES Illiger.

Californian Vulture. Brownish-black, lustrous above, paler below; secondary quills gray; greater coverts tipped with white; bill whitish; head and neck orange and red; "iris carmine." Most of the neck, as well as the head, naked, with scattered bristle-like feathers, and a feathered patch at base of the bill; plumage commencing on the neck, not with a downy ruff, as in the condor, but with lengthened lanceolate feathers continued on the breast; nostrils comparatively small; tail nearly even. Young covered with whitish down. Largest of the genus; length about 4 feet; extent 9 : wing $2\frac{3}{4}$; tail $1\frac{1}{2}$; thus approaching the condor in size. Egg white, granular, elliptical, $4\frac{1}{2}$ by $2\frac{3}{4}$ inches. General habits the same as those of the following species. Rocky Mountains to the Pacific, U. S. Aud., i, 12, pl. 1; Nutt., i, 39; ii, 557; Cass. in Bd., 5; Coop., 496. CALIFORNIANUS.

Turkey Buzzard. Blackish-brown; quills ashy-gray on their under surface; head red; feet flesh-colored; bill white. Skin of the head corrugated, sparsely beset with bristle-like feathers; plumage commencing in a circle on the neck; nostrils very large and open; tail rounded. Length about $2\frac{1}{2}$ feet; extent 6 : wing 2; tail 1. U. S., from Atlantic to Pacific, and somewhat northward; abundant in more southern portions; resident as far north as New Jersey. Nests on the ground, or near it, in hollow stumps and logs, generally breeding in communities; eggs commonly two, creamy white, blotched and speckled, $2\frac{3}{4}$ by $1\frac{7}{8}$. Wils., ix, pl. 75, f. 1; Nutt., i, 43; Aud., i, 15, pl. 2; Cass. in Bd., 4; Coop., 503. . . . . . AURA.

Carrion Crow. Blackish; quills very pale, almost whitish, on the under surface; head dusky; bill and feet grayish-yellow. Skin of the head as in the last species, but plumage running up the back of the neck to a point on the hind head; nostrils as before; tail square. Smaller than *aura*, in linear dimensions, but a heavier bird : length about 2 feet; wing $1\frac{1}{2}$; tail $\frac{6}{8}$. The difference in size and shape between this species and *aura* is strikingly displayed when the birds are flying together, as constantly occurs in the Southern States ; there is also a radical difference in the mode of flight, this species never sailing for any distance without flapping the wings. Nesting the same : eggs similar, but larger, or at any rate more elongate; $3\frac{1}{4}$ by 2. Chiefly South Atlantic and Gulf States, there very numerous, far outnumbering the turkey buzzard, and semi-domesticated in the towns; N. regularly to North Carolina, thence straggling even to Massachusetts (*Jillson*; Putnam, Proc. Essex Inst. 1856, 223) and Maine (Boardman, Am. Nat. iii, 498); Ohio (*Audubon*); not authenticated on the Pacific Coast. Wils., ix, pl. 75, f. 2; Nutt., i, 46; Aud., i, 17, pl. 3; Cass. in Bd., 5. . . ATRATUS.

OBS. *C. burrovianus* Cass. in Bd., 6; Elliot, pl. 36, a doubtful species, is said to inhabit Lower California. From various accounts, it seems probable that the king vulture really occurs on our southern border, but this remains to be determined. See Bartram, Travels in Florida, p. 150; Cassin in Bd., p. 6; Coues, Proc. Phila. Acad. 1866, p. 49; Allen, Bull. Mus. Comp. Zool. ii, 1871, p. 313.

Order COLUMBÆ. Columbine Birds.

An essential character of birds of this order is seen in the structure of the bill: horny and convex at the tip, somewhat contracted in the continuity, furnished at the base with a soft swollen membrane in which the nostrils open. There are four toes, three anterior, generally cleft, but occasionally with a slight basal web, and one behind, with few exceptions perfectly insistent or not obviously elevated. The feet are never lengthened; the tarsus is commonly shorter than the toes, either scutellate or extensively feathered anteriorly, reticulate on the sides and behind, the envelope rather membranous than corneous. The plumage is destitute of aftershafts. The syrinx has one pair of intrinsic muscles. There are two carotids. The sternum is doubly notched, or notched and fenestrate; there are other osteological characters. The regimen is exclusively vegetarian. Terrestrial progression gradient, never saltatory. As commonly accepted, the order is composed of three families. The strange dodo, *Didus ineptus*, recently extinct, represents one, *Dididæ*; another, *Didunculidæ*, consists of the only less singular tooth-billed pigeon, *Didunculus strigirostris*, of the Navigator Islands; the third is the *Columbidæ*. Some, like Lilljeborg, enlarge the order, under name of *Pullastræ*, to receive the *Cracidæ* (see beyond), and *Megapodidæ*, big-feet or mound-birds of the East Indies; mainly on account, it would appear, of the position of the hallux in these families; but the balance of evidence favors their reference to the gallinaceous birds. There is no question that the columbine are very closely related to the rasorial birds, but it seems best to draw the line between them as above indicated; and I shall accordingly close the great Insessorial series with the

Family COLUMBIDÆ. Pigeons.

The family may be framed simply by exclusion of the *Didunculidæ* and *Dididæ*. With one exception, all our species will be immediately recognized by their likeness to the familiar inmates of the dove-cot. One seemingly trivial circumstance is so constant as to become a good clue to these birds: the frontal feathers do not form antiæ by extension on either side of the culmen, but sweep across the base of the bill with a strongly convex outline projected on the culmen, thence rapidly retreating to the commissural point. The plumuleless plumage is generally compact, with thickened, spongy rhachis, the insertion of which will seem loose to one who skins a bird of this family. The head is remarkably small; the neck moderate; the body full, especially in the pectoral region. The wings are strong, generally lengthened and pointed, conferring a rapid, powerful, whistling flight; the peculiar aërial evolutions that these birds are wont to perform, have furnished a synonym for the family, *Gyrantes*. The tail varies in shape, from square to graduate, but is never forked; as a rule there are 12 rectrices, frequently increased to 14, rarely to 16. The feet show considerable modification when the strictly arboricole are compared with the more terrestrial species; their general character has just been indicated. The gizzard is large and muscular, particularly in the species that feed on seeds and other hard fruits; the gullet dilates to form a capacious circumscribed crop. This organ at times secretes a peculiar milky fluid, which, mixed with macerated food, is poured by regurgitation directly into the mouth of the young; thus the fabled "pigeon's milk" has a strong spice of fact, and in this remarkable circumstance we see probably the nearest approach, among birds, to the characteristic function of mammalia. "The voice of the turtle is heard in the land" as a plaintive cooing, so characteristic as to have afforded another name for the family, *Gemitores*. Pigeons are altricial, and monogamous—doubly monogamous, as is said when both sexes incubate and care for the young; this is a strong trait, compared with the præcocial and often polygamous nature of rasorial birds. They are amorous birds whose passion generally results in a tender and constant devotion, edifying to contemplate, but is often marked by high irascibility and pugnacity—traits at variance with the amiable meekness which doves are supposed to symbolize. The nest, as a rule, is a rude, frail, flat structure of twigs; the eggs are usually two in number, sometimes one, white.

"The entire number of pigeons known to exist is about 300; of these the Malay Archipelago already counts 118, while only 28 are found in India, 23 in Australia, less than 40 in Africa, and not more than 80 in the whole of America." They focus in the small district of which New Guinea is the centre, where more than a fourth of the species occur. Mr. Wallace accounts for this by the absence of fruit-eating forest mammals, such as monkeys and squirrels; and finds in the converse the reason why pigeons are so scarce in the Amazon valley, and there chiefly represented by species feeding much on the ground and breeding in the bushes lower than monkeys habitually descend. "In the Malay countries, also, there are no great families of fruit-eating *Passeres*, and their place seems to be taken by the true fruit-pigeons, which, unchecked by rivals or enemies, often form with the *Psittaci* the prominent and characteristic features of the Avifauna." (NEWTON.)

There are three prominent groups of pigeons. The *Treroninæ* are exclusively frugivorous and arboricole species, with short, soft, broad-soled and extensively feathered feet, 14 rectrices, and soft lustreless plumage, of which green is the characteristic color. These are all Old World; the genera are *Treron* and *Ptilonopus*, with their subdivisions: "54 species are confined to the Austro-Malayan, while 28 inhabit the Indo-Malayan, subregion; in India 14, and in Africa 6 species are found; 30 inhabit the Pacific Islands, and 8 occur in Australia or New Zealand, while New Guinea has 14 species." (WALLACE.) The *Gourinæ* are more or less terrestrial species, of both hemispheres, embracing a considerable number of more varied generic forms. In the New Guinean *Goura coronata* there are 16 rectrices, and the head is crested; in the singular *Calœnas nicobarica*, feathers of the upper parts are acuminate, elongate and even pendulous; each of these is sometimes made the type of a family. There are several other Old World forms, such as *Trugon*, *Phaps*, *Henicophaps*, *Geophaps*, *Lophophaps*, *Ocyphaps* and *Chalcophaps;* our genus *Starnœnas* is an interesting American one. The *Columbinæ* are the least specialized and most generally distributed group, comprising numerous species of which the domestic pigeon (*Columba livia*) is a type. Of these the Australian *Lopholæmus antarcticus*, if really belonging here, is one of the most peculiar; *Carpophaga* and *Turtur* are leading Old World genera. The North American genera, excepting the first two following, are probably *Gourinæ* in the current acceptation of that term; but in the uncertainty attending its precise limitation as compared with *Columbinæ*, I shall not attempt to distinguish subfamilies. In gen. 167-8 the tarsi are short and slightly feathered above, a characteristic of arboricole pigeons; in the rest, longer and entirely naked, as usual in the terrestrial species; and in many of these there is a naked space above the eyes. The males of nearly all our species show a beautiful iridescence on the neck; the sexes are distinguishable by color; the young resembles the female.

167. Genus COLUMBA Linnæus.

Band-tailed Pigeon. Ashy-blue, tinged with olive on the back; head, neck and under parts purplish, whitening on the belly; hind neck metallic golden, with a conspicuous white collar; tail with a dark bar, beyond this brownish-white; bill and feet yellow, former black-tipped: ♀ and young less or not purplish, the nuchal band often obscure or wanting; 15; wing 8½; tail 6¼, nearly even; tarsus 1, feathered above. Rocky Mountains to the Pacific, U. S. and southward; common. Bonap., Am. Orn. i, 77, pl. 8; Nutt., i, 624; Aud., iv, 312, pl. 279; Bd., 597; Coop., 506. . Fasciata.

Red-billed Pigeon. Slaty-blue, olive on the back and scapulars; head and neck all round, breast and wing-patch, chocolate-red; no nuchal iridescence; "bill, feet and eyes, purple;" 14; wing 8; tail 5¾. Mexico, to U. S. border; Cape St. Lucas. Lawr., Ann. Lyc. N. Y. 1851, 116; Bd., 598, pl. 61; Coop., 508. . . . . Flavirostris.

White-crowned Pigeon. Dark slaty-blue, paler below; crown pure white; hind neck purplish-brown, lower down metallic golden, each feather black-edged; iris white; bill and feet reddish, former blue-tipped; 13½; wing 7½; tail 5½; ♀ similar. West Indies and Florida Keys. Bonap., Am. Orn. ii, 11, pl. 15; Nutt., i, 625; Aud., iv, 315, pl. 280; Bd., 599. Leucocephala.

168. Genus ECTOPISTES Swainson.

Wild Pigeon. Adult ♂ dull blue with olivaceous tinge on back, below dull purplish-red whitening on vent and crissum; sides of neck golden and ruby; some wing coverts black-spotted; quills blackish, with slaty, whitish and rufous edging; middle tail feathers bluish-black, the others white or

Fig. 115. Wild Pigeon.

ashy, the inner webs basally black with a chestnut patch; bill black; feet yellow; ♀ and young duller and more brownish or olivaceous above, below dull grayish, with a tawny tinge anteriorly, or quite gray; very young have the feathers skirted with whitish; 15–17; wing 7–8; tail about the same, cuneate, of 12 narrow acuminate feathers. "Wanders continually in search of food throughout all parts of North America; wonderfully abundant at times in particular districts" (*Audubon*); chiefly, however, temperate Eastern North America; eminently gregarious. WILS., v, 102, pl. 44; NUTT., i, 629; AUD., v, 25, pl. 285; BD., 600. . . . . . MIGRATORIUS.

169. Genus ZENÆDURA Bonaparte.

Carolina Dove. Brownish-olive, glossed with blue on the crown and nape; below purplish-red, becoming tawny-white on the vent and crissum;

neck metallic golden; a velvety black spot on the auriculars, and others on the wing coverts and scapulars; middle tail feathers like the back, the rest ashy-blue at base, then crossed by a black bar, then white or ashy-white; bill very slender, black; feet carmine; ♀ and young differ as in the wild pigeon; 11–13; wing 5–6; tail 6–7, shaped as in the wild pigeon, but of

FIG. 116. Carolina Dove.

14 feathers; circumorbital space naked. Temperate North America, very abundant. WILS., v, 91, pl. 43; NUTT., i, 626; AUD., v, 30, pl. 286; BD., 604. . . . . CAROLINENSIS.

170. Genus ZENÆDA Bonaparte.

Zenaida Dove. Olive-gray with a reddish tinge, crown and under parts vinaceous-red, sides and axillars bluish; a velvety black auricular spot, and others on the wing coverts and tertiaries; secondaries tipped with white; neck with metallic lustre; middle tail feathers like the back, others bluish with whiter tips, a black band intervening; 10; wing 6; tail 4, rounded. West Indies and Florida Keys. BONAP., Am. Orn. ii, pl. 15, f. 2; NUTT., i, 625; AUD., v, 1, pl. 281; BD., 602. . . . . AMABILIS.

171. Genus MELOPELEIA Bonaparte.

White-winged Dove. A broad oblique white bar on the wing, formed by ends of greater coverts and alula. Tail feathers, except the middle, broadly tipped with white; general plumage resembling that of the Carolina dove; 11–12; wing 6–6½; tail 5, rounded. Southwestern U. S. and southward. *Columba trudeaui* AUD., vii, 352, pl. 496. BD., 603. . . LEUCOPTERA.

172. Genus CHAMÆPELEIA Swainson.

Ground Dove. Grayish-olive, glossed with blue on the hind head and neck, most feathers of the fore parts with darker edges, those of the breast with dusky centres; forehead, sides of head and neck, lesser wing coverts and under parts purplish-red of variable intensity, paler or grayish in the ♀; under surface of wings orange-brown or chestnut, this color suffusing the quills to a great extent, upper surface sprinkled with lustrous steel-blue spots; middle tail feathers like the back, others bluish-black; feet yellow; bill yellow with dark tip; diminutive; 6–6½; wing 3½, with inner secondaries nearly as long as the primaries; tail 2¾, rounded. Southern U. S., Atlantic to Pacific, but chiefly coastwise; N. to the Carolinas, and accidentally to Washington, D. C; common. WILS., iv, 15, pl. 46; NUTT., i, 635; AUD., v, 19, pl. 283; BD., 606; COOP., 516. Var. *pallescens* BD., Proc. Phila. Acad. 1859, 305; COOP., 517; Cape St. Lucas. . . PASSERINA.

173. Genus SCARDAFELLA Bonaparte.

Scaly Dove. General coloration much as in the ground dove, but all the body-feathers with sharp dark border producing a scaled appearance; tail long and cuneate, with (14?) narrow acuminate feathers, as in the common dove, broadly tipped with white, except the middle pair; wing shaded as in the ground dove. Small; 8; wing and tail about 4. Mexico to U. S. border. BD., 605; ELLIOT, pl. 37; COOP., 519. . SQUAMOSA var. INCA.

174. Genus GEOTRYGON Gosse.

Key West Dove. Above, vinaceous-red with highly iridescent lustre of various tints; below pale purplish fading to creamy; an infraocular stripe and the throat white; 11; wing and tail about 6, latter rounded. West Indies and Key West. *Columba montana*, AUD., v, 14, pl. 282. NUTT., i, 2d ed. 756; BD., 607. . . . . MARTINICA.

175. Genus STARNŒNAS Bonaparte.

Blue-headed Ground Dove. Crown rich blue bounded by black; a white stripe under the eye meeting its fellow on the chin; throat black, bordered with white; general color olivaceous-chocolate above, purplish-red below, lighter centrally; 11; wing 5½; tail 4½. West Indies and Florida Keys. A remarkable form, grading towards the gallinaceous birds in structure and habits; bill short; wings and tail very short, former rounded and concave, latter nearly even; legs very long and stout; tarsus bare, reticulate; hind toe not strictly insistent. AUD., v, 23, pl. 284; NUTT., i, 2d ed. 769; BD., 608. . . . . CYANOCEPHALA.

Subclass II. AVES TERRESTRES, or CURSORES.

TERRESTRIAL BIRDS.

This second series includes all living birds, between the *Columbæ* and the *Lamellirostres*, excepting, probably, the ostriches and their allies. Like the other two divisions called "subclasses" in the present work, it is insusceptible of definition by characters of more than the slightest morphological importance, and consequently has nothing of the taxonomic value commonly attaching to groups so named. It may be considered, however, to represent the teleological generalization, that a certain number of birds, differing greatly in structure, are collectively modified in a way that fits them for similar modes of life—that several different types of structure are bent to subserve a particular end. In a certain sense, therefore the *Cursores* may be said to hold together more by analogical relationship than by special morphological affinity; and among them there is certainly greater diversity of structure than that existing between some of them and the birds standing upon the confines of *Insessores* and *Natatores*. On the one hand, the gallinaceous birds shade directly into the columbine, while on the other, the *Grallatores* are perfectly linked with the *Natatores* by means of the flamingoes. As implied in their name, the birds of this series are especially terrestrial in habit, spending most of the time on the ground, not on trees or the water; although most of them fly vigorously, and some swim well. A character of general applicability is the combination of long or strong legs (as compared with *Insessores*), with the freedom of the knee and lower thigh from the body (as compared with *Natatores*). The hallux as a rule is reduced in length and elevated in position, and is often absent altogether—a modification rarely found outside this group; the front toes are generally webbed at base, often cleft, occasionally lobed or even full-webbed. Excluding the struthious birds, which cannot well be brought into this connection, the series represents two commonly received orders.

Order GALLINÆ. Gallinaceous Birds.

Equivalent to the old order *Rasores*, exclusive of the pigeons—this name being derived from the characteristic habit of scratching the ground in search of food; connecting the lower terrestrial pigeons with the higher members of the great plover-snipe group. On the one hand, it shades into the *Columbæ* so perfectly that Huxley has proposed to call the two together the "Gallo-columbine series;" on the other hand, some of its genera show a strong plover-ward tendency, and have even been placed in *Limicolæ*. The birds of this family are more or less perfectly

terrestrial ; the legs are of mean length, and stout ; the toes four (with rare exceptions), three in front, generally connected by basal webbing, but sometimes free,
and one behind, almost always short and elevated, occasionally absent. The tibiæ
are rarely naked below ; the tarsi often feathered, as the toes also sometimes are ;
but ordinarily both these are naked, scutellate and reticulate, and often developing
processes (*spurs*) of horny substance with a bony core, like the horns of cattle.
The bill as a rule is short, stout, convex and obtuse ; never cered, nor extensively
membranous ; the base of the culmen parts prominent antiæ, which frequently fill
the nasal fossæ ; when naked the nostrils show a superincumbent scale. The head is
frequently naked, wholly or partly, and often develops remarkable fleshy processes.
The wings are short, stout and concave, conferring power of rapid, whirring, but
unprotracted, flight. The tail varies extremely ; it is entirely wanting in some
genera, enormously developed in others ; the rectrices vary in number, but are
commonly more than twelve. The sternum, with certain exceptions, shows a
peculiar conformation ; the posterior notches seen in most birds, are inordinately
enlarged, so that the bone, viewed vertically, seems in most of its extent to be
simply a narrow central projection, with two long backward processes on each side,
the outer commonly hammer-shaped ; this form is modified in the tinamous, curassows, mound-birds and sand-grouse, and not at all shown in the hoazin. The palate
is schizognathous ; there are other distinctive osteological characters. As a rule, the
digestive system presents an ample special crop, a highly muscular gizzard, and
large cœca ; "the inferior larynx is always devoid of intrinsic muscles" (*Huxley*).
Excepting the *Pteroclidæ* (?), there are aftershafts, and a circlet around the oil-gland.
Gallinæ are præcocial. A part of them are polygamous — a circumstance shown in its
perfection by the sultan of the dunghill with his disciplined harem ; and in all such
the sexes are conspicuously dissimilar. The rest are monogamous, and the sexes of
these are as a rule nearly or quite alike. The eggs are very numerous, usually laid
on the ground, in a rude nest, or none. The order is cosmopolitan ; but most of its
groups have a special geographical distribution ; its great economic importance is
perceived in all forms of domestic poultry, and principal game-birds of various
countries ; and it is unsurpassed in beauty — some of these birds offer the most
gorgeous coloring of the class. The characters of the order have been ably
exposed by Blanchard, Parker, Huxley and other distinguished anatomists. I
will briefly recount the exotic families.

1. The tinamous, *Tinamidæ*, are so remarkably distinguished by certain cranial
characters that Huxley was induced to make them one of his four primary divisions
of carinate birds. The palate is "completely struthious ;" the sternum has a
singular conformation. An obvious external feature, in many cases, is the entire
lack of tail feathers (only elsewhere wanting among grebes) ; in others, however,
these are developed. Confined to Central and South America, and represented by
about forty species, of six or eight genera.

2. The wonderful hoazin of Guiana, *Opisthocomus cristatus*, is the sole representative of a family *Opisthocomidæ*, one of the most isolated and puzzling forms
in ornithology, sometimes placed near the *Musophagidæ*, but assigned by maturer
judgment to the fowls, which it resembles in most respects. The sternum and
shoulder-girdle are anomalous ; the keel is cut away in front ; the furcula anchylose
with the coracoids (very rare) and with the manubrium of the sternum (unique) ;
the digestive system is scarcely less singular.

3. The bush-quails of the Old World, *Turnicidæ*, differ widely from other

Gallinæ, resembling the sand-grouse and tinamous in some respects, and related to the plovers in others. A singular circumstance is a lack of the extensive vertebral anchyloses usual in birds, all the vertebræ remaining distinct (*Parker*). The crop is said to be wanting in some, as is also the hind toe. There are some twenty current species of the principal genus, *Turnix*, to which Gray adds the African *Ortyxelos meiffrenii*, and the Australian *Pediomomus torquatus;* the latter is placed, by some, with the *Grallæ*.

4. The sand-grouse, *Pteroclidæ*, inosculate with the pigeons, as the *Turnicidæ* do with the plovers. The digestive system is fowl-like; the sternum in *Pterocles* departs from the rasorial type to approach the columbine, the modification being even more marked than in the next family; the pterylosis is pigeon-like, lacking aftershafts (*Huxley*), or having small ones (*Nitzsch*). The wings are very long and pointed, the feet short, with reduced hallux, and variable feathering. Confined to Europe, Asia and Africa: the principal genus, *Pterocles*, has about a dozen species; the only other, *Syrrhaptes*, has two.

5. The mound-birds, *Megapodidæ*, as the name implies, have large feet, with little curved claws, and lengthened insistent hallux. They share this last feature with the *Cracidæ* (beyond); and the osseous structure of these two families, except as regards pneumaticity, is strikingly similar. Both show a modification of the sternum, the inner one of the two notches being less instead of more than half as deep as the sternum is long, as in typical *Gallinæ*. Confined to Australia and the East Indies; *Megapodius* is the principal genus, of a dozen or more species; there are three others, each of a species or two.

6. The guinea-fowl, *Numididæ*, of which a species, *Numida meleagris*, is commonly seen in domestication, are an African and Madagascan type. While the foregoing families are strongly specialized, this one, like the turkey family, more closely approaches the true fowl, and both may be only subfamilies of *Phasianidæ*. The bones of the pinion have a certain peculiarity; the frontal generally develops a protuberance; there are wattles, but no spurs; the tail is very short; the head naked. There are six or eight species of *Numida*, in some of which the trachea is convoluted·in an appendage to the furcula; *Acryllium vulturina*, *Agelastes meleagrides* and *Phasidus niger*, are the remaining ones.

7. Finally, we reach the *Phasianidæ*, or pheasants, a magnificent family of typical *Gallinæ*, of which the domestic fowl is a characteristic example. These birds do not show any of the foregoing special characters; the feet, nasal fossæ, and usually a part, if not the whole, of the head, are naked; the tarsi commonly develop spurs; the hallux is elevated; the tail, with or without its coverts, sometimes has an extraordinary development or a remarkable shape. There are fifty or sixty species, distributed in numerous modern genera, about twelve of which are well marked; they are all indigenous to Asia and neighboring islands, focussing in India. In the peacock, *Pavo cristatus*, the tail coverts form a superb train, capable of erection into a disk, the most gorgeous object in ornithology; in an allied genus, *Polyplectron*, there are a pair of spurs on each leg. The argus pheasant, *Argusianus giganteus*, is distinguished by the enormous development of the secondary quills, as well as by the length of the tail feathers and peculiarity of the middle pair. The combed, wattled and spurred barn-yard fowl, with folded tail and flowing middle feathers, are descendants of *Gallus bankiva*, type of a small genus. The tragopans, *Ceriornis*, are an allied form with few species; the macartneys, *Euplocomus*, with a dozen species, are another near form, as are the impeyans, *Lophophorus*,

with a slender aigrette on the head, like a peacock's. The naturalized English pheasant, *P. colchicus*, introduced into Britain prior to A. D. 1056, is the type of *Phasianus*, in which the tail feathers are very long and narrow; in one species, *P. reevesii*, the tail is said to attain a length of six feet. The golden and Amherstian pheasants, *Chrysolophus pictus* and *amherstiæ*, are singularly beautiful, even for this group. The other genera are *Crossoptilon* and *Pucrasia*. New species are still coming to light.

Family CRACIDÆ. Curassows. Guans.

This type is peculiar to America, where it may be considered to represent the *Megapodidæ*, though differing so much in habit and general appearance. The affinities of the two are indicated above, and some essential characters noted. According to the latest authority on the family, Messrs. Sclater and Salvin, it is divisible into three subfamilies: *Cracinæ*, curassows and hoccos, with four genera and twelve species; *Oreophasinæ*, with a single species, *Oreophasis derbianus*, and the

Subfamily PENELOPINÆ, Guans,

with seven genera and thirty-nine species, one of which reaches our border.

176. Genus ORTALIDA Merrem.

Texan Guan. Chiacalaca. Head crested, its sides, and strips on the chin, naked, but no wattles; tarsi naked, scutellate; hind toe insistent, about ⅓ the middle toe; tail graduated, longer than the wings, of 12 feathers. Length nearly 2 feet; wing 8½ inches; tail 11; tarsus 2⅓; middle toe the same. Dark olivaceous, paler and tinged with brownish-yellow below, plumbeous on the head; tail green, tipped with white except on the middle pair of feathers; bill and feet plumbeous. Mexico, to Texas. *O. vetula* Lawr., Ann. Lyc. N. Y. 1851, 116; *O. poliocephala* Cass., Ill. 267, pl. 44; *O. maccallii* Bd., 611. . . . . VETULA.

Family MELEAGRIDÆ. Turkeys.

Head and upper neck naked, carunculate; in our species with a dewlap and erectile process. Tarsi naked, scutellate before and behind, spurred in the ♂. Tail broad, rounded, of 14–18 feathers. Plumage compact, lustrous; in our species with a tuft of hair-like feathers on the breast. One genus, two species. *M. ocellatus* is a very beautiful species of Central America.

177. Genus MELEAGRIS Linnæus.

Turkey. Upper tail coverts chestnut, with paler or whitish tips; tail feathers tipped with brownish-yellow or whitish; 3–4 feet long, etc. Wild in Texas, New Mexico, Arizona and southward; domesticated elsewhere. There is reason to believe that the Mexican bird is the original of the domestic race; it was upon this form, imported into Europe, that Linnæus imposed the name *gallopavo* (Fn. Suec. No. 198; Syst. Nat. i, 1766, 268), which has generally been applied to the following feral variety. *M.*

mexicana GOULD, Proc. Zool. Soc. 1856, 61; BD., 618; ELLIOT, pl. 38; Coor., 523. . . . . GALLOPAVO.

Var. AMERICANA BART., Trav. 1791, 290. *Gallopavo sylvestris* LE CONTE, Proc. Phila. Acad. 1857, 179; *M. gallopavo* AUD., v, 42, pls. 287, 288; NUTT., i, 630; BD., 615. Upper tail coverts without light tips, and ends of tail feathers scarcely paler. This is the ordinary wild turkey of Eastern North America; N. to Canada, where it is said still to occur; apparently extirpated in New England. N.W. to the Missouri, and S.W. to Texas (*Audubon*). The slight differences just noted seem to be remarkably constant, and to be rarely, if ever, shown by the other form, although, as usual in domestic birds, this last varies interminably in color.

Family TETRAONIDÆ. Grouse, etc.

All the remaining gallinaceous birds are very closely related, and they will probably constitute a single family, although the term *Tetraonidæ* is usually restricted to the true grouse as below defined (*Tetraoninæ*), the partridges being erected into another family, *Perdicidæ*, with several subfamilies. But the grouse do not appear to differ more from the partridges than these do from each other, and they are all variously interrelated; so that no violence will be offered in uniting them. One group of the partridges is confined to America; all the rest to the Old World. The leading forms among the latter are *Perdix*, the true partridge; *Coturnix*, the true quail; *Francolinus*, the francolins; with *Rollulus* and *Caccabis*. In all, perhaps a hundred species and a dozen genera. Without attempting to frame a family diagnosis to cover all their modifications, I will precisely define the American forms, as two subfamilies.

Subfamily TETRAONINÆ. Grouse.

Head completely feathered, excepting, usually, a naked strip of skin over the eye. Nasal fossæ densely feathered. Tarsi more or less perfectly feathered, the feathering sometimes extending on the toes to the claws; the toes, when naked, with fringe-like processes. Tail variable in shape, but never folded, of 16–20 feathers. Sides of the neck frequently with lengthened or otherwise modified feathers, or a bare distensible skin, or both.

The true grouse are confined to the northern hemisphere, and reach their highest development, as a group, in North America, where singularly varied forms occur. The only Old World species are — the great *Tetrao urogallus*, or capercailzie of Europe, and its allied Asiatic species; *Tetrao tetrix*, the "black game" of Europe, with curiously curled tail feathers; *Tetrao falcipennis* of Siberia, the representative of our spruce partridge; *Bonasa betulina* of Northern Europe and Asia, like our ruffed grouse; and two or three species of ptarmigan (*Lagopus*).

178. Genus TETRAO Linnæus.

** No peculiar feathers on the neck; tarsus feathered to the toes; tail moderate, little rounded, of 16–20 broad feathers. Woodland birds of northerly or alpine distribution. Our species differ materially from the European capercailzie, *T. urogallus*, type of the genus, and might be properly separated.

* Tail normally of 16 (14–18) feathers. (*Canace.*)

Canada Grouse.   Spruce Partridge.   ♂ below mostly black with numerous white spots; above, vermiculated with blackish and slate, and usually some tawny, especially on the wings; quills variegated with tawny; tail with a terminal orange-brown band, its upper coverts plain; 15–17; wing 7; tail 5; ♀ rather less, no continuous black below, but variegated with blackish, white and tawny; above, much as in the ♂, but more tawny. N. Am., northerly; in Brit. Am., W. to Alaska; in U. S., W. to Rocky Mts.; S. into the northern tier of states; Maine, and casually to Massachusetts. NUTT., i, 667; AUD., v, 83, pl. 294; BD., 622. . . . . . . CANADENSIS.

FIG. 147. Canada Grouse.

Var. FRANKLINII. Tail less rounded, lacking the terminal orange-brown band, and its upper coverts conspicuously white-tipped. Rocky and Cascade Mts., U. S. BD., 623; COOP., 529.

** Tail normally of 20 (18–22?) feathers. (*Dendragapus*.)

Dusky Grouse. ♂ blackish, more or less variegated with slate-gray, or a peculiar slaty-black; throat and sides marked with white; breast black; belly slate; tail clouded with slate and black, and with a broad terminal slate bar; 18–20; wing 9–10; tail 7–8; ♀ smaller, not particularly different in color, but not so uniformly dark, having ochrey or reddish-brown variegation in places. Rocky and other Mountains, U. S. to the Pacific. NUTT., i, 666; AUD., v, 89, pl. 295; BD., 620; COOP., 526. . . . OBSCURUS.

Var. RICHARDSONII. Tail nearly square, entirely black, or with only a slight slate tipping. Central Rocky Mountains and northward. COOP., 582.

179. Genus CENTROCERCUS Swainson.

Sage Cock. *Cock of the Plains.* Tail very long, equalling or exceeding the wings, of twenty stiffened, graduated, narrowly acuminate feathers; sides of lower neck with a patch of peculiar sharp scaly feathers, the shafts of which terminate in bristly filaments, sometimes 3–4 inches long in the ♂; tarsi full feathered. Very large; two feet or more long, wing and tail each about a foot; ♀ much smaller. Above, variegated with black, gray and tawny; below, a large black abdominal patch in the adult. Confined to the sterile plains and sage-brush (*Artemisia*) tracts of Western U. S.; S. to about 35° (Mojave river; *Cooper*). Sw. and RICH., F. B.-A. ii, 358, pl. 58; NUTT., i, 666; AUD., v, 106, pl. 297; BD., 624. UROPHASIANUS.

180. Genus PEDIŒCETES Baird.

*** Neck without peculiar feathers; tail very short, of sixteen narrow, soft, true rectrices, and a middle pair, apparently developed coverts, projecting an inch beyond the rest; tarsi fully feathered. Length about 18; wing 8–9; tail 5–6. Below, white, with numerous dark marks; above, variegated with blackish and white, or tawny; quills dusky, with white or tawny spots on the outer web; central tail feathers like the back, others white on the inner web. Sexes alike.

Northern Sharp-tailed Grouse. The markings black, white and dark brown, with little or no tawny; spots on the under parts numerous, blackish, V-shaped; throat white, speckled. Arctic America; not S. to the U. S. *Tetrao phasianellus* LINN., Syst. Nat. i, 160; ELLIOT, Proc. Phila. Acad. 1862, 403. *P. kennicottii* SUCKLEY, *ibid.* 1861, 361. . PHASIANELLUS.

Var. COLUMBIANUS. *Common Sharp-tailed Grouse.* The markings black, white, and especially tawny; below, the spots fewer, brown, U-shaped; throat buff. *T. phasianellus* NUTT., i, 669; AUD., v, 110, pl. 298. BD., 626. *P. columbianus* ELLIOT, *l. c.*; COOP., 532. This is the ordinary U. S. bird, abundant on the prairies from Wisconsin and Kansas westward. It is accurately discriminated from the dark northern form by Dr. Suckley and Mr. Elliot, who, however, incorrectly suppose that the two forms are distinct species; they are geographical races differing from each other according to well known laws of climatic variation.

181. Genus CUPIDONIA Reichenbach.

Pinnated Grouse. Prairie Hen. Neck with a peculiar tuft of loose, lengthened, acuminate feathers, beneath which is a patch of bare, brightly colored skin, capable of great distension; tail short, rounded, of eighteen stiffish, not acuminate, feathers; tarsi barely feathered to the toes. Length 16–18; wing 8–9; tail about 5. Above, variegated with black, brown, tawny or ochrey, and white, the latter especially on the wings; below, pretty regularly barred with dark brown, white and tawny; throat tawny, a little speckled, or not; vent and crissum mostly white; quills fuscous, with white spots on the outer webs; tail fuscous, with narrow or imperfect white or tawny bars and tips; sexes alike in color, but ♀ smaller, with shorter

FIG. 118. Foot of Prairie Hen.

neck-tufts. This well known bird formerly ranged across the United States, in open country, from the Atlantic to the Eastern foothills of the Rocky Mountains, and now abounds on the prairies, from Illinois and Wisconsin, to Middle Kansas at least, if not found on the dryer plains westward. It has been almost extirpated in the Middle and Eastern States, though it still occurs sparingly in isolated localities in New York, New Jersey, Pennsylvania, Long Island, Nantucket and Martha's Vineyard, etc. Its abundance, and the excellence of its flesh, render it an object of commercial importance. Though there may be little probability of its extinction, legislation against its wanton or ill-timed destruction would be a measure of obvious propriety. WILS., iii, 104, pl. 27; NUTT., i, 662; AUD., v, 93, pl. 296; BD., 628. . . . . CUPIDO.

182. Genus BONASA Stephens.

Ruffed Grouse. Partridge; New England and Middle States. *Pheasant;* Southern States. Sides of the neck with a tuft of numerous (15–30), broad, soft, glossy-black feathers; head with a full soft crest; tail about as long as the wings, amply rounded, of (normally) eighteen soft broad feathers; tarsi naked below. Length 16–18; wing 7–8. Sexes nearly alike; variegated reddish- or grayish-brown, the back with numerous, oblong, pale, black-edged spots; below, whitish barred with brown; tail with a broad subterminal black zone, and tipped with gray. A woodland bird, like the species of *Tetrao,* abundantly distributed over Eastern North America, well known

FIG. 119. Ruffed Grouse.

under the above names in different sections; but it is neither a partridge nor a pheasant. The "drumming" sound for which this bird is noted, is not vocal, as some suppose, but is produced by rapidly beating the wings together, or against some hard object, as a fallen log. WILS., vi, 46, pl. 49; NUTT., i, 657; AUD., v, pl. 293, 72; BD., 630. . . . UMBELLUS.

Var. UMBELLOIDES. Pale; slaty-gray the prevailing shade. Rocky Mountain region. DOUGLAS, Linn. Trans. xvi, 1829, 148; BD., 925.

Var. SABINEI. Dark; chestnut-brown the prevailing shade. Pacific Coast region. DOUGLAS, *ibid.* 137; BD., 631; COOP., 540.

· 183. Genus LAGOPUS Vieillot.

*** No peculiar feathers on neck; tarsi and toes densely feathered; tail short, little rounded, normally of 14 broad feathers, with long upper coverts, some of which resemble rectrices. Boreal and alpine grouse, shaped nearly as in *Canace,* remarkable for the seasonal changes of plumage, becoming in winter snow-white. There are only five or six species, at most, and probably fewer; we certainly have the three here given.

Willow Ptarmigan. Tail black; no black stripe on head; bill very stout, culmen $\frac{3}{4}$, or more, its depth at base as much as the distance from nasal fossa to tip. In summer, the fore parts rich chestnut or orange-brown, variegated with blackish, the upper parts and sides barred with blackish, tawny and white; most other parts white. 15–17; wing 8; tail 5. British America, into northernmost U. S. NUTT., i, 674; AUD., v, 114, pl. 299; BD., 633. *L. salceti* Sw. and RICH., Fu. Bor.-Am. ii, 351. . . ALBUS.

Rock Ptarmigan. Tail black; ♂ with a black transocular stripe; bill slenderer, culmen about $\frac{2}{3}$, depth at base less than distance from nasal fossa to tip. In summer, the general plumage irregularly banded with black, reddish-yellow, and white. Rather smaller than the foregoing. Arctic

America. Sw. and Rich., Fn. Bor.-Am. ii, 354, pl. 64; Nutt., i, 610; Aud., v, 122, pl. 301; Bd., 635. *?L. americanus* Aud., v, 119, pl. 300; based on *L. mutus* Sw. and Rich., Fn. Bor.-Am. ii, 350. . . rupestris.

White-tailed Ptarmigan. Tail white at all seasons; in winter, no black anywhere; in summer, barred with dark brown and ochrey; bill slender, and other proportions nearly as in the last. A species of alpine distribution in western North America, from the Arctic regions to New Mexico (lat. 37°). Sw. and Rich., Fn. Bor.-Am. ii, 356, pl. 63; Nutt., i, 612; Aud., v, 125, pl. 302; Bd., 636; Coop., 542. . . . . leucurus.

Subfamily ODONTOPHORINÆ. *American Partridges.*

Head completely feathered, and usually crested, the crest frequently assuming a remarkable shape. Nasal fossæ not filled with feathers, the nostrils covered with a naked scale. Tarsi and toes naked, the latter scarcely or not fringed.

Our partridges may be distinguished, among American *Gallinæ*, by the foregoing characters, but not from those of the Old World; and it is highly improbable that, as a group, they are separable from all the forms of the latter by any decided peculiarities. I find that the principal supposed character, namely, a toothing of the under mandible, is very faintly indicated in some forms, and entirely wanting in others. Pending final issue, however, it is expedient to recognize the group, so strictly limited geographically, if not otherwise. Several beautiful and important

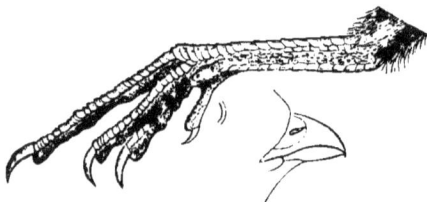

FIG. 150. Foot and bill of Partridge.

genera occur within our limits, but these partridges are most numerous in species in Central and South America. *Odontophorus* is the leading genus, with perhaps 15 species; *Eupsychortyx* and *Dendrortyx* are other extra-limital forms; and in all, some forty-odd species are known. In habits, they agree more or less completely with the well known bob-white. Our species are apparently monogamous, and go in small flocks, called "coveys," usually consisting of the members of one family; they are terrestrial, but take to the trees on occasion; nest on the ground, laying numerous white or speckled eggs; are chiefly granivorous, but also feed on buds, soft fruits, and insects; and are non-migratory.

184. Genus ORTYX Stephens.

Virginia Partridge, or *Quail.* *Bob-white.* *Quail;* New England and Middle States, wherever the ruffed grouse is called "partridge." *Partridge;* Southern States, wherever the ruffed grouse is called "pheasant." Coronal feathers somewhat lengthened, and erectile, but hardly forming a true crest. Forehead, superciliary line and throat, white, bordered with black; crown, neck all round, and upper part of breast, brownish-red, other under parts tawny-whitish, all with more or fewer doubly crescentic black bars; sides broadly streaked with brownish-red; upper parts variegated with chestnut, black, gray and tawny, the latter edging the inner quills. ♀ known by having the throat buff instead of white, less black about the fore parts, and general colors less intense; rather smaller than the ♂ . 9–10; wing 4½–5; tail 2½–3. Eastern United States to high central plains; the characteristic game bird of this country. Eggs white. WILS., vi, 21, pl. 47; NUTT., i, 647; AUD., v, 59, pl. 289; BD., 646. . . . . VIRGINIANUS.

Var. FLORIDANUS COUES, *n. v.* Rather smaller, the ♂ about the size of the ♀ *virginianus*, but bill relatively larger, and jet-black; colors darker, all the black markings heavier. Florida (*Allen*) ; an approach to the Cuban form (*O. cubanensis*).

Var. TEXANUS LAWR., Ann. Lyc. N. Y., vi, 1853, 1 ; BD., 641. Size of *floridanus;* colors paler, the prevailing shade rather gray than brown ; upper parts much variegated with tawny. Texas.

OBS. Among the thousands of bob-whites yearly destroyed, albinotic or melanotic, and other abnormally colored specimens, are frequently found; but the percentage of these cases is nothing unusual, and the sportsman must be cautioned against supposing that such birds have any status, in a scientific point of view, beyond their illustration of certain perfectly well known variations. Such specimens, however, are interesting and valuable, and should always be preserved.

185. Genus OREORTYX Baird.

Plumed Partridge. Mountain Quail of the Californians. With an arrowy crest of two slender keeled feathers, 3–4 inches long in the ♂ when fully developed, shorter in the ♀ . An elegant species, much larger than the bob-white, inhabiting the mountainous parts of California and Oregon. A foot long; wing over 5 inches; tail over 3; whole foot about 3 ; ♀ rather less. Hinder half of body above, with wings and tail, rich dark olive-brown, the inner edges of the inner quills brownish-white ; hinder half below purplish-chestnut, barred with white, black and tawny; fore parts above and below slaty-blue (above more or less glossed with olive, below finely marbled with black), the chin and throat purplish-chestnut, edged with black and bounded by a white stripe meeting its fellow under the bill; ♀ sufficiently similar. AUD., v, 69, pl. 291; NUTT., i, 2d ed. 791; BD., 642; COOP., 546. PICTUS.

FIG. 151. Plumed Partridge.

186. Genus LOPHORTYX Bonaparte.

** With an elegant crest, recurved helmet-wise, of several (6–10) keeled, clubbed, glossy black imbricated feathers, more than an inch long when fully developed; in the ♀ , smaller, of fewer feathers. Bulk of the bob-white, but longer; 10–11½ ; wing 4 or more; tail 3 or more. ♂ with the chin and throat jet-black, sharply bordered with white; a white line across the vertex and along the sides of the crown, bordered behind by black ; ♀ without these head-markings.

Californian Partridge. Valley Quail of the Californians. ♂ with a small white line from bill to eye; forehead whitish with black lines; occiput smoky-brown; nuchal and cervical feathers with very dark edging and shaft lines, and fine whitish speckling; general color of upper parts ashy with strong olive-brown gloss, the edging of the inner quills brownish-orange; fore breast slaty-blue; under parts tawny deepening centrally into rich golden-brown or orange-chestnut, all the feathers sharply edged with jet-black; sides like the back, with sharp white stripes; vent, flanks and crissum tawny, with dark stripes. Besides lacking the definite head-markings, the ♀ wants the rich sienna color of the under parts, which are whitish or tawny, with black semicircles as in the ♂; the breast is olive-gray. Lower portions of California and Oregon, East nearly to the Colorado River; abundant. Eggs of this and the next species speckled. AUD., v, 67, pl. 290; NUTT., i, 2d ed. 789; BD., 644; COOP., 549. . CALIFORNICUS.

Gambel's Partridge. Arizona Quail. ♂ without white loral line; fore-

head black with whitish lines; occiput chestnut; nuchal and cervical feathers with dark shaft lines, but few dark edgings or none, and no white speckling; general color of upper parts clear ash, the edging of the inner quills white; fore breast like the back; under parts whitish, middle of belly with a large jet-black patch; sides rich purplish-chestnut with sharp white stripes; vent, flanks and crissum white with dusky streaks. Besides lacking the definite head-markings, the ♀ wants the black abdominal area, where the feathers are whitish with dark lengthwise touches. New Mexico and Arizona, both in mountains and valleys, very abundant; E. to Pecos and San Elizario, Texas, beyond which replaced by the Massena partridge;

Fig. 152. Gambel's Partridge.

W. to Colorado R. and slightly beyond; N. to 35° and probably a little further; S. into Mexico. CASS., Ill. 45, pl. 9; COUES, Proc. Phila. Acad. 1866, 59, and Ibis, 1866, 46; BD., 645; COOP., 553.. . . . . GAMBELI.

187. Genus CALLIPEPLA Wagler.

Scaled Partridge. Blue Quail. With a short, full, soft crest. Grayish-blue, paler below, in places with a brownish shade, the sides with white stripes, nearly the whole plumage marked with semicircular black edging of the feathers, producing a scaled appearance; inner edges of inner quills, and end of crest, whitish; crissum rusty with dark streaks. ♀ not particularly different. 9–10; wing 5; tail 4. Texas, New Mexico, Arizona and southward. CASS., Ill. 129, pl. 19; BD., 646; COOP., 556. . . SQUAMATA.

188. Genus CYRTONYX Gould.

Massena Partridge. ♂ with the head singularly striped with black and white; the upper parts variegated with black, white and tawny, and with paired black spots on the wings; below velvety black, purplish-chestnut along the middle line, and with numerous sharp circular white spots; 9–10 long; wing 6; tail 2½; tarsus 1¼. ♀ smaller, and entirely different in color, but easily recognized by the peculiar generic characters; tail very short, soft, almost hidden by its coverts; wing coverts and inner quills highly developed; toes short; claws very large; head with a short, full, soft, occipital crest. Texas, New Mexico, Arizona and southward. N. at least to 35°. Cass., Ill. 21, pl. 4; Bd., 647; Coop., 558. . . . . MASSENA.

FIG. 153. Massena Partridge.

Obs. The Welcome Partridge, *Eupsychortyx cristatus* (*Ortyx neoxenus* Aud., v, 71, pl. 292) and several other species, have been admitted to our fauna upon unsatisfactory evidence, or erroneous reports. Some of them, however, may yet be found over our Mexican border.

Order GRALLATORES. Wading Birds.

A character of nearly unexceptional applicability is nakedness of the leg above the heel, or tibio-tarsal joint (*suffrago*). The bare space is generally of considerable length, but in several genera the ends of the feathers reach to the joint, while in others the tibiæ are completely feathered. The legs are usually long; as a rule the neck is lengthened *pari passu;* and the length of the bill is also in some measure correspondent. In its current acceptation, the order does not appear susceptible of further, or of any very exact, definition. Besides its several leading and characteristic groups, it contains a number of singular outlying forms, mostly represented each by a single genus, the location of which has not been satisfactorily determined. Present indications are, however, that all the grallatorial birds will fall in one or another of *three* groups, to be conventionally designated as suborders. All of these occur in this country; their nature may be approximately indicated, as follows : —

I. LIMICOLÆ. *Shore-birds.* Commonly known as the great "plover-snipe group," from the circumstance that the pluvialine and scolopacine birds form the bulk of the division. The species average of small size, with rounded or depressed (never extremely compressed) body, and live in open places on the ground, usually by the water's edge. With rare exceptions, the head is completely feathered; the general pterylosis is of a nearly uniform pattern. The osteological characters are shared to some extent by certain swimming birds, as gulls and auks; the palate is schizognathous; the carotids are double; the syringeal muscles, not more than one pair. The physiological nature is præcocial; the eggs, averaging four, as a rule are laid on the ground in a rude nest or bare depression; the young hatch clothed and able to run about. The food is insects, worms, and other small or soft animals, either picked up from the surface, or probed for in soft sand or mud, or forced to rise by stamping with the feet on the ground; from this latter circumstance, the birds have been named *Calcatores* (stampers). With a few exceptions, the wing is long, thin, flat and pointed, with narrow stiff primaries, rapidly graduated from 1st to 10th; secondaries in turn rapidly lengthening from without inward, the posterior border of the wing thus showing two salient points separated by a deep emargination. The tail, never long, is commonly quite short, and has from 12 (the usual number) up to 20 or even 26 feathers (in a remarkable group of snipe). The legs are commonly lengthened, sometimes extremely so, rarely quite short, and are usually slender; they are indifferently scutellate or reticulate, or both. The feathers rarely reach the suffrago. The toes are short (as compared with the case of herons and rails, of the next group), the anterior usually semipalmate, frequently cleft to the base, rarely palmate or lobate; the hinder is always short and elevated, or absent. The bill varies much in length and contour, but is almost always slender, contracted from the frontal region of the skull, and as long as, or much longer than, the head, representing the "pressirostral" and "longirostral" types of Cuvier. Furthermore, it is generally in large part, if not entirely, covered with softish skin, often membranous and sensitive to the very tip, and only rarely hard throughout. The nostril is generally a slit in the membranous part, and probably never feathered.

Most of the families of this division are well represented in this country, and will be found fully characterized beyond. The extra-limital ones are: — *Otididæ*, bustards, an important group of Europe, Asia and Africa, containing some 20 species; it has a certain gallinaceous bent, and stands, like the *Turnicidæ*, near the boundary line of the two orders. The remarkable genus *Chionis*, of two South American species, forms the family *Chionidæ* (or sheath-bills, so called because the bill is invested by a horny sheath forming a false cere), with some gallinaceous relationships, and appears to belong here, near the oyster-catchers. The *Thinocoridæ*, or "lark-partridges," as they are called, consisting of the South American genera *Thinocorus* and *Attagis*, of few species, appear to be plover-like birds, near the glareoline group of the latter. The singular African *Dromas ardeola*, representing a family *Dromadidæ*, of uncertain position, is sometimes placed near the avocets, sometimes with the herons, and is occasionally removed to another order.

II. HERODIONES. *Herons and their allies.* The species average of large size, some of them standing amongst the tallest of birds (excepting ostriches). The body is usually compressed; the legs, neck and bill are commonly extremely long. The general pterylosis is peculiar, in the presence, nearly throughout the group, of the remarkable powder down tracts, and in some other respects. A part, if not the whole of the head, is naked, as much of the neck also frequently is. The toes are long and slender; the hallux is long, and either not obviously elevated, or else perfectly insistent. A foot of insessorial character results, and the species frequently perch on trees, where the nest is usually placed. The physiological nature is altricial; the young hatch naked, unable to stand, and are fed in the nest. The food is fish, reptiles, mollusks and other animal matters, generally procured by spearing with a quick thrust of the sharp bill, given as the birds stand in wait, or stalk stealthily along; hence they are sometimes called *Gradatores* (stalkers). The bill represents the cultrirostral pattern; it is as a rule of lengthened, wedged shape, hard and acute at the end, if not hard throughout, with sharp cutting edges, and it enlarges regularly to the forehead, where the skull contracts gradually in sloping down to meet it. The palate is desmognathous. The wings normally show a striking difference from those of *Limicolæ*, being long, broad and ample, much as in the next group.

The herons (*Ardeidæ*, beyond), are typical of this group. The only extra-limital family is that of the *Ciconiidæ*, or storks ; these are birds standing very near the ibises and spoonbills (beyond), and distinguished from the herons, among other circumstances, by the absence of powder-down tracts. Excepting the jabiru of tropical America, *Mycteria americana*, the storks are all Old World, and chiefly inhabit warm countries ; there are only 8–10 species, representing nearly as many genera of authors ; among these, *Anastomus* and *Hiator* are remarkable for a wide interval between the cutting edges of the bill, which only come into apposition at base and tip. The singular African *Scopus umbretta*, type of a subfamily at least, is often placed among the herons, but its pterylosis is that of the storks. The cranes, which have been associated with *Herodiones* on account of their stature and other superficial resemblances, unquestionably belong to the next division, where also several doubtful forms appear to fall.

III. ALECTORIDES. *Cranes, Rails and their allies.* A portion of these birds, representing the *crane* type, have a general resemblance to the foregoing, but are readily distinguished by the technical characters given beyond under the head of *Gruidæ*, and in essential respects accord with the rest, representing the *rail* type. The latter are birds of medium and small size, with compressed body, and the head feathered. The neck and legs are not particularly lengthened, but as a rule the toes are remarkably long, enabling the birds to run lightly over the soft oozy ground and floating vegetation of the reedy swamps and marshes they inhabit. This length of the toes has given a name, *Macrodactyli*, to the group ; their shy retiring habit of skulking among the rushes has caused them to be sometimes called *Latitores* (skulkers). Their nature is præcocial ; the eggs are numerous, usually laid on the ground, in a rude nest. The nourishment is essentially the same as that of the *Limicolæ*, but it is simply picked up from the surface, not felt for in the mud, nor stamped out of the ground. The hallux is usually lengthened, and but little elevated ; the feet are conspicuously lobate in some forms. The wings are usually short, rounded and concave ; the tail is very short, few-feathered, often held cocked up, and wagged in unison with a bobbing motion of the head that occurs with each step taken. The *Alectorides* are schizognathous.

This country affords typical representatives of the two leading forms, that of the cranes, and of the rails, coots and gallinules, as given beyond ; there are, however, a number of remarkable outliers, that may be briefly mentioned, as follows ;— The kagu, *Rhinochætus jubatus* of New Caledonia, and the carle, *Eurypyga helias* of Guiana, are each the type and single representative of a family which seems near the cranes in principal osteological characters (*Huxley*), although pterylographically they are more like herons, both possessing powder-down tracts (*Bartlett*) ; and *Eurypyga*, in particular, resembles herons in other respects. More closely allied to the cranes are the trumpeters, *Psophiidæ*, of one genus and few species of South America, with the cariamas, *Cariamidæ*, of the same country, represented only by the *Cariama cristata* and the *Chunga burmeisterii*. The horned screamers, *Palamedeidæ*, of South America, consisting of three species, *Palamedea cornuta*, *Chauna chavaria* and *C. derbiana*, seem to be nearer the rails, and also closely approach the water birds ; one of them is by some considered the nearest living ally of the mesozoic *Archæopteryx*. Some gigantic extinct birds belong in the neighborhood of the rails and coots. Decidedly rail-like and better known birds are the jacanas, *Parridæ*, noted for the length of the toes, and especially of the claws; they have a sharp spur on the wing. There are less than 12 species, usually referred to several genera, of various parts of the world. Finally, the sun-birds, *Heliornithidæ*, are a small but remarkable family of one or two genera and about four species of tropical America, Africa, and southern Asia. They have been classed, on account of their lobate feet and a certain general resemblance, with the grebes; but the feet are like those of coots, and their whole structure shows that they belong with the ralliform birds. This completes an enumeration of the *Alectorides*.

Suborder LIMICOLÆ. Shore Birds. (See p. 239.)

Family CHARADRIIDÆ. Plover.

This is a large and important family of nearly a hundred species, of all parts of the world. Its limits are not settled, there being a few forms sometimes referred here, sometimes made the types of distinct families. I exclude from it the genera *Thinocorus, Attagis,* and *Chionis,* noted on a preceding page. The glareoles (*Glareolinæ* if not *Glareolidæ*)˙ are a remarkable Old World form, like long-legged swallows, with a cuckoo's bill; the tail is forked; there are four toes; the wings are extremely long and pointed; the tarsi are scutellate; the middle claw denticulate. The coursers, *Cursoriinæ*, are another Old World type, near the bustards, of one or two genera and less than ten species. In both of these the gape of the mouth is longer than in the true plovers; the hind toe, as usual for this family, is absent in the coursers. The thick-knees, *Œdicneminæ*, are truly plover-like birds, with one exception belonging to the Old World, comprising about eight species of the genera *Œdicnemus* and *Esacus*. All the remaining pluvialine birds appear to fall in the

Subfamily CHARADRIINÆ. True Plover.

Toes generally three, the hinder absent (excepting, among our forms, gen. 189, 193); tarsus reticulate, longer than the middle toe; toes with a basal web; tibiæ naked below. Bill of moderate length, much shorter or not longer than the head, shaped somewhat like that of a pigeon, with a convex horny terminal portion, contracted behind this; the nasal fossæ rather short and wide, filled with soft skin in which the nostrils open as a slit, not basal, and perforate. Gape very short, reaching little beyond base of culmen. Wings long and pointed, reaching, when folded, to or beyond the end of the tail, and sometimes spurred; crissal feathers long and full; tail short, generally nearly even and of 12 feathers; body plump; neck short and thick; head large, globose, sloping rapidly to the small base of the bill, usually fully feathered. Size moderate or small.

Our species (excepting *Aphriza*, if really belonging here) are very closely related, and will be readily recognized by the foregoing characters. There are in all perhaps sixty species. The most singular of them is the *Anarhynchus frontalis*, in which the bill is bent sideways. *Thinornis zelandiæ* of New Zealand, *Phegornis mitchellii* and *Oreophilus totanirostris* of Chili, are peculiar forms. Species of *Chettusia*, *Lobivanellus*, and *Hoplopterus* have fleshy wattles, or a tubercle, often developed into a spine, on the wing, or both; some of these, and others, are crested. These are all near *Vanellus* proper, and a part of them are 4-toed. Our species are found along the seashore, by the water's edge in other open places, and in dry plains and fields. They all perform extensive migrations, appearing with great regularity in the spring and fall, and most of them breed far northward. They are all more or less gregarious except when breeding. They run and fly with great rapidity; the voice is a mellow whistle; the food is chiefly of an animal nature. The eggs are commonly four in number, speckled, very large at one end and pointed at the other, placed with the small ends together in a slight nest or mere depression in the ground. The sexes are generally similar, but the changes with age and season are great.

Obs. The European lapwing, *Vanellus cristatus*, is reported by Mr. Dall from Alaska, where, however, specimens were not taken. (Alaska and its Resources, p.586.)

189.   Genus SQUATAROLA Cuvier.

Black-bellied Plover.   Beetle-headed Plover.   Whistling Field Plover. Bull-head.   Ox-eye. A small hind toe, hardly ¼ long; plumage speckled. Adult in breeding season (rarely seen in the U. S.) : face and entire under parts black, upper parts variegated with black, and white or ashy; tail barred with black and white; quills dusky, with large white patches. Adult at other times, and young : below white, more or less shaded with gray, the throat and breast speckled with dusky; above blackish, speckled with white or yellowish; the rump white with dark bars; legs dull bluish. Old birds changing show every grade, from a few

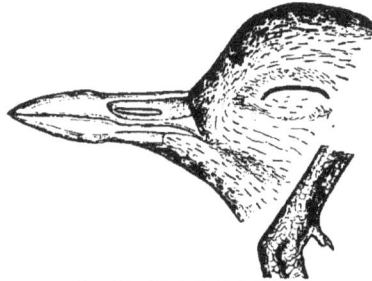

FIG. 154. Black-bellied Plover.

isolated black feathers on the under parts, to numerous large black patches. 11–12; wing 7 or more; tail 3; bill 1–1¼; tarsus 2; middle toe and claw 1¼. A bird commonly diffused over most parts of the world. WILS., vii, 41, pl. 57, f. 4 ; AUD., v, 199, pl. 315 ; NUTT., ii, 26 ; CASS. in BD., 697. HELVETICA.

FIG. 155. Golden Plover; winter plumage.

190.   Genus CHARADRIUS Linnæus.

Golden Plover.   Frost-bird.   Bull-head. No hind toe ; plumage speckled above, and in the breeding season black below, as in the last species, but much of the speckling bright yellow, and the rump and upper tail coverts like the back ; forehead, and a broad line over the eye to the nape, white ; tail feathers grayish-brown, with imperfect white or ashy bars ; axillars gray or ashy. At other times the under parts nearly as in the last species. 10–11 ; wing 7 or less ; tail under 3 ; bill 1 or less ; tarsus 1⅔ ; middle toe and claw 1¼. N. Am., abundant in the U. S. in great flocks in the fall, a well-known and highly-rated game bird. It is very near the European species, but seems distinct; the axillars are gray, not white. It appears to be a variety of the Asiatic; rather larger and with shorter toes. Wils., vii, 71, pl. 59, f. 5; Nutt., ii, 16; Aud., v, 203, pl. 316; Cass. in Bd., 690. . . . . . . fulvus var. virginicus.

191-2. Genus ÆGIALITIS Boie.

₊ Plumage not speckled; below, white; head and neck with black bands in the breeding season.
  * Tarsus about half as long again as the middle toe. (*Ægialitis*.)

Killdeer Plover. Rump and upper tail coverts tawny or orange-brown, most of the tail feathers white at base and tip, suffused with orange-brown

in a part of their length, and with 1–3 black bars; secondaries mostly white, and primaries with a white space; forehead white; a black bar across the crown, and *two* broad black bands on the neck and breast; bill black; feet pale grayish-blue. 9–10; wing 6 or more; tail 3½, much rounded; tarsus about 1½. North America, very abundant, especially on the Plains; breeds anywhere, but rarely in New England; name derived from its pecu-

Fig. 155. Killdeer Plover.

liar notes. Wils., vii, 73, pl. 59; Nutt., ii, 22; Aud., v, 207, pl. 317; Cass. in Bd., 692. . . . . vociferus.

Wilson's Plover. Pale ashy-brown, merging into fulvous on the nape; a black bar on the crown, and a broad black pectoral belt, grayish-brown in the ♀ and young; no bright ring round eye; legs flesh-colored; bill black, extremely large and stout, nearly as long as the head; 7–8; wing 4½–5; tail 2, nearly square. Seacoast of S. Atlantic and Gulf States, common; N. to the middle districts, and probably to New England; also on the Pacific side to California? Wils., ix, 77, pl. 73, f. 5; Nutt., ii, 21; Aud., v, 214, pl. 319; Cass. in Bd., 693. . . . . wilsonius.

Semipalmated Plover. Ring Plover. Ringneck. Dark ashy-brown with an olivaceous shade; very broad coronal and pectoral black bars, in the adult of both sexes, that on the breast grayish-brown in the young, but still evident; edges of eyelids bright orange; bill moderately short and stout, orange or yellow, black-tipped; legs yellowish; toes conspicuously semi-palmate. About 7; tail about ½ as long as the wings, rounded. North America, abundant. Breeds northward. Wils., vii, 65, pl. 59, f. 3; Nutt., ii, 24; Aud., v, 218, pl. 320; Cass. in Bd., 674. . . . semipalmatus.

Piping Plover. Ringneck. Very pale ashy-brown; the black bands narrow, often imperfect; bill colored as in the last, but very short and stumpy; edges of eyelids colored; no evident web between inner and middle toes, and only a slight one between middle and outer; size of the last, or rather less. Eastern and Middle North America; abundant on the Atlantic coast, breeding northward. WILS., v, 30, pl. 37, f. 3; NUTT., ii, 18; AUD., v, 223, pl. 321; CASS. in BD., 695. . . . . . . MELODUS.

Snowy Plover. With a general resemblance to the last, this species is rather darker (not so dark as in *semipalmatus*), and the hind head is tinged with fulvous, as in *wilsonius;* it may be at once recognized by its entirely black bill, slender, about ⅔ long; legs dark; several lateral tail feathers entirely white; 6½-7 long; wing 4-4¼; tail 2 or less. California coast, where I found it abundant in winter: Ibis, 1866, 274. It belongs to a different sub-group from the foregoing, and appears to be identical with the common Kentish plover of Europe, *A. cantiana;* but I have had no opportunity of a direct comparison. CASS. in BD., 696. . . CANTIANA.

** Tarsus about twice as long as the middle toe. (*Podasocys.*)

Mountain Plover. Bill black, slender, an inch long; middle toe and claw the same; tarsus 1⅔; tibiæ bare over ½; about 9; wing 6; tail 3, nearly square. Above brown, all the feathers skirted with rusty, which also shades the breast; other under parts, forehead, and short line over eye, white; a coronal (and pectoral?) black band in mature plumage; quills and tail blackish, former with white shafts, latter tipped with whitish. Middle Kansas to the Pacific, common on dry plains and even in deserts; independent of water; feeds on insects, especially grasshoppers. I found it in New Mexico in June, and abundantly in California in November. The eggs, hitherto undescribed, measure 1.40 to 1.50 in the greater axis, by 1.10 in the transverse; color an olive drab with a slight brown shade, thickly marked, especially toward the larger end, with small sharp speckling and fine dotting of blackish, dark brown and neutral tint. (Described from two specimens in the Smithsonian collection, taken by HAYDEN, July, 1859, in Nebraska). *Charadrius montanus* AUD., v, 213, pl. 318 ; *Ægialitis montanus* CASS. in BD., 693; *Podasocys montanus* COUES, Proc. Phila. Acad. 1866, 96; ELLIOT, pl. 39. . . . . ASIATICUS var. MONTANUS.

193. Genus APHRIZA Audubon.

Surf Bird. Dark ashy-brown, streaked with white on the head and neck, and in summer with chestnut and black on the back; upper tail coverts white; under parts white, often ashy-shaded, and variously marked with blackish; tail black and white; bases and shafts of primaries, most of the secondaries, and tips of greater coverts, white; bill black, flesh-colored at base below; legs dusky-greenish; 9-10; wing about 7; tail 3 or less; bill 1; tarsus 1¼, reticulate; hind toe present; front toes cleft to the base. Varies greatly in plumage with age and season. A remarkable bird, apparently a plover, connecting this family with the next, and also related to the sandpipers. Extensively dispersed on the coasts and islands of the Pacific. CASS. in BD., 698; *Aphriza townsendii* AUD., v, 228, pl. 322. . VIRGATA.

Family HÆMATOPODIDÆ. Oyster-catchers. Turnstones.

A small family of two genera and six or eight species, with the bill *hard*, and either acute or truncate, the nasal fossæ short, broad and shallow; the legs short, stout, brightly-colored. The two following genera differ much. *Hæmatopus* is 3-toed, with much basal webbing, the tarsi reticulate; the bill longer than the tarsus, stout, straight, constricted toward the base, compressed and truncate at the end, somewhat like a woodpecker's; it is an efficient instrument for prying open the shells of bivalve mollusks. *Strepsilas* is 4-toed, with no obvious basal webbing; the tarsi scutellate in front, the bill sharp-pointed, not longer than the tarsus; its scientific and vernacular names are both derived from its curious habit of turning over pebbles along the beach in search of food. There is but one species, cosmopolitan.

194. Genus HÆMATOPUS Linnæus.

Oyster-catcher.   Head and neck blackish tinged with brown or ashy; back ashy-brown; below from the breast, eyelid, rump, tips of greater wing coverts, most secondaries, and basal part of tail feathers, white; rest of tail, and quills, blackish; bill and edges of eyelids red or orange; legs flesh color; 17–18; wing 10; tail 4½; bill 3. Atlantic coast; California

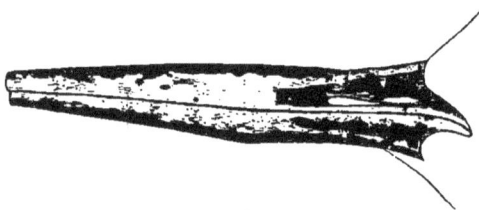

FIG. 157. Bill of Oyster-catcher.

(*Cooper*). WILS., viii, 15, pl. 64; NUTT., ii, 15; AUD., v, 236, pl. 324; CASS. in BD., 699. . . . . PALLIATUS.
Black Oyster-catcher. Nearly uniform blackish or sooty brown; the head and neck frequently with an ashy shade. Size of the foregoing. Pacific coast. CASS. in BD., 700. *H. townsendii* AUD., v, 243, pl. 325. NIGER.

OBS. *H. bachmani* AUD., v, 245, pl. 326; *H. ater* CASS. in BD., 700 (if really distinct from the last, which is doubtful), is a South American species improperly attributed to our fauna.

195. Genus STREPSILAS Linnæus.

Turnstone. Brant Bird. Calico-back. Adult in summer pied above with black, white, brown and chestnut-red, the latter color wanting in winter, and in young birds; below, from the breast (which is more or less completely black), throat, most of the secondaries, bases and shafts of primaries, and bases and tips of tail feathers, white; bill black; feet orange; 8–9; wing 5½–6; tail 2½; bill ⅞, almost recurved, with ascending gonys; tarsus 1; tibiæ bare but a little way. Both coasts, abundant during the migrations. WILS., vii, 32, pl. 57, f. 1; NUTT., ii, 30; AUD., v, 231, pl. 323; CASS. in BD.. 701. INTERPRES.

FIG. 158. Bill of Turnstone.

Var. MELANOCEPHALUS. *Black-headed Turnstone.* Without any of the chestnut coloration of the last, the parts that are pied in *interpres* being blackish ; the white parts, however, as before. Apparently a permanent melanism. Pacific Coast. CASS. in BD., 702.

Family RECURVIROSTRIDÆ.   Avocets.   Stilts.

Another small family, characterized by the extreme length of the slender legs, and the extreme slenderness of the long acute bill, which is either straight or curved upward. *Recurvirostra* is 4-toed, and full-webbed ; the bill is decidedly recurved, flattened, and tapers to a needle-like point ; the body is depressed ; the plumage underneath is thickened as in water birds. The species swim well. *Himantopus* is 3-toed, semipalmate, the bill nearly straight, and not flattened ; in relative length of leg it is probably not surpassed by any bird whatsoever. These two genera, each of three or four species of various parts of the world, with the *Cladorhynchus pectoralis* of Australia, compose the family.

196. Gen. RECURVIROSTRA Linnæus.

Avocet. Blue-stocking. White ; back and wings with much black ; head and neck cinnamon-brown in the adult, ashy in the young (*R. occidentalis* CASS., Ill. 232, pl.

FIG. 159. Bill and foot of Avocet.

40) ; bill black ; legs blue ; eyes red ; 16–18 ; wing 7–8 ; tail 3½ ; tarsus 3¼. Temperate N. Am. WILS., vii, 126, pl. 63, f. 2 ; NUTT., ii, 74 ; AUD., vi, 24, pl. 353 ; CASS. in BD., 703. AMERICANA.

197. Genus HIMANTOPUS Brisson.

Stilt. Longshanks. Lawyer. Glossy black ; forehead, sides of head and neck, rump and under parts, white ; tail white or ashy ; bill black ; legs carmine. Young with back and wings brown.   13–15 ; wing 8–9 ; tail 3 ; tarsus 4.   United States.   WILS., vii, 48, pl. 58, f. 2 ; AUD., vi, 31, pl. 354 ; NUTT., ii, 8 ; CASS. in BD., 704.   .   .   .   .   NIGRICOLLIS.

Fam. PHALAROPODIDÆ.   Phalaropes.

This is likewise a small family ; the three species comprising it resemble sandpipers, but are imme-

FIG. 160. Stilt.

diately distinguished by the lobate feet ; the toes are furnished with plain or scalloped membranes, like those of coots and grebes, but not so broad. The body is depressed, and the under plumage thick and duck-like to resist water, on which the birds swim with perfect ease and grace. The wings and tail are like those of ordinary sandpipers ; the tarsi are much compressed ; there is basal webbing of the toes besides the marginal membrane ; the bill, and some other details of form, differ in each of the three species. These birds inhabit the northern portions of both hemispheres, two of them at least breeding only in boreal regions, but they all wander far southward in winter.

198. Genus STEGANOPUS Vieillot.

Wilson's Phalarope.   Membranes straight-edged; bill very slender, subulate.   Length 9–10; wing 5; tail 2; bill, tarsus, and middle toe, each, over 1, black.   Adult ashy; upper tail coverts and under parts white; a black stripe from the eye down the side of the neck spreading into rich purplish-chestnut, which also variegates the back, and shades the throat; young lacking these last colors.   N. Am. WILS., ix, 72, pl. 73, f. 3; NUTT., ii, 245; AUD., v, 299, pl. 341; CASS. in BD., 705. . . . . . WILSONII.

FIG. 161. Wilson's Phalarope (head); Northern
Phalarope (foot).

199. Genus LOBIPES Cuvier.

Northern Phalarope.   Membranes scalloped; bill very slender, subulate.   Length about 7; wing 4¼; tail 2; bill, tarsus and middle toe, each, under 1, black.   Adult dark opaque ash or grayish-black, the back variegated with tawny; upper tail coverts and under parts mostly white; side of the head and neck with a broad stripe of rich chestnut, generally meeting on the jugulum; breast otherwise with ashy-gray; young lacking the chestnut. Northern N. Am., U. S. during the migration. BONAP., Am. Orn. iv, 82, pl. 25, f. 2; NUTT., ii, 239; AUD., v, 295, pl. 340; CASS. in BD., 706. . . HYPERBOREUS.

200. Genus PHALAROPUS.

Red Phalarope.   Membranes scalloped; bill comparatively stout, flattened, with lancet-shaped tip.   Length 7–8; wing 5; tail 2¾; bill 1, yellowish, black-tipped; tarsus ¾, greenish.   Adult with the under parts purplish-chestnut, of variable intensity, white in the young; above variegated with blackish and tawny. Northern N. Am., U. S. during the migrations. WILS., ix, 75, pl. 73, f. 4; NUTT., ii, 236; AUD., v, 291, pl. 339; CASS. in BD., 707. . . . . FULICARIUS.

Family SCOLOPACIDÆ.   Snipe, etc.

Snipe and their allies form a well-defined and perfectly natural assemblage, one of the two largest limicoline families, agreeing with plover in most essential respects, yet well distinguished from the pluvaline birds. In general, the bill is much elongated, frequently several times longer than the head, and in those cases in which it is as short as in plover, it does not show the particular, somewhat pigeon-like, shape described under *Charadriinæ*, being slender and soft-skinned throughout. It is generally straight, but frequently curved up or down. The nasal grooves, always long and narrow channels, range from one-half to almost the whole length of the bill; similar grooves usually occupy the sides of the under mandible; the interramal space is correspondingly long and narrow, and nearly naked. This length, slenderness, grooving, and peculiar *sensitiveness* are the prime characteristics of the scolopacine bill. The gape, never ample, is generally very short and narrow, reaching little, if any, beyond the base of the bill. The nostrils are short narrow slits, exposed. The head is completely feathered to the bill (except in one species), at the base of which the ptilosis stops abruptly, without forming projecting antiæ. The wings commonly show the thin pointed contour described under *Limicolæ*, but they are occasionally short and rounded. The tail, always short and soft, has as a rule 12 rectrices ; in one genus, however, there are from 12 to 26. The crura are rarely feathered to the suffrago. The tarsi are scutellate before and behind, and reticulate on the sides, except in the curlews, where they are scutellate only in front ; they are probably never entirely reticulate (the normal state in plover). The hallux is absent in only two or three instances ; the anterior toes commonly show one basal web, and often two, but in many species they are entirely cleft. The scolopacine birds are of medium and small size, ranking with plover in this respect ; none attain the average stature of *Herodiones*.

The general economy of these birds is similar to that of plover ; a chief peculiarity being probably their mode of procuring food, by feeling for it, in the majority of cases, in the sand or mud with their delicately sensitive, probe-like bill. The eggs are commonly four, parti-colored, pointed at one end and broad at the other, placed with the small ends together in a slight nest or mere depression on the ground ; the young run about at birth. The sexes, with very rare exceptions, are alike in color or nearly so, and the ♀ is usually a little larger than the ♂ ; but the sexual distinctions are very rarely strong enough to be perfectly reliable (remarkable exception in gen. 218). Color distinctions with age, likewise, are rarely marked ; but on the contrary, seasonal plumages are, in many cases, as throughout the sandpipers, very strongly indicated, the nuptial dress being entirely different from that worn the rest of the year. Excepting a few species that frequent dry open places like many plover, these birds are found by the water's edge where the ground is soft and oozy—in moist thickets, low rank meadows, bogs and marshes, by the riverside, and on the seashore. Some are solitary, but the majority are gregarious when not breeding, and many gather in immense flocks, especially during the extensive migrations that nearly all perform. The voice is a mellow pipe, a sharp bleat, or a harsh scream, according to the species. Few birds surpass the snipe in sapid quality of flesh, and many kinds rank high in the estimation of the sportsman and epicure. The family is cosmopolitan, but the majority inhabit the northern hemisphere, breeding in boreal regions. There are about ninety well-determined species of scolopacine birds, referable perhaps to fifteen tenable genera, although many more than this are often employed. Various attempts to divide the group into sub-families have met with little success, owing to the close intergradation of the several types. All the leading forms of the family, with most of the lesser genera, are represented in this country, and are indicated by the specific descriptions given beyond ; while its entire composition may be pointed out and rendered perfectly intelligible by a brief summary :—

a. In *woodcock* (gen. 201–2) and *true snipe* (203) the ear appears below and not behind the eye, which is placed far back and high up ; and if the brain be examined, it will be found curiously tilted over so that its anatomical base looks forward. The bill is perfectly straight and much longer than the head, deep-grooved to the very end, which is either knobbed, or widened just behind the tip, where there is a furrow in the flattened culmen. The membranous covering is abundantly supplied with nerves ; this organ constitutes a probe of delicate sensibility, an efficient instrument of touch, used to feel for food below the surface of the ground.

In the dried state, the soft skin shrinks tight like parchment to the bone, and becomes studded with small pits. The gape of the mouth is extremely short and narrow ; the toes are cleft ; the legs, neck and wings are comparatively short, and the body is rather full. There are no obvious seasonal or sexual differences in plumage. Not completely gregarious ; no such flights of woodcock and true snipe occur as are usually witnessed among sandpipers and bay-snipe ; they inhabit the bog and brake rather than the open waterside ; they cannot be treacherously massacred by scores, like some of their relatives ; they are knowing birds, if their brains are upset, and their successful pursuit calls into action all the better qualities of the true sportsman. There is but one species of *Philohela* ; two or three of *Scolopax*, and about twenty of *Gallinago*. The curious circumstance occurs, among the latter, that the tail feathers range from 12 to 26 in different species ; and in those with the higher numbers, several pairs are narrow and linear— a character upon which the genus *Spilura* rests.—The singular genus *Rhynchæa*, with two species, *R. capensis* (Africa) and *R. semicollaris* (S. America), may belong here.—*Macrorhamphus* (204), containing only our species, and one other, *M. semipalmatus* of the Old World, has the bill exactly as in *Gallinago*, but is distinguished by more pointed wings, and differently proportioned legs, with basal webbing of the toes. It stands exactly between the true snipe and

  b. The *godwits* (213), in which we find the same very long, wholly grooved, and extremely sensitive bill, which, however, is not dilated at the end, nor furrowed on the culmen, and is bent *slightly* upward ; the gape, as before, is exceedingly constricted. The toes show a basal web. These are rather large birds, with the colors and general aspect of curlews, but the bill is not decurved and the tarsi are scutellate behind. They frequent marshes, bays and estuaries, and are among the miscellaneous assortment of birds that are collectively designated "bay-snipe." There are only five or six species, of the single genus *Limosa*. The *Terekia cinerea* of various parts of the Old World, with the bill recurved almost as in an avocet, stands between the godwits and tattlers.

  c. The *sandpipers* (gen. 205–212) are a rather extensive group, notable for the variation in minor details of form, that it shows with almost every species—a circumstance that has caused the erection of a number of unwarranted genera. Here the bill retains much of the sensitiveness of a snipe's, and the gape likewise is much constricted ; but the bill is much shorter, averaging about equal to the head. One trivial circumstance affords a good clue to this group ; the tail feathers are plain colored, or with simple edgings, while in *almost* all the species of other groups these feathers are barred crosswise. In this group the seasonal changes of plumage are very great ; the proportions of the legs, and webbing of the toes, are variable with the species, but as a rule, the toes are cleft to the base (not so in 205, 206), and four in number (except 212). The sandpipers belong particularly to the northern hemisphere, and breed in high latitudes ; they perform extensive migrations, and in winter spread over most of the world. Among them are the most diminutive of waders. They are probably without exception gregarious, and often fleck the beach in vast multitudes ; they live by preference in *open* wet places, rather than in fens and marshes, and feed by probing, like snipe ; the voice is mellow and piping. They are pretty well distinguished from both the foregoing, though gen. 203 connects with the snipe through 204 ; but shade directly into the following group ; for instance, gen. 218–19–20, if not also 217, have been called *Tringa*, and "sandpiper." Nearly all the forms of sandpiper are described in detail beyond. There are in all about 20 species. The only generic forms not represented in this country are the *Limicola platyrhyncha*, the peculiarity of which is expressed in its name; and the *Eurinorhynchus pygmæus*, a wonderful and exceedingly rare species, in which the bill is expanded and flattened at the end, somewhat as in the spoonbill. The singular *Philomachus pugnax* should perhaps rather come here than among

d. The *tattlers*, with which it is ranged, beyond. In this, the largest and most varied group, the bill has comparatively little of the sensitiveness of that of all the foregoing, and the gape is longer, extending obviously beyond the base of the culmen, and sometimes to nearly below the eyes. It varies much in length and shape, but it is *usually* longer than the head, and very slender, not often grooved to the tip, and is either straight, or bent slightly upward. The body and its members are commonly more elongate than in the foregoing, the toes have a basal web or two, and the hinder is always present. The tail is usually barred. They are noisy, restless birds of the marshes and sand-flats and mud-bars of estuaries, and apparently do not probe for food to any extent ; they gain their name from their harsh voice. The yellowshanks is a typical example of the group ; most of the species cluster close about this type, and ought to go in the single genus *Totanus*. Gen. 217, 219, 220, are another slight group. The only extra-limital form is the *Prosobonia leucoptera*, of the Sandwich Islands, a curious species, apparently near 220. There are about 18 species in all, universally distributed. Finally,

e. The *curlews* (gen. 222) are distinguished by the downward curvature, extreme slenderness, and usually great length of the bill, with the slight scutellation of the tarsus. In size and general appearance they are near the godwits ; they inhabit all parts of the world. They all belong to the genus *Numenius*, which has about a dozen species — excepting the *Ibidorhyncha struthersii* of Asia, which is a three-toed curlew, not showing the coloration characteristic of the rest.

201. Genus PHILOHELA Gray.

American Woodcock. Bogsucker. First three primaries attenuate and falcate ; wings short ; when closed, the quills hidden by the coverts and

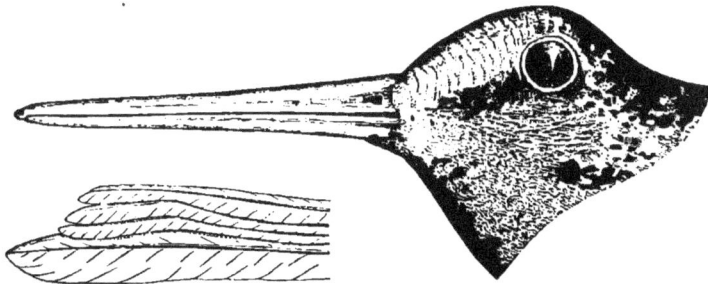

FIG. 162. Woodcock; with attenuate primaries.

tertiaries ; tibiæ feathered to the joint ; tarsi shorter than middle toe, scutellate before and behind ; toes slender, free to the base ; bill much longer than the head, stout and deep at base, grooved nearly its whole length, the tip knobbed ; gape very short and narrow ; ear under the eye, which is set in the back upper corner of the head ; colors above variegated and harmoniously blended black, brown, gray and russet ; below pale warm brown of variable shade; ♂ 10–11; ♀ 11–12; extent 16–18; wing 4½–5; bill 2½–3; tarsus 1¼; middle toe and claw 1½; weight 5–9 oz. Bogs, swamps, wet woodland and fields, Eastern U. S. and Canada. WILS., vi, 40, pl. 48, f. 2; NUTT., ii, 194; AUD., vi, 15, pl. 352; CASS. in BD., 709. . . . . . . MINOR.

202. Genus SCOLOPAX Linnæus.

European Woodcock. First primary alone attenuate; wings more pointed than in the last; one-third larger; weight 12–15 oz. This bird has not hitherto been formally introduced to our fauna in any systematic treatise; but there are several authentic instances of its capture in this country, and it is unquestionably entitled to a place here, as a straggler from Europe. See LEWIS, American Sportsman, ed. of 1868, p. 169, footnote (New Jersey); LAWR., Ann. Lyc. N. Y. 1866, 292 (Rhode Island and New Jersey); BAIRD, Am. Journ. Sci. xli, 1866, 25 (Newfoundland). Sportsmen who get a bird of this sort, will do well to report the fact at once. Of all the snipe-like birds of this country, called "Scolopax," the present is the only one to which the name is strictly applicable. . . . . RUSTICOLA.

203. Genus GALLINAGO Leach.

American Snipe. Wilson's Snipe. Bill much longer than the head, perfectly straight, soft to the end, where it is somewhat widened and grooved on top; gape narrow, not reaching beyond base of culmen; ear under eye; tibiæ feathered not quite to the joint; tarsus a little shorter than middle toe and claw; toes perfectly free. Crown black, with a pale middle stripe; back varied with black, bright bay and tawny, the latter forming two lengthwise stripes on the scapulars; neck and breast speckled with brown and dusky; lining of wings barred with black and white; tail usually of 16 feathers, barred with black, white and chestnut; sides waved with dusky; belly dull white; quills blackish, the outer white-edged. Length 9–11; wings 4½–5¼; bill about 2½; whole naked portion of leg and foot about 3. This is the genuine snipe, of all the birds loosely so called; its name of "English" snipe is a misnomer, as it is indigenous to this country, and distinct from any European species, though closely resembling one of them. Open wet places of North America, at large; migratory. WILS., vi, 18; pl. 47, f. 2; NUTT., ii, 185; AUD., v, 339, pl. 350; CASS. in BD., 700. *Scolopax drummondii* and *S. douglasii* Sw., F. B.-A., ii, 401; *S. leucurus* ID., *ibid.* 501. . . . . WILSONII.

FIG. 161. American Snipe.

204. Genus MACRORHAMPHUS Leach.

· *Red-breasted Snipe. Gray Snipe. Brown-back. Dowitcher.* A very snipe-like bird, with the bill exactly as in *Gallinago*, but readily distinguished generically: legs long; tibiæ bare upwards of ¾ of an inch; tarsus longer than middle toe and claw; outer and middle toes connected by an evident membrane; tail of 12 feathers. Tail and its coverts, at all seasons, conspicuously barred with black and white (or tawny); lining of wings, and axillars, the same; quills dusky, shaft of first primary, and tips of secondaries except the long inner ones, white; bill and feet greenish-black. In summer, brownish-black above, variegated with bay; below, brownish-red, variegated with dusky; a tawny superciliary stripe, and a dark one from bill to eye.

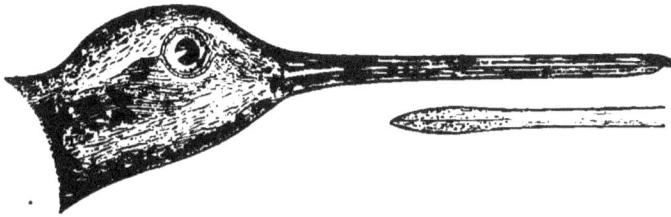

FIG. 164. Red-breasted Snipe; with end of bill, from above.

In winter, plain gray above, and on the breast, with few or no traces of black and bay, the belly, line over eye, and under eyelid, white. 10–11; wing 5–5½; tail 2½; bill about 2½; tarsus 1½; middle toe and claw 1¼. A variety of this bird is almost a foot long, the bill upward of 3 inches (*M. scolopaceus* LAWR., Ann. Lyc. 1852, 4, pl. 1; CASS. in BD., 712). North America, at large; abundant, migratory; it generally flies in large compact flocks, like the sandpipers and shore-birds generally, rather than singly or in wisps like the true snipe; and prefers the shores of bays and estuaries, instead of wet meadows. WILS., vii, 45, pl. 58, f. 1; NUTT., ii, 181; AUD., vi, 10, pl. 351; CASS. in BD., 712. . . . . . . GRISEUS.

205. Genus MICROPALAMA Baird.

Stilt Sandpiper. Bill much as in the last genus, but shorter, less evidently widened at the end and not so distinctly furrowed on top, sometimes perceptibly curved; legs very long; tibiæ bare an inch; tarsus as long as the bill, both 1½–1⅔; feet semipalmate, the front toes being connected by two evident webs; middle toe 1. Length 8–9; wing 5; tail 2¼; plumage resembling that of the last species, its changes the same. Adult in summer: above blackish, each feather edged and tipped with white and tawny or bay, which on the scapulars becomes scalloped; auriculars chestnut; a dusky line from bill to eye, and a light reddish superciliary one; upper tail coverts white with dusky bars; primaries dusky with blackish tips; tail feathers 12, ashy-gray, their edge and a central field white; under parts mixed reddish, black and whitish, in streaks on the jugulum, elsewhere in bars; bill and feet greenish-black. Young, and adult in winter: ashy-gray above, with or without traces of black and bay, the feathers usually with white edging; line over the eye and under parts white, the jugulum and sides suffused with the color of the back, and streaked with dusky; legs usually pale. N. Am., generally dispersed, but apparently not very common anywhere; West Indies in winter; U. S. during the migrations; breeds in high latitudes. AUD., v, 271, pl. 334; NUTT., ii, 138, 140, 141; Sw., F. B.-A. ii, 379, 380, pl. 66; CASS. in BD., 726; COUES, Proc. Phila. Acad. 1861, 174. HIMANTOPUS.

206. Genus EREUNETES Illiger.

Semipalmated Sandpiper. *Peep.* Bill, tarsus and middle toe with its claw, *about* equal to each other, an inch *or less* long, but bill very variable, and apt to be shorter — $\frac{2}{3}$–$\frac{7}{8}$; feet semipalmate, with two evident webs; length $5\frac{1}{2}$–$6\frac{1}{2}$; wing $3\frac{1}{4}$–$3\frac{3}{4}$; tail 2, doubly emarginate, the central feathers projecting. Adult in summer: above, variegated with black, bay, and ashy or white, each feather with a black field, reddish edge and whitish tip; rump, and upper tail coverts except the lateral ones, blackish; tail feathers ashy-gray, the central darker; primaries dusky, the shaft of the first white; a dusky line from bill to eye, and a white superciliary line; below, pure white, usually rufescent on the breast, and with more or less dusky speckling on the throat, breast and sides, in young birds usually wanting; in winter the upper parts mostly plain ashy-gray; but in any plumage and under any variation, the species is known by its small size and semipalmate feet. The extreme variation in the length of the bill is from $\frac{1}{2}$ to $1\frac{1}{4}$, or 86 per cent. of the average ($\frac{7}{8}$). N. Am., everywhere an abundant and well known little bird, thronging our beaches during the migrations. *Tringa semipalmata* WILS., vii, 131, pl. 63, f. 4; NUTT., ii, 136; AUD., v, 277, pl. 336; *Ereunetes petrificatus* CASS. in BD., 724; *E. pusillus* COUES, Proc. Phila. Acad. 1861, 177. Var. *occidentalis* LAWR., *ibid.* 1864, 107; ELLIOT, pl. 41. . . . . PUSILLUS.

207-11. Genus TRINGA Linnæus.

* Bill, tarsus, and middle toe with claw, of about equal length. (*Actodromas.*)

† Upper tail coverts (except the lateral series) black or very dark brown; jugulum with an ashy or brownish suffusion, and dusky streaks.

Least Sandpiper. *Peep.* Smallest of the sandpipers; $5\frac{1}{2}$–6; wing $3\frac{1}{4}$–$3\frac{1}{2}$; tail 2 or less; bill, tarsus, and middle toe with claw, about $\frac{3}{4}$; bill black; legs dusky-greenish; upper parts in summer with each feather blackish centrally, edged with bright bay, and tipped with ashy or white; in winter, and in the young, simply ashy; quills blackish, the shaft of the first white; tail feathers gray with whitish edges, the central blackish, usually with reddish edges; crown not conspicuously different from hind neck; chestnut edgings of scapulars usually scalloped; below, white, marked as above stated. North America, very abundant; this species and the last are usually confounded under the common name of " sandpeeps," and look much alike; but a glance at the toes is sufficient to distinguish them. *Tringa minutilla* VIEILLOT; COUES, Proc. Phila. Acad. 1861, 191; *T. pusilla* WILS., v, 32, pl. 37, f. 4; AUD., v, 280, pl. 337; *T. minuta* Sw., F. B.-A., ii, 385; NUTT., ii, 119; *T. wilsoni* NUTT., ii, 121; CASS. in BD., 721. MINUTILLA.

Baird's Sandpiper. Medium; 7–7½; wing 4½–4¾; tail 2¼; bill, tarsus, and middle toe with claw about ⅞; bill and feet black; colors almost exactly as in the last species; edgings of upper plumage rather tawny than chestnut; jugular suffusion pale, rather fulvous, the streaks small and sparse, sometimes almost obsolete. North and South America; rare on the Atlantic coast (Long Island, *Henshaw;* Am. Nat. vi, 306). *Tringa schinzii* WOODH., Sitgreaves Rep. 1853, 100. *T. bonapartei* CASS. in BD., 722 (in part). *T. maculata,* SCHLEGEL, Mus. Pays-Bas, *Scolopaces,* 39 (in part). *A. bairdii,* COUES, Proc. Phila. Acad. 1861, 194; 1866, 97; SCLATER, Proc. Zool. Soc. 1862, 369 (Mexico); 1867, 332 (Chili, etc.); DALL and BANN., Trans. Chicago Acad. i, 292 (Alaska); ALLEN, Bull. M. C. Z., 1872, 182 (Kansas); HARTING, Ibis, 1870, 151 (S. Africa!). . . . . . BAIRDII.

Pectoral Sandpiper. Jack Snipe. Grass Snipe. Large; 8½–9; wing 5–5½; bill, tarsus, and middle toe with claw about 1⅛; bill and feet greenish; crown noticeably different from cervix; edging of scapulars bright chestnut, straight-edged; chin whitish, definitely contrasted with the heavily ashy-shaded and sharply dusky-streaked jugulum. North America, abundant. NUTT., ii, 111; AUD., v, 259, pl. 359; CASS. in BD., 720. . MACULATA.

†† Upper tail coverts white, with or without dusky marks; jugulum sharply streaked, but with little or no ashy suffusion.

White-rumped Sandpiper. Medium; size of *bairdii;* feet black; bill black, light-colored at base below; plumage as in the foregoing species, excepting the above particulars. An ashy wash on the jugulum is hardly appreciable except in young birds, and then it is slight; the streaks are very numerous, broad and distinct, extending as specks nearly or quite to the bill, and as shaft lines along the sides; while the white upper tail coverts are a diagnostic feature. Eastern N. Am. to the Rocky Mountains. Western? An abundant species along the Atlantic Coast. *T. schinzii* Sw., Fn. Bor.-Am. ii, 384; NUTT., ii, 109; AUD., v, 275; *T. bonapartii* CASS. in BD., 275. . . . . BONAPARTEI.

Cooper's Sandpiper. Largest; 9½; wing 5¾; tail 2¾; bill 1¼; tarsus 1⅛; Like the last in color. Long Island; only one specimen known. It is uncertain whether this is a good species or an unusual state of *T. canutus* or *A. maculata.* BD., 716; COUES, Proc. Phila. Acad. 1861, 202. COOPERII.

** Bill, tarsus, and middle toe, obviously not of equal length.

‡ Tarsus shorter than middle toe; tibiæ feathered. (*Arquatella.*)

Purple Sandpiper. Bill little longer than head, much longer than tarsus, straight or nearly so; tibial feathers long, reaching to the joint, though the legs are really bare a little way above; tarsus shorter than middle toe; 8–9; wing 5; tail 2⅗, rounded; bill 1¼; tarsus ¾; middle toe 1 or a little more. Adult: above ashy-black with purplish and violet reflections, most of the feathers with pale or white edging; secondaries mostly white; line over eye, eyelids, and under parts white, the breast and jugulum a pale cast of the color of the back, and sides marked with the same. In winter, and most immature birds, the colors are similar but much duller; very young birds have tawny edgings above, and are mottled with ashy and dusky below. Atlantic coast, rather common. NUTT., ii, 115; AUD., v, 261, 330; CASS. in BD., 717. . . . . MARITIMA.

†† Tarsus not shorter than middle toe; tibiæ bare below.
+ Bill slightly decurved, much longer than tarsus. (*Pelidna*.) •

American Dunlin. Black-bellied Sandpiper. Red-backed Sandpiper. Ox-bird. Bill longer than head or tarsus, compressed at the base, rather depressed at the end, and usually appreciably decurved; 8–9; wing 4½–5; tail 2–2½; bill 1½–1¾; tibiæ bare about ½; tarsus 1 or rather more; middle toe and claw 1 or rather less. Adult in summer: above, chestnut, each feather with a central black field, and most of them whitish-tipped, rump and upper tail coverts blackish, tail feathers and wing coverts ashy-gray, quills dusky with pale shafts, secondaries mostly white, and inner primaries edged with the same; under parts white, belly with a broad, jet-black area, breast and jugulum thickly streaked with dusky; bill and feet black. Adult in winter, and young: above, plain ashy-gray, with dark shaft lines, with or without red or black traces; below white, little or no trace of black on belly; jugulum with few dusky streaks and an ashy suffusion. N. Am. WILS., vii, 25, pl. 56, f. 2; 39, pl. 57, f. 3; NUTT., ii, 106; AUD., v, 266, pl. 332; CASS. in BD., 719. ALPINA var. AMERICANA.

+ + Bill much decurved, slightly longer than tarsus. (*Ancylocheilus*.)

Curlew Sandpiper. Bill longer than head or tarsus, compressed throughout, decurved; size of the last; legs longer; tibiæ bare ¾; tarsus 1½; middle toe and claw under an inch; bill about 1½. Adult in summer: above, greenish-black, each feather tipped and edged with yellowish-red; below, deep brownish-red; upper tail coverts white with dusky bars; tail ashy-gray, with greenish gloss; wing coverts ashy with reddish edgings and dusky shaft lines; quills dusky, the shafts whitish along their central portion; bill and feet greenish-black. Adult in winter, and young: similar; above, duller blackish with little reddish; below, white, more or less buffy-tinged, the jugulum dusky-streaked. Atlantic coast, extremely rare, little more than a straggler; Europe; Asia; Africa. NUTT., ii, 104; AUD., v, 269, pl. 333; CASS. in BD., 718. . . . . SUBARQUATA.

+ + + Bill perfectly straight. (*Tringa*.)

Red-breasted Sandpiper. Ash-colored Sandpiper. Gray-back. Robin-snipe. Knot. Bill equalling or rather exceeding the head, straight, comparatively stout; toes evidently shorter than tarsus; large, 10–11; wing 6–6½; tail 2½, nearly square; bill about 1⅓ (very variable); tarsus 1¼; middle toe and claw 1; tibiæ bare ½ or more. Adult in summer: above, brownish-black, each feather tipped with ashy-white, and tinged with reddish on the

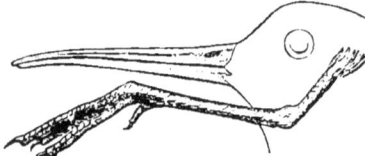

FIG. 166. Bill and foot of American Dunlin.

scapulars; below, uniform brownish-red, much as in the robin, fading into white on the flanks and crissum; upper tail coverts white with dusky bars; tail feathers and secondaries grayish-ash with white edges; quills blackish, gray on the inner webs and with white shafts; bill and feet blackish. Young: above clear ash, with numerous black-and-white semicircles; below white, more or less tinged with reddish, dusky-speckled on breast, wavy-barred on sides. Atlantic coast, abundant. WILS., vii, 36, 43, pl. 57, f. 2, 5; NUTT., ii, 125; AUD., v, 254, pl. 328; CASS. in BD., 715. . CANUTUS.

212. Genus CALIDRIS Cuvier.

Sanderling. Ruddy Plover. No hind toe; otherwise, form exactly as in *Tringa* proper; $7\frac{1}{2}$–8; wing $4\frac{1}{2}$–5; tail $2\frac{1}{4}$; bill about 1; tarsus 1 or rather less; middle toe and claw $\frac{3}{4}$. Adult in summer: head, neck and upper parts varied with black, ashy and bright reddish; below from the breast pure white; tail except central feathers light ash, nearly white; primaries gray with blackish edges and tips, the shafts of all and bases of most, white; secondaries white except a space at the end, and greater coverts broadly white-tipped; bill and feet black. Adult in winter, and young: little or no reddish; speckled with black and white, or ash and white, below white, sometimes tawny-tinged on the jugulum.

FIG. 167. Foot of Sanderling.

N. Am., coastwise, abundant. WILS., vii, 68, 129, pl. 59, f. 4; pl. 63, f. 3; NUTT.. ii, 4; AUD., v, 287, pl. 338; CASS. in BD., 723. . . ARENARIA.

213. Genus LIMOSA Brisson.

Great Marbled Godwit. Marlin. Tail barred throughout with black and rufous; rump and upper tail coverts like the back; no pure white anywhere. General plumage rufous or cinnamon-brown; below, nearly unmarked and of very variable shade, usually deepest on the lining of the wings; above, variegated with black and brown or gray; quills rufous and black; bill flesh colored, largely tipped with black; feet dark. Large; 16–22; wing about 9; tail about $3\frac{1}{2}$; bill 4–5, grooved nearly to the end, usually *slightly* recurved; tibiæ bare 1–$1\frac{1}{2}$; tarsus $2\frac{1}{2}$–$3\frac{1}{4}$; scutellate before and behind; toes $1\frac{1}{4}$, stout. Temperate North America, abundant; conspicuous by its size and coloration among the waders that throng the shores and muddy or sandy bars of bays and estuaries during

FIG. 168. Great Marbled Godwit.

the migration; breeds in the U. S. as well as northward. WILS., vii, 30, pl. 56, f. 1; NUTT., ii, 173; AUD., v, 331, pl. 348; CASS. in BD., 740. FEDOA.

White-tailed Godwit. Tail, its upper coverts and rump, white, barred throughout with black ; head, neck and under parts rusty-red in the breeding season, in winter whitish ; above, grayish-brown, the feathers with darker centres, and blackish shaft lines ; sides and crissum with sagittate black marks. About the size of the last.   A widely distributed Old World species, and a very near relative of *L. rufa* of Europe, lately discovered in Alaska (*Dall*). BD., Trans. Chicago Acad. i, 320, pl. 32. . . . . UROPYGIALIS.

Hudsonian or *Black-tailed Godwit. Ring-tailed Marlin.*   Tail black, largely white at the base, its coverts mostly white ; rump blackish ; lining of wings extensively blackish ; under parts in the breeding season intense rufous, variegated (chiefly barred) with dusky ; head, neck and upper parts brownish-black, variegated with gray, reddish, and usually with some whitish speckling ; quills blackish, more or less white at the base.   Young and apparently winter specimens much paler, tawny whitish below, more gray above.   Considerably smaller than either of the foregoing ; about 15 ; wing 8 or less ; bill 3½ or less ; tarsus 2½ or less.   North America, rather northerly, apparently not common in the United States ; a near relative of *L. ægocephala* of Europe. NUTT., ii, 175 ; AUD., v, 335, pl. 349 ; CASS. in BD., 741. . . . . HUDSONICA.

214-16. Genus TOTANUS Bechstein.

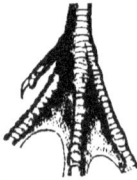

FIG. 169.   Base of toes of Willet.

* Toes with 2 subequal webs ; legs bluish or dark. (*Symphemia.*)

Willet. Semipalmated Tattler.   Bill straight, comparatively stout, grooved little if any more than half its length ; toes with two conspicuous basal webs ; 12-16 ; wing 7-8 ; tail 2½-3 ; bill or tarsus 2-2¾ ; tibiæ bare 1 or more ; middle toe and claw 1½-2.   In summer, gray above, with numerous black marks, white below, the jugulum streaked, the breast, sides and crissum barred or with arrow-shaped marks of dusky (in winter, and in young birds all these dark marks few or wanting, except on jugulum) ; upper tail coverts, most of the secondaries, and basal half of primaries, white ; ends of primaries, their coverts, lining of wings, and axillars, black ; bill bluish or dark.   Temperate N. Am., abundant ; resident in the U. S., conspicuous in the marshes of the Atlantic coast.   WILS., vii, 27, pl. 56, f. 3 ; NUTT., ii, 144 ; AUD., v, 324, pl. 347 ; CASS. in BD., 729.   SEMIPALMATA.

** Toes with inner web very small ; legs yellow or green. (*Glottis.*)

Greater Tell-tale. Greater Yellow-shanks. Stone-snipe. Tattler.   Bill straight or slightly bent upward, very slender, grooved half its length or less, black ; legs long and slender, yellow.   In summer, ashy-brown above varied with black and speckled with whitish, below white, jugulum streaked, and breast, sides and crissum speckled or barred, with blackish, these latter marks fewer or wanting in winter and in the young ; upper tail coverts white with dark bars ; tail feathers marbled or barred with ashy or white ; quills blackish.   Large ; length over 12 ; wing over 7 ; tail 3 or more ; bill 2 or more ; tarsus about 2¼ ; middle toe and claw 1½ ; tibiæ bare 1¼.   N. Am., abundant, migratory; like the last, a restless noisy denizen of the marshes, bays and estuaries. WILS., vi, 57, pl. 58; NUTT., ii, 148; AUD., v, 316, pl. 345; CASS. in BD., 731. . . . . MELANOLEUCUS.

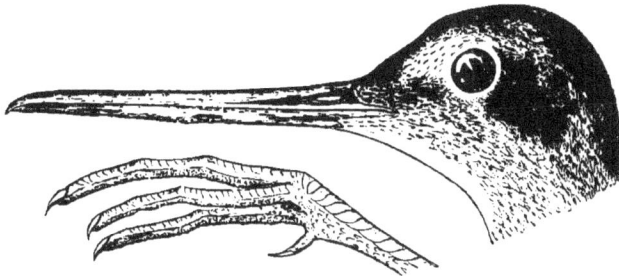

FIG. 170. Greater Tell-tale.

Lesser Tell-tale. Yellow-shanks. A miniature of the last; colors precisely the same; legs comparatively longer; bill grooved rather further. Length under 12; wing under 7; tail under 3; bill under 2; tarsus about 2; middle toe and claw, and bare tibia, each, 1¼. Eastern (and Western?) N. Am., abundant, in the same places as the last. WILS., vii, 55, pl. 57; NUTT., ii, 152; AUD., v, 313, pl. 344; CASS. in BD., 732. . . FLAVIPES.

Green-shanks. Size and form almost exactly as in the last species; bill longer, about 2¼; colors nearly the same, but bill and legs *greenish;* rump and lower back, as well as the tail and its coverts, white, with more or fewer dark marks. Florida. *T. glottis* AUD., v, 321, pl. 346; NUTT., ii, 68; *Glottis floridanus* CASS. in BD., 730. There is no reason to suppose that this bird is any thing more than a straggler to this country; Audubon's specimen is absolutely identical with European ones. . . . . CHLOROPUS.

*** Toes with inner web rudimentary; legs blackish. (*Rhyacophilus.*)

Solitary Tattler. Bill perfectly straight, very slender, grooved little beyond its middle; 8–9; wing 5; tail 2½; bill, tarsus, and middle toe, each about 1–1¼; tibiæ bare ⅔. Dark lustrous olive-brown, streaked on the head and neck, elsewhere finely speckled, with whitish; below, white, jugulum and sides of neck with brownish suffusion, and dusky streaks; rump and upper tail coverts like the back; tail, axillars and lining of wings beautifully barred with black and white; quills entirely blackish; bill and feet blackish; young duller above, less speckled, jugulum merely suffused with grayish-brown. N. America,

FIG. 171. Solitary Tattler.

abundant, migratory; a shy, quiet inhabitant of wet woods, moist meadows and secluded pools, rather than of the marshes; breeds in mountainous portions of the U. S., and northward. WILS., vi, 53, pl. 58; NUTT., ii, 159; AUD., v, 309, pl. 343; CASS. in BD., 733. . . . . . SOLITARIUS.

217. Genus TRINGOIDES Bonaparte.

Spotted Sandpiper. Bill short, straight, grooved nearly to tip; 7–8; wing about 4; tail about 2; bill, tarsus and middle toe, each, about 1. Above, olive (quaker-color; exactly as in a cuckoo) with a coppery lustre, finely varied with black; line over eye, and entire under parts, pure white, with numerous sharp circular black spots, larger and more crowded in the ♀ than in the ♂, entirely wanting in very young birds; secondaries broadly white-tipped and inner primaries with a white spot; most of the tail feathers like the back, with subterminal black bar and white tip; bill pale yellow, tipped with black; feet flesh-color. N. Am., extremely abundant everywhere near water, and breeding throughout the country; famil-

FIG. 172. Spotted Sandpiper.

iarly known as the sandlark, peetweet, teeter-tail, tip-up, etc., these last names being given in allusion to its habit (shared by allied species) of jetting the tail as it moves; a custom as marked as the continual bobbing of the head of the solitary tattler and others. Nest a slight affair of dried grasses, on the ground, often in a field or orchard, but generally near water; eggs 4, pointed, creamy or clay colored, blotched with blackish and neutral tint. WILS., vii, 60, pl. 59, f. 1; NUTT., ii, 162; AUD., v, 303, pl. 342; CASS. in BD., 735. . . . . MACULARIUS.

218. Genus PHILOMACHUS Mœhring.

Ruff (♂). *Reeve* (♀). Bill straight, about as long as the head, grooved nearly to tip; gape reaching behind culmen; outer and middle toe webbed at base, inner cleft; tail barred; ♂ in the breeding season with the face bare and beset with papillæ, and the neck with an extravagant ruff of elongated feathers; plumage endlessly variable in color; about 10; wing 6½–7; tail 2½–3; bill 1¼; tarsus 1¾; middle toe and claw 1¼; ♀ smaller, the head fully feathered, and no ruff. A widely distributed bird of the Old World, noted for its pugnacity; occasionally killed on the coast of New England and the Middle States; some half dozen instances are recorded. NUTT., ii, 131; CASS. in BD., 737; LAWR., Ann. Lyc. N. Y. 1852, 220 (Long Island); BREWSTER, Am. Nat. vi, 306 (Massachusetts). . PUGNAX.

219. Genus ACTITURUS Bonaparte.

Bartramian Sandpiper. Upland Plover. Field Plover. Bill straight, about as long as the head, grooved ⅔ its length, the gape very deep, reaching nearly to below the eyes, the feathers extending on the upper mandible beyond those on the lower, which do not fill the interramal space; tail very long, more than half the wing, graduated; tarsi much longer than the middle toe and claw; tibiæ bare nearly the length of the latter; length 11–13; wing 6–7; tail 3–4; bill 1–1¼; middle toe and claw the same;

tarsus about 2. Above blackish, with a slight greenish reflection, variegated with tawny and whitish; below, pale tawny of varying shade, bleaching on throat and belly; jugulum with streaks, breast and sides with arrowheads and bars, of blackish; axillars and lining of wings pure white, black-barred; quills blackish, with white bars on the inner webs; tail varied with tawny, black and white, chiefly in bars; bill and legs pale, former black-tipped. N. Am., abundant, migratory; a highly esteemed game bird found usually in flocks, in fields, not necessarily near water; feeds chiefly on insects. WILS., vii, 63, pl. 59, f. 2; AUD., v, 248, pl. 327; NUTT., ii, 168; CASS. in BD., 737. . . . . BARTRAMIUS.

220. Genus TRYNGITES Cabanis.

Buff-breasted Sandpiper. Bill extremely small and slender, appearing the more so because of the extension of the feathers on its base—on the upper mandible, quite to the nostrils, nevertheless not reaching nearly so far as on the sides of the lower, and the interramal space completely filled; gape reaching beyond base of culmen; basal webbing of toes rudimentary, hardly noticeable; tail rounded, with projecting central feathers; 7–8; wing 5–5½; tail 2¼; tarsus 1¼; middle toe and claw, and bill, under an inch. Quills largely white on the inner web, and with beautiful black marbling or mottling, best seen from below; tail unbarred, gray, the central feathers darker, all with subterminal black edging and white tips; crown and upper parts blackish, the feathers with whitish and tawny edging, especially on the wings; sides of the head, neck all round, and under parts, pale rufous, or fawn color, speckled on the neck and breast with dusky; bill black; feet greenish-yellow. N. Am., generally distributed in open country, but apparently not abundant; a remarkable bird both in form and coloration, in the latter

FIG. 173. Buff-breasted Sandpiper.

respect somewhat resembling the foregoing, with which it shares many habits. NUTT., ii, 113; AUD., v, 264, pl. 331; CASS. in BD., 739. RUFESCENS.

221. Genus HETEROSCELUS Baird.

Wandering Tattler. Bill straight, stout, compressed, grooved about ⅔ its length, gape reaching beyond base of culmen; legs rather short, rugous, reticulate, scutellate only in front of the tarsus; outer toe with an evident basal web, inner with a rudimentary one; 10½; wing 6½; tail 3½; bill 1½; tarsus 1¼; tail unbarred; plumage variable, generally uniform plumbeous-gray above, below white shaded on breast and sides, or barred on the latter, with the color of the upper parts. A species of almost universal distribution on the coasts and islands of the Pacific, described under at least twelve different names, without counting its various generic appellations. *H. brevipes* CASS. in BD., 734. . . . . INCANUS.

222. Genus NUMENIUS Linnæus.

Long-billed Curlew. Sickle-bill. Bill of extreme length and curvature, measuring from 5 to 8 or 9 inches; total length about 2 feet; wing a foot or less; tail about 4 inches; tarsus 2½-2¾, scutellate only in front. Plumage very similar to that of the godwit; prevailing tone rufous, of varying intensity in different birds and on different parts of the same bird, usually more intense under the wing than elsewhere; below, the jugulum streaked, and the breast and sides with arrow-heads and bars, of dusky; above, variegated with black, especially on the crown, back and wings; tail barred

FIG. 171. Long-billed Curlew.

throughout with black and rufous; secondaries rufous; primaries blackish and rufous; no pure white anywhere; bill black, the under mandible flesh colored for some distance; legs dark. Temperate N. Am., abundant; breeds in the U. S. WILS., viii, 23, pl. 64; NUTT., ii, 94; AUD., vi, 35, pl. 355; CASS. in BD., 743. . . . LONGIROSTRIS.

Hudsonian Curlew. Jack Curlew. Bill medium, 3 or 4 inches long; length 16–18; wing 9; tail 3½; tarsus 2¼-2½. Plumage as in last species in pattern, but general tone much paler; quills barred. N. Am., abundant; breeds in British America; U. S. chiefly during the migrations. *Scolopax borealis* WILS., vii, 92, pl. 56; *N. intermedius* NUTT., ii, 100. AUD., vi, 42, pl. 356; NUTT., ii, 97; CASS. in BD., 744. . . . . . HUDSONICUS.

Esquimaux Curlew. Dough-bird. Bill small, under 3 inches long; length 12–15 inches; wing under 9; tail 3; tarsus 2. Plumage in tone and pattern almost exactly as in the last species, but averaging more rufous, especially under the wings, and primaries not barred. N. Am., abundant; distribution much as in the last species. SWAINS., F. B.-Am., ii, 378, pl. 65; NUTT., ii, 101; AUD., vi, 45, pl. 357; CASS. in BD., 744. BOREALIS.

Suborder HERODIONES. Herons and their Allies.

The character of this group has been indicated on p. 240.

Family TANTALIDÆ. Ibises. Spoonbills.

Under this head I associate the genera *Tantalus, Platalea* and *Ibis*, with its subdivisions; all of these, especially the first, are very nearly related to the storks (*Ciconiidæ*); the last two agree more closely with each other, in the remarkable smallness of the tongue, and other characters. In all, the pterylosis is more or less completely stork-like. The head is more or less perfectly bare of feathers in the adult state, downy in young. Birds of medium and very large size, long-legged, long-necked and small-bodied, like the cranes, storks and herons, with ample, more or less rounded wings, of which the inner quills are very large; tail very short, usually, if not always, of 12 broad rectrices; tibiæ bare for a long distance; tarsi reticulate, or scutellate in front only; toes four, the anterior webbed at base, the hinder lengthened and inserted low down, as in storks and herons (not cranes) ; middle claw not pectinate as in the herons.   Chiefly lacustrine and palustrine inhabitants of the warmer parts of the globe, feeding on fish, reptiles and other animals.   The sexes are alike ; the young different.   The manifest modification of the bill is the principal external character of the three subfamilies into which the group is divisible.

Subfamily TANTALINÆ.   Wood Ibises.

Bill long, extremely stout at base, where it is as broad as the face, gradually tapering to the decurved tip, without nasal groove or membrane, the nostrils directly perforating its substance.   One genus and three or four species of America, Africa, Southern Asia, and part of the East Indies.

FIG. 175. Wood Ibis.

224. Genus TANTALUS Linnæus.

Wood Ibis.   Adult with head and part of the neck naked, corrugate, bluish ; legs blue ; bill pale greenish ; plumage entirely white, excepting the quills, tail, primary coverts and alula, which are glossy black ; young with the head downy-feathered, the plumage dark gray, the quills and tail blackish ; about 4 feet long ; wing 18–20 inches ; bill 8–9 ; tarsus 7–8.   Wooded swampy places in the Southern states, N. to Ohio and the Carolinas, W. to the Colorado, abundant ; gregarious ; nests in trees and bushes.   WILS., viii, 39, pl. 66 ; NUTT., ii, 82 ; AUD., vi, 64, pl. 361 ; BD., 682. . . . . LOCULATOR.

Subfamily IBIDINÆ.   Ibises.

Bill long, very slender, curved throughout, and grooved nearly or quite to the tip (thus closely resembling a curlew's).   There are about twenty species of ibises, among which minor details of form vary considerably, nearly every one of them having been made the type of some genus.   They probably form two genera, *Ibis*, with the tarsi scutellate in front, and *Geronticus*, with the legs entirely reticulate.   Our species vary in the nakedness of the head, which in one is little more than in the herons, and in none is it complete, as in the preceding and following genus.

225-6. Genus IBIS Moehring.

Glossy Ibis.   Plumage rich dark chestnut, changing to glossy dark green with purplish reflections on the head, wings and elsewhere ; bill dark ; young similar, much duller, or grayish-brown, especially on the head and neck, which are white-streaked.   Claws slender, nearly straight ; head bare only about the eyes and between the forks of the jaw.   Length about 2 feet ; wing 10–11 ; tail 4 ; bill 4½ ; tarsus 3½ ; middle toe and claw .3.   U. S., generally but irregularly distributed, chiefly southerly and especially coastwise ; N. casually to New England.   BONAP., Am. Orn. iv, 23, pl. 23 ; NUTT., ii, 88 ; AUD., vi, 50, 358 ; BD., 685. .   FALCINELLUS var. ORDII.

White Ibis. Plumage pure white, outer primaries tipped with glossy black; bill and feet reddish; young dull brown or gray, the legs bluish, the bill yellowish. Claws curved; face and throat bare in the adult. Size of the last or rather larger; bill 7; tarsus 4. South Atlantic and Gulf States, casually N. to Long Island (*Lawrence*). WILS., viii, 43, pl. 66; NUTT., ii, 86; AUD., vi, 54, pl. 360; BD., 684. . . . . ALBA.

Scarlet Ibis. Plumage rich scarlet, outer primaries tipped with glossy black; bill and feet reddish. Young ashy-gray, darker above, paler or whitish below. Size and proportions nearly as in the last species. Tropical America; accidental in the U. S. (Louisiana; seen at a distance, not procured, *Audubon;* Rio Grande, fragment of a specimen examined, *Coues.*) WILS., viii, 41, pl. 66; NUTT., ii, 84; AUD., vi, 53, pl. 359; BD., 683. RUBRA.

Subfamily *PLATALEINÆ.*   Spoonbills.

Bill long, perfectly flat, remarkably widened, rounded and spoon-shaped at the end. Birds of this group are known at a glance, by the singularity of the bill; they closely resemble the foregoing in structure and habit. One genus, with five or six species of various countries.

227. Genus PLATALEA Linnæus.

Roseate Spoonbill. In full plumage rosy-red, whitening on neck; lesser wing coverts, tail coverts, and lower throat crimson; tail brownish-yellow; legs pale carmine; bare head yellowish-green, with a dark stripe; bill mostly grayish-blue. Young with the head mostly feathered, colors much less vivid (no crimson); tail rosy; in an early stage probably grayish. Length about 30; wing 14–15; tail 4–5; tarsus 4; bill 6–7. South Atlantic and Gulf states, N. casually to the Carolinas and Natchez (*Audubon*); common; gregarious; breeds on trees and bushes in the wooded swamps. WILS., vii, 123, pl. 62; NUTT., ii, 79; AUD., ii, 72, pl. 362; BD., 686. . . AJAJA.

Family ARDEIDÆ.   Herons.

It is in this family that powder-down tracts (p. 4, § 6) reach their highest development; and although these peculiar feathers occur in some other birds, there appears to be then only a single pair; so that the presence of two or more pairs is probably diagnostic of this family. In the genus *Ardea* and its immediate allies there are three pairs, the normal number; one on the lower back over the hips, one on the lower belly under the hips, and one on the breast, along the track of the furcula. In the bitterns, the second of these is wanting. In the boat-billed heron, *Cancroma cochlearia,* there is still another pair, over the shoulder blades. There are other pterylographic characters; in general, the tracts (p. 5, § 9) are extremely narrow, often only two feathers wide; there are lateral neck tracts; the lower neck is frequently bare behind. More obvious characters are, the complete feathering of the head (as compared with storks, etc.) except definite nakedness of the lores alone—the bill appearing to run directly into the eyes; a general looseness of the plumage (as compared with *Limicolæ*), and especially the frequent development of remarkably lengthened, or otherwise modified, feathers, constituting the beautiful crests and dorsal plumes that ornament many species, but which, as a rule, are worn only during the breeding season. These features will suffice to determine the *Ardeidæ,* taken in connection with the more general ones indicated under head of *Herodiones,* and the details given beyond.

The boat-billed heron of Central America, with a singular shape of the bill that has suggested the name, and the four pairs of powder-down tracts, constitutes one subfamily, *Cancrominæ.* The still more remarkable *Balæniceps rex,* of Africa, with an enormous head and bill, thick neck, and one pair of such tracts, is probably assignable here as a second subfamily, *Balænicepinæ ;* but it approaches the storks, and may form a separate intermediate family. The disputed cases of *Rhinochetus, Eurypyga* and *Scopus* have been already mentioned ; these five forms aside, the herons all fall in the single

.

Subfamily *ARDEINÆ.* True Herons.

Bill longer than head, straight, or very nearly so, more or less compressed, acute, cultrate (with sharp cutting edges) ; upper mandible with a long groove ; nostrils more or less linear, pervious. Head narrow and elongate, sloping down to the bill, its sides flattened. Lores naked, rest of head feathered, the frontal feathers extending in a rounded outline on the base of the culmen, generally to the nostrils. Wings broad and ample ; the inner quills usually as long as the primaries, when closed. Tail very short, of twelve (usually), or fewer soft broad feathers. Tibiæ naked below, sometimes for a great distance. Tarsi scutellate in front, and sometimes behind, generally reticulate there and on the sides. Toes long and slender ; the outer usually connected with the middle by a basal web, the hinder very long (for this order), inserted on the level of the rest. Hind claw larger and more curved than the middle one (always?) ; the middle claw *pectinate.*

The group thus defined offers little variation in form ; all the numerous genera now in vogue have been successively detached from *Ardea,* the typical one, with which most of them should be reunited. The night herons (235-6) differ somewhat in shortness and especially stoutness of bill ; while the bitterns (237, and the South American genus *Tigrisoma*) are still better marked. There are about seventy-five species, very generally distributed over the globe, but especially abounding in the torrid and temperate zones. Those that penetrate to cold countries in summer, are regular migrants ; the others are generally stationary. They are maritime, lacustrine and paludicole birds, drawing their chief sustenance from animal substances taken from the water, or from soft ground in its vicinity ; such as fish, reptiles, testaceans and insects, captured by a quick thrust of the spear-like bill, given as the bird stands in wait or wades stealthily along. In conformity with this, the gullet is capacious, but without special dilation, the stomach is small and little muscular, the intestines are long and extremely slender, with a large globular cloaca, and a cœcum. Herons are altricial, and generally nest in trees or bushes (where their insessorial feet enable them to perch with ease) in swampy or other places near the water, often in large communities, building a large flat rude structure of sticks. The eggs vary in number, coincidently, it would seem, with the *size* of the species ; the larger herons generally lay two or three, the smaller kinds five or six ; the eggs are somewhat elliptical in shape, and usually of an unvariegated bluish or greenish shade. The voice is a rough croak. The sexes are nearly always alike in color (remarkable exception in gen. 238) ; but the species in which, as in the bittern, the plumage is nearly unchangeable, are very few. Indeed, probably no birds show greater changes of plumage, with age and season, than nearly all the herons. Their beautiful plumes are only worn during the breeding season; the young invariably lack them, and there are still more remarkable changes of plumage in many cases. Thus, the young may be pure white while the adults are dark colored, as in the small blue heron; and sometimes even, as in the remarkable case of our reddish egret, most individuals change from white to a dark plumage after two years, while others appear to remain white their whole lives,

FIG. 176. Great Blue Heron.

and others again are dark from the nest. Many species are pure white at all times, and to these the name of "egret" more particularly belongs; but I should correct a prevalent impression that an egret is anything particularly different from other herons. The name, a corruption of the French word "aigrette," simply refers to the plumes that ornament most of the herons, white or otherwise, and has no classificatory meaning; its application, in any given instance, is purely conventional. The colors of the bill, lores and feet are extremely variable, not only with age or season, but as individual peculiarities; sometimes the two legs of the same specimen are not colored exactly alike. The ♀ is commonly smaller than the ♂. The normal individual variability in stature and relative length of parts is very great; and it has even been noted that a specimen may have one leg larger than the other, and the toes of one foot longer than those of the other — a circumstance perhaps resulting from the common habit of these birds, of standing for a long time on one leg. *

228-34. Genus ARDEA Linnæus.

* Species of large size, and varied dark colors. (*Ardea*.)

Great Blue Heron. Back without peculiar plumes at any season, but scapulars lengthened and lanceolate; an occipital crest, two feathers of which are long and filamentous; long loose feathers on the lower neck. Length about 4 feet; extent 6; bill 5½ inches; tarsus 6½; middle toe and claw 5; wing 18–20; tail 7. ♀ much smaller than ♂. Adult of both sexes grayish-blue above, the neck pale purplish-brown with a white throat-line, the head black with a white frontal patch, the under parts mostly black, streaked with white; tibiæ, edge of wing, and some of the lower neck feathers, orange-brown; bill and eyes yellow, culmen dusky, lores and legs greenish. The young differ considerably, but are never white, and cannot be confounded with any of the succeeding. Entire temperate North America, abundant; migratory in northerly portions. WILS., viii, 68, pl. 65; NUTT., ii, 42; AUD., vi, 122, pl. 369; BD., 668. . . . . HERODIAS.

Florida Heron. Similar; larger; bill 6½; tarsus 8 or more; tibiæ bare nearly ⅓ their length; middle toe not ⅔ the tarsus; below, white, the sides streaked with black; neck ashy; head, with the crest, white, the forehead streaked with black. Southern Florida (*Wurdemann*). BD., 669. It seems improbable that this is anything more than a special state of the last species, but as it is useless to exchange one doubtful opinion for another, I retain it, pending final determination. . . . . . . WURDEMANNII.

** Species (large or small) white at all times. (*Audubonia, Herodias* and *Garzetta*.)

Great White Heron. Size and form nearly as in the foregoing; no greatly elongated occipital feathers nor lengthened scapulars; bill 6½; tarsus 8½; tibiæ bare 6. Color entirely pure white; bill and eyes yellow; culmen greenish at base; lores bluish; legs yellow, greenish in front. Southern Florida. AUD., vi, 110, pl. 368; NUTT., ii, 39; BD., 670. OCCIDENTALIS.

Great White Egret. *White Heron.* No obviously lengthened feathers on the head at any time; in the breeding season, back with very long plumes of decomposed feathers drooping far beyond the tail; neck closely feathered; plumage entirely white at all seasons; bill, lores and eyes, yellow; legs and feet black. Length 36–42 (not including the dorsal train); wing 16–17; bill nearly 5; tarsus nearly 6; rather larger specimens constitute var. *californica* BD., 667. Distribution the same as that of the snowy heron; abundant. WILS., vii, 106, pl. 61, f. 4; NUTT., ii, 47; AUD., vi, 132, pl. 370; BD., 666. . . . . EGRETTA.

Little White Egret. *Snowy Heron.* Adult with a long occipital crest of decomposed feathers, and similar dorsal plumes, latter *recurved* when perfect; similar, but not recurved plumes on the lower neck, which is bare behind; lores, eyes and toes yellow; bill and legs black, former yellow at base, latter yellow at the lower part behind. Plumage always entirely white. Size of the little blue heron. S. States; Cala.; Middle States, in summer; N. occasionally to New England; abundant. WILS., vii, 120, pl. 62, f. 4; NUTT., ii, 49; AUD., vi, 163, pl. 374; BD., 665. . . . CANDIDISSIMA.

*** Species under 3 feet long, of varied dark colors when adult, in some cases white when young. (*Hydranassa, Florida* and *Butorides*.)

Louisiana Egret. Adult slaty-blue on the back and wings, mostly white below and along the throat-line; crest and most of the neck reddish-purple, mixed below with slaty; the longer narrow feathers of the crest white; lower back and rump white, but concealed by the dull purplish-brown feathers of the train, which whiten towards the end; bill black and yellow; lores yellow; legs yellowish-green, dusky in front. Young variously different, but never white. Length about 24 (exclusive of the long train); wing 10-11; bill 4-5; tarsus 4; middle toe and claw 3. S. Atlantic and Gulf States, chiefly maritime, very rarely N. to the Middle districts. *A. ludoviciana* WILS., viii, 13, pl. 64, f. 1; NUTT., ii, 51; AUD., vi, 156, pl. 373; BD., 663. . . . . LEUCOGASTRA var. LEUCOPRYMNA.

Reddish Egret. Adult grayish-blue, rather paler below, head and neck lilac-brown, ends of the train yellowish; bill black on the terminal third, the rest flesh-colored, like the lores; iris white; legs blue, the scales of the tarsus blackish; length about 30; wing 14-15; bill 4; tarsus 5½. Young usually entirely white, for a year or two; some individuals permanently so; bill as in the adult; legs greenish, with yellowish soles; in this state the bird *A. pealei* BONAP., Am. Orn. iv, 96, pl. 26, f. 1; NUTT., ii, 49; GAMBEL, Proc. Phila. Acad. 1848, 127; BD., 661. Florida and Gulf States, strictly; maritime, abundant. *Ardea rufescens* AUD., vi, 139, pl. 371; *Demiegretta rufa* BD., 662. . . . . RUFA.

Little Blue Heron. Head of the adult with lengthened decomposed feathers, those of the lower neck, and scapulars, elongate and lanceolate; no dorsal plumes; neck bare behind below; length about 24; wing 11-12; bill 3; tarsus 3½-4. Adult slaty-blue, becoming purplish on the head and neck; bill and loral space blue, shading to black at the end; eyes yellow; legs black. Young *pure white*, but generally showing blue traces, by which it is distinguished from the snowy heron, as well as by the color of the bill and feet, though at first the legs are greenish-blue with yellowish traces. S. Atlantic and Gulf States, abundant; N. casually to New England in summer. WILS., vii, 117, pl. 62, f. 3; NUTT., ii, 58; · AUD., vi, 148, pl. 372; BD., 671. . . . . CÆRULEA.

Green Heron. Adult in the breeding season with the crown, long soft occipital crest, and lengthened narrow feathers of the back lustrous dark green, sometimes with a bronzy iridescence, and on the back often with a glaucous cast; wing coverts green, with conspicuous tawny edgings; neck

purplish-chestnut, the throat-line variegated with dusky or whitish; under
parts mostly dark brownish-ash, belly variegated with white; quills and tail
greenish-dusky with a glaucous shade, edge of the wing white; some of the
quills usually white-tipped; bill greenish-black, much of the under mandible
yellow; lores and iris yellow; legs greenish-yellow; lower neck with
lengthened feathers in front, a bare space behind. Young with the head less
crested, the back without long plumes, but glossy greenish, neck merely
reddish-brown, and whole under parts white, variegated with tawny and
dark brown. Length 16–18; wing about 7; bill 2½; tarsus 2; middle toe
and claw about the same; tibiæ bare 1 or less. U. S., and a little beyond,
abundant in summer; resident in the South. WILS., vii, 97, pl. 61, f.
1; NUTT., ii, 63; AUD., vi, 105, pl. 367; BD., 676. . . . VIRESCENS.

235-6. Genus NYCTIARDEA Swainson.

Night Heron. Qua-bird. Squawk. No peculiar feathers excepting 2–3
very long filamentous plumes springing from the occiput, generally imbricated
in one bundle; bill very stout; tarsi reticulate below in front; length about
2 feet; wing 12–14 inches; bill, tarsus and middle toe about 3. Crown,
scapulars and interscapulars, very dark glossy green; general plumage
bluish-gray, more or less tinged with lilac; forehead, throat-line and most
under parts, whitish; occipital plumes white; bill black; lores greenish;
eyes red; feet yellow. Young very different; lacking the plumes; grayish-
brown, paler below, extensively speckled with white; quills chocolate-brown,
white-tipped. U. S., and somewhat northward, abundant; resident in the
South, migratory elsewhere. *A. nycticorax* WILS., vii, 101, pl. 61, f.
2, 3; AUD., vi, 82, pl. 363; *A. discors* NUTT., ii, 54; *N. gardeni* BD.,
678; *A. nævia* BODD., Planches Enluminées, 939. .. GRISEA var. NÆVIA.

Yellow-crowned Night Heron. Adult with the head crested, some of the
feathers extremely long, and back with long loose feathers, some of which
reach beyond the tail; bill short, extremely stout; tarsi mostly reticulate,
longer than middle toe; about as large as the last; bill 2¾; tarsus 3¼;
middle toe 3. Grayish-plumbeous, darker on the back, where the feathers
have black centres and pale edges, and rather paler below, the head and
upper neck behind black, with a cheek-patch, the crown, and most of the
crest, white, more or less tinged with tawny; bill black, eyes orange, lores
greenish, feet black and yellow. Young speckled, as in the last, but showing the different proportions of the bill and feet. WILS., viii, 26, pl. 65,
f. 1; NUTT., ii, 52; AUD., vi, 89, pl. 364; BD., 679. . . . VIOLACEUS.

237. Genus BOTAURUS Stephens.

Bittern. Indian Hen. Stake-driver. Bog-bull. Plumage of the upper
parts singularly freckled with brown of various shades, blackish, tawny and
whitish; neck and under parts ochrey or tawny white, each feather marked
with a brown dark-edged stripe, the throat-line white, with brown streaks,
a velvety black patch on each side of the neck above; crown dull brown,

with buff superciliary stripe; tail brown; quills greenish-black, with a glaucous shade, brown-tipped; bill black and yellowish, legs greenish, soles yellow; 23–28 long; wing 10–13; tail 4½, of only 10 feathers; bill about 3; tarsus about 3½. Temperate N. Am., abundant. Not gregarious; nests on the ground; eggs 4–5, drab-colored. WILS., viii, 35, pl. 65, f. 3; NUTT., ii, 60; AUD., vi, 94, pl. 365; BD., 674. ENDICOTT, Am. Nat. iii, 169. MINOR.

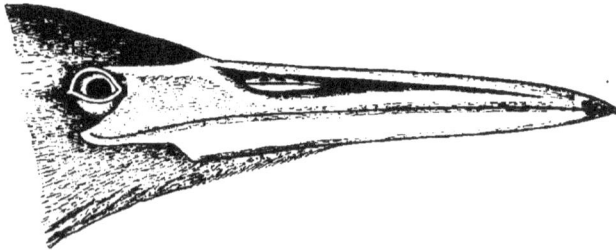

FIG. 177. Bittern.

238. Genus ARDETTA Gray.

Least Bittern. No peculiar feathers, but those of the lower neck long and loose, as in the bittern; size very small; 11–14 inches long; wing 4–5; tail 2 or less; bill 2 or less; tarsus about 1⅜. Sexes dissimilar. ♂ with the slightly crested crown, back and tail, glossy greenish-black; neck behind, most of the wing coverts, and outer edges of inner quills, rich chestnut, other wing coverts brownish-yellow; front and sides of neck, and under parts, brownish-yellow, varied with white along the throat-line, the sides of the breast with a blackish-brown patch; bill and lores mostly pale yellow, the culmen blackish; eyes and soles yellow; legs greenish-yellow; ♀ with the black of the back entirely, that of the crown mostly or wholly, replaced by rich purplish-chestnut, the edges of the scapulars forming a brownish-white stripe on either side. U. S., common. WILS., viii, 37, pl. 65, f. 4; NUTT., ii, 66; AUD., vi, 100, pl. 366; BD., 673. . . EXILIS.

Family GRUIDÆ. Cranes.

As already intimated, cranes are related to rails in essential points of structure, though more resembling herons in their general aspect. They are all large birds, some being of immense stature; the legs and neck are extremely long, the wings ample, and the tail short, usually of twelve broad feathers. The head is generally, in part, naked and papillose or wattled in the adult, with a growth of hair-like feathers, or, in some cases, an upright tuft of curiously bushy plumes. The general plumage is compact, in striking contrast to that of herons; but the inner remiges, in most cases, are enlarged and flowing. In some species, the sternum is enlarged and hollowed to receive a fold of the windpipe, as in swans. Bill equalling or exceeding the head in length, straight, rather slender but strong, compressed, contracted opposite the nostrils, obtusely pointed; nasal fossæ short, broad, shallow; nostrils near the middle of the bill, large, broadly open and completely pervious;

tibiæ naked for a great distance; tarsi scutellate in front; toes short, webbed at base; hallux very short, highly elevated. About 14 species of various parts of the world; only 2 of them American. Most of them fall in the genus *Grus;* the elegant "demoiselle" cranes of the Old World, *Anthropoides virgo* and *paradisæ*, and the African *Balearica pavonina*, are the principal exceptions.

223. Genus GRUS Linnæus.

White or Whooping Crane. Adult with the bare part of the head extending in a point on the occiput above, on each side below the eyes, and very hairy. Bill very stout, gonys convex, ascending, that part of the under mandible as deep as the upper opposite it. Adult plumage pure white, with black primaries, primary coverts and alula; bill dusky greenish; legs black; head carmine, the hair-like feathers blackish. Young with the head feathered; general plumage gray? varied with brown. Length about 50 inches; wing 24; tail 9; tarsus 12; middle toe 5; bill 6. Temperate N. Am., but apparently of irregular distribution, not well made out; said to be common in the South Atlantic and Gulf States. WILS., viii, 20, pl. 64, f. 3 : NUTT., ii, 34; AUD., v, 188, pl. 313; BD., 654. . AMERICANUS.

Brown or Sandhill Crane. Adult with the bare part of the head forking behind to receive a pointed extension of the occipital feathers, not reaching on the sides below the eyes, and sparsely hairy. Bill moderately stout, with nearly straight and scarcely ascending gonys, that part of the under mandible not so deep as the upper at the same place. Adult plumage plumbeous-gray, never whitening; primaries, their coverts, and alula, blackish. Young with head feathered, and plumage varied with rusty brown. Rather smaller than the last. Temperate N. Am., rare or irregular in the east, very abundant in the south and west. NUTT., ii, 38; AUD., v, 188 (in part), pl. 314; BD., 655. Also, *G. fraterculus* CASS. in BD., 656 (young). CANADENSIS.

Family ARAMIDÆ. Courlan.

Consisting of a single genus, with probably only one species, of the warmer portions of America; closely allied to the rails in all essential points of structure, and perhaps only forming a subfamily of *Rallidæ*. Bill twice as long as the head, slender but strong, compressed, grooved for about half its length, contracted opposite the nostrils, the terminal portion enlarged, and decurved; nostrils long, linear, pervious; head completely feathered to the bill; tibiæ half bare; tarsus scutellate anteriorly, as long as the bill; toes cleft, the hinder elevated; wings short, rounded, with falcate 1st primary; tail short, of 12 broad feathers.

239. Genus ARAMUS Vieillot.

Scolopaceous Courlan. Crying-bird. Chocolate-brown with a slight olivaceous or other gloss, paler on the face, chin and throat, most of the plumage sharply streaked with white; 24–28 long; wing 12–14; tail 6–7; bill and tarsus, each, about 5. Florida. BONAP., Am. Orn. iv, 111, pl. 26; NUTT., ii, 68 : AUD., v, 181, pl. 312; BD., 657. . SCOLOPACEUS var. GIGANTEUS.

Family RALLIDÆ, Rails, etc.

This is a large and important family, abundantly represented in most parts of the world. They are birds of medium and small size, generally with compressed body and large strong legs (the muscularity of the thighs is very noticeable), enabling them to run rapidly and thread with ease the mazes of the reedy marshes to which they are almost exclusively confined; while by means of their very long toes they are prevented from sinking in the mire or the floating vegetation. The wings are never long and pointed as among *Limicolæ*, being in fact of the shortest, most rounded and concave form found among waders; and the flight is rarely protracted to any great distance. The tail is always very short, generally of 10 or 12 soft feathers. Details of the bill and feet vary with the genera; but the former is never sensitive at the tip, and the latter have the hallux longer and lower down than it is in the shore-birds. The nostrils are pervious, of variable shape. The head is completely feathered; the general plumage is ordinarily of subdued and blended coloration, lacking much of the variegation commonly observed in shore-birds; the sexes are usually alike, and the changes of plumage not great with age or season. The food, never probed for in the mud, but gathered from the surface of the ground or water, consists of a variety of aquatic animal and vegetable substances. The nest is a rude structure, placed on the ground, or in a tuft of reeds or other herbage; the eggs are numerous, generally variegated in color; the young are hatched clothed. The general habit is gregarious, and migratory; many species occur in vast multitudes, though their skulking ways, and the nature of their resorts, withdraw them from casual observation. Some species swim habitually.

There appear to be upward of 150 species of the family, falling in several well marked groups. The *Ocydrominæ* are an Old World type of some 35 species, ranking with some authors as a distinct family. Mr. Gray makes the African *Himantornis hæmatopus* the type and single representative of another subfamily. Excluding the *Parridæ* and *Heliornithidæ* (see p. 241), both of which are sometimes brought under *Rallidæ*, as subfamilies, the three remaining groups are represented in this country.

Subfamily RALLINÆ. Rails.

This is the largest, and central or typical, group, to which most of the foregoing paragraph is especially applicable. The species are strictly paludicole; the compression of the body is at a maximum; the form is blunt and thick behind, with a very short tip-up tail, and tapers to a point in front; the whole figure being thus adapted to wedge through narrow places. The wings are extremely short and rounded, and the ordinary flight appears feeble and vacillating, though the migrations of many species are very extensive. The flank-feathers are commonly enlarged and conspicuously colored; the thighs are very muscular; the tibiæ are generally if not always naked below; the toes are long, completely cleft, without lobes or any obvious marginal membranes. The bill occurs under two principal modifications: in *Rallus* proper it is longer than the head, slender, compressed, slightly curved, long-grooved, with linear nostrils; in most genera, however, it is shorter or not longer than the head, straight, rather stout, with short broad nasal fossæ, and linear-oblong nostrils—altogether somewhat as in gallinaceous birds. The culmen more or less obviously parts antial extension of the frontal feathers, but never forms a frontal shield, as in the coots and gallinules. Of the 35

American species (*Sclater* and *Salvin*) only 7 occur in this country, one of which is merely a straggler. There are some 25 Old World species.

240. Genus RALLUS Linnæus.

Clapper Rail. Salt-water Marsh-hen. Mud-hen. Above, variegated with dark olive-brown and pale olive-ash, the latter edging the feathers; below, pale dull ochrey-brown, whitening on the throat, frequently ashy-shaded on the breast; flanks, axillars and lining of wings, fuscous-gray, with sharp white bars; quills and tail plain dark-brown; eyelids and short superciliary line whitish; young birds are mostly soiled whitish be-low; when just from the egg, en-tirely sooty black.

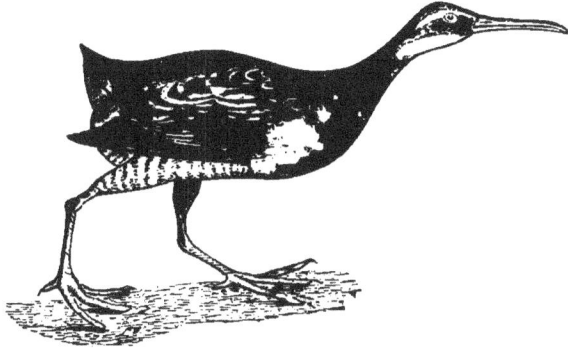
Fig. 178. Clapper Rail.

14–16 long; wing 5–6; tail 2–2½; bill 2–2½; tarsus 1⅔–2; middle toe and claw 2–2⅓; ♀ smaller than the ♂. Salt marshes of Atlantic States, extremely abundant southerly; N. regularly to middle districts, sometimes to Massachusetts; Great Salt Lake (*Allen*). Eggs 6 to 10, 1⅔ by 1¼, whitish, creamy or buff, variously speckled and blotched with reddish-brown, with a few obscure lavender marks. *R. crepitans* Wils., vii, p. 112; Nutt., ii, 201; Aud., v, 165, pl. 310; Cass. in Bd., 747.   longirostris.

King Rail. Fresh-water Marsh-hen. With a general resemblance to the last species, but much more brightly colored; above, brownish-black, variegated with olive-brown, becoming rich chestnut on the wing coverts; under parts rich rufous or cinnamon-brown, usually paler on the middle of the belly, and whitening on the throat; flanks and axillars blackish, white-barred. Rather larger than the last. U. S., chiefly in fresh-water marshes. Wils., vii, pl. 62, f. 2; Aud., v, 160, pl. 309; Cass. in Bd., 746. elegans.

Virginia Rail. Coloration exactly as in *elegans*, of which it is a perfect miniature. Length 8½–10½; wing about 4; tail about 1½; bill 1½–1⅔; tarsus 1¼–1½; middle toe and claw 1½–1¾. Temperate N. Am., common, migratory; many winter in the S. states. Wils., vii, 109, pl. 62, f. 1; Nutt., ii, 205; Aud., v, 174, pl. 311; Cass. in Bd., 748. . virginianus.

241. Genus PORZANA Vieillot.

Carolina Rail. Common Rail. Sora. "Ortolan." Above, olive-brown, varied with black, with numerous sharp white streaks and specks; flanks, axillars and lining of wings, barred with white and blackish; belly whitish;

crissum rufescent. Adult with the face and central line of throat black, the rest of the throat, line over eye, and especially the breast, more or less intensely slate-gray, the sides of the breast usually with some obsolete whitish barring and speckling; young without this black, the throat whitish, the breast brown. Length 8–9; wing $4–4\frac{1}{2}$; tail about 2; bill $\frac{2}{3}–\frac{3}{4}$; tarsus $1\frac{1}{3}$; middle toe and claw $1\frac{2}{3}$. Temperate N. Am., exceedingly abundant during the migration in the reedy swamps of the Atlantic states. WILS., vi, 27, pl. 48, f. 1; NUTT., ii, 209; AUD., v, 145, pl. 306; CASS. in Bd., 749. . . . . CAROLINA.

FIG. 179. Carolina Rail.

Yellow Rail. Above, varied with blackish and ochrey-brown, and thickly marked with narrow white semicircles and transverse bars; below, pale ochrey-brown, fading on the belly, deepest on the breast, where many feathers are dark-tipped; flanks dark with numerous white bars; crissum varied with black, white and rufous. Small, about 6 long; wing $3\frac{1}{4}$; tail $1\frac{1}{2}$; bill $\frac{1}{2}$; tarsus $\frac{7}{8}$; middle toe and claw $1\frac{1}{8}$. Eastern N. Am., not abundant. BONAP., Am. Orn. iv, 136, pl. 27, f. 2; SW. and RICH., Fn. B.-A. ii, 402; NUTT., ii, 215; AUD., v, 152, pl. 307; CASS. in BD., 750. NOVEBORACENSIS.

Black Rail. Blackish; head and under parts dark slaty, paler or whitening on the throat; above, speckled with white, the cervix and upper back varied with dark chestnut; lower belly, crissum, flanks and axillars, white-barred; quills with white spots. Very small; about $5\frac{1}{2}$; wing $2\frac{3}{4}–3$; tail $1\frac{1}{3}$; tarsus $\frac{3}{4}$. S. and Cent. Am. and West Indies; rarely observed in the U. S. Washington, D. C., (*Coues* and *Prentiss*); Kansas, (*Allen*). AUD., v, 157, pl. 308; CASS. in BD., 749. . . . . JAMAICENSIS.

242. Genus CREX Bechstein.

Corn Crake. Yellowish-brown, varied with black; below, cinereous-whitish, palest on the throat and belly; wings extensively rufous both above and below; flanks and crissum barred with the same; 10–11; wing $5\frac{1}{2}–6$; tail 2; bill 1; tarsus $1\frac{2}{3}$. Europe; Greenland; accidentally on the Atlantic Coast, U. S. CASS., Proc. Phila. Acad. 1855, 265 (New Jersey), and in BD., 751; BAIRD, Am. Journ. Sci. 1866, xli, 25. . . . . PRATENSIS.

Subfamily GALLINULINÆ. Gallinules.

Forehead shielded by a broad, bare, horny plate, a prolongation and expansion of the culmen. Bill otherwise much as in the shorter-billed rails, like *Porzana;* general form much the same, though the body is not so compressed; toes slender, simple or slightly margined. The two following genera differ considerably, and each becomes the type of a subfamily with those who elevate the gallinules to the rank of a family; but this does not seem to be required. In *Gallinula*, the nostrils are linear, and the toes have an evident marginal membrane. *Porphyrio* (not "Porphyrula") has broadly oval nostrils and no obvious digital membranes; the legs are notably longer, with shorter toes; the bill is thicker, etc. There are about thirty species of gallinules, of various parts of the world.

243. Genus GALLINULA Brisson.

Florida Gallinule. Head, neck and under parts, grayish-black, darkest on the former, paler or whitening on the belly; back brownish-olive; wings and tail dusky; crissum, edge of wing, and stripes on the flanks, white; bill, frontal plate, and ring round tibiæ, red, the former tipped with yellow; tarsi and toes greenish. 12–15 long; wing 6½–7½; tail 3–3½; gape of bill about 1½; tarsus about 2. S. Atlantic and Gulf states, N. sometimes to Massachusetts. BONAP., Am. Orn. iv, 128, pl. 27, f. 1; NUTT., ii, 223; AUD., v, 132, pl. 304; CASS. in BD., 752. . (CHLOROPUS var?) GALEATA.

244. Genus PORPHYRIO Temminck.

Purple Gallinule. Head, neck and under parts beautiful purplish-blue, blackening on the belly, the crissum white; above, olivaceous-green, the cervix and wing coverts tinted with blue; frontal shield blue; bill red, tipped with yellow; legs yellowish. Young with the head, neck and lower back brownish, the under parts mostly white, mixed with ochrey. 10–12; wing 6½–7; tail 2½–3; bill from gape about 1¼; tarsus about 2¼; middle toe and claw about 3. S. Atlantic and Gulf States, N. casually to New England (*Maine*, BOARDMAN, Am. Nat. iii, 498). WILS., ix, 69, pl. 73; f. 2; NUTT., ii, 221; AUD., v, 128, pl. 303; CASS. in BD., 753. MARTINICA.

Subfamily FULICINÆ. Coots.

Bill and frontal plate much as in the gallinules; body depressed, the under plumage thick and duck-like, to resist water; feet highly natatorial; tarsus shorter than middle toe; toes, including the hinder, *lobate*, being furnished with large semicircular membranous flaps. The coots are eminently aquatic birds, swimming with ease, by means of their lobate feet, like phalaropes and grebes. There are about ten species, of both hemispheres, all referable to the

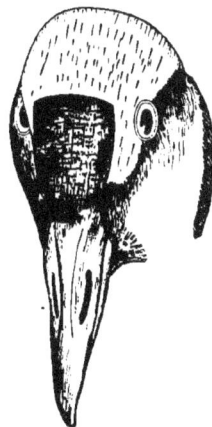

FIG. 180. Frontal plate of a species of Coot.

245. Genus FULICA Linnæus.

Coot. Dark slate, paler or grayish below, blackening on the head and neck, tinged with olive on the back; crissum, whole edge of wing and tips of secondaries, white; bill white or flesh color, marked with reddish-black near the end; feet dull olivaceous; young similar, paler and duller. About 14; wing 7–8; tail 2; bill from the gape 1¼–1½; tarsus about 2; middle toe and claw about 3. Temperate N. Am., abundant. WILS., ix, 61, pl. 73, f. 1; NUTT., ii, 229; AUD., v, 138, 305; CASS. in BD., 751. . . . . AMERICANA.

Subclass III.   AVES AQUATICÆ, or NATATORES.

AQUATIC BIRDS.   SWIMMERS.

Tuis, the third and last series, containing all remaining carinate birds of the present geologic epoch, is a group that may readily be defined upon the principles of adaptive modification already explained under head of *Aves Terrestres;* although as in the cases of the other two "subclasses," it does not rest upon characters of much morphological significance.   The birds composing it are aquatic in a strict sense, fitted to progress upon or through the water, and to derive the greater part of their sustenance from the same source; many of them are absolutely independent of land, except for the purpose of reproduction.   Manifest indications to be fulfilled in adaptation to an aquatic mode of life, are such a configuration of the body as will enable the bird to rest upright on the water, boat-like; and such conformation of the legs as will render them a pair of paddles rather than simple pillars of support, together with water proof clothing of the body.   Accordingly, all swimming birds have a more or less broad and depressed shape, especially flattened underneath.   The coat of feathers is compact and impervious to water, either by its close imbrication, or its thickening with broad tracts and abundant down-feathers, or its plentiful lubrication with oil from the well-developed gland on the rump; in general, these three circumstances conspire to the single result.   The modifications of the legs are especially interesting.   In general, these limbs are transformed into oars by means of webs stretching from tip to tip of the front toes, and sometimes also from the inner toe to the hallux.   This complete palmation is so nearly universal that it alone would characterize the Swimmers, were it not that in one family the same result is effected by means of broad lobes instead of plain webs, and for the fact that a very few genera of waders are more or less completely palmiped.   Since these broad webs would interfere in passing each other were the legs as close together and as parallel as they are in higher birds, another feature is introduced.   The limbs are widely separated, in swimming, not only by the unusual width of the body, but by an outward obliquity of the members themselves; divergence begins at the hip-joint in the direction of the axis of the femur, and increases thence to the terminal segments.   Greater power being required to *push* the body through the water than is needed to simply support it, first on one leg and then on the other, as in ordinary walking, the femur is shortened to become rather a fulcrum for advantageous application of power, than a lever for increase of motion. This shortening is generally so marked that the knee is entirely withdrawn within the general skin of the body—a special characteristic of swimming birds ; and in the best swimmers, the whole limb is thus buried almost to the heel-joint. Finally, the natatorial limb becomes a rudder as well as an oar, serving to steer the bird's course through the water, as much as the tail guides flight through the air. This is accomplished by the backward set of the legs—they project so far posteriorly, in many cases, that in standing on land, the birds necessarily assume a nearly upright position. The wings, tail and bill differ according to families, as noticed under the several heads, beyond, while the more important points of the osseous and digestive systems are similarly diversified.

The Natatorial plan is primarily carried out in *four* different ways, affording as many orders. All of these, and all their families excepting one, are represented within our limits. The missing family is that of the *Spheniscidæ*, or penguins.

Order LAMELLIROSTRES. Anserine Birds.

Bill lamellate: that is, both mandibles furnished along their tomial edges with series of laminar or teeth-like projections, alternating and fitting within each other. Covering of bill membranous, wholly or in greatest part. Tongue fleshy, usually with horny tip, and serrate or papillate edges corresponding to the denticulations of the bill. Feet palmate ; hallux elevated, free, simple or lobed (rarely absent). Wings never exceedingly long, rarely very short. Tail generally short and many-feathered. Œsophagus narrower than in the lower flesh-eating orders, usually with a more or less specially formed crop ; gizzard strongly muscular ; intestines and their cœca long ; cloaca capacious. Legs near centre of equilibrium ; position of body in walking horizontal or nearly so. Reproduction præcocial. Sexual habit frequently polygamous. Diet various, commonly rather vegetarian than animal. There are two remarkably diverse families of lamellirostral birds.

Family PHŒNICOPTERIDÆ. Flamingoes.

Bill very large and thick, entirely invested with membrane (without the horny terminal nail of the *Anatidæ*) which extends around the eyes, and abruptly bent downward at the middle. Legs and neck exceedingly long. Tibiæ largely bare below ; tarsus broadly scutellate, much longer than the toes. Front toes completely webbed ; hallux very small, or wanting. Wings moderately long, ample.

This is a small but very peculiar group of about six species, inhabiting various warm parts of the world. The external characters are so nicely balanced between those of wading and swimming birds, that the flamingoes have been placed indifferently in both groups ; but nearly the whole organization corresponds essentially with that of the duck tribe, the grallatorial relationship, in form and habits, though so evident, being rather of analogy than of affinity. In length of legs and neck these birds exhibit even an exaggeration of the characters of cranes, storks and herons. The bill is unique in shape ; its abrupt bend brings the upper surface in contact with the ground in the act of feeding. The nest is a heap of earth and other material, which the birds bestride in an ungainly attitude; but it is not high enough to permit their long legs to dangle, as represented in some popular accounts and pictorial efforts. The young are said, on good authority, to take to the water as soon as hatched.

246. Genus PHŒNICOPTERUS Linnæus.

American Flamingo. Adult plumage scarlet; most of the quill feathers black; legs lake-red; bill orange-yellow, black-tipped. Length about 4 feet; wing 16 inches; tail 6; bill 5; tarsus 12; middle toe and claw 3½. Florida and Gulf coast; N. casually to S. Carolina (*Audubon*). WILS., viii, 45, pl. 66; NUTT., ii, 70; AUD., vi, 169, pl. 375; BD., 687.   RUBER.

Family ANATIDÆ.   Geese, Ducks, etc.

Bill lamellate, stout, more or less elevated and compressed at base, widened or flattened at tip, invested with soft, tough, leathery membrane, except at the end, which is furnished with a hard, horny "nail," generally somewhat overhanging, sometimes small and distinct, sometimes large and fused; that is, changing insensibly into the general covering. (This soft covering is regarded by some as a prolonged cere; but this is purely theoretical.) Body full, heavy, flattened beneath; neck of variable length; head large; eyes small. No antiæ, the frontal feathers encroaching on the culmen with a convex or pointed outline, and forming other projections on the sides of the bill, and in the interramal space, which latter is broad and long, the mandibular crura being united only at the end by a broad short bridge; no culminal ridge nor keel of gonys. Nostrils subbasal, median or subterminal, usually broadly oval.

FIG. 181. Wild Duck.   Wings of moderate length (rarely very short), stiff, strong, pointed, conferring rapid, vigorous, whistling flight; a wild duck at full speed is said to make ninety miles an hour. Tail of variable shape, but usually short and rounded, never forked, sometimes cuneate, of 12-24 feathers, usually 14-16, the under coverts very long and full, forming a conspicuous crissal tuft. Feet short; knees buried in the general integument; tibiæ feathered nearly or quite to the suffrago; tarsi reticulate or scutellate, or both; toes palmate, the hinder always present and free, simple or lobate. Wing occasionally spurred.

Like the gallinaceous, the anserine type is a familiar one, comprising all kinds of "water-fowl," among which are the originals of all our domestic breeds of swans, geese and ducks, that vie with poultry in point of economic consequence, ornament our parks, or furnish exquisite material for wearing apparel. But additional information respecting the structure of this, the largest and most important family of swimming birds, may be desirable. It is definitely characterized by many important points besides those external features just stated. In palatal structure, the *Anatidæ* are desmognathous; "the lachrymal region of the skull is remarkably long [the lachrymal bone itself is large]. The basisphenoidal nostrum has oval sessile basipterygoid facets. The flat and lamellar maxillo-palatines unite and form a bridge across the palate. The angle of the mandible is produced and greatly recurved" (*Huxley*). The interorbital septum is more or less completely ossified, and the orbits are better defined than in many birds, by well developed processes. The premaxillary is large, and its three prongs are so extensively fused that only a slight nasal aperture remains. Sometimes the top of the skull shows crescentic depressions for lodgment of the supraorbital gland, the secretion of which lubricates the nasal passages; but this feature is never so marked as in most of the pisciv- orous swimmers. The sternum is both long and broad, more or less transverse posteriorly, with a simple notch or fenestra on each side; sometimes its keel is curiously hollowed out for a purpose stated beyond. The vertebræ vary a good deal in number, owing to the variability of the cervicals, which run up to 23 in some swans. The pelvis is ample, arched and extensively ossified, with small foramina, showing nothing of the straight, constricted, largely fenestrated figure prevalent among lower water-birds.

The tongue is large and fleshy; its main bone (*glosso-hyal*) is highly developed; its sides show a fringe of processes corresponding to the lamellæ of the bill. The gullet is not so ample as in the flesh-eating swimmers. The gizzard resembles that of a fowl in its shape and great muscularity; the muscles are deep-colored, and well show the typical disposition of large hemispherical lateral masses converging to central tendons. The cœca vary with the genera according to food; they are very long—12 or 15 inches—in some of the herbivorous species. The male genital armature merits special notice. "In some *Natatores* which copulate on the water there is provision for more efficient coitus than by simple contact of everted cloacæ; and in the *Anatidæ* a long penis is developed. It is essentially a saccular produc- tion of a highly vascular part of the lining membrane of the cloaca. * * * In the passive state it is coiled up like a screw by the elasticity of associated ligamentous structure. * * * A groove commencing widely at the base follows the spiral turns of the sac to its termination; the sperm ducts open upon papillæ at the base of this groove. This form of penis has a muscle by which it can be everted, protruded and raised." (*Owen.*) Among the most interesting structures of the *Anatidæ* are the curious modifications of the windpipe, prevailing almost throughout the family. In a number of swans, this organ enters a cavity in the keel of the sternum, doubles on itself and then emerges to pass to the lungs, forming either a horizontal or a vertical coil. In some geese the windpipe coils between the pectoral muscles and the skin. These vagaries of the windpipe are not, however, confined to the present family, occurring in some of the cranes, certain *Gallinæ*, and also, it is said, in the curious snipe, *Rhynchæa capensis*. In most of the ducks, furthermore, and in the mergansers, the lower larynx is a singularly enlarged and complicated affair; several of the lower rings of the trachea being soldered together and greatly magnified to produce a large irregularly shaped capsule. Its use is not known; in some sense it is a sexual character, since it is only fully developed in the male; it varies greatly in size and shape in different species. Finally, it should be added, that the pterylosis of the family is perfectly definite, a certain type of tract-formation prevailing throughout, with very slight minor modifications.

It is not easy to overrate the economic importance of this large family. It is true that the mergansers, some of the sea ducks, and certain maritime geese, that feed chiefly upon animal substances, are scarcely fit for food; but the great majority afford a bounteous supply of sapid meat, a chief dependence, indeed, with the population of some inhospitable regions. Such is the case, for example, in the boreal parts of this continent, whither vast bands of water-fowl resort to breed during the fleeting arctic summer. Their coming marks a season of comparative plenty in places where hunger often pinches the belly, and their warm downy covering is patched into garments almost cold-proof.

The general traits of the anserine birds are too well known to require more than passing notice. They are salacious to a degree remarkable even in the hot-blooded, passionate class of birds ; a circumstance rendering the production of hybrids frequent, and favoring the study of this subject. If we recall the peculiar actions of geese nipping herbage, and of ducks "dabbling" in the water, and know that some species, as the mergansers, pursue fish and other live prey under water, we have the principal modes of feeding. Nidification is usually on the ground ; sometimes in a hollow tree ; the nest is often warmly lined with live feathers ; the eggs are usually of some plain pale color, as greenish or creamy ; the clutch varies in number, commonly ranging from half a dozen to a dozen and a half. The young are clothed with stiffish down, and swim at once. Among the ducks and mergansers, marked sexual diversity in color is the rule ; the reverse is the case with swans and geese. A noteworthy coloration of many species, especially of ducks, is the *speculum;* a brightly colored, generally iridescent, area on the secondary quills. Most of the species are migratory, particularly those of the northern hemisphere ; the flight is performed in bands, that seem to preserve discipline as well as companionship ; and with such regularity, that no birds are better entitled to the claim of weather-prophets.

There are upward of 175 species of this family, inhabiting all parts of the world. They differ a good deal in minor details, and represent a number of peculiar genera aside from the ordinary types, though none are so aberrant as to endanger the integrity of the group. It is difficult to establish divisions higher than generic, because the swans, geese and ducks, if not also the mergansers, are closely united by intermediate genera. But the five groups presented as subfamilies in the following pages, and representing the whole of the family, may be conveniently recognized, and are readily distinguished, so far as our species are concerned, by the characters assigned.

Subfamily CYGNINÆ. Swans.

A strip of bare skin between the eye and bill; tarsi reticulate. In the swans, the neck is of extreme length and flexibility ; the movements and attitudes on the water are proverbially elegant and graceful. The bill equals or exceeds the head in length ; it is high and compressed at base (where sometimes tuberculate), flatter and widened at the end ; the nostrils are median. Some of the inner remiges are usually enlarged, and when elevated in a peculiar position of the wing, they act as sails to help the course of the bird over the water. The legs are placed rather far back for this family, so that the gait is awkward and constrained. The tail is short, of 20 or more feathers. Although the voice is sonorous at times, an habitual reticence of swans contrasts strongly with the noisy gabbling of geese and ducks ; it is hardly necessary to add, that their fancied musical ability, either in health or at the approach of death, is not confirmed by examination of their vocal apparatus ; this is in many cases convoluted as already described, but there are no syringeal muscles nor other apparatus for modulating the voice. There are eight or ten species, of various countries, among them the celebrated black swan of Australia, *Chenopsis atratus,* the black-necked swan of South America, *Cygnus nigricollis;* and the *Coscoroba anatoides* of the same country, a species with feathered lores ; in none of these does the trachea enter the breast-bone. Our two species belong to the subgenus *Olor,* distinguished from *Cygnus* proper by absence of a tubercle at the base of the bill. The sexes are alike throughout the group.

247. Genus CYGNUS Linnæus.

*** Adult plumage entirely white; younger, the head and neck washed with rusty brown; still younger, gray or ashy. Bill and feet black. Length 4-5 feet. *Trumpeter Swan.* Tail (normally) of 24 feathers. No yellow spot on bill, which is rather longer than the head, the nostrils fairly in its basal half. Mississippi Valley, westward and northward; Canada (*C. passmorei* HINCKS). Sw. and RICH., Fn. Bor.-Am. ii, 464; NUTT., ii, 370; AUD., vi, 219, pls. 382, 383; BD., 758. . . . . BUCCINATOR.

FIG. 182. American Swan.

Whistling Swan. Tail (normally) of 20 feathers. A yellow spot on bill, which is not longer than the head; nostrils median. N. Am. *C. bewickii* Sw., Fn. Bor.-Am., 465; *C. ferus* NUTT., ii, 366; *C. bewickii* NUTT., ii, 372; *C. americanus* AUD., vi, 226, pl. 384; BD., 758. . . AMERICANUS.

Subfamily ANSERINÆ. Geese.

Lores completely feathered; *tarsi entirely reticulate.* Neck in length between that of swans and of ducks; cervical vertebræ about 16; body elevated and not so much flattened as in the ducks; legs relatively longer; tarsus generally exceeding, or at least not shorter than, the middle toe; bill generally rather short, high and compressed at base, and tapering to tip, which is less widened and flattened than is usual among ducks, and almost wholly occupied by the broad nail. The species as a rule are more terrestrial, and walk better, than ducks; they are generally herbivorous, although several maritime species (gen. 249, and an allied South American group) are animal-feeders, and their flesh is rank. Both sexes attend to the young. A notable trait, shared by the swans, is their mode of resenting intrusion by hissing with outstretched neck, and striking with the wings. With some exceptions the plumage is not so bright and variegated as that of ducks, and the speculum is wanting; there is only an annual moult, and no seasonal change of plumage; the sexes are generally alike. Most of the geese fall in or very near gen. 248 and 250, and are modelled in the likeness of the domestic breeds. The more notable exotic forms are: — the Australian *Anseranas melanoleuca* and *Cereopsis novæ-hollandiæ*, the former having the feet little more than semipalmate, the latter scarcely aquatic, with very long legs, much bare above the suffrago, and the bill small, very membranous; the African *Plectropterus gambensis*, a purplish-black bird with spurs on the wings and a tubercle at the base of the bill; the Asiatic *Cynopsis cygnoides*, frequently domesticated, a true goose with a swan-like aspect; the Egyptian goose, *Chenalopex œgyptiaca*. The geese appear to pass directly into the ducks through the rather large shieldrake group, the species of which resemble the latter in many external features, but are more essentially like geese. Characteristic examples of this group are the European *Tadorna vulpanser* and *Casarca rutila;* there are several others in the southern hemisphere; our long-legged arboricole genus *Dendrocygna* belongs in the immediate vicinity, while the domesticated musk duck, *Cairina moschata*, is not far removed. Through such forms as these we are brought directly among the ducks proper.

248. Genus ANSER Linnæus.

*** Bill and feet light or bright colored; plumage *white*, or much variegated.

American White-fronted Goose. Bill smooth; the laminæ moderately exposed; tail normally of 16 feathers. Under parts white or gray, extensively blotched with black; back dark gray, with paler or brownish edgings of the feathers; upper tail coverts white; head and neck grayish-brown, the forehead conspicuously pure white (in the adult; dark in some states); bill pale lake; feet orange, with pale claws. About 27 long; wing 16–18; tail 5–6; tarsus 2¾–3; middle toe and claw about the same. North America; only differs from the European in an average longer bill (1¾–2, instead of 1½–1¾). Sw. and Rich., Fn. B.-A. ii, 466; Nutt., ii, 346; Aud., vi, 209, pl. 380; Bd., 761; *A. frontalis* Bd., 762 (young). Albifrons var. Gambelii.

? Blue Goose. With nearly the size, and exactly the form, of the next species, but the plumage ashy, varied with dark brown, the head, upper neck, tail coverts and most of the under parts white, the wing coverts silvery-ash. Questionably the young of the snow goose. Wils., viii, 89, pl. 69, f. 5; Cass., Proc. Phila. Acad. 1856, 12; Ell. pl. 43. . . cærulescens.

Snow Goose. Bill smooth; the laminæ very prominent, owing to arching of the edges of the bill. Adult plumage pure white, but in most specimens the head washed with rusty-red; primaries broadly black-tipped; bill lake-red with white nail; feet the same, with dark claws. "Young, dull bluish or pale lead colored on the head and upper parts of the body" (*Cassin*). Length about 30; wing 17–19; tail 5½–6½; bill 2½; tarsus 3½. North America; U. S. in winter; extremely abundant in the West, much less so in the East. Wils., viii, 76, pl. 68, f. 5; Sw. and Rich., Fn. B.-A. ii, 467; Nutt., ii, 344; Aud., vi, 212, pl. 381; Bd., 760. . hyperboreus.

Var. albatus. *Lesser Snow Goose.* Smaller; "length about 25 inches; wing 15½; tail 5¾; bill 2; tarsus 3." Western N. Am. Cass., Proc. Phila. Acad. 1856, 41; 1861, 73; Bd., 760, 925; Elliot, pl. 42.

Ross' Goose. Bill studded at the base with numerous elevated papillæ. Color white, with black-tipped quills, exactly as in the snow goose, but less than 24 long; wing 14–15; tail 5; bill 1¼; tarsus 2½. Arctic regions (U. S. in winter?). "Horned Wavey" of *Hearne*, Journ. 442: *A. rossii* Bd.; Cass., Proc. Phila. Acad. 1861, 73; *Exanthemops rossii* Elliot, pl. 44. rossii.

249. Genus PHILACTE Bannister.

Painted Goose. Emperor Goose. Wavy bluish-gray, with lavender or lilac tinting, and sharp black crescentic marks; head, nape and tail white, former often washed with amber-yellow; throat black, white-speckled; quills varied with black and white; 25–28; wing 15–17; tail 5–6; bill 1½; tarsus 3. N. W. coast; abundant at mouth of Yukon. *Chloephaga canagica* BD., 768; ELL., pl. 45; DALL., Trans. Chicago Acad. i, 296; *Philacte canagica* BANN., Proc. Phila. Acad. 1870, 131.   .   .   . CANAGICA.

FIG. 183. Emperor Goose.

250. Genus BRANTA Scopoli.

*** Bill and feet black; head and neck black, with white spaces.

Barnacle Goose. Blackish; tail coverts, sides of rump, forehead, sides of head, and throat, white; interscapulars and wing coverts bluish-gray; under parts plumbeous-white; 28; wing 17; tarsus 2¾; bill 1½.   Europe; very rare or merely casual in N. Am. BD., Am. Nat. ii, 49 (Hudson's Bay); LAWR., *ibid.* v, 10 (North Carolina). NUTT., ii, 355; AUD., vi, 200, pl. 378; BD., 768.   .   .   . LEUCOPSIS.

Brant Goose. Head, neck, body anteriorly, quills and tail, black; a small patch of white streaks on the middle of the neck, and usually white touches on the under eyelid and chin; upper tail coverts white; back brown-ish-gray; under parts the same, but paler, and fading into white on the lower

FIG. 184. *a,* Brant Goose; *b,* var. *nigricans.*

belly and crissum; black of jugulum well defined against the color of the breast; 2 feet long; wing 13; tail 5; bill 1⅜; tarsus 2¼. Hudson's Bay; Arctic and Atlantic (and Pacific?) Coast, S. in winter to Carolina or further; common. WILS., viii, 131, pl. 72, f. 1; Sw. and RICH., F. B.-A. ii, 469; NUTT., ii, 359; AUD., vi, 203, pl. 379; BD., 767. . . . . . BERNICLA.

Var. NIGRICANS. *Black Brant.* Similar; black of jugulum extending over most of the under parts, gradually fading behind; white neck patches usually larger and meeting in front. Both coasts; very abundant on the Pacific; not common on the Atlantic. LAWR., Ann. Lyc. N. Y. 1846, 171; CASS., Ill. 52, pl. 10; BD., 767.

FIG. 185. *a,* Canada Goose; *b,* var. *leucoparcia.*

Canada Goose. Common Wild Goose. Tail normally 18-feathered. Grayish-brown, below paler or whitish-gray, bleaching on the crissum, all the feathers with lighter edges; head and neck black, with a broad white patch on the throat mounting each side of the head; tail black, with white upper coverts. About 36; wing 18-20; tail 6½-7½; bill 1¾-2; tarsus usually over 3. N. Am., abundant; U. S. chiefly in winter, but also occasionally in summer, breeding sparingly. WILS., viii, 52, pl. 67, f. 4; Sw. and RICH., Fn. B.-A. ii, 468; NUTT., ii, 349; AUD., vi, 178, pl. 376; BD., 764. . . . . CANADENSIS.

Var. LEUCOPAREIA. Black of neck bounded below by a white jugular collar; under parts rather darker than is usual in the Canada goose, well defined against the white of the jugulum and crissum. Size of the last; tail feathers 18. CASS., Ill. 272, pl. 45; BD., 765. *B. occidentalis* BD., 766 (in text).

Var. HUTCHINSII. Tail usually 16-feathered. Colors exactly as in the Canada goose, but size less. About 2½ feet long; wing 15-17; tail 5-6; bill 1¼-1¾; tarsus rather under 3. N. Am., but chiefly northern and western. Sw. and RICH., F. B.-A., ii, 470; NUTT., ii, 362; AUD., vi, 198, pl. 377; BD., 766.

251. Genus DENDROCYGNA Swainson.

*** Duck-like arboricole geese, with the bill longer than the head, terminated by a prominent nail, the legs very long with the tibiæ extensively denuded below, the hind toe lengthened, more than one-third as long as the tarsus. In addition to the following species, a third, *D. arborea,* of the West Indies, may occur in the South.

Fulvous Tree Duck. Pale cinnamon or yellowish-brown, darker on the crown, the nape with a black line, the bend of the wing chocolate-brown; rest of the wing, rump and tail, black, its upper and under coverts white; scapulars and fore back dark with pale cinnamon edgings; bill and feet blackish; 20; wing 9½; tail 3¾; bill 1½; tarsus 2¼. S. and Cent. Am. and Mexico; Southwestern U. S., not common. Fort Tejon, Cal. (*Xantus*), BD., 770; Fort Whipple, Ariz., COUES, Proc. Phila. Acad. 1866, 98; Galveston, Tex. (*Dresser;* breeding); New Orleans, La. (*Moore*). . FULVA.

Autumnal Tree Duck. Blackish, including a nuchal stripe; crown, most of neck and fore breast, middle of back and scapulars, reddish-chocolate;

a large white wing-patch; bill and legs reddish. Size of the last. South and Central America and Mexico, to Texas (*Schott*). LAWR., Ann. Lyc. N. Y. 1851, 117; BD., 770. . . . . AUTUMNALIS.

Subfamily *ANATINÆ*. River Ducks.

Tarsi scutellate in front; hind toe simple. This expression separates the present group from all the North American examples of the foregoing and succeeding subfamilies, although not a perfect diagnosis. The neck and legs are shorter than they average in geese, while the feet are smaller than in the sea-ducks, the toes and their webs not being so highly developed. None of the *Anatinæ* are extensively maritime, like most of the *Fuligulinæ;* yet they are not by any means confined to fresh waters, and some species constantly associate with the sea-ducks. They feed extensively, like most geese, upon succulent aquatic herbage, but also upon various animal substances; their flesh is, almost without exception, excellent. They do not dive for their food. The moult is double; the sexes are almost invariably markedly distinct in color; the young resemble the ♀ ; the wing has usually a brilliant speculum, which, like the other wing-markings, is the same in both sexes. Unlike geese, these and other ducks are not doubly monogamous, but simply so if not polygamous; the male pays no attention to the young. Excluding the shield-rake group, already mentioned as pertaining rather to the geese than the ducks, there are about fifty species, generally distributed over the world. They are split into a large number of modern genera, most of which indicate little more than specific characters; the majority are represented in this country. Of those here following, only two, *Spatula* and *Aix*, represent any decided structural peculiarity; the rest might all be referred to *Anas*, type of the group. The *Malacorhynchus membranaceus*, of Australia, is a notable exotic form.

252. Genus ANAS Linnæus.

Mallard. ♂ with the head and upper neck glossy green, succeeded by a white ring; breast purplish-chestnut; tail feathers mostly whitish; greater wing coverts tipped with black and white, the speculum violet, black-bordered; bill greenish-yellow; feet orange-red; ♀ with the wing as in the ♂ ; head, neck and under parts pale ochrey, speckled and streaked with dusky. Length about 24; wing 10–12. N. Am., abundant; rare or casual in New England and further eastward. WILS., viii, 112, pl. 70, f. 7 ; NUTT., ii, 378; AUD., vi, 236, pl. 385; BD., 774. . . . . BOSCHAS.

OBS. This is the well-known original of the common tame duck. An anomalous duck, with the general aspect of this species, but nearly as large as a goose, is occasionally taken on the Atlantic coast. It is unquestionably part mallard, but the balance of its parentage is unknown—supposed to be muscovy. (*A. maxima* GOSSE, Birds of Jamaica, 399 ; *Fuligula viola* BELL, Ann. Lyc. N. Y. 1852, 219.) *A. glocitans* or *A. breweri* of AUD., vi, 252, pl. 387 (*A. audubonii* of BONAP.) is supposed to be a hybrid between the mallard and gadwall. The mallard is known to cross with various other species. Upwards of fifty kinds of hybrid ducks are recorded ; some of them have proved *fertile*, contrary to an assumed rule.

Dusky Duck. Black Duck. Size of the mallard, and resembling the ♀ of that species, but darker and without decided white anywhere except under the wings. Tail 16–18-feathered. Eastern N. Am., abundant, especially in New England and eastward. Wils., viii, 141, pl. 72, f. 5; Nutt., ii, 392; Aud., vi, 244, pl. 386; Bd., 775. . . . . OBSCURA.

253. Genus DAFILA Leach.

Pintail. Sprigtail. Tail cuneate, when fully developed the central feathers much projecting and nearly equalling the wing in length; much shorter and not so narrow in the ♀ and young; 4 to 9 inches long; wing 11; total length about 24. Bill black and blue, feet grayish-blue; head and upper neck dark brown, with green and purple gloss, sides of neck with a long white stripe; lower neck and under parts white, dorsal line of neck black, passing into the gray of the back, which, like the sides, is vermiculated with black; speculum greenish-purple, anteriorly bordered by buff tips of the greater coverts,

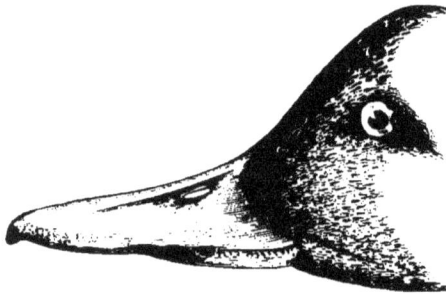

FIG. 186. Female Pintail.

elsewhere by black and white; tertials and scapulars black and silvery; ♀ and young with the whole head and neck speckled or finely streaked with dark brown and grayish or yellowish-brown; below, dusky-freckled; above, blackish, all the feathers pale-edged; only a trace of the speculum between the white or whitish tips of the greater coverts and secondaries. N. Am., abundant. Wils., viii, 72, pl. 68, f. 3; Nutt., ii, 386; Aud., vi, 266, 390; Bd., 776. *Anas caudacuta* Sw. and Rich., F. B.-A. ii, 441. ACUTA.

254. Genus CHAULELASMUS Gray.

Gadwall. ♂ with most of the plumage barred or half-ringed with black and white, or whitish; middle wing coverts *chestnut*, greater coverts *black*, *speculum white;* ♀ known by these wing-marks; 19–22; wing 10–11; N. Am., common. Sw. and Rich., F. B.-A., ii, 440; Wils., viii, 120, pl. 71, f. 1; Nutt., ii, 383; Aud., vi, 254, pl. 388; Bd., 782. STREPERUS.

255. Genus MARECA Stephens.

*** Bill shorter than head, grayish-blue like the feet; tail 14–16-feathered, pointed, but hardly or not half as long as the wing; top of head white or nearly so, plain or speckled, its sides, and the neck, more or less speckled; fore breast light brownish-red; belly pure white; crissum abruptly black; middle and greater coverts white, latter black-tipped; speculum green, black-bordered; 20–22; wing 11; tail 5; tarsus 2; bill 1½–1½; ♀ known by the wing-markings.

European Widgeon. Head and neck reddish-brown, scarcely varied; top of head creamy, or brownish-white, its sides with mere traces of green.

Europe; casually on the Atlantic coast, Greenland to Florida; California (*Cooper*). GIRAUD, Birds Long Island, 307; BD., 784. . . PENELOPE. *American Widgeon. Baldpate.* Head and neck grayish, dusky-speckled; top of head white (in full plumage), its sides with a broad green patch. N. Am., abundant. Scarcely distinct from the last. Sw. and RICH., F. B.-A. ii, 445; WILS., viii, 86, pl. 69, f. 4; NUTT., ii, 389; AUD., vi, 259, pl. 389; BD., 783 . . AMERICANA.

256-7. Genus QUERQUEDULA Stephens.

* Subcrested; head and upper neck chestnut, with a broad glossy green band on each side, whitish-bordered, uniting and blackening on the nape; under parts white, the fore breast with circular black spots; upper parts and flanks closely waved with blackish and white; crissum

FIG. 187. American Widgeon.

black, varied with white or creamy; speculum rich green, bordered in front with buffy tips of the greater coverts, behind with white tips of the secondaries; no blue on the wing; bill black; feet gray. ♀ differs especially in the head markings, but those of the wings are the same. Small; 11-15; wing 7½; tail 3¼; bill 1½; tarsus 1¼. (*Nettion.*)

FIG. 188. Green-winged Teal.

English Teal. No white crescent in front of the wing; long scapulars black externally, creamy internally. Europe; accidental on the Atlantic Coast. COUES, Proc. Phil. Acad. 1861, 238 (Labrador); BD., 778. . . . . CRECCA.

Green-winged Teal. A conspicuous white crescent on the side of the body just in front of the bend of the wing; scapulars plain. N. Am., abundant. WILS., viii, 101, pl. 70, f. 4; NUTT., ii, 400; AUD., vi, 281, pl. 392; BD., 777. . CAROLINENSIS.

** Wing-coverts in both sexes sky-blue, the greater white-tipped; speculum green, white-tipped; axillars and most under wing coverts white; scapulars striped with tawny and blue (not in the ♀) or dark green; fore back barred; rump and tail dark, plain; crissum dark or black; bill black; feet not dark. (*Querquedula.*)

Blue-winged Teal. Head and neck of the ♂ blackish-plumbeous, darkest on the crown, usually with purplish iridescence; a white crescent in front of the eye; under parts thickly dark-spotted; ♀ with head and neck altogether different; under parts much paler and obscurely spotted; but known by the wing-markings from any species except the next one. 15-16; wing 7; tail 3; tarsus 1¼; bill 1½-1¾. Eastern N. Am. to the Rocky Mountains, abundant; also, Alaska (*Dall*). Sw. and RICH., F. B.-A. ii, 444; WILS., viii, 74, pl. 68, f. 4; NUTT., ii, 397; AUD., vi, 287, pl. 393; BD., 779. DISCORS.

Cinnamon Teal. ♂ with head, neck, and whole under parts, rich purplish-chestnut, darkening on crown, chin and crissum, and blackening on middle of belly; rather larger than the last; bill longer, 1⅜-1¾. ♀ with the chestnut replaced by mottled brown and tawny, and difficult to distinguish from ♀ *discors;* but darker, usually with some chestnut traces; head, and especially chin, more spotted; bill longer. A generally distributed S. Am. species, now abundant in the U. S. west of the Rocky Mountains; of casual occurrence in the Gulf States (Louisiana, *Pilaté;* Florida, *Maynard*).

CASS., Proc. Phila. Acad. 1848, 195, and Ill. 82, pl. 25; LAWR., Ann. Lyc. N. Y. 1852, 220; BD., 780, and Stansbury's Rep. 322. CYANOPTERA.

258. Genus SPATULA Boie.

Shoveller. Broad-bill. Bill twice as wide at the end as at the base; with very numerous .and prominent laminæ. Head and neck of ♂ green; fore breast white; belly purplish-chestnut; wing coverts blue; speculum green, bordered with black and white; some scapulars blue, others green, all white-striped; bill blackish; feet red. ♀ known by bill and wings. 20; wing 9½; tarsus 1⅜; bill 2½–2¾. N. Am., abundant. WILS., viii, 65, pl. 67, f. 7; NUTT., ii, 375; AUD., vi, 293, pl. 394; BD., 781. . . CLYPEATA.

259. Genus AIX Swainson.

Summer Duck. Wood Duck. Crested; head iridescent green and purple, with parallel curved, white superciliary and postocular stripes, and a broad, forked, white throat patch; 18–20; wing 8½–9½; tail

FIG. 189. Summer Duck.

4½–5; tarsus 1¼–1½; bill 1⅜; ♀ with the head mostly gray. N. Am., abundant, breeding in most sections, nesting in trees. WILS., viii, 97, pl. 70, f. 3; NUTT., ii, 394; AUD., vi, 271, pl. 391; BD., 785. . SPONSA.

Subfamily *FULIGULINÆ.* Sea Ducks.

Tarsi scutellate in front; hind toe lobate. The large membranous flap depending from the hind toe distinguishes this group from the preceding, probably without exception. While the general form is the same as that of the *Anatinæ*, the feet are notably larger, with relative shorter tarsi, longer toes, and broader webs, and placed somewhat further back, in consequence of which the gait is still more awkward and constrained than the "waddle" of ordinary ducks; but swimming powers are enhanced, and diving is facilitated. A large number of the species are exclusively maritime, but this is no more the case with all of them, than is the reverse with the river ducks. These birds feed more upon mollusks and other animal substances (not, however, upon fish, like the mergansers) than the river ducks do, and their flesh, as a rule, is coarser, if not entirely too rank to be eaten; there are, however, single exceptions to this, as in the case of the canvas-back. The sexes are unlike, as among the *Anatinæ;* and besides the difference in color, the ♀ is often distinguished by the absence or slight development of certain tuberosities of the bill that the ♂ of several species, as of scoters and eiders, possesses. A large majority of the species inhabit the Northern Hemisphere; there are some forty in all, exhibiting a good deal of diversity in minor details, but to no such extent as the number of current genera would imply. Among notable exotics, we have the soft-billed *Hymenolæmus malacorhynchus* of New Zealand, and the short-winged *Micropterus cinereus* of South America, both related to our gen. 264; there are but very few others. The genus *Erismatura* is the type of a small remarkable group, as noticed beyond, sometimes considered as a subfamily. *Biziura lobata* of Australia, with a fleshy appendage under the bill, the African *Thalassornis leuconota*, the *Nesonetta aucklandica*, and several species of *Erismatura*, compose the subgroup.

260-1. Genus FULIGULA Stephens.

* ♂ with the head, neck, and body anteriorly, black, the former glossy; lower back, rump, tail and its coverts, blackish; below, white, with fine black waving on the sides and lower belly; ♀ with the head and anterior parts brown, with or without pure white around the bill, and other black parts of the ♂ rather brown; ♂ ♀ bill black and blue, or dusky; feet livid. (*Fulix.*)

Greater Scaup Duck. Big Black-head. Blue-bill. Raft Duck. Flocking Fowl. Shuffler. No ring round neck; speculum white; back and sides whitish, finely waved in zigzag with black; gloss of head green; bill dull blue with black nail; legs plumbeous; ♀ with the face pure white, the black-and-white vermiculation less distinct. About 20 long; wing 9. N. Am. WILS., viii, 84, pl. 69, f. 3; NUTT., ii, 437 (includes next species); AUD., vii, 355, pl. 498 (not of vi, 316); BD., 791. . . . . . MARILA.

?Lesser Scaup Duck. Little Black-head (with other names of the foregoing). Extremely similar; smaller, about 16; wing 8; gloss of head chiefly purple; flanks and scapulars less closely waved with black? It is very difficult to define this bird specifically, and it may be simply a small southern form; but it appears to preserve its characters, although constantly associated with the last. *F. marila* AUD., vi, 316, pl. 397; *F. minor* GIRAUD, Birds of Long Island, 323; *F. affinis* BD., 791. . . AFFINIS.

Ring-necked Duck. An orange-brown ring round the neck; speculum gray; back nearly uniform blackish; bill black, pale at base and near tip; ♀ with head and neck brown, and no collar, but loral space and chin whitish, as is a ring round eye; bill plain dusky. In size between the foregoing. N. Am. WILS., viii, 60, pl. 67, f. 5; NUTT., ii, 439; AUD., vi, 320, pl. 398; BD. 792. . . COLLARIS.

** ♂ with the head and neck chestnut, pure or obscured, in the ♀ plain brown; body anteriorly, rump and tail coverts, black, in the ♀ dark brown; back, scapulars and sides plumbeous-white, finely waved with black, less distinct in the ♀; speculum bluish-ash. Length about 20; wing 9-10; tarsus 1¾-1¾. (*Aythya.*)

FIG. 190. Canvas-back.

Red-head. Pochard. Bill dull blue with a black belt at the end, broad and depressed, shorter than head (2 or less), the nostrils within its basal half; color of head rich pure chestnut, with bronzy or red reflections ; of back, mixed silvery-gray and black in about equal amount, the dark waved lines unbroken. N. Am., abundant. WILS., viii, 110, pl. 70, f. 6 ; NUTT., ii, 434 ; AUD., vi, 311, pl. 396 ; BD., 793. FERINA var. AMERICANA.

Canvas-back. Bill blackish, high at the base and narrow throughout, not shorter than head (2½, or more), the nostrils at its middle ; head much obscured with dusky ; black waved lines of the back sparse and much broken up into dots, the whitish thus predominating. N. Am., especially abundant along the middle Atlantic Coast in winter, where from feeding on the wild celery (*Vallisneria*) its flesh acquires a peculiar flavor, though not particularly excellent under other circumstances. WILS., viii, 103, pl. 70, f. 5 ; NUTT., ii, 430 ; AUD., vii, 299, pl. 395 ; BD. 794. . . . . . . VALLISNERIA.

262. Genus BUCEPHALA Baird.

*** ♂ with the head puffy, dark colored, iridescent, with large white patches ; lower neck all around, under parts, including sides, most of the scapulars, wing coverts and secondaries, white ; lining of wings and axillars dark ; most of upper parts black ; no waving on back and sides. ♀ with the head less puffy, brown or dark gray, with traces of the white patches, or not ; somewhat less white on the wings ; fore breast and sides with gray, the feathers paler-edged. Bill much shorter than head, very high at the base, tapering, with median nostrils.

Golden-eye. Garrot. ♂ with the head and upper neck glossy dark green, and a white oval or rounded loral spot, not touching the base of the bill throughout ; white continuous on outer surface of wing ; bill black with pale or yellow end, with nostrils in anterior half ; feet orange ; webs dusky ; eyes yellow ; head uniformly puffy ; ♀ with head snuffy-brown, and no white patch in front of the eye. Length 16–19 ; wing 8–9. N. Am., abundant. Our bird does not appear to differ in the least from the European. WILS., viii, 62, pl. 67, f. 6 ; NUTT., ii, 441 ; AUD., vi, 362, pl. 406 (describes the next species as summer plumage) ; BD., 796. . . . . . . CLANGULA.

Barrow's Golden-eye. Rocky Mountain Garrot. Very similar ; gloss of head purplish and violet ; the loral spot larger, triangular or crescentic, applied against the whole side of the bill at base ; white on surface of wing divided by a dark bar ; rather larger than the last ; 19–22 ; wing 9–10 ; occipital feathers lengthening into a slight crest ; bill shorter ; ♀ probably not distinguishable with certainty from that of the foregoing, unless by the dark bar on the wing. Arctic America to the N. States in winter, not common. Also N. Europe. It is doubtfully distinct from the last, with which, however, I am not prepared to unite it. Sw. and RICH., F. B.-A. 456, pl. 70 ; NUTT., ii, 444 ; BD., 796 ; ELLIOT, pl. 46, and Ann. Lyc. N. Y. 1862. . ISLANDICA.

Buffle-headed Duck. Butter-ball. Spirit Duck. Dipper. ♂ with the head particularly puffy, of varied rich iridescence, with a large white auricular patch confluent with its fellow on the nape ; small ; 14–16 ; wing 6–7 ; bill 1, with nostrils in basal half ; ♀ still smaller, an insignificant looking duck, with head scarcely puffy, dark gray, with traces of the white auricular patch. N. Am., abundant. WILS., viii, 51, pl. 67, f. 2, 3 ; NUTT., ii, 445 ; AUD., vi, 369, pl. 408 ; BD., 797. . . . . ALBEOLA.

263. Genus HARELDA Leach.

Long-tailed Duck. South-southerly. Old-wife. Tail of 14 narrow pointed feathers, in the ♂ in summer the central ones very slender and much elongated, nearly or quite equalling the wing; nail of bill occupying the whole tip; seasonal changes remarkable. ♂ in summer with the back and the long narrowly lanceolate scapulars varied with reddish-brown, wanting in winter, when this color is exchanged for pearly-gray or white; general color blackish or very dark brown, below from the breast abruptly white; no white on the wing; sides of head plumbeous-gray; in winter the head, neck and body anteriorly, white, but the gray cheek-patch persistent, and a large dark patch below this; bill at all seasons black, broadly orange-barred. ♀ without lengthened scapulars or tail feathers, the bill dusky greenish, and otherwise different; but recognized by presence of head- and neck-patches, and absence of white on the wing. Length 15–20, or more, according to tail; wing 8–9. N. Am., northerly, coastwise; U. S. only in winter; common. Also Northern Europe. WILS., viii, 93, pl. 70, f. 1, 2; NUTT., ii, 453; AUD., vi, 379, pl. 410; BD., 800. . . . . GLACIALIS.

264. Genus CAMPTOLÆMUS Gray.

Labrador, or Pied Duck. Bill enlarged towards end by membranous expansion, the nostrils in its basal third; cheek feathers rigid; ♂ with the body and primaries black; rest of the wing, with neck and head, white, with a black collar and lengthwise coronal stripe; ♀ plumbeous gray; about 2 feet long; wing 9. N. Atlantic Coast, to middle districts in winter; formerly common, now apparently rare. WILS., viii, 91, pl. 69, f. 6; NUTT., ii, 428; AUD., vi, 329, pl. 400; BD., 803. . . . LABRADORIUS.

265. Genus HISTRIONICUS Lesson.

Harlequin Duck. Bill very small and short, rapidly tapering to tip, which is wholly occupied by the nail, and with a membranous lobe at its base; tertiaries curly; plumage singularly patched with different colors; ♂ deep leaden-bluish, browner below; sides of head, and of body posteriorly, chestnut; coronal stripe and tail black; a white patch at base of bill, another on side of occiput, of breast and of tail, two transverse ones on side of neck, forming a nearly complete ring, and several on the wings; a white jugular collar; speculum violet and purple; ♀ dark brown, paler below, whitening on belly; a white patch on auriculars and before eye. 15–18; wing 8; Northwestern Europe; N. Am., northerly, and entirely coastwise, U. S. only in winter, not abundant. WILS., viii, 139, pl. 72, f. 4; NUTT., ii, 448; AUD., vi, 374, pl. 409; BD., 799. . . TORQUATUS.

266-8. Genus SOMATERIA Leach.

* Bill without frontal process, not feathered to the nostrils. (*Polysticta*.)

Steller's Eider. Head white, with a pearly gray tinge, a green occipital band, and a black chin-patch and eye-ring; collar round neck, and upper parts, lustrous velvety black, the lengthened curly scapulars and tertiaries silvery-white on the inner webs, the lesser and middle wing coverts white, the greater coverts and secondaries white-tipped, enclosing the violet speculum; under parts rich reddish-brown, blackening on the belly and crissum, fading through buff to white on the breast and sides, where there are black spots. ♀ reddish-brown, blackening below, varied with darker on the head, neck and fore parts; tips of greater coverts and secondaries alone white, enclosing the speculum. Length about 18; wing 8. Northwest Coast. NUTT., ii, 451; AUD., vi, 368, pl. 407; BD., 801. STELLERI.

** Bill without frontal processes, feathered to the nostrils. (*Lampronetta.*)

Spectacled Eider. ♂ black or blackish, the throat, most of neck, fore back, wing coverts, scapulars, tertials and flank-patch, white; nape and occiput green; a whitish space round eye, bounded by black; ♀

FIG. 191. Spectacled Eider.

said to be brown, varied with darker, the chin and throat whitish, the eye patch obscurely indicated; after the summer moult the ♂ is said to be like the ♀. Length about 2 feet. Northwest Coast, common about St. Michaels. DALL, Trans. Chicago Acad. i, 299; ELLIOT, pl. 47; BD., 803. FISCHERI.

*** Bill with frontal processes, not feathered to the nostrils. (*Somateria.*)

Eider Duck. Bill with long club-shaped processes extending in a line with the culmen upon the sides of the forehead, divided by a broad feathered interspace. ♂ in breeding attire white, creamy-tinted on breast and washed with green on head; under parts from the breast, lower back, rump, tail, quills, and large forked patch on the crown, black. ♀ with the bill less developed, general plumage an extremely variable shade of reddish-brown or ochrey-brown, speckled, mottled and barred with darker; ♂ in certain stages resembling the ♀. Length about 2 feet; wing 11–12 inches. Arctic and N. Atlantic Coasts, abundant, S. in winter to New England commonly, to the Middle States rarely. This celebrated bird, semi-domesticated in some places, yields most of the prized eider-down of commerce, which the parent plucks from the breast to cover the eggs; eggs commonly 3–4, pale dull greenish. WILS., viii, 122, pl. 71, f. 2, 3; NUTT., ii, 407; AUD., vi, 349, pl. 405; BD., 809. The American bird has lately been separated from the

European under name of *S. dresseri*, by Mr. Sharpe, but I doubt the exclusive pertinence of the assigned characters. . . MOLLISSIMA (var?).

Pacific Eider. Precisely like the last, excepting a V-shaped black·mark on the chin; may require to be treated as merely a variety. Arctic and North Pacific coast, common. BD., 810; ELLIOT, pl. 48. . . . V-NIGRA.

King Eider. Bill with broad squarish nearly vertical frontal processes bulging angularly out of line with culmen. ♂ in breeding attire black, including a forked chin-patch, a frontal band, and small space round eye; the neck and fore parts of the body, part of interscapulars, of wing coverts and of lining of wings, and a flank patch, white, creamy on the jugulum, greenish on sides of head; crown and nape fine bluish-ash. ♀ resembling that of the common eider,

FIG. 192. Eider Ducks. Upper fig., ♂; lower fig., ♀.

but bill different. Size of the last, or rather less. Both coasts, arctic and northerly; S. in winter sometimes to New York. NUTT., ii, 414; AUD., vi, 347, pl. 404; BD., 810. . . . . SPECTABILIS.

269. Genus ŒDEMIA Fleming.

✱✱ Embracing the black sea-ducks, surf-ducks, scoters or "coots" as they are variously called: maritime mollusk-eating species, scarcely fit for food; ♂ black, relieved or not by definite white patches on head or wings, or both, with brightly parti-colored bill, very broad at the end, singularly gibbous at base, but of different form in each of the following species, unnecessarily causing their separation into the three genera, mentioned below; ♀ sooty-brown. etc., bill simply turgid at base, much widened at end; but may be known by having the nostrils at the middle of the bill or beyond it, the nail broad, fused, occupying all the tip, the frontal feathers reaching further on culmen than on sides of upper mandible, and forming no reëntrance at its back upper corner; young ♂ resembling the ♀. Our three species inhabit both coasts, and sometimes the larger inland waters, breeding northward; they occur abundantly in winter along the whole length of the U. S.

American Black Scoter. Bill scarcely encroached upon by the frontal feathers, shorter than the head, black, the gibbosity superior, circumscribed, orange (♂); nostrils at its middle; tail normally 16-feathered. (*Œdemia.*) Plumage of ♂ entirely black. ♀ sooty-brown, paler below, becoming grayish-white on the belly, there dusky-speckled, on the sides and flanks dusky-waved; throat and sides of head mostly continuous whitish; bill all black; feet livid olivaceous, with black webs. ♂ nearly 2 feet long; wing about 10 inches: ♀ 18–19 inches; wing 8–9; gape 2; culmen 1¾. Differs from the European in the shape and coloration of the protuberance on the bill. Wils.,

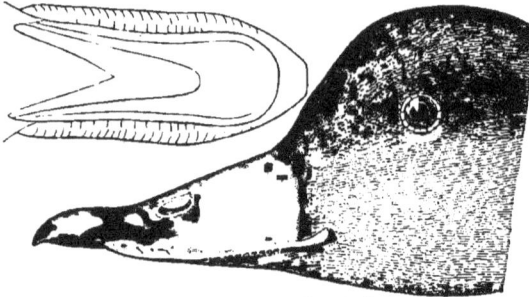

viii, 135, pl. 72, f. 2; Nutt., ii, 422 and 423; Aud., vi, 343, pl. 403; Bd., 807. . americana. *Velvet Scoter. White-winged Surf-duck.* Bill broadly encroached upon by the frontal feathers, on the culmen nearly

Fig. 193. *Female Black Scoter, with outline of bill viewed from below.*

or quite to the nostrils, and on its sides to a less extent, shorter than head, black, broadly orange-tipped (♂); nail broad and truncate; gibbosity superior, circumscribed. (*Melanetta.*) ♂ black, with a large white wing-patch, and another under the eye; feet orange-red, with dusky webs. Size of the last, or rather larger; ♀ smaller, sooty-brown, pale grayish below, with much whitish about head, but showing white speculum; bill all black. Said to differ from the European by greater encroachment of feathers on bill, but the ascribed feature is not tangible. Wils., viii, 137, pl. 72, f. 3; Nutt, ii, 419; Aud., vi, 332, pl. 401; *M. velvetina* Bd., 805. Also, *Fuligula bimaculata* Herbert, Field Sports, ii, 2d ed. 366; *O. bimaculata* Bd., 808 (*immature*). . . . . fusca (var?)

Surf Duck. Sea Coot. Bill narrowly encroached upon by the frontal feathers, on the culmen nearly or quite to the nostrils, but not at all upon its sides, about as long as head, with the nail narrowed anteriorly, the swelling lateral as well as superior; nostrils beyond its middle; bill of ♂ orange-red, whitish on the sides, with a large circular black spot on each side at the base; tail normally 14-feathered.

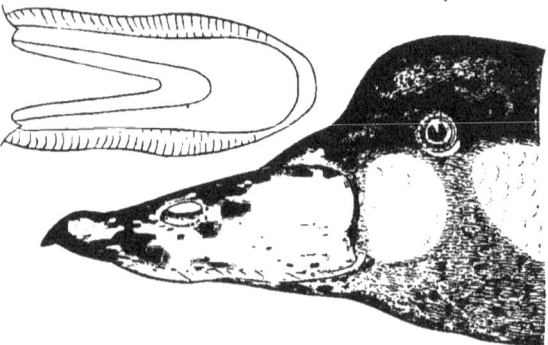

Fig. 194. *Young male Surf Duck, with outline of bill viewed from below.*

(*Pelionetta.*) ♂ black, with a triangular white patch on the forehead and another on the nape; no white on wings; feet orange, with dusky webs.

Size of the first; gape of bill about 2½; ♀ smaller; bill black, shorter, gape about 2¼; feathers of culmen hardly or not reaching nostrils; feet dark, tinged with dusky-reddish; webs black; plumage sooty-brown, below silvery-gray, sides of head with much whitish, chiefly in two patches, one loral, the other auricular. WILS., viii, 49, pl. 67, f. 1; NUTT., ii, 416; AUD., vi, 337, pl. 402; BD., 806. . . . . PERSPICILLATA.

Var. TROWBRIDGEI. With the bill longer, exceeding the head, and of slightly different shape; feathers falling short of nostrils; gape about 2⅓; white frontal patch small, its posterior border anterior to a line between the eyes, instead of reaching or passing beyond this. Cala. BD., 806; ELLIOT, Introd. B. A., No. 64.

270. Genus ERISMATURA Bonaparte.

₊ Remarkably distinguished from other *Fuligulinæ* by the stiffened, linear-lanceolate tail feathers (16–20 in number) exposed to the base, by reason of extreme shortness of the coverts; bill broad, flattened, the nail large, overhanging.

Ruddy Duck. The ♂ in perfect plumage with the neck all round and the upper parts brownish-red, the lower parts silky silvery-white watered with dusky, the chin and sides of the head dead-white, the crown and nape black; but not often seen in this condition in the U. S. As generally observed, and the ♀ at all times, brown above finely dotted and waved with dusky, paler and duller below with darker undulations and sometimes a slight tawny tinge, as also occurs on the sides of the head; crown and nape dark brown; bill dusky; crissum always white. Length 14–17; wing 5–6; tarsus 1¼. N. Am., abundant. WILS., viii, 128, 130, pl. 71, f. 5, 6; NUTT., ii, 426; AUD., vi, 324, pl. 399; BD., 811. . . . . RUBIDA.

St. Domingo Duck. ♂ head anteriorly and chin black; hind-head, neck and breast deep ferruginous; above brownish-red, blotched with black; below lighter ferruginous; speculum white. ♀ similar, but less strongly marked. 13½; wing 6¼; tail 3¾; bill 1¼, smaller and less expanded than in the preceding. S. Am. and W. Indies, accidental in U. S. The only known instances are Lake Champlain (CABOT, Proc. B. S. N. H., vi, 375); Wisconsin (KUMLEIN; *ibid.* xiv, 154; Am. Nat. v, 441). *E. dominica* BD, 925; *E. ortygoides* GOSSE, Birds of Jamaica, 405. . . DOMINICA.

Subfamily *MERGINÆ. Mergansers.*

Bill more or less nearly cylindrical, the nail hooked and overhanging, the lamellæ highly developed into prominent retrorse serrations. Excepting these characters of the bill, the fishing-ducks are simply *Fuligulinæ*, somewhat modified in adaptation to a more exclusively animal regimen; the principal point in their economy is ability to pursue fish under water, like cormorants, loons and other birds of lower orders. The nature of their food renders their flesh rank and unpalatable. The gizzard is rather less muscular than in most ducks; the intestines and their cœca are shorter; the laryngeal capsule of the males is very large, irregular, and partly membranous; the trachea has other dilations. Birds of this group inhabit fresh as well as salt water, and are abundant in individuals if not in species. There are only about eight species, chiefly of the Northern Hemisphere; but several occur in South America.

OBS. The smew, or white nun, *Mergellus albellus*, of Europe, has been attributed to N. Am. upon insufficient evidence, though very possibly occurring. WILS., viii, 126, pl. 71, f. 4; NUTT., ii, 467; AUD., vi, 408, pl. 414; BD., 817.

271-2. Genus MERGUS Linnæus.

* Bill not shorter than head, mostly red. (*Mergus.*)

Merganser. *Goosander.* *Fish Duck.* Nostrils nearly median; frontal feathers reaching beyond those on sides of bill; ♂ with the head scarcely crested, glossy green; back and wings black and white, latter crossed by one black bar; under parts salmon-colored; about 24; wing 11; ♀ smaller; occipital crest better developed, but still flimsy; head and neck reddish-brown; black parts of the ♂ ashy-gray; less white on the wing; under parts less tinted with salmon. N. Am., common. WILS., viii, 68, pl. 68, f. 1, 2; NUTT., ii, 460; AUD., vi, 387, pl. 411; *M. americanus* BD., 813. MERGANSER.

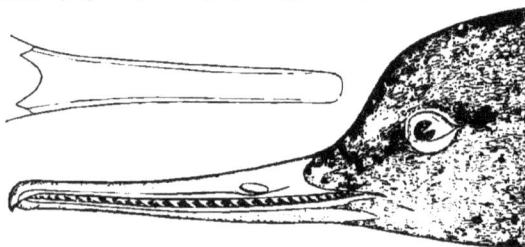

FIG. 195. Red-breasted Merganser, with outline of bill from above.

Red-breasted Merganser. *Fish Duck.* Nostrils sub-basal; frontal feathers not reaching beyond those on sides of bill; a long thin pointed crest in both sexes. Smaller than the last; wing 8–9; general coloration, and sexual differences, the same, but the ♂ with the jugulum rich reddish-brown, black-streaked, the sides conspicuously finely waved with black, a white, black-bordered mark in front of the wing, and the wing crossed by two black bars. N. Am., abundant. WILS., viii, 91, pl. 69, f. 2; NUTT., ii, 463; AUD., vi, 395, pl. 412; BD., 814. . . SERRATOR.

** Bill shorter than head, mostly or entirely black. (*Lophodytes.*)

Hooded Merganser. Nostrils sub-basal; frontal feathers reaching beyond those on sides of bill; a compact, erect, semicircular, laterally compressed crest in the ♂, smaller and less rounded in the ♀; ♂ black, including two crescents in front of wing, and bar across speculum; under parts, centre of crest, speculum, and stripes on tertials, white; sides chestnut, black-barred; 18–19; wing 8; ♀ smaller; head and neck brown; chin whitish; back and sides dark brown, the feathers with paler edges; white on the wing less; bill reddish at base below. N. Am., common. WILS., viii, 79, pl. 69, f. 1; NUTT., ii, 465; AUD., vi, 402, pl. 413; BD., 816. . . . . CUCULLATUS.

Order STEGANOPODES. Totipalmate Birds.

Feet totipalmate, with three full webs; hind toe semi-lateral, larger and lower down than in other water birds, *connected with the inner toe by a complete web reaching from tip to tip.* Nostrils minute, rudimentary or entirely abortive. A gular pouch. Bill not membranous nor lamellate, but tomia sometimes serrate.

This is a definite and perfectly natural group, which will be immediately recognized by the foregoing characters, one of which, the complete webbing of the hallux, is not elsewhere observed among birds. It is represented by six genera, all North American, each the type of a family.

The nature is altricial throughout the order. The eggs are very few, frequently only one, usually if not always plain-colored, and encrusted with a peculiar white chalky substance; they are deposited in a rude bulky nest on the ground, on rocky ledges, or on low trees and bushes in the vicinity of water. The dietetic regimen is exclusively carnivorous, the food being chiefly fish, sometimes pursued under water, sometimes plunged after, sometimes scooped up. In accordance with this, we find the alimentary canal to consist of a capacious distensible œsophagus not developing a special crop, a large proventriculus with numerous solvent glands, a small and very moderately muscular gizzard, rather long and slender intestines, with small cœca, if any, and an ample globular cloaca. The tongue is extremely small, a mere knob-like rudiment (as we have seen in the piscivorous kingfishers). The characteristic gular pouch varies greatly in development. The condition of

Fig. 196. Totipalmate Foot.

the external nostrils is a curious and unexplained feature; they appear to be open at first, and in some species, like the tropic-bird, they remain so; but they are generally completely obliterated in the adult state. There are probably no intrinsic syringeal muscles in any birds of this order. But the most notable fact in connection with the respiratory system is the extraordinary pneumaticity of the body, which reaches its height in the pelicans and gannets; it is described by Nitzsch substantially as follows: The interior air receptacles are of an ordinary character, but the anterior of these cells are more subdivided than usual; from them, the air gets under the skin through the axillary cavities, and diffuses over the entire pectoral and ventral regions, in two large parallel inter-communicating cells on each side, over which the skin does not fit close to the body, but hangs loosely. It is further remarkable that the skin itself does not form a wall of these cavities, a very delicate membrane being stretched from the inwardly projecting bases of the contour-feathers. Thus there is yet another, although a very shallow, interval between this membrane and the skin, this also containing air, admitted from the larger spaces by numerous minute orifices close to the roots of the feathers. This subcutaneous areolar tissue is that which, in ordinary birds and mammals, holds the deposit of fat, no trace of which substance is found in these birds.

The pterylosis of *Steganopodes* adheres throughout to one marked type, there being little variation except in the density of the plumage, which would seem to accord with temperature, the tropical forms being the more sparsely feathered. Excepting one genus, the gular sac is wholly or in part bare. The contour feathers appear to always lack aftershafts. The remiges are from 26 to 40 in number, of which 10 are always long, strong, pointed primaries. There are usually 22–24 tail feathers in the pelicans, but 12, 14 or 16 in the other genera. All have the oil gland large, with a circlet of feathers and more than one orifice; sometimes, as in the pelicans, it is protuberant, heart-shaped, and as large as pigeons' eggs, with two sets of six orifices; in the gannets it is flat and disc-like.

The palatal structure is desmognathous; there are no basipterygoids; the maxillo-palatines are large and spongy; the mandibular angle is truncate; other cranial characters appear under two aspects, one peculiar to the pelicans, the other common to the rest of the order. (HUXLEY.) The sternum is short and broad, with transverse, entire or emarginate, posterior border; the apex of the furculum commonly, if not always, anchyloses with the sternal keel. The upper arm bones are very long; the tibia does not develop the long proximal apophysis seen in many *Pygopodes*. The carotids are double.

The species of this order are few — apparently not over fifty, of which the cormorants represent half — very generally distributed over the world.

Family SULIDÆ. Gannets.

Bill rather longer than the head, cleft to beyond the eyes, very stout at the base, tapering and a little decurved toward the tip, which however is not hooked, the tomia irregularly serrate, or rather lacerate. Nostrils abortive. Gular sac little developed, but naked. Wings rather long, pointed. Tail long, stiff, wedge-shaped, 12-14-feathered. Feet more nearly beneath centre of equilibrium than in some other families of this order. General configuration somewhat that of a goose; body stout; neck rather long; head large, uncrested; plumage compact.

Gannets are large heavy sea-birds of various parts of the world. There are only five or six well established species, of which the two following, with the *S. piscator* of the Indian Ocean, and the Australian *S. cyanops*, are the principal ones. They are piscivorous, and feed by plunging on their prey from on high, when they are completely submerged for a few moments; but they do not appear to dive from the surface of the water, like cormorants. The gait is firm; the flight vigorous and protracted, performed with alternate sailing and flapping. Although so heavy, they swim lightly, owing to the remarkable pneumaticity of the body, already noticed. They are highly gregarious; the common gannet congregates to breed in almost incredible numbers on rocky coasts and islands, of high latitudes, while the booby similarly assembles on the low shores of warmer seas. The nest is a rude bulky structure of sticks and seaweed, placed on the rock or in low thick bushes; the egg, generally single, is plain in color and encrusted with calcareous matter. Both sexes appear to incubate; they are alike in color, the young being different.

273. Genus SULA Brisson.

Common Gannet. Solan Goose. White, with black primaries, the head washed with amber-yellow; bill not yellow; lores, sac and feet blackish. Young: dark brown speckled with white, below from the neck grayish-white, each feather darker-edged; quills and tail blackish. Length about 31; extent 60; wing 17–21; tail about 10; bill 4. Atlantic Coast; swarming in summer at certain northern breeding places, S. to the Gulf of Mexico in winter. NUTT., ii, 495; AUD., vii, 44, pl. 425; LAWR. in BD., 871. . BASSANA.

Booby Gannet. Brown; below from the neck white; bill and feet yellow. Young: grayish-brown, merely paler below; bill dusky. Rather smaller than the last. S. Atlantic and Gulf States, very abundant. NUTT., ii, 500; AUD., vii, 57, pl. 426; LAWR. in BD., 872. . . . . FIBER.

Family PELECANIDÆ. Pelicans.

Bill several times as long as the head, comparatively slender but strong, straight, broad, flattened, ending with a distinct claw-like hook. Mandibular rami joining only at their apex; the long broad interramal space, and the throat, occupied by an enormous membranous sac. Nostrils abortive. Wings extremely long, in the upper and fore-arm portions, as well as the pinion, with very numerous remiges. Tail very short, of 20 or more feathers. Feet short, very stout. Size large.

The remarkable pneumaticity of the body (shared however by the gannets) has been already described. A principal osteological character is, that "the inferior edge of the ossified interorbital septum rises rapidly forward, so as to leave a space at the base of the skull, which is filled by a triangular crest formed by the union of the greatly developed ascending processes of the palatines" (*Huxley*). The tongue is a mere rudiment. But the most obvious peculiarity of these birds is the immense skinny bag hung to the bill, capable of holding several quarts when distended; its structure is as follows: The covering is ordinary skin, but very thin; the lining is skin modified somewhat like mucous membrane; between these "is interposed an equally thin layer, composed of two sets of very slender muscular fibres, separated from each other, and running in opposite directions. The outer fibres run in fascicles from the lower and inner edge of the mandible, those from its base passing downward, those arising more anteriorly passing gradually more forward, and reach the middle line of the pouch. The inner fibres have the same origin, and pass in a contrary direction, backwards and downwards. From the hyoid bone to the junction of the two crura of the mandible, there extends a thin band of longitudinal muscular fibres, in the centre of which is a cord of elastic tissue. By means of this apparatus, the sac is contracted, so as to occupy but little space. When the bill is opened, the crura of the lower mandible separate from each other to a considerable extent [in their continuity — not at the symphysis], by the action of muscles inserted into their base, and the sac is expanded" (*Audubon*). This organ is used like a dip-net, to catch fish with; when it is filled, the bird closes and throws up the bill, contracts the pouch, letting the water run out of the corners of its mouth, and swallows the prey. Pelicans feed in two ways; most of them, like our white one, scoop up fish as they swim along on the water; but the brown species plunges headlong into the water from on wing, like a gannet, and makes

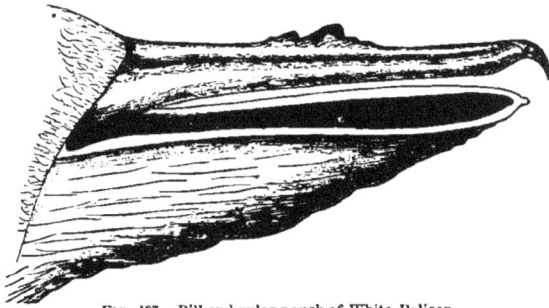

FIG. 197. Bill and gular pouch of White Pelican.

a grab, often remaining submerged for a few seconds. Neither species often catches large fish; they prefer small fry of which several hundred may be required for a full meal. The prevalent impression that the pouch serves to convey live fish, swimming in water, to the little pelicans in the nest, is untrue; the young are fed with partially macerated fish disgorged by the parents from the crop. As Audubon remarks, it is doubtful whether a pelican could fly at all with its burden so out of trim.

The gular pouch varies in size with the different species, reaching its greatest development in the brown pelican, where it extends half-way down the neck in front, is a foot deep when distended, and will hold a gallon. Besides this singular adjunct, the bill of our white pelican has another curious structure, not found in other species. The culmen is surmounted near the middle by a high thin upright comb or crest, the use of which is not known. It is supposed to be a weapon of attack or defence in the combats that occur at the breeding season between rival males, being found only in this sex, and during the breeding season alone. It appears to be shed and renewed in a manner analogous to the casting of deer's horns—a remarkable circumstance first noticed, I believe, by Mr. Ridgway. Its structure explains how this can be: "The crest-like excrescence on the ridge of the upper mandible is not formed of bone, nor otherwise connected with the osseous surface, which is smooth and continuous beneath it, than by being placed upon it, like any other part of the skin; and when softened by immersion in a liquid may be bent a little to either side. It is composed internally of erect slender plates of a fibrous texture, externally of horny fibres, which are erect on the sides, and longitudinal on the broadened ridge; these fibres being continuous with the cutis and cuticle" (*Audubon*).

Pelicans are found in most temperate and tropical countries, both coastwise and inland; they are gregarious birds at all times, and gather in immense troops to breed. A large rude nest is prepared on the ground, or built of sticks in a low bush near the water; the eggs appear to be one to three, plain dull whitish, with a thick roughened shell. The gait of these cumbersome birds is awkward and constrained; but their flight is easy, firm and protracted, and they swim lightly and gracefully, buoyed up by the interior air-sacs. The sexes are alike; the young different; most species are white, with yellow or rosy hue at times, and a crest or lengthened feathers, at the breeding season; while nearly every one of them has a peculiar contour of the feathering at the base of the bill, by which it may be known. There are only six unquestionable species, although some authors admit eight or nine. The four exotic ones are: *P. onocrotalus* of Europe, Asia and Africa (including the *P. minor* and *javanicus* of authors), with the frontal feathers extending in a point on the culmen; *P. crispus* of the same countries, the largest of the genus, and *P. rufescens* (with *philippinus*) of various parts of the Old World, in both of which the frontal outline is concave on the base of the culmen; and, finally, the Australian *P. conspicillatus*, in which a strip of feathers cuts off the naked circumocular region from the base of the bill. This is an entirely peculiar feature; and our white pelican shows another, having the sides of the under mandible feathered at base for a short distance. Excellent accounts of the genus have been given by Dr. Sclater and Mr. Elliot (Proc. Zool. Soc. 1868, 264, and 1869, 571).

274. Genus PELECANUS Linnæus.

American White Pelican. White; occiput and breast yellow; primaries, their coverts, bastard quills, and many secondaries, black; bill, sac, lores and feet yellow. About 5 feet long; expanse 7–9; wing 2; bill 1 or more; tail ½, normally 24-feathered. N. Am.; N. to 61°; very abundant in the west; only accidental in the Middle and Eastern States. Rich. and Sw., F. B.-A., ii, 472; Nutt., ii, 471; Aud., vii, 20, pl. 422; Lawr. in Bd., 868. . . . . TRACHYRHYNCHUS.

Brown Pelican. Dark-colored, variegated; neck of the adult mostly reddish-brown, head mostly white; bill dark, varied with red; sac blackish; feet black; rather smaller than the last; tail normally 22-feathered. S. Atlantic and Gulf Coasts and California, abundant, strictly maritime. Nutt., ii, 476; Aud., vii, 32, pls. 423, 424; Lawr. in Bd., 870. FUSCUS.

Family GRACULIDÆ. Cormorants.

Bill about as long as head, stout or slender, more or less nearly terete, always strongly hooked at the end; tomia generally found irregularly jagged, but not truly serrate; a long, narrow, nasal groove, but nostrils obliterated in the adult state; gape reaching below the eyes, which are set in naked skin. Gular pouch small, but forming an evident naked space under the bill and on the throat, variously encroached upon by the feathers. Wings short for the order, stiff and strong, the 2d primary usually longer than the 3d, both these exceeding the 1st. Tail rather long, large, more or less fan-shaped, of 12–14 very stiff, strong feathers, denuded to the base by extreme shortness of the coverts; thus almost "scansorial" in structure, recalling that of a woodpecker or creeper, and used in a similar way, as a support in standing, or an aid in scrambling over rocks and bushes. The body is compact and heavy, with a long neck; the general configuration, and especially the far backward set of the legs, is much like that of pygopodous birds. While other *Steganopodes* can stand with the body more or less nearly approaching a horizontal position, the cormorants are forced into a nearly upright posture, when the tail affords with the feet a tripod of support. They also, like the birds just mentioned, dive and swim under water in pursuit of their prey, using their wings for submarine progression, which is not the case with the other families, excepting *Plotidæ*.

Among osteological characters, aside from the general figure of the skeleton, a long bony style in the nape, in the position of the *ligamentum nuchæ* of many animals, and ossified with the occiput, is the most remarkable. It occurs in the anhinga also, but is there much smaller. The desmognathous structure is seen in its highest development; the palatines being not only soldered, but sending down a keel along their line of union; the interorbital septum is very defective, with horizontal inferior border (a general character of the order except in the pelicans). The pterylosis agrees essentially with the ordinal pterylographic characters, but the plumage is peculiar in certain details. Excepting a few speckled species, and some others that are largely white below, the plumage is glossy or lustrous black, often highly iridescent with green, purple and violet tints, commonly uniform on the head, neck and under parts, but on the back and wing coverts, where the feathers are sharp-edged and distinct, the shade is more apt to be coppery or bronzy, each feather with well defined darker border. This concerns, however, only the adult plumage, which is the same in both sexes; the young are plain brownish or blackish. The cormorants have other special featherings, generally of a temporary character, assumed at the breeding season and lost soon after; these are curious long filamentous feathers (considered by Nitzsch filoplumaceous), on the head and neck, and even, in some cases, on the upper and under parts too. These feathers are commonly white, as is also a large silky flank-patch acquired by several species. Many cormorants are also crested with ordinary long slender feathers; the crest is often double, and when so, the two crests may be either one on each side of the head, or they may follow each other on the middle line of the hind head and nape. Our species illustrate all these various featherings. The naked parts about the head vary with the species and afford good characters, especially considering the shape of the pouch, as noted by Mr. Lawrence and Prof. Schlegel; the skin is usually brightly colored, and sometimes carunculate. The eyes, as a rule, are green—a color not common among birds.

Twenty-five species of cormorants may be considered established. Their study is difficult, owing to the great changes in plumage, the high normal variability in size, and their close inter-relation, which is such that the single genus *Graculus* does not appear capable of well founded division. Species are found all over the world, excepting the uttermost polar regions, and are usually very abundant in individuals; they are all very much alike in their habits. Many are maritime, but others range over fresh waters as well. They are eminently gregarious, especially in the breeding season, when they congregate by thousands—the boreal kinds generally on rock-begirt coasts and islands, those of warm countries in the dense fringes of shrubbery. They often migrate in large serried ranks. The nest is rude and bulky; the eggs are commonly two, of elliptical form and pale greenish color, overlaid with a white, chalky substance. They feed principally upon fish, and their voracity is proverbial, though probably no greater than in the cases of allied birds. Under some circumstances they have shown an intelligent docility; witness their semi-domestication by the Chinese, who train them to fish for their masters, a close collar being slipped around the neck to prevent them from swallowing the booty.

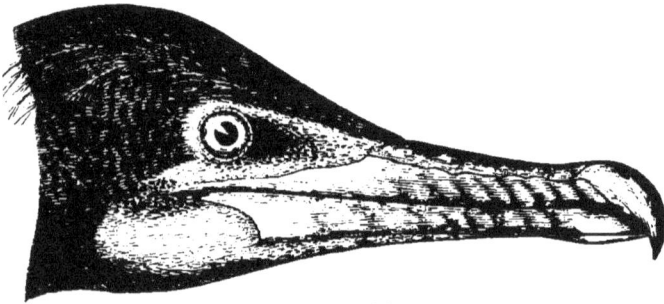

Fig. 198. Double-crested Cormorant.

275. Genus GRACULUS Linnæus.

* *Tail of* 14 *feathers* (and gular sac heart-shaped behind).

Common Cormorant. Shag. Glossy greenish-black, feathers of back and wing coverts bronzy-gray, black-edged; quills and tail grayish-black; gular sac yellow, white-bordered; feet black; in summer a white flank patch, numerous long thready white plumes on head and neck, and a small black occipital crest; length 36; wing 12–14; tail 6–7; tarsus over 2; bill 4 along the gape. Atlantic Coast of Europe and North America; breeds in great numbers in Labrador and Newfoundland; S. to the Middle States in winter. NUTT., ii, 479; AUD., vi, 412, pl. 415; LAWR. in BD., 876. CARBO.

** *Tail of* 12 *feathers.*

† Gular sac convex, or nearly straight-edged, behind.

White-tufted Cormorant. Glossy greenish-black, the back and wing coverts with the feathers gray, black-edged; *lateral* crests, of a superciliary bundle of long curly filamentous feathers, *white.* Size of the last. Alaska. I have never seen this bird, and do not know of any specimen in this country: description compiled from the original account. BRANDT, Bull. Imp. Acad. St. Petersburg, iii, 55; BONAP., Consp. Av. ii, 168; SCHLEGEL, Mus. Pays-Bas, iv, 22; LAWR. in BD., 877; ELLIOT, pl. 51. CINCINNATUS.

Double-crested Cormorant. Glossy greenish-black; feathers of the back and wings coppery-gray, black-shafted, black-edged; adult with curly black *lateral* crests, and in the breeding season other filamentous white ones, over the eyes and along the sides of the neck; white flank-patch not observed in the specimens examined, but probably occurring; gular sac and lores orange. Length 30–33 inches; wing 12 or more; tail 6 or more; bill along gape 3½; tarsus a little over 2. Young plain dark brown, paler or grayish (even white on the breast) below, without head-plumes. N. Am., at large, the commonest species. Sw. and Rich., F. B.-A. ii, 473; Nutt., ii, 483; Aud., vi, 423, pl. 416; Lawr. in Bd., 877. . . . . . . dilophus.

Var. FLORIDANUS. *Florida Cormorant.* Similar, smaller (wing 12 or less; tail 6 or less; tarsus a little under 2), but bill as large if not larger; gape nearly 4. The plumage is exactly the same, excepting, probably, that white plumes are not developed. There are said to be certain differences in the life-colors of the bills (blue instead of yellow on under mandible and edges of upper — *Audubon*), but none show in my specimens. This is simply a localized southern race of *dilophus*, smaller in general dimensions, with relatively larger bill, as usual in such cases; the sac seems to be more extensively denuded. Resident on the Floridan and Gulf coast, breeding by thousands on the mangrove bushes; in summer, ranging up the Mississippi valley to Ohio (*Audubon*) and along the coast to North Carolina (*Coues*). Aud., vi, 430, pl. 417; Lawr. in Bd., 879.

Mexican Cormorant. Resembling the last; lustre more intense, rather violet-purplish than green; long filamentous white feathers on head and neck (but no definite black lateral crests?); sac orange, *white-edged. Small;* length about 24; wing about 10; tail 6, thus relatively long; tarsus under 2; gape of bill under 3. The sac is not strongly convex in outline behind, the feathers passing across in a straight or even convex line. Central America and West Indies; Texas; up the Mississippi to Illinois (*Ridgway*). Brandt, *l. c.* 56; Lawr. in Bd., 879. . . . mexicanus.

† † Gular sac heart-shaped behind, owing to a narrow pointed forward extension of the feathers on the middle line.

Brandt's Cormorant. Deep lustrous green, changing to violet or steel-blue on the neck, the back proper like the under parts, but the scapulars and wing coverts showing narrow dark edgings of the individual feathers (much less conspicuous than in any of the foregoing species: nothing of the sort is seen in any of the following ones). *Sac dark blue, surrounded by a gorget of fawn-colored or mouse-brown plumage,* largely naked, the feathers extending on it little if any in advance of those on the lower mandible. White filamentous plumes, 2 inches or more long, straight and stiffish, spring in a series down each side of the neck; a few others are irregularly scattered over the back of the neck; many others, still longer, grow on the upper part of the back. No black crests, nor white flank-patch, observed. Wing nearly 12; tail scarcely or not 6, thus relatively very short; bill along culmen 2¾; tarsus 2½. Does not particularly resemble any other species here described. *Young:* blackish-brown, rustier below, the belly grayish; scapulars and wing coverts with edges of the feathers paler than the centres ; gorget fawn-colored, as in the adult (*Phalacrocorax townsendii!* AUD., vi, 438, pl. 418).   Pacific Coast, U. S., common. BRANDT, *l. c.* 55 ; GAMBEL, Journ. Phila. Acad. 1849, 227 ; LAWR. in BD., 880. . . . . PENICILLATUS.

Pallas's Cormorant. Deep lustrous green, above and below, with blue gloss on the neck, and rich purplish on the scapulars and wing coverts, the latter not edged ; *shafts of tail feathers* (said to be) *white; if* this holds, it is a unique character among our species.   Adult with coronal and occipital crests (not lateral paired crests) ; a white flank-patch in the breeding season ; face and neck with long sparse straw-yellow plumes ; sac orange.   Large ; 36 ; wing 13 ; tail 7 ? 9 ? tarsus 3 ; bill (along gape ?) 4, very stout, ⅔ deep at base.   N. Pacific Coast.   I have not seen this species, which seems to be well marked.   PALLAS, Zoog. R.-A. ii, 305 ; GOULD, Voy. Sulphur, 49, pl. 32 ; SCHLEGEL, *l. c.* 17 ; BONAP., Consp. Av. ii, 167 ; LAWR. in BD., 877 ; ELLIOT, pl. 50. . . . . PERSPICILLATUS.

Red-faced Cormorant.   Frontal feathers not reaching base of the culmen, the bill being entirely surrounded by naked *red* skin which also encircles the eyes, somewhat carunculate, forming a kind of wattle on each side of the chin ; base of under mandible *blue;* feet black, blotched with yellow. Crown with a median black crest, and nape with another, in the same line. In the specimen examined, a large white flank-patch, but no white plumes on neck.   Plumage richly iridescent, mostly green, but violet and steel-blue on the neck, purplish, violet and bronzy on the back and wings, the feathers there *without* definite dark edgings.   Length 33 ; extent 48 ; wing 12 ; tarsus 2⅔ ; gape of bill 3.   Kadiak, Alaska ; described from the single recognized specimen, No. 52, 512, Mus. Smiths. Inst., the same noticed by BAIRD, Trans. Chicago Acad. i, 321, pl. 33, believed to represent the *Phalacrocorax bicris-latus* of PALLAS, Zoog. R.-A. ii, 183.   Probably the "red-faced cormorant," *Pelecanus urile,* of Pennant, Latham and Gmelin, but as this point cannot be decided, I accept Baird's identification. . . . . BICRISTATUS.

Violet-green Cormorant.   Frontal feathers reaching culmen ; gular sac inconspicuous, very extensively feathered, the feathers reaching on the sides of the under mandible to below the eyes, and running in a point on the sac far in advance of this.   Small ; length 24–28 ; wing 10–11 ; tail 6 *or less;* tarsus 2 *or less;* bill along gape 3 *or less,* very slender, and smooth on the sides, its depth at base about ⅓.   Deep lustrous green, including the back ; the scapulars, wing coverts and sides of the body iridescent with purplish or coppery, the neck with rich violet and blue ; gular sac orange ; feet black ; Two median lengthwise crests as in the last two species.   Among the speci- mens before me, one has no white flank-patch, but a few white scattered plumes on the neck ; another, marked ♀ , has none of these, but a large snowy tuft on the flanks.   A third, labelled *"bairdii,* ♂ , Farallones, Apr.'61," has both the flank tufts and the neck plumes ; it is very small, the wing being under 10, the tarsus 1⅔, the gape 2⅔, and the bill is extremely slender ; it possibly represents a small southern race, bearing somewhat the relation to *violaceus* that *floridanus* does to *dilophus*. Pacific Coast, N. A.— *Pelecanus violaceus* GM., i, 575? *Graculus violaceus* LAWR. in BD., 881; SCHLEGEL, *l. c.* 17; *Urile bicristatus* BONAP., Consp. Av. ii, 175 (ucc Pall.); *Phalacrocorax resplendens* AUD., vi, 430, pl. 419; *G. bairdii* COOP., Proc. Phila. Acad. 1865, 5, 6; ELLIOT, pl. 49. . . . . VIOLACEUS.

Family PLOTIDÆ. Darters.

Bill about twice as long as the head, straight, slender, very acute, paragnathous, the tomia with fine serratures. Gular sac moderate, naked. Nostrils minute, entirely obliterated in the adult. Wings moderate, the 3d quill longest. Tail rather long, stiff, broad and fan-shaped, of 12 feathers widening towards the end, the outer web of the middle pair curiously crimped (in our species).

There is an occipital style, as in cormorants, but it is very small. The digestive system shows a remarkable feature; instead of the lower part of the œsophagus being occupied by the proventricular glands, these are placed in a small distinct sac on the side of the gullet. As in other *Steganopodes*, the gizzard develops a special pyloric cavity. There are no proper cœca, but there is a small rounded termination of the rectum (*Audubon*).

The darters are birds of singular appearance, somewhat like a cormorant but much more slightly built, and with exceedingly long slender neck and small constricted head that seems to taper directly into the bill. As in the cormorants, there are long slender feathers on the neck; the sexes are commonly distinguishable, but the ♀ is said sometimes to resemble the ♂. Other changes of plumage appear to be considerable, but not well made out. The feet are short, and placed rather far back, but the birds perch with ease. Unlike most of the order, they are not maritime, shunning the seacoast, dwelling in the most impenetrable swamps of warm countries. They fly swiftly, and dive with amazing ease and celerity. They are timid and vigilant birds; when alarmed they drop from their perch into the water below, noiselessly and with scarcely a ripple of the surface, and swim beneath the surface to a safe distance before reappearing. When surprised on the water, they have the curious habit of sinking quietly backward, like grebes; and they often swim with the body submerged, only the head and neck in sight, looking like some strange kind of water serpent. They feed on fish, which they do not dart down upon, but dive for and pursue under water like cormorants and loons. The eggs are three or four, pale bluish, with white chalky incrustation. There are only three or four species: the African *P. levaillantii;* the *P. melanogaster* of Southern Asia, with the Australian *P. nova-hollandiæ,* if distinct from the last; with the following:

276. Genus PLOTUS Linnæus.

Darter. Anhinga. Snake-bird. Water-turkey. Glossy greenish-black; a broad gray wing-band formed by most of the coverts; lower neck behind and scapulars speckled with grayish-white; tertiaries striped with silvery ash; tail pale-tipped; filamentous feathers of neck purplish-ash; ♀ with parts of the head, neck and back brown, the jugulum and breast fawn-color sharply margined with rich brown. Length about 36; extent nearly 4; wing 14; tail 11; bill 3⅓; tarsus 1⅓. S. Atlantic and Gulf States, common; in summer to North Carolina (*Audubon*), and up the Mississippi to S. Illinois

(*Kennicott*) ; Fort Thorn, New Mexico (*Henry*). NUTT., ii, 507 ; AUD., vi, 443, pl. 420 ; LAWR. in BD., 883 ; *P. melanogaster* WILS., ix, 79, 82, pl. 74, f. 1, 2. .   .   .   . ANHINGA.

Family TACHYPETIDÆ. Frigates.

Bill longer than the head, stout, straight, wider than high at the base, thence gradually compressed to the strongly hooked extremity. Nostrils very small, linear, almost entirely closed, in a long narrow groove. Gular sac small, but capable of considerable distension. Wings exceedingly long and pointed, of about 34 remiges, of which the 10 primaries are very powerful, with stout quadrangular shafts ; upper and middle portion of the wings greatly lengthened. Tail very long, deeply forked, of 12 strong feathers. Feet exceedingly small, the tarsus, in particular, extraordinarily short (§ 75, p. 45), feathered ; middle claw pectinate. Bulk of body slight compared with the great length of the wings and tail. Here only in this order is found the *os uncinatum*, a peculiar skull-bone occurring in nearly all the petrels, the turacous (*Musophagidæ*, p. 178) and many cuckoos.

The frigates are maritime and pelagic birds of most warm parts of the globe. Their general contour is unique among water-birds, in the immense length and sweep of the wings, length of the forked tail and extreme smallness of the feet. In command of wing they are unsurpassed, and but few birds approach them in this respect. They are more nearly independent of land than any other birds excepting albatrosses and petrels, being often seen hundreds of miles at sea, and delight to soar at an astonishing elevation. They cannot dive, and scarcely swim or walk ; food is procured by dashing down on wing with unerring aim, and by harassing gulls, terns and other less active or weaker birds until they are forced to disgorge or drop their prey. Their habit is gregarious, especially during the breeding season, when thousands congregate to nest in low thick bushes by the water's edge. The nest is a shallow flat structure of sticks ; the eggs, two or three in number, are greenish-white with a thick smooth shell. "The young are covered with yellowish-white down, and look at first as if they had no feet. They are fed by regurgitation, but grow tardily, and do not leave the nest until they are able to follow their parents on wing" (*Audubon*). The following is the principal if not the only species.

277. Genus TACHYPETES Vieillot.

Frigate. Man-of-war Bird. ♂ brownish-black, glossed with green or purplish, duller on the belly, wings showing brown and gray ; ♀ with white on neck and breast. Length about 3½ feet ; extent about 8 ; wing 2 ; tail 1½ ; bill 5 or 6 inches ; tarsi 1 inch or less. S. Atlantic and Gulf Coast. NUTT., ii, 491 ; AUD., vii, 10, pl. 421 ; LAWR. in BD., 873.   ·   AQUILUS.

FIG. 199. Frigate.

Family PHAETHONTIDÆ. Tropic Birds.

Bill about as long as the head, stout, straight, compressed, tapering, acute, paragnathous. Gular sac rudimentary, almost completely feathered. Nostrils small, linear, but remaining patulous. Tail with the two middle feathers in the adult filamentous and extraordinarily prolonged, the rest short and broad.

The tropic birds resemble a large, stout tern in their general figure; the bill, especially, being almost exactly like that of a tern. The principal external peculiarity is the development of the middle tail-feathers; the feathering of the gular sac and the permanent patulence of the nostrils are other features. They are graceful birds on the wing, capable of protracted flight, venturing far from land. They are gregarious at all times, and nest in communities along coasts and on islands, in rocky places or among low trees and bushes. As implied in their name, they are birds of the .torrid zone, though in their extensive wanderings they visit Southern seas, and have even been reported from latitude 40° N. There are but three well determined species: *P. flavirostris* (below); *P. æthereus*, and *P. rubricauda*.

278. Genus PHAETHON Linnæus.

Tropic Bird. White, satiny, rosy-tinted; long tail feathers reddened, black-shafted; sides of head, wings and flanks varied with black; bill orange; tarsi yellow; toes and webs black; young with more black on upper parts. Wing 11; bill 1¾–2; tarsus 1; tail 4–5, its middle feathers up to 15–20. Gulf Coast, rare or casual. *P. æthereus* NUTT., ii, 503; AUD., vii, 64, pl. 427; *P. flavirostris* BRANDT; LAWR. in BD., 885. . . . . FLAVIROSTRIS.

Order LONGIPENNES. Long-winged Swimmers.

Wings long, pointed, reaching when closed beyond the base, in many cases beyond the end, of the tail, which is usually lengthened and of less than 20 rectrices (oftenest 12). Legs more or less perfectly beneath centre of equilibrium when the body is in the horizontal position; the crura more nearly free from the body than in other *Natatores*, if not completely external. 'Anterior toes palmate; hallux never united with the inner toe, highly elevated, directly posterior, very small, rudimentary or absent; tibiæ naked below. Bill of variable form, but never extensively membranous nor lamellate, the covering horny throughout, sometimes dis-. continuous. Nostrils variable, but never abortive. No gular pouch. Altricial.

This order, which may be recognized among web-footed birds by the foregoing external characters, is less substantially put together than either of the two preceding — not that its components are not sufficiently related to each other, but because the essential points of structure are shared to a considerable extent by other groups. Thus the osteological resemblances of longipennine birds with loons, auks, and plover, are quite close, as noted by Huxley; while the digestive system agrees in general characters with that of other fish-eating birds. In some of the lower members of the order, the tibia develops an apophysis, as in the loons; while even in external characters, one genus at least, *Halodroma*, resembles the *Alcidæ*. It is not certain, that the order must not be broken up, or rather enlarged and differently defined, to include some of the genera now ranged under *Pygopodes*.

The palate has the schizognathous structure; "the maxillo-palatines are usually lamellar and concavo-convex, but in the *Procellariidæ* they become tumid and spongy" (*Huxley*); basipterygoid processes may be often wanting, but they are certainly present in many more cases than Huxley supposed. There is apparently one pair of syringeal muscles throughout the order. The œsophagus is capacious and distensible; there is no special crop; the proventriculus is a bulging o the gullet; the gizzard is small and little muscular; the cœca are variable; the cloaca is large. Certain genera offer peculiarities of this general type of alimentary canal. According to Nitzsch, the pterylosis of the gulls "approaches very closely that of the *Scolopacidæ*, and can hardly be distinguished therefrom with certainty by any character." In the terns, "in consequence of the slender and elegant form of the body, the tracts are very narrow, and perfectly scolopacine." The jaëgers differ "in having the outer branch of the inferior tract united with the main stem in the first part of its course, and all the tracts still broader and stronger than in" the gulls; while in the petrels, "the tract formation of the jaëgers is elevated into the type of a group, undergoing scarcely any change in the form of the inferior tract, but showing some little modification of the dorsal tract."

As here constituted, the order embraces two families, to be known by the character of the nostrils; both are well represented in this country.

Family LARIDÆ. Gulls, Terns, etc.

Nostrils not tubular (linear, linear-oblong, oval or drop-shaped), sub-basal or median, lateral, pervious. The hallux, though very small and elevated, with its tip hardly touching the ground, is, except in one instance, better developed than in the petrels. The habitat is fluviatile, lacustrine and maritime, rather than pelagic. The family contains four leading genera, each of which may be assumed as the basis of a subfamily; all four occur in North America.

Subfamily LESTRIDINÆ. Jaëgers, or Skua Gulls.

Covering of bill discontinuous, the upper mandible being saddled with a large horny "cere," beneath the edges of which the nostrils open (unique, among water-birds); bill epignathous. Tail nearly square, but the middle pair of feathers abruptly long-exserted. Feet strong, the podotheca granular or otherwise roughened behind, scutellate in front; webs full. Certain pterylographic characters have been already noted. A leading anatomical peculiarity in the large size of the cœca, as compared with the cases of the other subfamilies. There is but one genus, and only four species are well determined. They belong more particularly to the northern hemisphere, although some also, inhabit southern seas; they mostly breed in boreal regions, but wander extensively at other seasons. They inhabit sea coasts, and also large inland waters; the nidification resembles that of the gulls; eggs, 2-3, dark-colored, variegated. The sexes are alike; the young different, excepting one species; there is also a particular melanotic plumage, apparently a normal transient condition. At first the central tail feathers do not project, and they grow tardily. The skua gulls are eminently rapacious, whence their name of "jaëger" (hunter); they habitually attack and harass terns and the smaller gulls, until these weaker and less spirited birds are forced to drop or disgorge their prey. Their flight is vigorous; lashing the air with the long tail, they are able to accomplish the rapid and varied evolutions required for the successful practice of piracy. Thus in their leading traits they are marine Raptores; whilst the cered bill furnishes a curious analogy to the true birds of prey.

279-80. Genus STERCORARIUS Brisson.

* Bill shorter than middle toe without claw; tarsus shorter than middle toe and claw; central rectrices little projecting, broad to the tip. (*Buphagus*.)

Skua Gull. Length about 2 feet; wing 17 inches; tail 6; tarsus 2¾, middle toe and claw 3; bill about 2, its depth at base ¾. Above, blackish-brown, varied with chestnut and whitish; throat and sides of neck yellowish-brown, streaked with white; below, fusco-rufous, with an ashy shade; quills blackish, with white shafts and a conspicuous large white area at base; tail feathers blackish, white at the base; very old birds are much darker and more uniform brown, almost blackish above, rather smoky brown below. Northern N. Am., rare or casual; "California." *Lestris cataractes* NUTT., ii, 312; *Stercorarius cataractes* LAWR., Ann. Lyc. N. Y. 1853, 71, and in BD., 838; *Buphagus skua* COUES, Proc. Phila. Acad. 1863, 125. . SKUA.

** Bill and tarsi relatively longer than in the foregoing; central rectrices finally projecting far beyond the rest. Smaller and less robust. (*Stercorarius.*)

Pomarine Jaëger. Middle tail feathers finally projecting about 4 inches, *broad to the tip.* Length about 20 inches; wing 14; bill 1½-1¾; tarsus about 2. *Adult:* back, wings, tail, crissum and lower belly blackish-brown, deepening on the top of the head and slight occipital crest to brownish-black; below, from bill to belly, and neck all round, pure white, excepting acuminate feathers of sides of neck, which are pale yellow; quills whitish basally, their shafts largely white; tarsi above blue, below, with the toes and webs, black. *Not quite adult:* as before, but breast with dark spots, sides of the body with dark bars, blackish of lower belly interrupted; feet black. *Younger:* whole under parts, with upper wing and tail coverts, variously marked with white and dark; feet blotched with yellow. *Young:* whole plumage transversely barred with dark brown and rufous; feet mostly yellow. *Dusky stage* (coming next after the barred plumage just given?); fuliginous, unicolor; blackish-brown all over, quite black on the head, rather sooty brown on the belly; sides of the neck slightly gilded. Northern N. Am., ranging S. to the Middle States in winter. Sw. and RICH., F. B.-A. ii, 429; NUTT., ii. 315; AUD., vii, 186, pl. 451; LAWR. in BD., 838; COUES, *l. c.* 129. . . . . POMATORHINUS.

Parasitic, or *Richardson's Jaëger.* Middle tail feathers finally projecting about 4 inches, *tapering, acuminate;* smaller; wing 12-13; tarsus 1¾-1⅞; bill 1¼-1½; tail 5-6, the long feathers up to 9 inches. *Adult:* upper parts, including top of the head and slight occipital crest, and crissum, blackish-brown, deeper on wings and tail; chin, throat, sides of head, neck all round and under parts to the vent, white, the sides of the neck pale yellow; quills and tail feathers with whitish shafts; feet blue and black. *Younger:* clouded below with dusky in variable pattern and amount. *Young:* barred cross-wise with rufous and dusky; feet mostly yellow. There is a fuliginous stage, precisely as in the last species. Northern N. Am.; U. S. in winter. *Lestris richardsonii* Sw., F. B.-A. ii, 433, pl. 73 (in dusky plumage); NUTT., ii, 319 (dusky); AUD., vii, 190, pl. 452; *Stercorarius richardsonii* COUES, *l. c.* 135; *Lestris cepphus* NUTT., ii, 318 (adult); *Stercorarius parasiticus* LAWR. in BD., 839; COUES, *l. c.* 132. . . . . . PARASITICUS.

Arctic, Long-tailed, or *Buffon's Jaëger.* Middle tail feathers finally projecting 8 or 10 inches, very slender and almost filamentous for a great part of their length; smaller still; wing about 12; tail about 6; tarsus $1\frac{1}{2}$–$1\frac{2}{3}$; bill 1–$1\frac{1}{6}$; plumage as in the last. Same habitat. *Lestris parasiticus* Sw. and RICH., F. B.-A. ii, 430; NUTT., ii, 317; AUD., vii, 192, pl. 453; *S. cepphus* LAWR. in BD., 840; *S. buffonii* COUES, *l. c.* 136.   .   .   . BUFFONII.

Subfamily *LARINÆ*. *Gulls.*

Covering of bill continuous, horny throughout: bill more or less strongly epignathous, compressed, with more or less protuberant gonys; nostrils linear-oblong, median or sub-basal, pervious. Tail even or nearly so, rarely forked or cuneate, without projecting middle feathers. Certain of the smaller slenderer-billed species alone resemble terns, but may be known by the not forked tail (except *Xema*); in all the larger species, the hook of the bill is distinctive. Gulls average much larger than terns, with stouter build; the feet are larger and more ambulatorial, the wings are shorter and not so thin; the birds winnow the air in a steady course unlike the buoyant dashing flight of their relatives. They are cosmopolitan; species occur in abundance on all sea coasts, and over large inland waters; in general, large numbers are seen together, not only at the breeding places, but during the migrations, and in winter, when their association depends upon community of interest in the matter of food. This is almost entirely of an animal nature, and consists principally of fish; the birds seem to be always hungry, always feeding or trying to do so. Many kinds procure food by plunging for it, like terns; others pick up floating substances; some of the smaller kinds are adroit parasites of the pelicans, snatching food from their very mouths. They all swim lightly—a circumstance explained by the smallness of the body compared with its apparent dimensions with the feathers on. The voice of the larger species is hoarse, that of the smaller shrill; they have an ordinary note of several abrupt syllables during the breeding season, and a harsh cry of anger or impatience; the young emit a querulous whine. The nest is commonly built on the ground; the eggs, 2–3 in number, are variegated in color.

Several circumstances conspire to render the study of these birds difficult. With few exceptions, they are almost identical in form; while in size they show an unbroken series. Individual variability in size is high; northerly birds are usually appreciably larger than those of the same species hatched further south; the ♂ exceeds the ♀ a little (usually); very old birds are likely to be larger, with especially stouter bill, than young or middle aged ones. There is, besides, a certain plasticity of organization, or ready susceptibility to modifying influences, so marked that the individuals hatched at a particular spot may be appreciably different in some slight points from others reared but a few miles away. One pattern of coloration runs through nearly all the species: they are *white*, with a darker mantle (*stragulum; § 38, p. 17*), and in most cases with black crossing the primaries near the end, the tips of the quills white. The shade of the mantle is very variable in the same species, according to climate, action of the sun, friction and other causes; the pattern of the black on the quills is still more so, since it is *continually* changing with age, at least until a final stage is reached. Incredible as it may appear, species and even genera have been based upon such shadowy characters. One group of species has the head enveloped in a dark hood in the breeding season, the under parts tinted with peach-blossom hue. The sexes are always alike; the moult appears to be twice a year, so that a winter plumage more or less different from that of summer results; while the young are never like the old. The change is slow, generally requiring

2–3 years ; in the interim, birds are found in every stage. They are always *darker* than the old, often quite dusky ; usually with black or flesh-colored bill ; and of those with black on the primaries when adult, the young usually have these quills all black. There being no peculiar extra-limital species, those of our country give a perfect idea of the whole group. Some seventy-five species are currently reported ; there are certainly not over fifty, and I doubt that there are over forty unquestionable species. For these, thirty! generic names have been invented, nine-tenths of which are simply preposterous.

N. B. In using the following descriptions, understand that the color is *white*, unless otherwise stated.

281-5. Genus LARUS Linnæus.

A. Species of largest to medium size, of robust form, with a stout bill, more or less strongly hooked, and protuberant at the symphysis ; *the white of the under parts never rosy-tinted, nor the head enveloped in a dark-colored hood.*

   a. Hind toe well developed, bearing a perfect claw.

      * Tail *of the adult* entirely white.

       † Feet not black ; and with full webs.

        ‡ Primaries without any black.

Glaucous Gull. Ice Gull. Burgomaster. Primaries entirely white, or palest possible pearly-blue fading insensibly into white at some distance from the end, their shafts straw-yellow ; mantle palest pearly-blue ; bill yellow, with vermilion spot on lower mandible ; feet flesh colored or pale yellowish. In winter, head and hind neck lightly touched with dusky. *Young:* impure white, with or without traces of pearly on the mantle ; head, neck and upper parts mottled with pale brownish (sometimes quite dusky on the back), the under parts a nearly uniform but very faint shade of the same, the quills and tail often imperfectly barred with the same ; bill flesh-colored or yellowish, black-tipped. Very large ; length about 30 ; extent 60 ; wing 18, or more ; bill 2¾ *or more;* tarsi 3 *or more.* Arctic America ; S. coastwise in winter to the Middle States. Rich., F. B.-A. ii, 416 ; Nutt., ii, 306 ; Aud., vii, 170, pl. 449 ; Lawr. in Bd., 842.—*L. hutchinsii* Rich., F. B.-A. ii, 419? Coues, Proc. Phila. Acad. 1862, 294, and Proc. Essex Inst. v, 306 ; Elliot, pl. 53 (young). . . . . GLAUCUS.

White-winged Gull. Precisely like the last, but smaller ; length about 24 (rather less than more) ; wing 16–17 ; bill 1¾–2 ; tarsus 2–2¼. Same habitat. Rich., F. B.-A. ii, 418 ; Nutt., ii, 305 ; Aud., vii, 159, pl. 447 ; Lawr. in Bd., 843. . . . . LEUCOPTERUS.

Glaucous-winged Gull. Primaries of the color of the mantle to the very tips, which are occupied by definite small white spots ; the 1st also with a large white sub-apical spot. Mantle average "gull-blue ;" bill yellow with red spot ; feet flesh-colored ; in winter, the head and hind neck clouded with dusky. *Young:* gray, more or less variegated with whitish, chiefly in bars on the back and wings ; bill black, or pale with dark tip. Size and shape of *argentatus;* the adult is exactly like that species, excepting that the primaries have the color of the mantle, instead of black ; the young are much paler than young herring-gulls. I have seen no specimens not instantly distinguishable from the foregoing. Pacific Coast, common; breeding northerly; U. S. in winter. *L. glaucescens* and *L. chalcopterus* (younger) LAWR. in BD., 842, 843; COUES, *l. c.* 295; BONAP., Consp. Av. ii, 216; *Laroides glaucopterus* BRUCH. . . . . GLAUCESCENS.

‡‡ Primaries crossed with black (adult), or all black (young).

Great Black-backed Gull. Saddle-back. Coffin-carrier. Cobb. Feet flesh-colored; bill yellow with red spot. Mantle *blackish slate-color;* 1st primary with the end white for 2–3 inches; 2nd primary with a white sub-apical spot, and, like the remaining ones that are crossed with black, having the tip white (when not quite mature, the 1st with small white tip and sub-apical spot, the 2nd with white tip alone). In winter, head and neck streaked with dusky. *Young:* whitish, variously washed, mottled and patched with brown or dusky; quills and tail black, with or without white tips; bill black. Very large; equalling or even exceeding *L. glaucus.* N. Atlantic; S. along the U. S. coast in winter; Florida (*Audubon*). NUTT., ii, 308; AUD., vii, 172, pl. 450; LAWR. in BD., 844. . . . . MARINUS.

OBS. *L. fuscus,* a European species bearing the same relation to *marinus* that *leucopterus* does to *glaucus,* has been attributed to this country, upon insufficient evidence. BONAP., Synopsis U. S. Birds, No. 298; NUTT., ii, 302.

Herring Gull. Common Gull. Feet flesh color; bill yellow with red spot; mantle pale dull blue (darker than in *glaucus,* but nothing like the deep slate of *marinus*—much the same as in all the rest of the species); primaries marked as in *marinus* (but the great majority of specimens will be found to have the not quite mature or final condition); length 22–27; wing 15–18; tarsus 2¼–2¾; bill about 2¼ long, about ⅔–¾ deep at the base, and about the same at the protuberance. In winter: head and hind neck streaked with dusky. *Young:* at first almost entirely fuscous or sooty-brown, the feathers of the back and wings with paler edges; bill black; quills and tail black, white-tipped or not; size at the minimum above given. As it grows old, it gradually lightens; the head, neck and under parts are usually quite whitish, before the markings of the quills are apparent, and before the blue begins to show, as it does in patches, mixed with brown; the black on the tail narrows to a bar, at the time the primaries are assuming their characters, but this bar disappears before the primaries gain their perfect pattern. At one time the bill is flesh colored or yellowish, black-tipped. The American bird proves to average larger than the European in all its parts, as observed in several other water-birds: whence *L. smithsoni-anus* COUES, *l. c.* 296. N. Am., abundant, both coastwise and in the interior, breeding northward, generally distributed at other seasons. *L. argentatoides* BONAP., Syn. No. 229; RICH., F. B.-A. ii, 417. NUTT., ii, 304; AUD., vii, 163, pl. 448; LAWR. in BD., 844. . . ARGENTATUS.

Var. OCCIDENTALIS. Mantle notably darker, rather slaty-blue than grayish-blue; bill stouter, especially towards the end, the depth at the protuberance usually rather greater than at the base; greatest depth ⅞; at the nostrils ¾. Pacific Coast, abundant. AUD., vii, 161; LAWR. in BD., 845; COUES, *l. c.* 296; ELLIOT, pl. 52. The ordinary Californian bird is distinguished by the above particulars; but connects directly with *argentatus* by the North Pacific strain (*L. borealis* BRANDT; BAIRD, Trans. Chicago Acad. i, 324), and the Siberian bird (*L. cachinnans* PALL.; *L. argentatus* var., MIDDENDORF, SCHRENK).

Ring-billed Gull. Adult plumage precisely like that of the last species, and its changes substantially the same; bill *greenish*-yellow, encircled with a *black band* near the end, usually complete, sometimes defective, the tip and most of the cutting edges of the bill yellow; in high condition, the angle of the mouth and a small spot beside the black, red; *feet olivaceous*, obscured with dusky or bluish, and partly yellow; the webs bright chrome. (Observe the coloration of the feet in this and in *californicus*, as compared with *argentatus*.) Notably smaller than *argentatus;* length usually 18–20 inches; extent about 48; wing about 15; bill *under* 2, and only about ⅓ deep at the protuberance; tarsus about 2, obviously longer than the middle toe. N. Am., abundant and generally distributed. *L. delawarensis* ORD, Guthrie's Geog. 2d Am. ed. ii, 319; LAWR. in BD., 846; *L. canus* NUTT., ii, 299; *L. zonorhynchus* RICH., F. B.-A. ii, 421; NUTT., ii, 300; AUD., vii, 152, pl. 446. COUES, *l. c.* 302. . . . . DELAWARENSIS.

Var. CALIFORNICUS. Apparently larger than ordinary *delawarensis*, and sometimes nearly equalling *argentatus*, averaging perhaps 22 inches; bill about 2, the black band probably never perfect, the red spot more obvious; *feet colored as in the last;* tarsus 2¼. yet not, or not obviously, longer than the middle toe and claw. In all the adult birds observed, the white spot on the 1st primary had enlarged to occupy the whole end of the feather for about 2 inches; while the subapical spot on the 2d was large — a state I have not observed in typical *delawarensis*. Arctic and Western America, abundant. LAWR., Ann. Lyc. N. Y. 1854, 79, and in BD., 846; COUES, *l. c.* 300 (excl. syn.). (Type specimen examined.)

American Mew Gull. Small; length 16–18 inches; extent about 40; wing 13–14; bill 1⅓, slender, its depth hardly or not over ⅓; tarsus about equal to the middle toe and claw, both about 1¾. Bill bluish-green, yellow-tipped, without any red or black; feet dusky bluish-green, webs yellow. Mantle considerably darker than in *delawarensis*. Arctic and Western N. Am., in the interior and along the Pacific Coast to California; I am not aware that it occurs on the Atlantic, or anywhere in the United States east of the Rocky Mountains; Nuttall and Bonaparte seem to refer to the preceding species in giving this range. It will be seen at once to be different from any of the foregoing: and it appears to show constantly some slight discrepancies from the European *L. canus*. *L. canus* (adult) and *L. brachyrhynchus* (young — type specimen examined) RICH., F. B.-A. ii, 420, 422; NUTT., ii, 299, 301; COUES, *l. c.* 302; *Rissa septentrionalis* (adult) and *L. suckleyi* (young — types of both examined) LAWR., Ann. Lyc. N. Y. vi, 265, 264, and in BD., 854, 848. CANUS var. BRACHYRHYNCHUS.

†† Feet black, stout, rough, with short tarsi and excised webs. (*Pagophila*.)

Ivory Gull. Adult plumage entirely pure white, the shafts of the primaries yellow; bill yellow, more or less extensively greenish or dusky toward the base; feet black. *Young:* more or less spotted and patched with blackish, and bill often black. Length 16–20 inches; wing 11–13; bill 1⅛–1½; tarsi about the same, and rather shorter than the middle toe and claw. Quite different from any other species. Arctic America and Europe, coastwise, rarely S. to U. S. in winter. Sw. and Rich., F. B.-A. ii, 419; Nutt., ii, 301; Aud., vii, 150, pl. 445; *Pagophila eburnea* and *P. brachytarsi* Lawr. in Bd., 856; Coues, *l. c.* 308, 309. . . . . . . eburneus.

** Tail of the adult almost entirely black. (*Blasipus*.)

White-headed Gull. Adult with the head white, gradually merging on the neck and under parts into pale ash; mantle dark plumbeous; upper tail coverts whitish; ends of secondaries and tertiaries white; primaries and tail feathers black, some of the former usually with white specks at the end, the latter white at extreme tip and base; bill red, black-tipped; feet dark. This is the final plumage; but the blanching is very gradual and tardy, a more usual condition being leaden-gray all over, the mantle slate-gray, the quills and tail black. Very young birds are fuliginous brown, paler or grayish below, the feathers of the upper parts with lighter brown edges, the bill pale with dark tip. Length 16–20 inches; wing 13–14. Pacific Coast, U. S. and southward, abundant. *L. belcheri* Vigors, Zool. Voy. Blossom, iv, 358; *L. fuliginosus* Gould, Zool. Voy. Beagle, Birds, 141; *L. heermanni* Cass., Proc. Phila. Acad. vi, 1852, 157, and Ill. 28, pl. 5; *B. heermanni* Lawr. in Bd., 848; Coues, *l. c.* 304. . . . . . belcherii.

b. Hind toe rudimentary or minute, usually without perfect claw. (*Rissa.*)

Kittiwake Gull. Hind toe only appearing as a minute knob, its claw abortive. Mantle rather dark grayish-blue; 1st primary with the whole outer web, and the entire end for about 2 inches, black; next one, with the end black about as far, but outer web elsewhere light, and a white speck at extreme tip; on the rest of the primaries that have black, this color decreases in extent proportionally to the shortening of the quills, so that the base of the black on all is in the same line when the wings are closed (a pattern peculiar to the species of *Rissa*); and these all have white apex. Bill yellow, usually clouded with olivaceous; feet dusky olivaceous. Rather small; 16–18; wing 12 long; bill 1⅛–1½; tarsus about the same; middle toe and claw longer; tail usually slightly emarginate. In winter, nape and hind neck shaded with the color of the mantle. Young: bill black; a black bar on the tail, another across the neck behind; wings and back variously patched with black; dark spots before and behind the eyes; quills mostly black. Arctic America and Europe, chiefly coastwise, very abundant; in winter, commonly S. to the Middle States; breeds from New England northward. Sw. and Rich., F. B.-A. ii, 423; Nutt., ii, 298; Aud., vii, 146, pl. 444; Lawr. in Bd., 854; Coues, *l. c.* 304. . tridactylus.

Var. kotzebui. It is a curious fact, that the common kittiwake of the North Pacific usually has the hind toe better formed — sometimes nearly if not quite as long as in ordinary gulls, with a nearly or quite perfect, though small, claw. But I cannot predicate a specific character on this score, since the development of the toe is by insensible degrees. See Coues, Proc. Phila. Acad. 1869, 207 (footnote). Bonap., Consp. Av. ii, 226; Coues, Proc. Phila. Acad. 1862, 305; Elliot, pl. 54.

Short-billed Kittiwake.   Red-legged Kittiwake.   Bill very short, stout, wide and deep at the base, with very convex culmen; its color *clear yellow; feet coral-red, drying yellow;* tarsus only about two-thirds as long as the middle toe and claw; hind toe very small (little if any larger than in an Atlantic kittiwake, smaller than in the best marked var. *kotzebui*), its rudimentary claw showing as a little black speck.   I do not know the young bird, in which the color of the bill and feet is probably materially different. Adult with the mantle leaden-gray, much darker than in the common kittiwake; pattern of the primaries essentially the same as in that species.   Wing 13; bill 1⅛-1¼, its depth at base ⅓, at angle little less; tarsus 1¼; middle toe and claw nearly 2.   North Pacific Coast, abundant.   This is unquestionably a different bird from the foregoing, and in adult plumage it would seem impossible to mistake it.   Here belong the following names: — *Rissa brevirostris* BRANDT; LAWR. in BD., 855; DALL and BANN., Trans. Chicago Acad. i, 305 (breeding by thousands about St. George's, Alaska); *Larus brachyrhynchus* GOULD, Proc. Zool. Soc. 1843, p.   , and Zool. Voy. Sulphur, 50, pl. 34; *Rissa brachyrhyncha* BONAP., Consp. Av. ii, 226; COUES, Proc. Phila. Acad. 1862, 306; *R. brevirostris* and *R. nivea* LAWR. in BD., 855; *R. nivea* ELLIOT, pl. 54 (not *Larus niveus* PALL.).   BREVIROSTRIS.

B.   Species of medium to smallest size, of less robust form and slenderer bill than most of the foregoing; in the breeding season *the white of the under parts rosy-tinted, and the head enveloped in a dark-colored hood.* (*Chrœcocephalus.*)

Black-headed, or *Laughing Gull.*   Tarsus one-fourth longer than middle toe and claw.   Large; 16-19; wing 12-13; tarsus 2; middle toe and claw 1½; bill about 1¾, the tip elongated and decurved, so that the point comes down nearly or quite to the level of the small, acute prominence of the gonys. Mantle grayish-plumbeous; hood dark plumbeous; eyelids white; black on primaries taking in nearly all the 1st quill, but rapidly decreasing to the 6th; the white tips very small, few, or wanting; bill and feet dusky carmine.   In

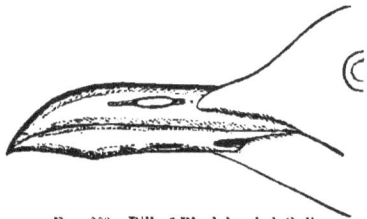

FIG. 200.   Bill of Black-headed Gull.

winter: not rosy, and unhooded; head white, with dusky or grayish patches on the nape and auriculars.   *Young:* quite brown, paler, grayish or whitish below and on the upper tail coverts; feathers of the back dark with paler edges; quills and tail black, or latter white or partly grayish-blue, with a black bar; bill and feet dusky or brownish.   United States, chiefly coastwise, breeding northward to Bay of Fundy (*Boardman*), but more abundantly southward; extremely numerous along the South Atlantic coast. New Mexico and Arizona (*Coues*); Pacific Coast (*Xantus*).   *Larus ridibundus* WILS., ix, 89, pl. 74, f. 4; *L. atricilla* NUTT., ii, 291; AUD., vii, 136, pl. 443; LAWR. in BD., 850.   . . . . ATRICILLA.

Franklin's Rosy Gull.   Tarsus about equal to the middle toe and claw. Medium; 14-16; wing about 11; bill 1¼-1⅛; tarsus 1⅔; bill and feet carmine, former usually with a black mark near the end; mantle bluish-plumbeous, the ends of the secondaries white nearly an inch; hood blackish-plumbeous, with white eyelids.   Final pattern of primaries :—shaft of 1st entirely white, of next 5 white except in the portion of the quill occupied by black; 1st with its outer web and a bar on the inner web, black, leaving the tip wholly white an inch or more, rest of the feather pearly white; next 5 crossed by a black bar on both webs, 2–3 inches wide on the 2nd quill, narrowing to a mere spot on the 6th; tips of all these broadly white. Younger birds have much more black on the wing, in a different pattern, and the tail washed with bluish (*Ch. cucullatus* LICHT.; LAWR. in BD., 851, pl. 95; COUES, Proc. Phila. Acad. 1862, 309).   Central America and Mexico in winter, migrating in the interior, west of the Mississippi, to the Arctic regions; abundant; has not been observed in the Atlantic States. *Larus franklini* RICH., F. B.-A., ii, 424, pl. 71; NUTT., ii, 293; AUD., vii, 145; *Ch. franklini* LAWR. in BD., 851. . . . . . . FRANKLINII.

Bonaparte's Gull. Tarsus about equal to middle toe and claw. Small; 12–14; wing 9½–10½; tarsus 1⅜; bill 1⅛–1¼, very slender, like a tern's. Adult in summer: *bill black;* mantle pearly blue, much paler than in the foregoing; hood slaty-plumbeous, with white touches on the eyelids; many wing coverts white; feet chrome yellow, tinged with coral red; webs vermilion. Primaries finally :—the first 5–6 with the shafts white except at tip; 1st white, with outer web and extreme tip black; 2d white, more broadly crossed with black; 3d to 6th–8th with the black successively decreasing. In winter, no hood, but a dark auricular spot. *Young:* mottled and patched above with brown or gray, and usually a dusky bar on the wing; the tail with a black bar, the primaries with more black, the bill dusky, much of the lower mandible flesh-colored or yellowish, as are the feet. N. Am.; breeds in the Arctic regions; very abundant in the U. S. during the migration. *Sterna philadelphia* ORD, Guthrie's Geog. 2d Am. ed. ii, 319; *Ch. philadelphia* LAWR. in BD., 852; *L. bonapartei* RICH., F. B.-A., ii, 425, pl. 72; NUTT., ii, 294; AUD., vii, 131, pl. 442; COUES, *l. c.* 310.   PHILADELPHIA.

OBS. The sexes of this gull are alike, as in all other cases. Audubon is wrong in figuring the ♀ with a brown hood. But it is a question whether the "brown-headed gull," *Larus capistratus* of BON., Syn. p. 358, No. 293—NUTT., ii, 290, should be considered as this species, or as the true European bird, *L. ridibundus*, erroneously attributed to this country. The European Least Gull, *L. minutus*, has been introduced to our fauna upon erroneous information, the single authority (SABINE) for its occurrence having doubtless mistaken the last species for it. RICH., F. B.-A. ii, 426; NUTT., ii, 289; LAWR. in Bd. 853. See COUES, *l. c.* 311.

286. Genus RHODOSTETHIA Macgillivray.

Wedge-tailed, or *Ross' Rosy Gull.*   Adult : white, rosy-tinted; a black collar, but no hood; mantle pearly-blue; primaries marked with black; bill black; feet vermilion; length 14; wing 10½; "bill along the ridge ¾," very slender; tarsus little over 1; tail 5½, *cuneate*, the graduation being one inch. Arctic America, apparently very rare; I have never seen a specimen, and do not know of any in this country. Rich., F. B.-A. ii, 427; Nutt., ii, 295; Aud., vii, 130; Lawr. in Bd., 856; Coues, *l. c.* 311. . rosea.

287-8. Genus XEMA Leach.

Fork-tailed Gull. Adult: white, including inner primaries, most of secondaries, and greater coverts: head enveloped in a slate-colored hood, succeeded by a velvety-black collar; mantle slaty-blue, extending quite to the tips of the tertiaries; whole edge of the wing, and first 5 primaries, black, their extreme tips, and the outer half of their inner webs to near the end, white; bill black, tipped with yellow; feet black; length 13–14; wing 10–11; bill 1; tarsus $1\frac{1}{4}$; tail 5, *forked* an inch or more. The changes of plumage are correspondent with those of *L. philadelphia;* in the young the tail is often simply emarginate. Arctic America, both coastwise and in the interior, common, but still rare in collections; in winter, S. occasionally to New York (*Audubon*) and Utah (*Allen*). Rich., F. B.-A. ii, 428; Nutt., ii, 296; Aud., vii, 127, pl. 441; Lawr. in Bd., 857; Coues, *l. c.* 311. sabinei.

Swallow-tailed Gull. Head and nearly all the neck grayish-brown; a white spot on each side of the forehead; mantle grayish-white; lesser wing coverts white, greater slate, white-bordered; bill black at the base, white at the end, much bent; eyes and feet red; eyelids orange; claws black; tail white, very much forked. Length about 2 feet. "California." This bird appears to be exceedingly rare; no one in this country has seen it. The description is compiled from the original account. *Larus furcatus* Neboux, Rev. Zool. 1840, 290; Prevost and Des Murs, Voy. Venus, pl. 10; *Creagrus furcatus* Bonap.; Lawr. in Bd., 857; Coues, *l. c.* 312. . . . . . . furcatum.

Fig. 201. Roseate Tern.

Subfamily STERNINÆ. Terns.

Covering of bill continuous (no cere), hard and horny throughout; bill *paragnathous*, relatively longer and slenderer than in the gulls, very acute, the commissure straight or nearly so to the very end; nostrils generally linear. Tail never square, almost invariably forked (often deeply forficate), in one group double-rounded. Wings extremely long, thin and pointed. Feet small, weak, scarcely ambulatorial.

The terns are not distinguished from the gulls by any strong structural peculiarities, but they invariably show a special contour, in the production of which the longer, slenderer and acutely paragnathous bill is a conspicuous element. Only one species has the bill in any noticeable degree like that of a gull. A few of the terns are as large as middle-sized gulls, but the normal stature is much less; and they are invariably of a slenderer build, more trim in shape, with smoother, closer-fitting plumage. The great length and sharpness of the wing relative to the bulk of the body confer a dash and buoyancy of flight wanting in the gulls; in flying over the water in search of food, they hold the bill pointing straight downward, which makes them look curiously like colossal mosquitoes; and they secure their prey by darting impetuously upon it, when they are usually submerged for a moment. The larger kinds feed principally upon little fish, procured in this way; but most of the smaller ones are insectivorous, and flutter about over marshy spots like swallows or nighthawks. The general appearance and mode of flight have suggested the name of "sea-swallow," the equivalent of which is applied in nearly all civilized languages. A forking of the tail is an almost universal character. In the Caspian and marsh terns, the black tern and its allies, and some others, the forking is moderate, and not accompanied by attenuation of the lateral feathers; but ordinarily, these are remarkably lengthened and almost filamentous, as in the barn swallow. It should be observed that in all such cases the narrowing elongation is gradual, and consequently less evident in the young; and that it is very variable in its development. The noddies offer the peculiarity of a tail lightly forked centrally, but rounded laterally. The feet are small and relatively weak throughout the group; the terns walk but little, and scarcely swim at all. Ordinarily the webbing is rather narrow, and excised, particularly that between the middle and inner toe; in *Hydrochelidon*, this occurs to such extent that the toes seem simply semipalmate. The webs are fullest in *Anous*, where also the hallux is unusually long; in some species, this toe is slightly connected with the tarsus by a web. The inner toe is shorter than the outer, and much less than the middle, which, especially in *Hydrochelidon*, is much lengthened, and has the inner edge of its claw dilated, or even slightly serrate. The coloration is very constant, almost throughout the subfamily. Most of the species are white (often rosy-tinted below), with a pearly-blue mantle, a black cap on the head, and dark-colored primaries, along the inner web of which *usually* runs a white stripe. These dark-colored quills, when new, are beautifully frosted or silvered over; but this hoariness being very superficial, soon wears off, leaving the feathers simply blackish. The black cap is often interrupted by a white frontal crescent; it is sometimes prolonged into a slight occipital crest; in a few species, it is replaced by a black bar on each side of the head. One species, *Inca mystacalis*, has a curious bundle of curly white plumes on each side of the head. Another, *Gygis alba*, is pure white all over; *Procelsterna cinerea* is wholly ashy: the noddies are all fuliginous; the upper parts of *Haliplana* are dark; the species of *Hydrochelidon* are largely black. These are the principal if not the only exceptions to the normal coloration just given. The sexes are never distinguishable, either by size or color; but nearly all the species, in the progress toward maturity, undergo changes of plumage, like gulls; while the seasonal differences are usually considerable. As a rule, the black cap is imperfect in young and winter specimens, and the former show gray or brown patching instead of the pure final color of the mantle. In all those species in which the bill is red, orange or yellow, it is more or less dusky in the young. The changes are probably greatest in the black terns.

The general economy is much the same throughout the group. The eggs are laid in a slight depression on the ground—generally the shingle of beaches, or in a tussock of grass in a marsh, or in a rude nest of sticks in low thick bushes; they are 1-3 in number, variegated in color. Most of the species are maritime, and such is particularly the case with the noddies; but nearly all are also found inland. They are noisy birds, of shrill penetrating voice; and no less gregarious than gulls, often assembling in multitudes to breed, and generally moving in company. Species occur near water in almost every part of the world, and most of them are widely distributed; of those occurring in North America, the majority are found in corresponding latitudes in the Old World. About seventy species are currently reported; these must be reduced nearly one-half; the true number is apparently just about that of the gulls. Some twenty "genera" have been imposed upon the terns — three-quarters of these are of no account whatever.

N. B. Understand *white, the pileum black, the quills silvered-dusky with long white stripe*, unless the descriptions state otherwise.

289-92. Genus STERNA Linnæus.

* Bill remarkably short, stout and obtuse, hardly or not half as long again as the tarsus. (*Gelochelidon.*)

Gull-billed, or *Marsh Tern*. Bill and feet black; mantle pearly grayish-blue, this color extending on the rump and tail; primaries with the white stripe restricted to their base, their shafts white. Length 13–15; extent about 34; wing 10–12; tail 4, forked only 2 or less, the lateral feathers little narrowed; tarsi 1–1¼; bill 1⅜. Eastern United States; apparently not abundant in this country. Europe, etc. *S. aranea* WILS., viii, 143, pl. 72, f. 6; LAWR. in BD., 859; *S. anglica* NUTT., ii, 269; AUD., vii, 81, pl. 430; *Gelochelidon anglica* COUES, Proc. Phila. Acad. 1862, 536. . . ANGLICA.

** Bill of an ordinary sternine character.

† Occiput slightly crested. Feet black. Size large. (*Thalasseus.*)

Caspian Tern. Bill red. Mantle pearly grayish-blue; cap extending below the eyes, but the under eyelid white; primaries *without* any white band. In winter, black of the cap chiefly restricted to the occiput; young, with the bill dusky and yellowish, the back, wings and tail patched with brown or blackish. Much the largest of the terns; length 20 or more; wing 15–17; tail 5–6, moderately forked, without narrowed feathers; bill 2¼–2¾, very stout, ¾ or more deep at base, ½ wide opposite nostrils; tarsus 1⅜–1¾; middle toe and claw rather less. Arctic America and Europe, S. in winter to the Middle States; apparently not abundant in this country. LAWR. Ann. Lyc. N. Y., 1851, v, 37; COUES, *l. c.* 537 (var. *imperator*), and Proc. Essex Inst. v, 308; ELLIOT, pl. 56. . . . . CASPIA.

Royal Tern. Bill orange. Mantle pearly grayish-blue. In winter, bill duller colored; cap mostly restricted to occiput; rump and tail shaded with the color of the mantle. Young, with the crown much like that of the adults in winter; upper parts without bluish, or this only showing in patches, and variously spotted with dusky. Scarcely *shorter* than the last, owing to length of tail, but much less bulky; length 18–20; wing 14–15; tail 6–8, deeply forked, with narrowed lateral feathers; tarsus about 1¼, middle toe and claw rather more than less; bill 2½–2¾ (in the young sometimes only ˙2¼), ½–¾ deep at base, the gonys about 1 long. Atlantic Coast, U. S., to New York (*Lawrence*), abundant southerly; California? *S. cayana* NUTT., ii, 268; AUD., vii, 76, pl. 429; *S. regia* GAMBEL, Proc. Phila. Acad. 1848, 128; LAWR. in BD., 859; *Thalasseus regius* COUES, *l. c.* 538. REGIA.

Elegant Tern. Similar to the last; mantle very pale; under parts rosy-tinted in high plumage. Smaller and somewhat differently proportioned; bill much slenderer; tarsus obviously longer than middle toe and claw. Length about 17; wing 12–13; tail 6–7; bill 2½, under ½ deep at base, the gonys about 1½ long; tarsus rather over 1; middle toe and claw under 1.

California, Mexico, Cent. and S. Am., and Africa. *S. galericulata* Licht., Verz. 1823, 81; *S. elegans* Gamb., Proc. Phila. Acad. 1848, 129; Lawr. in Bd. 860; *Thalasseus elegans* Coues, *l. c.* 540. . . . GALERICULATA.

· *Sandwich Tern.* Bill black, tipped with yellow. Plumage as in *regius* or *galericulata*, but mantle extremely pale; smaller; length 15–16; wing about 12; tail 5–6; bill $2–2\frac{1}{3}$, the yellow part from $\frac{3}{4}$ of an inch to a mere point; tarsus 1; middle toe and claw $1\frac{1}{8}$. Atlantic and Gulf Coast of U. S., abundant. Europe. *S. boysii* Nutt., ii, 276; *S. cantiaca* Aud., vii, 87, pl. 431; *S. acuflavida* Cabot, Proc. Bost. Soc. Nat. Hist. 1847, 257; Lawr. in Bd., 860; *Thalasseus acuflavidus* Coues, *l. c.* 540. . . . CANTIACA.

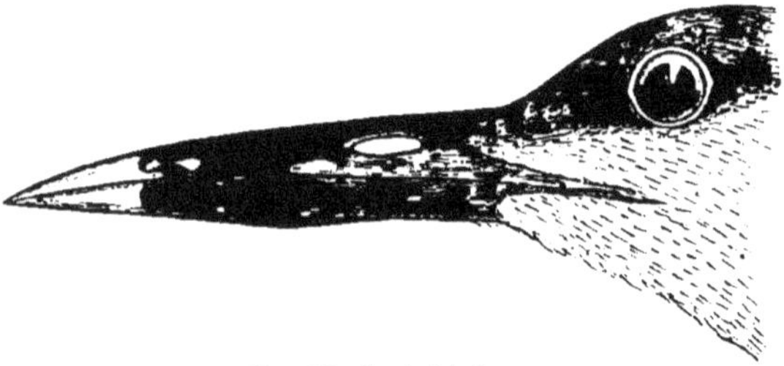

Fig. 202. Sandwich Tern.

†† Occiput not crested. Feet not black. Medium and small. (*Sterna.*)

Common Tern. Wilson's Tern. Sea Swallow. Bill red, blackening on the terminal third, the very point usually light; feet coral-red. Mantle pearly grayish-blue; primary shafts white except at the end; below white, washed with pale pearly plumbeous, blanching on throat and lower belly. Tail mostly white, the *outer* web of the outer feather darker than the inner web of the same. Length of ♂ $14\frac{1}{2}$ (13 to 16); extent 31 (29–32); wing $10\frac{1}{2}$ ($9\frac{3}{4}$–$11\frac{3}{4}$); tail 6 (5–7); tarsus $\frac{3}{4}$ ($\frac{2}{3}$–$\frac{7}{8}$); bill $1\frac{2}{3}$ ($1\frac{1}{4}$–$\frac{1}{2}$); whole foot averaging $1\frac{3}{4}$; ♀ rather less; averaging toward these minima: young birds may show a little smaller, in length of tail particularly, and so of total length; length 12+; wing 9+; tail 4+; bill $1\frac{1}{8}$+. In winter, this species does not appear to lose the black cap, contrary to a nearly universal rule. *Young:* bill mostly dusky, but much of the under mandible yellowish; feet simply yellowish; cap more or less defective; back and wings patched and barred with gray and light brown, the bluish showing imperfectly if at all, but this color shading much of the tail; usually a blackish bar along the lesser coverts, and several tail feathers dusky on the *outer* web; below, pure white, or with very little plumbeous shade. N. Am., abundant; breeds at various points along the Atlantic States, and northward. It does not differ in the least from the European. *S. hirundo* Wils., vii, 76, pl. 60, f. 1; Aud., vii, 97, pl. 433; Nutt., ii, 271; *S. wilsoni* Lawr. in Bd., 861. . HIRUNDO.

Forster's Tern. Like the last; larger, tail longer and wings shorter. Wing of adult $9\frac{1}{2}$–$10\frac{1}{2}$; tail $6\frac{1}{2}$–8, thus often beyond the extreme of *hirundo*, and nearly as in *macroura;* bill $1\frac{2}{3}$ ($1\frac{1}{2}$–$1\frac{3}{4}$), and about $\frac{2}{5}$ deep at base (in *hirundo* rarely if ever so deep); tarsus seldom down to $\frac{7}{8}$; whole foot about 2. Little or no plumbeous wash below; *inner* web of the outer tail feather darker than the outer web of the same. Young and winter birds may be distinguished from *hirundo* at gunshot range; the black cap is almost entirely wanting, and in its place is a broad black band on each side of the head through the eye; several lateral tail feathers are largely dusky on the *inner* webs; their outer webs are white. (*Sterna havelli!* AUDUBON, vii, 103, pl. 334.) N. Am., at large, abundant. *S. hirundo* Sw. and RICH., F. B.-A. ii, 412; *S. forsteri* NUTT., ii, 274; LAWR. in BD., 862. See COUES, Proc. Phila. Acad. 1862, 543, 544; ID., *ibid.* 1871, 44. . . . FORSTERII.

Arctic Tern. Bill carmine or lake-red throughout; feet vermilion. Plumage like that of *hirundo*, but much darker below, the plumbeous wash so heavy that these parts are but little if any paler than the mantle; crissum pure white in marked contrast; the throat and sides of the neck pale or white. In winter, cap defective; in young, the same; upper parts patched with gray, brown or rufous; under parts paler or white; a dark bar on the wing; outer

FIG.203. Foot of Forster's Tern.

webs of several tail feathers dusky; bill blackish or dusky-red with yellow on the under mandible; feet dull orange. Smaller than *hirundo*, but tail much longer. Length 14–17; extent 28–30; wing 10–12; tail 5–8; bill $1\frac{1}{4}$-$1\frac{2}{3}$; *tarsus only* $\frac{1}{2}$-$\frac{2}{3}$; whole foot about $1\frac{1}{4}$. Europe; N. Am., especially coastwise and northerly; breeds plentifully in New England and northward; abundant in Alaska. *S. arctica* Sw. and RICH., F. B.-A. ii, 414; NUTT., ii, 275; AUD., vii, 107, pl. 436; *S. macroura* LAWR. in BD., 862; COUES, *l. c.* 549. . . . . MACROURA.

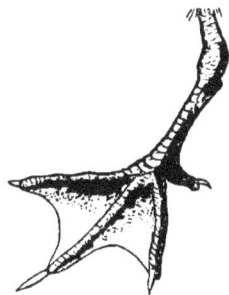

Pike's Tern. Bill black, or reddish-black, the point often whitish. Plumage resembling, that of *hirundo*, and size about the same; wings and tail relatively longer; bill $1\frac{1}{4}$-$1\frac{1}{4}$, very slender, $\frac{1}{4}$-$\frac{2}{3}$, high at the base; tarsi $\frac{1}{2}$-$\frac{2}{3}$. Pacific Coast, N. Am. I have never seen an adult, nor indeed any authentic specimen of this bird; but the type of *Sterna pikei* (a young bird, in poor condition) which I have examined, seems almost unquestionably referable here; if not this species, it is a young *macroura*. *S. longipennis* NORDMANN, Verz. 1835, 17; MIDDENDORF, Sibirische Reise, 246, pl. 25, f. 4; SCHLEGEL, M. P.-B. *Sternæ*, 23. *S. pikei* LAWR., Ann. Lyc. 1853, 3 and in BD., 863; COUES, *l. c.* 550. . . . . LONGIPENNIS.

Roseate Tern. Bill black, usually orange at base below. Mantle very pale pearly blue; primaries with the white band broad and usually extending to the very tip; below, pure white, or rosy-tinted; feet coral-red. Changes of plumage as in other species. Length 12–16; wing 9–10; *tail* 5–8; bill $1\frac{1}{4}$-$1\frac{2}{3}$, very slender; tarsus $\frac{3}{4}$-$\frac{4}{5}$. Atlantic Coast, U. S., abundant. *S. dougallii* NUTT., ii, 278; AUD., vii, 112, pl. 437; *S. paradisea* LAWR. in BD., 863; COUES, *l. c.* 551. . . . . PARADISÆA.

Least Tern. Bill yellow, usually tipped with black. Mantle pale pearly grayish-blue, extending unchanged on the rump and tail; *a white frontal crescent*, separating the black cap from the bill, bounded below by a black loral stripe reaching the bill; shafts of two or more outer primaries *black* on the upper surface, white underneath; feet orange. Young; cap too defective to show the crescent; bill dark, much of the under mandible pale; feet obscured. Very small, only 8–9; wing 6–6½; tail 2–3½; bill 1–1¼; tarsus ⅔. U. S. and somewhat northward, chiefly coastwise, abundant. Appears to be perfectly distinct from the European bird. *S. minuta* WILS., vii, 80, pl. 60, f. 2; AUD., vii, 119, pl. 439. *S. superciliaris* VIEILLOT, Dict. Deterv. 1819, xxxii, 176. *S. argentea* MAXIM., Voy. i, 67; NUTT., ii, 280. *S. antillarum* and *melanorhyncha* LESSON, 1847; COUES, *l. c.* 552. *S. frenata* GAMB., Proc. Phila. Acad. 1848, 128; LAWR. in BD., 864.   SUPERCILIARIS.

Trudeau's Tern. Bill orange, crossed by a blackish band, the tip yellow. Entire plumage pearly grayish-blue, little if any paler below than above but whitening on the head; a black band through the eye; no black cap. Size and proportions precisely as in *forsteri* (excepting shorter tail?). South and Central America, rare or only casual on the Atlantic Coast (New Jersey and Long Island, *Trudeau*). AUD., vii, 105, pl. 435; LAWR. in BD., 861; COUES, *l. c.* 542. . . . . TRUDEAUI.

††† No occipital crest. Feet and bill *black;* colors darker than in any of the foregoing. Size medium. (*Haliplana.*)

FIG. 201. Aleutian Tern.

Aleutian Tern. Top of the head black, with a white frontal crescent; back very dark ash, or dull slaty-blue; under parts similar, paler; tail white; chin and sides of head, edge and lining of wings, and shafts of primaries, white. Length about 14; wing 10½; tail 7½, forked nearly 4; bill 1½; tarsus .55; middle toe and claw 1½. Alaska; one specimen known. A remarkable species, entirely different from any other known to me; it stands exactly between *Sterna* proper and *Haliplana*, and appears related to *S. lunata* PEALE (CASS., U. S. Expl. Exp. 1858, 382). BAIRD, Trans. Chicago Acad., i, 1869, 321, pl. 31, f. 1. . . . . ALEUTICA.

Sooty Tern. Brownish-black, continuous from head to tail; under parts, outer web of outer tail feather, and a frontal crescent, white. The frontal lunule is short and wide, its horns not reaching beyond the eyes; the black loral stripe does not quite reach the bill. Very young birds are fuliginous, speckled with white. Length 15–17; wing 11–12; tail 6–8; bill 1½–2; tarsus ⅞. Atlantic and Gulf Coast, southerly; breeds by thousands in Florida, with the noddies. WILS., viii, 145, pl. 72, f. 7; NUTT., ii, 284; AUD., vii, 90, pl. 432; LAWR. in BD., 861. . . . . FULIGINOSA.

Bridled Tern. Slaty-gray, blackening on crown and quills, the color of the head separated from that of the back by an ashy-gray interval on the cervix; white frontal crescent very narrow, with long horns reaching beyond the eyes, involving the upper eyelid and forming a superciliary line; black loral stripe reaching the bill; under parts, and most of 2–3 outer tail feathers, white; smaller than the last, and easily distinguished. Central America, and various warmer parts of the world; I introduced it to our fauna upon the strength of a specimen from Audubon's collection, now in Mr. Lawrence's cabinet, labelled "Florida." *S. anosthætus* SCOPOLI (tide

Gray) ; *S. panayensis* of AUTHORS : *Haliplana discolor* COUES, Ibis, 1864, 392 ; LAWRENCE, Ann. Lyc. N. Y. viii, 105 ; ELLIOT, pl. 57. ANOSTHÆTA.

293. Genus HYDROCHELIDON Boie.

Black, or *Short-tailed Tern.* Adult in breeding plumage : head, neck and under parts, uniform jet-black ; back, wings and tail, plumbeous ; primaries unstriped ; crissum pure white ; bill black. In winter and young birds, the black is mostly replaced by white on the forehead, sides of head and under parts, the crown, occiput and neck behind, with the sides under the wings, being dusky gray ; a dark auricular patch and another before the eye ; in a very early stage, the upper parts are varied with dull brown. Small ; wing 8–9, little less than the whole length of the bird ; tail 3½, simply forked ; bill 1–1½ ; tarsus ⅔ ; middle toe and claw 1⅛. N. Am., chiefly inland, breeding in marshy places. *S. plumbea* WILS., vii, 83, pl. 60, f. 3 (young) ; *H. plumbea* LAWR. in BD., 864 ; *S. nigra* NUTT., ii, 282 ; AUD., vii, 116, pl. 438 ; *H. fissipes* COUES, *l. c.* 554. . . FISSIPES.

294. Genus ANOUS Leach.

Noddy Tern. Frontal feathers in convex outline on the bill (the antiæ, shown by all the foregoing, here wanting) ; webs remarkably full ; tail graduated laterally, emarginate in the middle, the feathers broad and stiffish. Plumage fuliginous, blackening on quills and tail, with a plumbeous cast on the head and neck, the crown more or less purely white ; bill black ; length 15–17 ; wing 10–11 ; tail 6–7 ; bill 1½–1¾ ; tarsus 1 ; middle toe and claw 1⅔–1⅜. S. Atlantic and Gulf Coasts, breeding in vast multitudes ; the nest is placed on bushes. NUTT., ii, 285 ; AUD., vii, 123, pl. 440 ; LAWR. in BD., 865. *A. stolidus* and *A. frater* COUES, *l. c.* 558. . . . STOLIDUS.

FIG. 205. Foot of Black Tern.

Subfamily RHYNCHOPINÆ. Skimmers.

Bill hypognathous. Among the singular bills of birds that frequently excite our wonder, that of the skimmers is one of the most anomalous. The under mandible is much longer than the upper, compressed like a knife-blade ; its end is obtuse ;

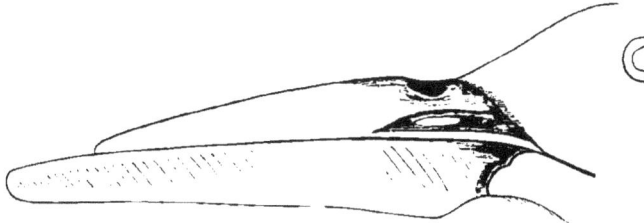

FIG. 206. Bill of Skimmer.

its sides come abruptly together and are completely soldered ; the upper edge is as sharp as the under, and fits a groove in the upper mandible ; the jawbone, viewed apart, looks like a short-handled pitch-fork. The upper mandible is also compressed, but less so, nor is it so obtuse at the end; its substance is nearly hollow, with light cancellated structure, much as in a toucan; it is freely movable by means of an elastic hinge at the forehead. There are cranial peculiarities. Conformably with the shape of the mouth, the tongue differs from that of other *Longipennes* in being very short and stumpy, as in kingfishers, and the *Steganopodes*. The wings are exceedingly long, and the flight more measured and sweeping than that of terns; the birds fly in close flocks moving simultaneously, rather than in straggling companies. They seem to feed as they skim low over water, with the fore parts inclined downward, the under mandible probably grazing or cutting the surface; but they are also said to use their odd bill to pry open weak bivalve mollusks. The voice is very hoarse and raucous, rather than strident. They are somewhat nocturnal or at least crepuscular; their general economy is the same as that of terns, as are all points of structure excepting those above specified. Besides the following, there are only two species: *R. flavirostris* and *R. albicollis*, of Asia.

295. Genus RHYNCHOPS Linnæus.

Black Skimmer.   Cut-water.   Glossy black, the forehead, sides of head and neck and all under parts pure white, or rosy-tinted; tail ashy and white; bill red, black-tipped; feet orange.   Young: grayish-black or dull brown above, varied with white; bill yellow, dusky-tipped.   Length 16–20 inches; extent 3–4 feet; wing 13–15; tail 4–5, forked; under mandible $3\frac{1}{4}$–$4\frac{1}{2}$, upper $2\frac{1}{2}$–3.   Coast of South Atlantic and Gulf States, very abundant; frequently to the Middle States, and even straying to New England.   WILS., vii, 85, pl. 60, f. 4; NUTT., ii, 264; AUD., vii, 67, pl. 428; LAWR. in BD., 866. NIGRA.

Family PROCELLARIIDÆ.   Petrels.

Nostrils tubular.   Bill epignathous; its covering discontinuous, consisting of several horny pieces separated by deep grooves.   Hallux small, elevated, functionless, appearing merely as a sessile claw, often minute, absent in two genera. These are oceanic birds, rarely landing except to breed, unsurpassed in powers of flight, and usually strong swimmers; excepting the sea-runners, none of them dive.   With the same exception, the wings are long, strong and pointed, of 10 stiff primaries and numerous short secondaries; the humeral and ante-brachial portions are sometimes extremely lengthened.   The tail is short or moderate, of less than 20 feathers, of variable shape.   The feet are usually short, with long full-webbed front toes, and a rudimentary hallux, as above stated, or none.   In size, these birds vary remarkably, ranging from that of a swallow up to the immense albatrosses, probably unsurpassed by any birds whatever in alar expanse, and yielding to few in bulk of body.   The plumage is compact and oily, to resist water; the sexes appear to be always alike, and no seasonal changes are determined; but some color variation with age, or according to individual peculiarities, certainly occurs in most cases, and in the *Puffini*, for instance, in which some currently admitted species are uniformly fuliginous, it is not proven that this feature is not temporary, as in the jaegers.   The food is entirely of an animal nature, and fatty substances, in particular, are eagerly devoured; when irritated, many species eject an oily fluid from the mouth or nostrils, and some are so fat they are occasionally used for lamps, by running a wick through the body.   The eggs are few, or only one, laid in a rude nest, or none, on the ground or in a burrow.   Petrels are silent birds, as a rule, contrasting with gulls and terns in this particular; many or most are gregarious, congregating by thousands at their breeding places or where food is plenty.

Birds of this family abound on all seas; but the group is yet imperfectly known. Bonaparte gave 69 species, in 1856; my memoirs upon the subject (1864-6) present 92, of which 17 are marked as doubtful or obscure; last year, Gray recorded 112; there are probably about seventy good species. They are sharply divided by the character of the nostrils into three groups; two represented in North America, as beyond, and the *Halodrominæ*. These last, consisting of one genus and three species or varieties, are remarkably distinguished from the rest, resembling auks in external appearance and habits; the wings and tail are very short; there is no hind toe; the skin of the throat is naked and distensible; the tubular nostrils, in fact, are the principal if not the only petrel-mark, and these organs are unique in opening directly upward, the nasal tube being vertical instead of horizontal as in all the rest.

Subfamily DIOMEDEINÆ. *Albatrosses.*

Nostrils disconnected, placed one on each side of the bill near the base. No hind toe. Of largest size in this family. There are eight unquestionable species, with two or three doubtful or obscure ones. Only three have proven their right to a place here. As Mr. Lawrence observes (BD., 821), there is no well authenticated instance of the occurrence of the great wandering albatross, *D. exulans*, off our coasts; but it has been taken in Europe, and is liable to appear at any time. It is distinguished from the first species following by its great size, and the outline of the frontal feathers: deeply concave on the culmen, strongly convex on the sides of the bill to a point nearly opposite the nostrils. The yellow-nosed albatross, *D. chlororhyncha* of AUDUBON, vii, 196; LAWR. in BD., 822, is the *D. culminata*, a species of Australian and other Southern seas, *said* to have been taken "not far from the Columbia river," but there is no reason, as yet, to believe it ever comes within a thousand miles of this country. It has the bill black with the culmen and under edge yellow. Other well known species of Southern seas are *D. chlororhyncha, cauta* and *melanophrys*.

296-7. Genus DIOMEDEA Linnæus.

* Sides of under mandible smooth; bill very stout, moderately compressed, with rounded culmen, the feathers running nearly straight around its base. Wing three or more times as long as the rounded tail. (*Diomedea*.)

Short-tailed Albatross. Bill 5 or 6 inches long, with moderately concave culmen and prominent hook. Tail very short, contained about $3\frac{1}{2}$ times in the wing. Length about

FIG. 207. Short-tailed Albatross.

3 feet; extent 7; wing 20 inches; tail $5\frac{1}{2}$; tarsi $3\frac{3}{4}$. Adult plumage white, with a yellowish wash on the head and neck; primaries black; other quills, the wing coverts and tail feathers, marked with blackish; bill and feet pale. Young dark colored, resembling the next species. Off the Pacific Coast, abundant. CASS., Ill. 289, pl. 50; LAWR. in BD., 822. . . BRACHYURA.

Black-footed Albatross. Bill about 4 (never 5) inches long, extremely stout, with the culmen almost perfectly straight to the hook, which is comparatively small and weak; the horny piece forming the culmen very broad, especially at base, where it overlaps the lateral piece; depth of bill at base 1⅓, its width there 1¼. Tail contained about 3 times in the wing. General dimensions of the last species, or rather less; tail longer. Adult plumage dark brown, paler and grayer, or rather plumbeous below, lightening or even whitening about the head; quills black with yellow shafts; bill dark; feet black. A final plumage may be lighter than as described, but is never white; and other characters seem to prove the validity of the species. Pacific Coast, very abundant. AUD., vii, 198; SCHLEGEL, M. P.-B., *Procellariæ*, 33; SWINHOE, Ibis, 1863, 431; COUES, Proc. Phila. Acad. 1866, 178; CASS., Ill. 210, pl. 35. *D. gibbosa* GOULD? . NIGRIPES.

** Sides of under mandible with a long colored groove; bill comparatively slender, strongly compressed, with sharp culmen; frontal feathers forming a deep reëntrance on the culmen, a strong salience on the sides of the lower mandible. Wing about twice as long as the cuneate tail. (*Phœbetria*.)

FIG. 208. Sooty Albatross.

Sooty Albatross. Fuliginous brown, nearly uniform, in some cases lightening on various parts; quills and tail blackish with white shafts; eyelids white; bill black, the groove yellow; feet yellow. Length about 3 feet; wing 20–22 inches; tail 10–11, its graduation 3½–4½; tarsi 3; bill 4–4½, at base 1½ deep, but only ¾ wide. *D. fusca* AUD., vii, 200, pl. 454; *D. fuliginosa* LAWR. in BD., 823; *Phœbetria fuliginosa* COUES, *l. c.* 186. FULIGINOSA.

Subfamily *PROCELLARIINÆ*. *Petrels.*

Nostrils united in one double-barrelled tube laid horizontally on the culmen at base. Hallux present, though it may be minute. Five groups of petrels may be distinguished, although they grade into each other; four of them are abundantly represented on our coasts. The *fulmars* are large gull-like species (one of them might be taken for a gull were it not for the nostrils), usually white with a darker mantle, the tail large, well formed (of 14–16 feathers), the nasal case prominent, with a thin partition. They shade into the group of which the genus *Estrelata* is typical, embracing a large number of medium sized species, chiefly of Southern seas, in which the bill is short, stout, very strongly hooked, with prominent nasal case; the tail rather long, usually graduated. The *shearwaters* have the bill longer than usual, comparatively slender, with short low nasal case, obliquely truncate at the end, and the partition between the nostrils thick; the tail short and rounded;

the wings extremely long; the feet large. The elegant little "Mother Carey's chickens" or "stormy petrels" (genus *Thalassidroma* of authors; gen. 303-8, beyond) are a fourth group, marked by their small size, slight build, and other characters; their flight is peculiarly airy and flickering, more like that of a butterfly than of ordinary birds; they are almost always seen on wing, appear to swim little if any, and some, if not all, breed in holes in the ground, apparently like bank-swallows. Like other petrels they gather in troops about vessels at sea, often following their course for many miles, to pick up the refuse of the cook's galley. Some of them, like gen. 307, have remarkably long legs, with fused scutella, flat obtuse claws. and the hallux exceedingly minute; in the rest, the feet are

FIG. 209. Stormy Petrel (Leach's).

of an ordinary character. The exotic genus *Prion* typifies a fifth group. of five or six species; here the bill is expanded, and furnished with strong laminæ, like a duck's; the colors are bluish and white.

298-300. Genus FULMARUS Leach.

* Tail 16-feathered; bill longer than the tarsus. (*Ossifraga.*)

Giant Fulmar. The largest of the petrels, equalling most of the albatrosses in size; length 3 feet; extent 7; wing 20 inches; tail 8; bill 4, the nasal case nearly 2. Plumage dark dingy gray, paler below, often whitening in places; bill and feet yellow. Pacific Coast; "common off Monterey" (*Cooper*). NUTT., ii, 329; AUD., vii, 202; LAWR. in BD., 825. GIGANTEUS.

** Tail 12-14-feathered; bill not longer than the tarsus. (*Fulmarus* and *Priocella.*)

Fulmar. Bill obviously shorter than the tarsus. Adult white, the mantle pale pearly blue, frequently extending on the neck and tail; quills blackish-brown; usually a dark spot before the eye; bill yellow, feet the same tinged with greenish. Young: smoky gray, paler below, the feathers of the back and wings dark-edged; colors of bill and feet obscured. Length usually about 16½, but from 15 to 18; wing 11–13; tail 4–5; tarsus about 2; bill 1⅓-1⅔, about ¾ deep and almost as wide at base; nasal tubes ⅔. Extraordinarily abundant in the North Atlantic; S. to U. S. in winter. NUTT., ii. 331; AUD., vii, 204, pl. 455; LAWR. in BD., 825. . . . . . GLACIALIS.

Var. PACIFICUS will probably average considerably darker on the mantle, with a weaker bill. N. Pacific Coast. AUD., vii, 208; LAWR. in BD., 826; COUES, Proc. Phila. Acad. 1866, 28.

Var. RODGERSII. The mantle dark, as in *pacificus*, but much restricted, most of the wing coverts and inner quills being white; primaries mostly white on the inner webs, their shafts yellow. A particular condition of the last variety? N. Pacific Coast. CASS., Proc. Phila. Acad. 1862. 290; COUES, *ibid.* 1866, 29; BD., Trans. Chicago Acad. i, 323, pl. 34, f. 1.

Slender-billed Fulmar. Bill little if any shorter than the tarsus. Adult white, with pearly blue mantle; primaries pearly whitish basally, white-tipped, crossed with definite black, much as in a herring gull; usually a small dark spot before the eye; feet yellow; bill yellow, obscured on the tube, at tip, and often at base. Changes of plumage as in the foregoing; size the same, but bill 2 long, scarcely $\frac{3}{4}$ wide or high at the base, the tube about $\frac{3}{8}$ long. Pacific Coast; only casual? *P. tenuirostris* AUD., vii, 210; LAWR. in BD., 826; *P. glacialoides* SMITH; *Thalassoica glacialoides* REICH.; COUES, *l. c.* 30; *P. smithii* SCHLEGEL; *Priocella garnoti* HOMB. and JACQ. . . . TENUIROSTRIS.

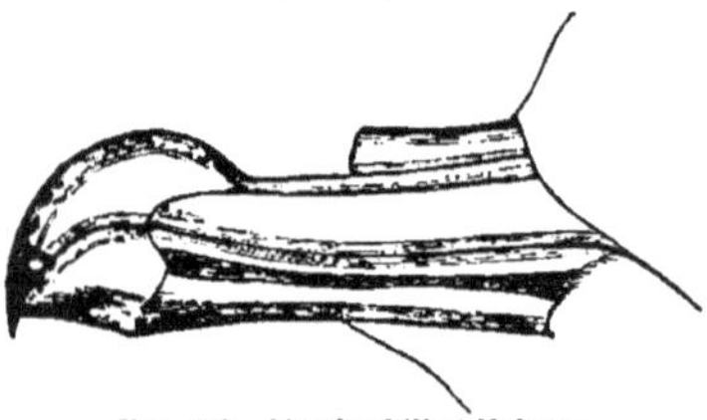

FIG. 210. Slender-billed Fulmar.

301. Genus DAPTION Stephens.

Pintado Petrel. Cape Pigeon. Speckled above with blackish and white; white below; tail black-barred; bill black; 15; wing 11; tail $4\frac{1}{2}$; bill $1\frac{1}{3}$; tarsus $1\frac{2}{3}$. Accidental on the Coast of California. LAWR., Ann. Lyc. N. Y. 1853, 6, and in BD., 828. . . . . CAPENSIS.

302. Genus ÆSTRELATA Bonaparte.

Black-capped Petrel. Adult: forehead, sides of head, neck all round, upper tail coverts, base of tail and all under parts, white; back clear bistre-brown (nearly uniform, but the feathers often with paler or ashy edges), deepening on the quills and terminal half of tail; crown with an isolated blackish cap, and sides of head with a black bar (younger birds with the white of the head and neck behind restricted, so that these dark areas run together); bill black; tarsi and base of toes and webs, flesh-colored (drying yellowish); rest of toes and webs black. Young extensively dark below? Length 16; wing 12; tail $5\frac{1}{4}$, cuneate, its graduation $1\frac{1}{2}$: tarsus $1\frac{2}{5}$; middle toe and claw $2\frac{1}{8}$; bill $1\frac{2}{5}$, $\frac{3}{5}$ deep at base, $\frac{2}{5}$ wide; tube $\frac{1}{3}$. Of casual occurrence on the Atlantic Coast, U. S. *P. hæsitata* KUHL, Monog. 142, No. 11; TEMM., Pl. Col. No. 416; NEWTON, Zool. x, 1852, p. 3691; SCHLEG., M. P.-B. 13; *Æ. hæsitata* COUES, Proc. Phila. Acad. 1866, 139; *P. meridionalis* LAWR., Ann. Lyc. N. Y. iv, 475; v. 220, pl. 15; in BD., 827. HÆSITATA.

303. Genus HALOCYPTENA Coues.

Wedge-tailed, or *Least Petrel.* Blackish, more fuliginous below, the greater wing coverts more grayish, the quills, tail, bill and feet black; no white anywhere. Length $5\frac{3}{4}$; wing $4\frac{3}{4}$; tail $2\frac{1}{2}$, *cuneate*, the graduation $\frac{1}{3}$; bill $\frac{1}{2}$, its height at base $\frac{1}{5}$; tarsus .90; tibiæ bare $\frac{1}{3}$. Lower California; one specimen known (No. 11, 420, Mus. Smiths. Inst.). COUES, Proc. Phila. Acad. 1864, 79; ELLIOT, pl. 61. . . . . MICROSOMA.

304. Genus PROCELLARIA Linnæus.

Stormy Petrel. Mother Carey's Chicken. Coloration of the last species, but upper tail coverts *white*, with black tips, and usually some white under the tail and wings; no yellow on the webs; tail a little *rounded*. About the size of the last. Common off the Atlantic Coast. NUTT., ii, 327; AUD., vii, 228, pl. 461; LAWR. in BD., 831. . . . . PELAGICA.

305. Genus CYMOCHOREA Coues.

Leach's Petrel. Coloration as in the last species, with conspicuous white upper tail coverts, but apt to be lighter — rather of a grayish or even ashy tint on some parts. Much larger: length about 8; wing $6-6\frac{1}{2}$; tail $3-3\frac{1}{2}$, *forked;* tarsus about 1; middle toe and claw the same; bill $\frac{2}{3}$, strong. Both coasts; abundant on the Atlantic. *P. leucorrhoa* VIEILL., Nouv. Dict. xxv, 422; *C. leucorrhoa* COUES, *l. c.* 76. *Thalassidroma leachii* NUTT., ii, 326; AUD., vii, 219, pl. 459; LAWR. in BD., 830. . . LEUCORRHOA.

FIG. 211. Black Petrel.

Black Petrel. Coloration as in the last species, but no white anywhere. Very large; 9; extent $18\frac{1}{2}$; wing $6\frac{3}{4}$; tail 4, *forked* an inch or more; tarsus $1\frac{1}{4}$; bill $\frac{3}{5}$. Coast of California. *Procellaria melania* BONAP., Compt. Rend. 1854, 662; *C. melania* COUES, *l. c.* 76 (described from No. 13,025, Mus. S. I.). . . . . . MELANIA.

Ashy Petrel. Somewhat similar to the last, like it having no white anywhere, but plumbeous rather than fuliginous, and *much* smaller. Length about $7\frac{1}{4}$; wing about 5; tail $3\frac{1}{4}$, *forked* about $\frac{1}{2}$ an inch; tarsus under an inch; bill $\frac{1}{2}$. California. *Thalassidroma melania* LAWR. in BD., 829, pl. 90 (nec BONAP.). *C. homochroa* COUES, *l. c.* 77; ELLIOT, pl. 87. HOMOCHROA.

306. Genus OCEANODROMA Reichenbach.

Fork-tailed Petrel. Bluish-ash, paler below and on the greater wing coverts, dusky around the eye; quills and tail brownish, outer web of the external tail feather white; bill black; feet dark. Length about 8; wing 6; tail $3\frac{3}{4}$, *forked;* bill $\frac{2}{3}$; tarsus $\frac{7}{8}$. N. Pacific Coast, common. CASS., Ill. i, 294, pl. 47; LAWR. in BD., 829. . . . . FURCATA.

Hornby's Petrel. "Front, cheeks, throat, collar round hind part of neck, breast and abdomen, pure white," quills black, other parts dark gray. Size of the last. N.W. coast. I have never seen this species, of which there are not to my knowledge any specimens in this country. *Thalassidroma hornbyi* GRAY, Proc. Zool. Soc. 1853, 62; LAWR. in BD., 829. . . . HORNBYI.

307. Genus OCEANITES Keys. and Blas.

Wilson's Petrel. Dark sooty brown, pale gray on the wing coverts; the upper tail coverts, and frequently the crissum and sides of rump and base of tail, *white;* bill and feet black, but webs with a *yellow* spot. Legs very long; tibiæ bare an inch or more; tarsi "booted," much longer than the toes; claws flat, obtuse; bill small and weak; hind toe very minute, liable to be overlooked. Length 7–8; wing about 6; tail 3, *nearly even;* tarsus $1\frac{1}{3}$; middle toe and claw $1\frac{1}{8}$; bill only $\frac{1}{2}$. Atlantic Coast, common. *Proc. oceanica* KUHL, Monog. 136, pl. 10, f. 1; *Oceanites oceanica* COUES, *l. c.* 82; *Proc. pelagica* WILS., vii, 90, pl. 60, f. 6; *Thalassidroma wilsoni* NUTT., ii, 322; AUD., vii, 223, pl. 460; LAWR. in BD., 831. . OCEANICA.

308. Genus FREGETTA Bonaparte.

White-bellied Petrel. Blackish-gray of variable intensity, blackening on the quills and tail, the whole under parts from the breast, the upper tail coverts, most of the under wing coverts, and bases of all the tail feathers except the middle pair, white; bill and feet black. Length about 8; wing 6–6½; tail 3, about even, with very broad, square-tipped feathers; bill ½; tarsus 1⅓; longest toe (outer) and claw 1 or less; tibiæ bare 1 or more. Florida, accidental, one instance (LAWR. Ann. Lyc. N. Y. v, 117). *Procellaria grallaria* VIEILL. *Procellaria fregetta* KUHL, and many authors. *Thalassidroma leucogastra* GOULD. *Fregetta lawrencii* BONAP.; LAWR., in BD., 832 (unquestionably this species). . . . . GRALLARIA.

309-10. Genus PUFFINUS Brisson.

* Nasal tubes vertically truncate, with thin septum. (*Priofinus.*)

Black-tailed Shearwater. Upper parts cinereous, nearly uniform, but some of the feathers with paler edges; under parts white, without line of demarcation from the color of the upper parts; *tail, crissum and vent blackish;* lining of wings, axillars, and some feathers on the sides of the body, brownish-cinereous; quills blackish-cinereous on outer webs and tips, paler internally and basally, with brown shafts; bill yellow, the nasal case, culmen as far as the hook, cutting edge and groove of lower mandible, black; feet (dried) dingy greenish with yellow webs. Large; 19; wing 13; tail 5–5¾, wedge-shaped, 12-feathered, the outer feathers an inch or more shorter than the middle; bill 1⅘, ⅔ high at base, the nasal tubes nearly ⅓; tarsus 2⅖; middle toe and claw 2⅞. Accidental off the coast of California. A peculiar species, very different from any of the following, approaching the fulmars. *Proc. cinereus* GM.? *Proc. melanura* BONN. *Proc. hæsitata* FORST., Descr. Anim. 1844, 208; GOULD, B. Aust. pl. 67; *Puffinus hæsitatus* LAWR., Ann. Lyc. N. Y. vi, 5. *Proc. adamastor* SCHLEGEL. *Adamastor typus* BONAP. *Puffinus cinereus* LAWR. in BD., 835; *Adamastor cinereus* COUES, Proc. Phila. Acad. 1864, 119; *Priofinus cinereus* COUES, Proc. Essex Inst. v, 303. *Puffinus kuhlii* CASS., Proc. Phila. Acad. 1862, 327 (err.). . MELANURUS.

** Nasal tubes obliquely truncate, with thick septum. (*Puffinus.*)

† Below, white or nearly so, the upper parts different.

Cinereous Shearwater. Above, pale brownish-ash, interrupted by paler or white edges of the feathers, most of the upper tail coverts white; below, entirely pure white, except some slight gray touches on the flanks; on the sides of the head and neck the ash and white gradually mingling; lining of wings and axillars white; quills dark with large white spaces on the inner webs; bill and feet mostly yellowish. Younger birds are darker, the bill and feet obscured. Length about 18; wing 13; tail 5½, outer feathers nearly an inch shorter; bill 1⅞, ⅜ high at base, nasal tube only about ⅕ its length; tarsus 1⅞; middle toe and claw 2½. A common bird of the North Atlantic, not hitherto introduced into our fauna. *Proc. kuhlii* BOIE, Isis, 1835,

257. *Puffinus kuhlii* BONAP., Consp. Av. ii, 202; COUES, Proc. Phila. Acad. 1864, 128. (*Proc. cinereus* GM.?) . . . . KUHLII.

Greater, or *Wandering Shearwater*. Dark bistre-brown, somewhat plumbeous on the head, most feathers of the back and wings with pale edges, most upper tail coverts partly white; below, white, with a plain line of demarcation from the color of the upper parts on the side of the head and fore neck, and dark flank patches; quills and tail blackish, paler or whitish at bases of inner webs; lining of wings mostly white; crissum mostly dark; *bill dark;* outside of tarsi and outer toe dark, rest of feet pale. Length 18–20; extent 45; wing 13; tail $5\frac{3}{4}$, outer feathers an inch less; bill 2; tarsus $2\frac{3}{5}$; middle toe and claw $2\frac{4}{5}$. Whole Atlantic coast, abundant. *P. cinereus* NUTT., ii, 334; AUD., vii, 212, pl. 456; *P. major* LAWR. in BD., 833. . MAJOR.

Flesh-footed Shearwater. Similar to the last; no white on upper or under tail coverts or bases of quills; bill yellowish flesh color, with dark tube, culmen and hook, short, very stout at base, with turgid tube; *feet flesh color*. Size of the last, but bill only $1\frac{5}{8}$, height or width at base nearly $\frac{3}{8}$; tarsus $2\frac{1}{8}$; middle toe and claw $2\frac{3}{8}$. San Nicholas Island, Cal. (No. 31,964, Mus. Smiths. Inst.). A doubtful species; I have little faith in its validity, but cannot refer it to any species known to me; it looks like one of the following section (††) passing to a bicolor plumage. COUES, *l. c.* 131. . CREATOPUS.

Manks Shearwater. Blackish, this color extending below the eyes, leaving the under eyelid white; under parts, including crissum and lining of wings, white; bill greenish-black; outside of foot mostly blackish, inner side dingy orange; about 15; extent 33; wing $9\frac{1}{2}$; tail 4, graduated $\frac{3}{4}$; bill $1\frac{1}{3}$–$1\frac{1}{2}$, but nearly $\frac{1}{2}$ deep at base; tarsus under 2; middle toe and claw 2 or rather less. Very distinct from the rest. N. Atlantic Coast, common. NUTT., ii, 336; AUD., vii, 214, pl. 457; LAWR. in BD., 834. ANGLORUM.

Dusky Shearwater. Resembling the last, but rather grayish- or plumbeous-black, the dark color not reaching below the eyes; crissum mostly white; bill dark leaden-blue; much smaller and otherwise distinct. Length about a foot; extent 26 inches; wing $7\frac{1}{2}$–8; tail $4\frac{1}{4}$, graduated an inch; bill $1\frac{1}{4}$; tarsus $1\frac{3}{5}$; middle toe and claw under 2. S. Atlantic Coast, as far as the Middle States, common. NUTT., ii, 337; AUD., vii, 216, pl. 458; LAWR. in BD., 835. . . . . OBSCURUS.

Black-vented Shearwater. Like the last; crissum and lining of wings mostly blackish; sides of head dark below the eyes; rather larger; bill $1\frac{3}{5}$; wing 9; tail $3\frac{3}{4}$, thus shorter; tarsus $1\frac{3}{4}$; middle toe and claw $2\frac{1}{3}$. Cape St. Lucas (Nos. 16,990–1, Mus. Smiths. Inst.). Seems to be distinct from the last, but may be the same as an exotic species of prior name. COUES, Proc. Phila. Acad. 1864, 139. . . . . OPISTHOMELAS.

††Below, dark, much like the upper parts.

Sooty Shearwater. Dark sooty brown, blackening on the quills and tail, paler and grayish below, usually with some whitish on the lining of the wings; bill dark; feet dark outside, pale on the inner aspect. Length 18; extent 40; wing 12; tail 4; bill $1\frac{3}{4}$–2; tarsus $2\frac{1}{5}$–$2\frac{1}{4}$; middle toe and claw

2⅜. North Atlantic, abundant; S. at least to Carolina (*Coues*). A special state of *P. major?* STRICKLAND, Proc. Zool. Soc. 1832, 129; DEKAY, New York Birds, 287, pl. 136, f. 298; LAWR. in BD., 834; COUES,

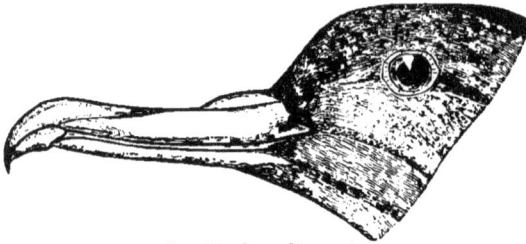

FIG. 212. Sooty Shearwater.

Proc. Phila. Acad. 1864, 123. . FULIGINOSUS. *Dark-bodied Shearwater.* Similar to the last; feet flesh color, slightly obscured outside; lining of wings mostly white; smaller; wing 11, etc. Cape St. Lucas. Very doubtful. I allow this, and several others, to stand, because it is still uncertain what reduction of the species of this genus will prove necessary. *Nectris amaurosoma* COUES, *l. c.* 124. . . . . AMAUROSOMA.

Slender-billed Shearwater. Plumage as in the foregoing; size less; tail shorter; bill smaller. Bill dusky-greenish, with yellow; feet yellowish, blackish behind and under the webs. Length about 14; wing 10; tail 3½, graduated an inch; bill 1¼; tarsus under 2; middle toe and claw 2¼. N. Pacific Coast. TEMM., Planches Color. No. 587; TEMM. and SCHL., Fn. Japon. 131, pl. 86; BONAP., Consp. Av. ii, 202; COUES, *l. c.* 126; BAIRD, Trans. Chicago Acad. i, 1869, 322, pl. 34, f. 2. *P. tristis, curilicus* and *brevicauda* of authors? . . . . TENUIROSTRIS.

Order PYGOPODES. Diving Birds.

In the birds of this order the natatorial plan reaches its highest development. All the species swim and dive with perfect ease; many are capable of remaining long submerged, and of traversing great distances under water, progress being effected by the wings as well as by the feet. Few other birds, such as cormorants and anhingas, resemble the *Pygopodes* in this respect. The legs are so completely posterior, that in standing the horizontal position of the axis of the body is impossible; the birds rest upright or nearly so, the whole tarsus being often applied to the ground, while the tail affords additional support; progression on land is awkward and constrained, only accomplished, in most cases, with a shuffling motion, when the belly partly trails on the ground. The penguins, and one species of auk, cannot fly — the former, because the wings are reduced to mere flippers with scaly feathers, the latter because the wings, although perfectly formed, are too small to support the body. The rest of the order fly swiftly and vigorously, with continuous wing-beats. The rostrum varies in shape with the genera; but it is never extensively membranous, nor lamellate, nor furnished with a pouch. The nostrils vary, but are neither tubular nor abortive. The wings are short, never reaching when folded to the end of the tail, and often not to its base. The tail is short, never of peculiar shape, generally of many feathers; there are, however, no perfect rectrices in the grebes. The crura are almost completely buried, and feathered nearly or quite to the heel. The tarsus is usually compressed, sometimes, as in the loons, extremely so; in the penguins, on the contrary, it is much broader across than in the opposite direction, being nearly as wide as long. The front toes are completely palmate in the loons,

auks and penguins, lobate with basal webbing in the grebes ; the hallux is present
and well formed, with a membranous expansion, in loons and grebes, very minute
and lateral in position in the penguins, wanting in the auks. The plumage is thick
and completely water-proof; once observing some loons under peculiarly favorable
circumstances in the limpid water of the Pacific, I saw that bubbles of air clung to
the plumage whilst the birds were under water, giving them a beautiful spangled
appearance. The pterylosis shows both contour and down-feathers, both after-
shafted ; in the penguins the feathers are implanted evenly over the whole skin ; in
the rest there are definite apteria ; the auks have free outer branches of the inferior
tract-bands, wanting in the loons and grebes. The oil-gland is large with several
orifices. Among osteological characters should be particularly mentioned the long
apophysis of the tibia found in the loons (fig. 8) and grebes, but not in the auks
and penguins : in the latter, the patella is of great size, and it is stated to develop
from two centres. In penguins and auks, the elbow has two sesamoids ; among the
former, there is a free ossicle in the heel joint. The thoracic walls are very extensive ; the long jointed ribs grow all along the backbone from the neck to the pelvis,
and form with the long broad sternum a bony box enclosing much of the abdominal
viscera as well as those of the chest, perhaps to prevent their undue compression
under water. The top of the skull has a pair of crescentic depressions for lodgment of a large gland ; the palate is schizognathous. The sternum has a different
shape in each of the families. There are two carotids, except among the grebes.
The digestive system shows minor modifications, but accords in general with the
piscivorous regimen of the whole order. The sexes are alike ; the young different ;
the seasonal changes often great. A part of the order are altricial, the rest
præcocial. There are four families of *Pygopodes*, sharply distinguished by external
characters ; three of them are represented in this country. The penguins (*Spheniscidæ*) are confined to the seashores of the southern hemisphere. This group is
well marked by the solidity of the skeleton, and the flatness of most of the bones,
with many peculiar osseous details ; by a very special ptilosis, both in the lack of
tracts, and the structure of the feathers themselves, many of which are curiously
scale-like ; by the completely posterior set of the legs with extremely short tarsus,
and especially, among external features, by the reduction of the wings to mere
paddles, lacking specially formed remiges, unserviceable for flight, but highly
efficient as fins to aid progress under water. There are twelve species of penguins,
referable to three or four genera. One of the most singular facts in ornithology is,
that some species of penguin do not lay their egg in a nest in the ordinary way,
but carry it about with them in a pouch temporarily formed by a fold of the
abdominal integument (*Verreaux*) ; thus affording a wonderful analogy to marsu-
pial mammals. The author's monograph of the *Spheniscidæ* will be found in the
Proceedings of the Philadelphia Academy, of the present year.

Family COLYMBIDÆ. Loons.

Bill stout, straight, compressed, tapering, acute, paragnathous, entirely horny.
Nostrils narrowly linear, their upper edge lobed. Head completely feathered, the
antiæ prominent, acute, reaching the nostrils ; no crests nor ruffs. Wings strong,
with stiff primaries and short inner quills. Legs completely posterior, buried,
feathered on to the heel-joint ; tarsi entirely reticulate, extremely compressed, the
back edge smooth ; toes four, the anterior palmate, the posterior semilateral and
having a lobe connecting it with the base of the inner. Tail short, but well formed,

of many feathers. Back spotted. Loons are large heavy birds with broad flattened body and rather long sinuous neck, abundant on the coasts and larger inland waters of the Northern Hemisphere; they are noted for their powers of diving, being able to evade the shot from a gun by disappearing at the flash, and to swim many fathoms under water. They are migratory, breeding in high latitudes, generally dispersed further south in winter. They are præcocial, and lay two or three dark-colored spotted eggs in a rude nest of rushes by the water's edge. The voice is extremely loud, harsh and resonant. The ♀ is smaller than the ♂. There is but one genus, with only three well-determined species.

311. Genus COLYMBUS Linnæus.

Great Northern Diver, or *Loon*. Black; below from the breast white, with dark touches on the sides and vent; back with numerous square white spots; head and neck iridescent with violet and green, having a patch of sharp white streaks on each side of the neck and another on the throat;

bill black. Young: dark gray above, the feathers with paler edges; below white from the bill, the sides dusky; bill yellowish-green and dusky. Length 2½–3 feet; extent about 4; wing about 14 inches; tarsus 3 or more; longest

Fig. 213. Great Northern Diver.

toe and claw 4 or more; bill 3 or less, at base 1 deep and ½ wide; the culmen, commissure and gonys all gently curved. N. Am., abundant; the whole U. S. in winter. Wils., ix, 84, pl. 74. f. 3; Nutt., ii, 513; Aud., vii, 282, pl. 476; Lawr. in Bd., 888. . . . . . torquatus.

Var. adamsii. *Yellow-billed Loon.* Similar; larger; spots on the back larger, not so nearly square; gloss of the neck rather steel-blue, the white patches smaller, but the individual streaks larger; *bill mostly yellowish-white*, nearly 4 long, higher and comparatively narrower at the base, the gape straight, the culmen and gonys nearly so (fig. 213 shows the shape exactly, although intended for the common species). Northwestern America, chiefly; England; Asia. Gray, Proc. Zool. Soc. 1859, 167; Coues, Proc. Phila. Acad. 1862, 227; Elliot, pl. 63.

Black-throated Diver. Back and under parts much as in the last species; upper part of head, and hind neck, *bluish-ash* or hoary gray; fore neck purplish-black with a patch of white streaks, the dark color ending abruptly; bill black. The young resemble those of that species, but will be known by their inferior size. Length under 2½ feet; extent about 3; wing 13 or less; tarsus 3; bill about 2¼. N. Am. and N. Europe; said to be common and generally dispersed throughout the U. S. in winter, which is contrary to my experience. Sw. and Rich., F. B.-A. ii, 475; Nutt., ii, 517; Aud., vii, 295, pl. 477; Lawr. in Bd., 888. . . . arcticus.

Var. PACIFICUS. Colors the same; size less; length about 2 feet; wing about 11; tarsus 2¼; bill 2-2¼, very weak and slender. Northwestern Am., abundant on the Pacific Coast of the U. S. in winter. LAWR. in BD., 889; COUES, *l. c.* 228.

Red-throated Diver. Blackish; below white, dark along the sides and on the vent and crissum; most of head and fore neck bluish-gray, the throat with a large *chestnut* patch; hind neck sharply streaked with white on a blackish ground; bill black. Young have not these marks on the head and neck, but a profusion of small, sharp, circular or oval white spots on the back. Size of the last, or rather less. N. Am. and N. Europe, common; dispersed over most of the U. S. in winter. Sw. and RICH., F. B.-A. ii, 476; NUTT., ii, 519; AUD., vii, 299, pl. 478; LAWR. in BD., 890.. SEPTENTRIONALIS.

Family PODICIPIDÆ. Grebes.

Bill of much the same character as that of loons, but generally weaker, in one genus only·quite stout and somewhat hooked. Nostrils linear, linear-oblong or oval, not lobed. Head incompletely feathered, with definitely *naked lores*, the feathers not reaching the nostrils; commonly adorned in the breeding season with lengthened gayly-colored crests, ruffs, or ear-tufts. Back not spotted; under plumage peculiarly silky and lustrous, usually white. Wings very short and concave, the primaries often attenuated at the end, covered by the large inner quills when closed. Tail a mere tuft of downy feathers, without perfectly formed rectrices. *Feet lobate*, the front toes also semipalmate; tarsi compressed, scutellate, their hinder edge rough with a double row of protuberant scales; toes flattened; claws short, broad, flat, obtuse, something like human nails.

The grebes are strongly marked by the foregoing characters, especially of the feet and tail, though they agree closely with the loons in general structure and economy. Principal internal characters are the absence, it is said, of one carotid, the greater number of cervical vertebræ (19 instead of 13) and shortness of the sternum, with lateral processes reaching beyond the transverse main part (the reverse of the case in loons). The gizzard has a special pyloric sac. These birds are expert divers, and have the curious habit of sinking back quietly into the water when alarmed, like anhingas. Owing to the virtual absence of the tail the general aspect is singular, rendered still more so by the almost grotesque parti-colored ruffs and crests that most species possess. These ornaments are very transient; old birds in winter, and the young, are very different from the adults in breeding attire. The eggs are more numerous than in other pygopodous birds, frequently numbering 6–8; elliptical, of a pale or whitish color, unvariegated; commonly covered with chalky substance. The nest is formed of matted vegetation, close to the water, or even, *it is said*, floating among aquatic plants; the young swim directly. Grebes are the only cosmopolitan birds of the order, being abundantly distributed over the lakes and rivers of all parts of the world, though they are less maritime than the species of either of the other families. There are not over twenty well determined species, for which fifteen generic, and about seventy specific, names are recorded. The genera requiring recognition are only two. In *Podilymbus*, the bill is short, stout, and bent at the end, the lores are broadly naked, the frontal feathers are bristly and there are no ruffs or crests; in all the rest of the grebes the bill is slender, straight and more or less acutely paragnathous, the naked loral strip is narrow, and the soft feathers of the head form lengthened tufts of various kinds.

312-4. Genus PODICEPS Latham.

* Large, with very long neck; bill very slender and sharp-pointed, longer than the head, straight or almost recurved; tarsus as long as the middle toe and claw. No colored ruffs at any season? (*Æchmophorus.*)

Western Grebe. Length about 30; extent 36; wing 8–9; bill and tarsus, each about 3. Above, blackish-gray, with paler edges of the feathers, blackening on the hind neck and top of the head, the loral region gray; quills ashy-brown, bases of the primaries and most of the secondaries white; below, from bill to tail, pure silky white, with dark touches on the sides; bill obscurely olivaceous, brighter along the edges and at tip. Adult in the breeding season with a short occipital crest, and slight indications of a ruff; but no brightly colored feathers on the head or neck as yet observed. Pacific Coast, U. S., abundant. *P. occidentalis* LAWR. in BD., 894; *Æchmophorus occidentalis* COUES, *l. c.* 229. OCCIDENTALIS.

FIG. 211. Western Grebe.

Var. CLARKII. Similar; loral region white; bill bright yellow, the ridge black, shorter, slenderer, extremely acute and almost recurved; smaller; length 2 feet or less; wing 7; tarsus $2\frac{3}{4}$; bill $2\frac{1}{2}$. Same habitat. LAWR. in BD., 895; COUES, *l. c.* 229 and 404. *** The foregoing species has been united with the ordinary bird of Central and South America (*P. major, cayennensis, bicornis* and *leucopterus*) by Dr. Schlegel. This seems premature, but it may be required if *occidentalis* proves to assume the red neck and other coloration of *major*.

** Medium, with moderately long neck; bill not longer than the head, shorter than the tarsus, moderately stout and acute; tarsus shorter than the middle toe and claw. Conspicuous crests, ruffs or tufts, in the breeding season. (*Podiceps.*)

Crested Grebe. Tarsus equal to the middle toe without its claw; bill equal to the head, about $\frac{4}{5}$ the tarsus; crests and ruff highly developed. About 24; extent 34; wing $7\frac{1}{2}$–$8\frac{1}{2}$; bill 2–$2\frac{1}{4}$; tarsus $2\frac{1}{2}$–$2\frac{3}{4}$. Adult: throat and sides of head white changing to brownish-red on the ruff, which is tipped with black; fore part and sides of neck like the ruff; top of head and long occipital tufts dark brown, as are the upper parts generally, the feathers of the back pale-edged; primaries brown, part of them and nearly all the secondaries white; under parts silky silvery white, *without* dark mottling, but the sides dark-marked. Young: without any lengthened colored feathers on the head or neck. N. Am. at large; U. S. in winter, but not nearly so common as the next species. SW. and RICH., F. B.-A, ii, 419; NUTT., ii, 250; AUD., vii, 308, pl. 479; LAWR. in Bd., 893. *P. cooperi* ID., *ibid.*; COUES, Proc. Phila. Acad. 1862, 230. ?*P. affinis* SALVADORI; ELLIOT, Introd. No. 98, with figure of head. . . . . CRISTATUS.

Red-necked Grebe. Tarsus about $\frac{4}{5}$ the middle toe and claw; bill little shorter than tarsus; crests and ruff moderately developed. Medium; length about 18; wing 7–8; bill $1\frac{7}{8}$, to nearly 2; tarsus 2; middle toe and claw $2\frac{3}{8}$. Adult: front and sides of the neck rich brownish-red; throat and sides of head ashy, whitening where it joins the dark color of the crown, the feathers *slightly* ruffed; top of head with its *slight* occipital crests, upper parts generally, and wings, as in the last species, but much less white on the inner quills; lower parts pale silvery-ash, with dark sides (*not* pure white, but watered or obscurely mottled, sometimes obviously speckled, with dusky) ; bill black, more or less yellow at base. The young will be recognized by these last characters, joined with the peculiar dimensions and proportions. N. Am. ; common in the U. S. in winter. *P. rubricollis* Sw. and Rich., F. B.-A. ii, 411 ; Nutt., ii, 253 ; Aud., vii, 312, pl. 480 ; *P. griseigena* Lawr. in Bd., 892 ; *P. holbölli* Reinhardt ; Coues, *l. c.* 231. Our bird appears to differ constantly from the European in being larger, with the bill dispro-portionately large, and differently colored. . Griseigena var. holbölli.

Horned Grebe. Tarsus about equal to the middle toe without its claw ; bill much shorter than head, little more than half the tarsus, *compressed,* higher than wide at the nostrils, rather obtuse ; crests and ruffs highly devel-oped. Small ; length about 14 ; extent 24 ; wing 6 or less ; bill about $\frac{3}{4}$; tarsus 1¼. Adult : above, dark brown, the feathers paler-edged ; below, silvery-white, the sides mixed dusky and reddish ; most of the secondaries white ; fore neck and upper breast brownish-red ; head glossy black, including the ruff ; a broad band over the eye, to and including the occipital crests, brownish-yellow ; bill black, yellow-tipped. The young differ as in other species, but are always recognizable by the above measurements and propor-tions of parts. N. Am., abundant. Sw. and Rich., F. B.-A. ii, 411 ; Nutt., 254 ; Aud., vii, 316, pl. 481 ; Lawr. in Bd., 895. . . . . . cornutus.

Eared Grebe. Proportions substantially the same as in the last species ; size rather less ; bill shorter and more acute, *depressed,* wider than high at the nostrils. Adult : above, blackish-brown, the feathers with scarcely or not paler edges ; below silky-white, reddish along the sides ; all the prima-ries chocolate-brown, most of the secondaries white ; head and neck all round black, the auriculars lengthened into a rich golden-brown tuft, but no obvious crests or ruff. Young : known from the last by the different shape of the bill. Arctic America, chiefly western ; *common* in the Pacific States in winter ; has not been observed in the Atlantic States. *P. auritus* Nutt., ii, 256 ; Aud., vii, 322, pl. 482 ; *P. californicus* Heermann, Proc. Phila. Acad. 1855, 179, and Pac. R. R. Rept. x, Cala. Route, pl. 8 (young) ; Lawr. in Bd., 896 ; Coues, *l. c.* 231. All the American specimens I have seen, differ from the European ones examined, in having less white on the wings. . . . . auritus var. californicus.

*** Very small ; bill much shorter than the head, $\frac{2}{3}$ or less the tarsus ; tarsus about $\frac{3}{4}$ the middle toe and claw. No colored crests or ruffs. (*Sylbeocyclus.*)

St. Domingo Grebe. Adult : top of head deep glossy steel-blue ; rest of head and neck ashy-gray, deepest behind, the throat with whitish ; upper parts brownish-black, with greenish gloss ; primaries chocolate-brown, a great part of most of them, and all the secondaries, pure white ; under parts silky-white thickly mottled with dusky. Length 9½ ; wing 3⅔ ; bill ⅝ ;

tarsus 1¼; middle toe and claw 1⅜. Central America, West Indies and Mexico, Texas, Southern Colorado, Lower California. *Sylbeocyclus dominicus* COUES, *l. c.* 232. . . . . DOMINICUS.

315. Genus PODILYMBUS Lesson.

Pied-billed Grebe. *Dab-chick.* *Dipper.* *Diedapper.* *Water-witch.* Length 12–14; wing about 5; bill 1 or less; tarsus 1½. Adult: bill bluish, dusky on the ridge, encircled with a black bar; throat with a long black patch; upper parts blackish-brown; primaries ashy-brown, secondaries ashy and white; lower parts silky-white, more or less mottled or obscured with dusky; the lower neck in front, fore breast and sides, washed with rusty. Young: lacking the throat patch and peculiar marks of the bill, otherwise not particularly different; in a very early plumage with the head curiously striped. N. Am., very abundant. NUTT., ii, 259; AUD., vii, 324, pl. 483; LAWR. in BD., 898. . . . . PODICEPS.

Family ALCIDÆ. Auks.

Feet three-toed, palmate. Bill horny, non-lamellate, of extremely variable shape, often curiously appendaged; nostrils variable, but not tubular. Wings and tail short; tarsi shorter than the middle toe and claw. Form heavy, thickset.

Birds of this family will be immediately recognized by the foregoing circumstances, taken in connection with general pygopodous characters. Agreeing closely in essential respects, they differ among themselves to a remarkable degree in the form of the bill, with every genus and almost every species; this organ frequently assuming an odd shape, developing horny processes, showing various ridges and furrows, or being brilliantly colored. It is the rule that any *soft* part that may be observed on the bill will finally become hard, or form an outgrowth, or both; and such processes, in some cases at least, are temporary, appearing only during the breeding season. The bill, besides, varies greatly with age, in size and shape, often showing at first little trace of its adult character. In gen. 316-7 the bill is high, compressed, with curved vertical colored grooves, the nostrils densely feathered; in 318-23, the feathers are remote from the nostrils, and the bill reaches its maximum of diversity and singularity of contour; in the rest, the bill is of simpler shape, usually conico-elongate, with more or less

FIG. 215. Great Auk.

perfectly feathered nostrils. The general coloration is simple; but many species develop very remarkable frontal or lateral crests; the sexes are alike; the young different; seasonal changes are almost always strongly marked.

The family is confined to the Northern Hemisphere, where it represents the penguins of the Southern; several species occur in the North Atlantic, in almost incredible numbers, or are of circumpolar distribution; but the majority, including all the stranger kinds, inhabit the North Pacific; some range as far south, in winter, at least as the Middle States and Lower California. They are all marine; feed on fish and other animal substances, exclusively; lay 1–3 eggs on bare ledges, in rifts of rocks, or in burrows; and are altricial. The voice is hoarse; the flight swift and firm, performed with vigorous rapid wing-beats; one species is deprived of flight owing to the shortness of the wings, although these members are well formed with perfect remiges; all swim and dive with great facility. They are eminently gregarious, and mostly migratory. All the species are represented in this country. The number of species given by Brandt in 1837 (Bull. Acad. Sci. St. Petersburg), by Cassin in 1858 (Baird's B. N. A.) and by myself in 1868 (Proc. Phila. Acad.), must be materially reduced, as Brandt himself has since shown (op. cit. 1869), and as I now admit. Only twenty-one are unquestionably valid.

316. Genus ALCA Linnæus.

Great Auk. Coloration as in the next species, but a large white area before the eye; length about 30; wing 6; tail 3; bill 3, along gape 4, its depth 1⅔. Nutt., ii, 553; Aud., vii, 245, pl. 465; Cass. in Bd., 900. Special interest attaches to this bird, which is now on the point of extinction, largely through human agency. It formerly inhabited this coast from Massachusetts northward, as attested by earlier observers, and by the plentiful occurrence of its bones in shell-heaps; also, Greenland, Iceland, and the N. W. shores of Europe, to the Arctic Circle. On our shores it was apparently last alive at the Funks, a small island off the S. Coast of Newfoundland; while in Iceland, its living history has been brought down to 1844. Of late years, it has been currently, but, as it appears, prematurely, reported extinct. Mr. R. Deane has recently recorded (Am. Nat. vi, 368) that a specimen was "found dead in the vicinity of St. Augustine, Labrador, in November, 1870;" this one, though in poor condition, sold for $200, and was sent to Europe. I know of only four specimens in this country—in the Smithsonian Institution, in the Philadelphia Academy, the Cambridge Museum, and in Vassar College, Poughkeepsie (the latter the original of Audubon's figures). There is an egg in each of the first two mentioned collections. About 60 skins appear to be preserved in various museums. See Steenstrup, Viddensk. Meddel., Copenhagen, 1856–7, 33–116; Newton, Ibis, 1862, p. —; Coues, Proc. Phila. Acad. 1868, 15; Orton, Am. Nat., iii, 539. . . . . Impennis.

317. Genus UTAMANIA Leach.

Razor-billed Auk. Tinker. Brownish-black, browner on the head and throat; under parts from the throat (in summer; from the bill in winter, and in *young*), tips of secondaries, and sharp line from bill to eye, white; bill black with a white curved line; mouth yellow; 16–19; wing 7–8; tail 3–3½, graduated 1 or more; tarsus 1–1¼; bill 1⅛, along gape 2¼, nearly 1 deep. N. Atlantic, extremely abundant on rocky shores and islands with murres, puffins and gannets; egg generally single, and deposited in a rift of rocks; 3×2, white or whitish variously speckled and blotched with brown. Comes S. in winter to the Middle States. N. Pacific, casually. NUTT., ii, 547; AUD., vii, 247, pl. 466; CASS. in BD., 901; COUES, *l. c.* 18, and *op. cit.* 1861, 249. . . . . TORDA.

318-9. Genus FRATERCULA Brisson.

*Not crested; eyelids appendaged; under mandible sulcate, like the upper, the grooves convex forward; culmen simple, with one curve; base of bill bossed; corners of mouth callous. Blackish, including the throat, the sides of the head ashy-gray, with dusky maxillary patches (whole face dusky in the young); below, white; bill red, blue and yellow; feet red. (*Fratercula.*)

Horned Puffin. A slender sharp spur on upper eyelid. Black of throat reaching the bill. 14½; wing 7¼; tail 2¾; bill 2; tarsus 1½. N. Pacific; *not* authentic on our Atlantic Coast. *Mormon glacialis* AUD., vii, 236, pl. 463; *M. corniculata* CASS. in BD., 902; COUES, *l. c.* 24. . CORNICULATA.

Common Puffin. *Sea Parrot.* A thick blunt excrescence on eyelids. Black of throat not reaching the bill. 13½; wing 6½; tarsus 1; bill 2, depth at base 1½. N. Atlantic, breeding in vast numbers, in burrows; egg 2½×1¾,

FIG. 216. Common Puffin.

broadly ovoid, rough-granular, white or whitish, more rarely brownish, obsoletely or not at all variegated. S. in winter to the Middle States (to Georgia, *Audubon*). NUTT., ii, 542; AUD., vii, 238, pl. 464; CASS. in BD., 903; COUES, *l. c.* 21, and *op. cit.* 1861, 251. . . . . . ARCTICUS.

Var. GLACIALIS, from the Arctic Coasts, is rather larger, especially the bill, which is about 2½ long, 1½-1¾ deep at base. CASS. in BD., 903; COUES, *l. c.* 23.

** Adult with a long flowing crest of filamentous feathers on each side of the head; eyelids simple; under mandible smooth, upper sulcate, the grooves concave forward; culmen with two curves, the basal part bossed. (*Lunda.*)

Tufted Puffin. Blackish, duller and more fuliginous below; face white; crests straw-yellow; bill red and livid; feet red. Young not crested, face not white, and at an early age the undeveloped bill has a different shape, represented in fig. 217 (*this is Saqmatorhina lathami* Br., Proc. Zool. Soc. 1851, 202, pl. 44; COUES, *l. c.* 31; *S. labradora* CASS. in BD., 904; ELL., pl. 66. See Brandt, *l. c.* 244). Length 15–16; wing 7½–8; tail 2; tarsus 1⅓; bill 2⅓, nearly 2 deep. N. Pacific, abundant; S. to Cala. in winter; rare or casual on the Atlantic (Maine, *Aud.*). *Mormon cirrhatus* Nutt., ii, 539; Aud., vii, 234, pl. 462; Cass. in Bd., 902. . . cirrhata.

320. Genus CERATORHINA Bonaparte.

Horn-billed Auk. Glossy blackish, below ashy-gray, breast and belly white. Adult with two series of stiffish lanceolate white feathers on each

Fig. 217. Undeveloped bill of very young Tufted Puffin.

side of the head, and a stout upright horn at base of culmen; immature birds without these lateral crests, and with soft membrane, more or less bulging, in place of the horn; some specimens (♀ ?) in perfect plumage have no trace of a horn (*C. suckleyi* Cass. in Bd., 906; *Sagmatorrhina suckleyi* Coues, *l. c.* 32; see Elliot, Introd. No. 102, with figs.; Brandt,

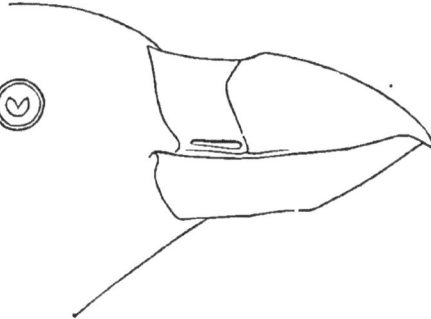

Figs. 218-21. Various stages of the bill of Horn-billed Auk.

l. c. 239. Figs. 218-21 show several conditions). Length 15½; wing 7¼; tail 2½; tarsus 1¼; bill 1⅓, including horn; nostril to top of horn sometimes ¾. Pacific Coast to Cala.; breeds S. to the Farallones. *Uria occidentalis* Aud., vii, 264, pl. 471; Nutt., ii, 538; *Cerorhina monocerata* Cass. in Bd., 905; *Ceratorhyncha monocerata* Coues, *l. c.* 28. . . monocerata.

321. Genus PHALERIS Temminck.

Parroquet Auk. Bill smooth; upper mandible oval, under falcate, rictus recurved. Blackish, below paler, gray, white, or varied; adult with a series of filamentous white feathers behind each eye; bill red, yellow-tipped.

9; wing 5½; tarsus 1; bill ⅔. N. Pacific. The curious bill is used to pry open bivalve mollusks (*Brandt*). NUTT., ii, 534; *Ombria psittacula* CASS. in BD., 910; *Simorhynchus psittaculus* COUES, *l. c.* 36; ELLIOT, pl. 70. . . . . . PSITTACULA.

322. Genus SIMORHYNCHUS Merrem.

Crested Auk. Adult in summer: blackish, paler and grayish below. A recurved frontal crest of 12–20 narrow feathers, dark; a bundle of lengthened filamentous feathers over and behind each eye, white. Bill red, yellow-tipped, with singularly irregular rictus, sides of lower mandible wholly naked, and a horny development at the commissural angle. *Phaleris cristatella* AUD., vii, 253, pl. 467; CASS. in BD., 906; *Simorhynchus cristatellus* COUES, *l. c.* 38.

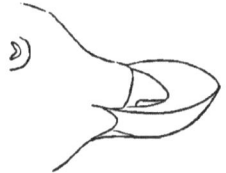

FIG. 222. Parroquet Auk.

FIG. 223. Crested Auk: adult in summer.   FIG. 224. Crested Auk: immature.

Fig. 223. In winter, bill dark, without the horny plate: *Uria dubia* PALL.; *Simorhynchus dubius* COUES, *l. c.* 40. Younger birds with a white spot under the eye, base of lower mandible feathered, gape straighter, and no horny plate at the angle. *Alca tetracula* PALL.; *Phaleris tetracula* Cass. in BD., 907; *Simorhynchus tetraculus* COUES, *l. c.* 43; ELLIOT, pl. 67. Fig. 224. See BRANDT, *l. c.* 224. All the foregoing stages show the crest, but it is wanting in very young birds. Length about 9; wing 5½; tail 1½; tarsus under 1; bill ½. N. Pacific; not observed in U. S. CRISTATELLUS.

Whiskered Auk. Similar; smaller; *two* series (postocular and maxillary) of filamentous white feathers on each side of head; bill smaller, never irregular. Very young: blackish-plumbeous, paler below, no crest, bill dark (*S. cassini* COUES, *l. c.*

FIG. 225. Whiskered Auk: adult.   FIG. 226. The same, young.

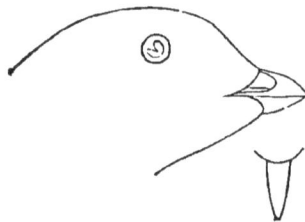

45; BD., Tr. Chicago Acad. i, 324, pl. 31, f. 2; *Alca pygmæa* GM.?). N. Pacific. *Phaleris camtschatica* CASS. in BD., 908; *Simorhynchus camtschaticus* COUES, *l. c.* 41; *Uria mystacea* PALL; *Mormon superciliosum* LICHT. CAMTSCHATICUS.

Knob-billed, or *Least Auk*. Very small; under 7; wing 4 or less; tarsus $\frac{3}{3}$; bill $\frac{2}{3}$. No crest, but white hair-like feathers on forehead and often about eyes. Blackish, with more or less white on scapulars; below white, pure or much varied with dusky; bill of adult in summer with a little knob at base. N. Pacific.

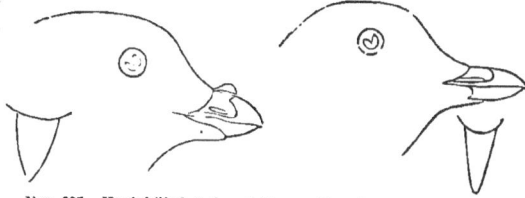

Phaleris nodirostris AUD. vii, 255; pl. 468; *P. microceros* BRANDT; CASS. in BD., 908; *S. microceros* COUES, *l. c.* 46. *Uria pusilla* PALL; *P. pusilla* CASS. in BD., 909; *S. pusillus* ELL., pl. 68; COUES, *l. c.* 49 (young or winter specimens). PUSILLUS.

FIG. 227. Knob-billed Auk; adult.   FIG. 228. The same, young.

323. Genus PTYCHORHAMPHUS Brandt.

Aleutian Auk. Blackish-cinereous, paler below, white on breast and belly; no long feathers about head; bill conic, acute, about $\frac{2}{3}$ the head, wrinkled at base, nostrils scaled; 9; wing 5; tarsus 1; bill $\frac{3}{4}$, $\frac{2}{3}$ deep at base, $\frac{1}{4}$ wide. Pacific Coast to L. Cala. CASS. in BD., 910; ELLIOT, pl. 69; COUES, *l. c.* 52. *Mergulus cassini* GAMBEL. . . . . . ALEUTICUS.

324. Genus MERGULUS Auctorum.

Sea Dove, or *Dovekie*. Glossy blue-black, below from the breast (in winter, and in young, from the bill) white; scapulars white-striped; secondaries white-tipped; white speck over eye; bill black, short, obtuse, turgid. $8\frac{1}{2}$; wing $4\frac{3}{4}$; tarsus $\frac{4}{5}$; bill $\frac{1}{2}$, about $\frac{1}{4}$ deep or wide at base. N. Atlantic, abundant, S. in winter to New Jersey (to Florida, *Maynard*). WILS., ix, 94, pl. 74, f. 5; NUTT., ii, 531; AUD., vii, 257, pl. 469; CASS. in BD., 918. . ALLE.

FIG. 229. Sea Dove.

325. Genus SYNTHLIBORHAMPHUS Brandt.

*** Tarsi much compressed, broadly scutellate in front and on the sides, not shorter than middle toe without its claw; bill compressed, shorter than head or tarsus; nostrils broadly oval, reached by feathers. Length $9\frac{1}{2}$–11; extent $16\frac{1}{2}$–$18\frac{1}{2}$; wing 5–$5\frac{1}{2}$; tail $1\frac{3}{4}$; bill $\frac{3}{4}$–$\frac{2}{3}$; tarsus 1. Head and neck black or blackish, with white stripe over eye and numerous others on nape and side of neck; upper parts and sides under the wings black or blackish-plumbeous; other under parts white, from the throat in summer, from the bill in winter, and in young birds.

FIG. 230. Black-throated Guillemot.

Black-throated Guillemot. Not crested. White superciliary stripe not running in advance of the eye. Bill stout, obtuse, at base $\frac{1}{2}$ *or more* of the length of culmen, pale, culmen and base black. N. Pacific. *Uria senicula* PALL. *Mergulus cirrhocephalus* VIGORS. *Uria antiqua* AUD., vii, 263, pl.

470, f. 1 (fig. 2, of supposed young, is *B. kittlitzii*) ; *Brachyrhamphus antiquus* CASS. in BD., 916 ; *S. antiquus* COUES, *l. c.*, 56. *Brachyrhamphus brachypterus* BRANDT ; CASS. in BD., 917 ; COUES, *l. c.* 67. . . ANTIQUUS.

Temminck's Guillemot. Adult in the breeding season crested. White superciliary stripe advancing far in front of the eye. Bill slenderer and more acute, scarcely *or not* ½ as deep at base as long, yellow, with black on culmen.

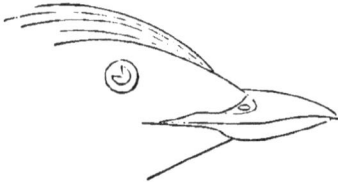

FIG. 231. Temminck's Guillemot. Adult.   FIG. 232. Temminck's Guillemot. Young.

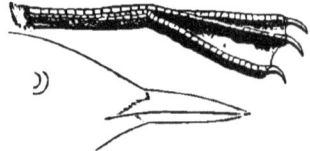

Young (and adult each winter?), uncrested, bill black ; above nearly uniform cinereous, below entirely white except along the sides : this is *Brachyrhamphus hypoleucus* XANTUS ; COUES, *l. c.* 64 ; ELLIOT, pl. 72 ; *Uria craveri* SALVADORI ; COUES, *l. c.* 66 ; ELLIOT, Introd. No. 172, with fig. of head. Whole Pacific Coast to Cape St. Lucas, abundant. *B. temminckii* CASS. in BD., 916 ; *S. wurmizusume* COUES, *l. c.* WURMIZUSUME.

326. Genus BRACHYRHAMPHUS Brandt.

₊ Tarsi little compressed, entirely reticulate, obviously shorter than the middle toe without its claw ; bill shorter than head, very slender and acute, with inflected tomia ; nostrils minute, overlaid by feathers. No crest.

Marbled Guillemot, or *Murrelet.* Adult in summer blackish, singularly variegated with chestnut or rusty, and white ; bill black ; adult in winter plumbeous, the feathers with darker centres, the scapulars and entire under parts, excepting some dark touches on the flanks, pure white. Length 9–10 ; wing 5 ; tail 1½ ; tarsus and bill ¾ or less. Pacific Coast to Cala. *Uria townsendii* AUD., vii, 278, pl. 475 (winter and summer, not old and young as supposed) ; *Uria marmorata* NUTT., ii, 525 ; *B. marmoratus* CASS. in BD., 915 ; COUES, *l. c.* 61 ; *B. wrangeli* BRANDT ; CASS. in BD., 917 ; COUES, *l. c.* 63 (winter, not different species as supposed) . MARMORATUS.

Kittlitz's Murrelet. Described as differing from the foregoing by its much shorter bill (only 1 along gape), deeper at base, and lateral tail feathers white, black-striped lengthwise. N. Pacific. Unknown to me ; no recognized specimens in American collections. *B. kittlitzii* BRANDT, 1837, 346 ; 1869, 213 ; CASS. in BD., 917. Brandt holds that Audubon's fig. 2 of pl. 470 represents this species. *?Uria brevirostris* VIGORS, Zool. Journ. vi, 1827, 357 ; Voy. Blossom, 32. . . . . KITTLITZII.

327. Genus URIA Brisson.

₊ Tarsi entirely reticulate, little, if any, shorter than middle toe without its claw ; bill straight, smooth, about equal to tarsus ; nostrils incompletely feathered ;

outer claw smooth; tail contained less than 3 times in length of wing. In summer, black, with white on wings or head; in winter, largely white; bill black, feet red. Length 12–15; wing 5½–7½; tarsus 1⅓–1⅛; bill 1⅓–1⅔.

Black Guillemot. Sea Pigeon. A large continuous white area on both upper and under surface of the wing, (rarely imperfect or wanting); head and neck with greenish gloss; tail feathers 12; wing 5½–6¼; bill rather acute.

FIG. 233. Black Guillemot.

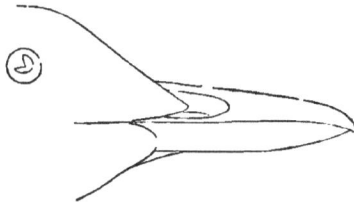

N. Atlantic, very abundant, S. in winter to New Jersey. Eggs laid in fissures of rock, 2 (3?) in number, 2⅜×1⅝ in size, nearly elliptical in shape, greenish-white, variously blotched with brown and purplish. Arctic Seas; rare or casual in the N. Pacific? NUTT., ii, 523; AUD., vii, 272, pl. 474; CASS. in BD., 911. COUES, *l. c.* 68; and *op. cit.* 1861, 255. . . . . . GRYLLE.

Pigeon Guillemot. A large white area on upper surface of wings only, partly

FIG. 234. Pigeon Guillemot.

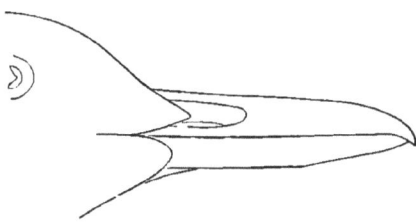

divided by a black line; head and neck with opaque ashy shade; tail feathers 14 (always?); bill rather obtuse; size rather greater, wing about 7. N. Pacific. CASS. in BD., 912; COUES, *l. c.* 72. . . . . COLUMBA.

Sooty Guillemot. No white on the wings, but usually whitish patches on the head. Larger; wing nearly 8; bill 1½–1¾. North Pacific. CASS. in BD., 913, pl. 97; COUES, *l. c.* 73. . . . CARBO.

328. Genus LOMVIA Brandt.

*** Tarsi scutellate in front, much

FIG. 235. Sooty Guillemot.

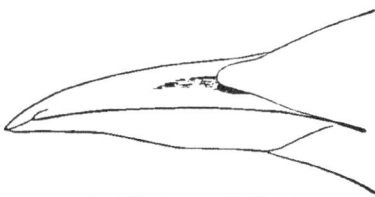

shorter than middle toe without claw; bill with decurved commissure, much longer than tarsus, its end, and the outer claw, grooved; nostrils feathered; tail graduated, contained more than 3 times in the length of wing. Size and coloration almost exactly as described under *Utamania*, but no white line from bill to eye.

Common Guillemot, or *Murre*. Depth of bill at nostrils not nearly ½ the length of culmen; tomia of upper mandible at base feathered, not noticeably dilated, nor brightly colored. In some cases, edges of eyelids, and line behind eye, white. N. Atlantic, Arctic and Pacific coasts, S. to New Jersey and California; breeding in myriads on rocky islands. Egg single, on bare ledges; 3–3½ long, by 1⅜–2½ broad; the ground color varying from white

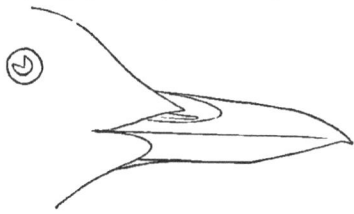

FIG. 236. Common Guillemot.

to dark green; sometimes perfectly plain, usually fantastically streaked or blotched in interminably varying pattern. Sw. and Rich., F. B.-A. ii, 477; Nutt., ii, 526; Aud., vii, 267, pl. 473; *Uria lomvia* and *ringvia* Cass. in Bd., 913, 914; *Lomvia troile* and *L. ringvia* Coues, *l. c.* 75, 78. Pacific specimens have a somewhat differently shaped bill, constituting var. *californica* Bryant, Proc. Bost. Soc. N. H. 1861, p. 11; Coues, *l. c.* 79. TROILE.

Thick-billed, or *Brünnich's Guillemot.* Depth of bill at nostrils about ½ the length of culmen; tomia of upper mandible at base dilated, denuded, usually yellowish. Distribution as in the last species. *Uria brünichii* Sw. and Rich.,

FIG. 237. Thick-billed Guillemot.

F. B.-A. ii, 477; Nutt., ii, 529; Aud., vii, 265, pl. 472; *U. arra* Cass. in Bd., 914; *Lomvia svarbag* Coues, *l. c.* 80. . . . . ARRA.

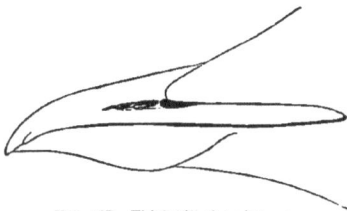

FIG. 238. Murres.

SYNOPSIS OF THE FOSSIL FORMS.

There is at present no satisfactory evidence of the existence of Birds in this country earlier than the Cretaceous period. The footprints in the sandstone of the Connecticut Valley, which have been attributed to Birds, were probably all made by Dinosaurian Reptiles.

The species which have been described from the Cretaceous formation are nearly all known only from fragmentary remains. Those from the Tertiary and Post-tertiary are generally represented by better preserved specimens.

The following synopsis has been revised by the accomplished palæontologist who has described nearly all the known species. Through his courteous attentions, the list has been completed to the date on which these pages go to press. This first connected account of the Extinct Birds of North America will unquestionably be very largely supplemented by future discoveries. Work in this field of research was never more actively and successfully prosecuted than at present, and new species are almost continually being brought to light.

PICARIÆ.

UINTORNIS LUCARIS Marsh.
American Journal of Science, iv, 259. Oct., 1872.

This bird was about as large as a robin, and apparently related to the Woodpeckers. The only known remains are from the Lower Tertiary formation of Wyoming Territory. They are preserved in the museum of Yale College.

RAPTORES.

AQUILA DANANA Marsh.
American Journal of Science, ii, 125. August, 1871.

This species was nearly as large as the Golden Eagle (*A. chrysaëtos*). The only known remains were found in the Pliocene of Nebraska, and are now in the Yale museum.

BUBO LEPTOSTEUS Marsh.
American Journal of Science, ii, 126. August, 1871.

A species about two-thirds as large as the Great Horned Owl (*B. virginianus*). The remains were discovered in the Lower Tertiary beds of Wyoming, and are also in the Yale museum.

GALLINÆ.

MELEAGRIS ANTIQUUS Marsh.
American Journal of Science, ii, 126. August, 1871.

This species was nearly as large as the Wild Turkey (*M. gallopavo*). The remains representing it were found in the Miocene of Colorado, and are preserved in the Yale museum.

(347)

MELEAGRIS ALTUS Marsh.

Proceedings of the Philadelphia Academy, 11. March, 1870. — American Naturalist, iv, 317. July, 1870.—American Journal of Science, iv, 260. Oct., 1872. (*M. superbus* Cope. Synopsis Extinct Batrachia, etc., 239.)

"Represented by portions of three skeletons, of different ages, which belonged to birds about the size of the Wild Turkey, although proportionally much taller. The tibiæ and tarso-metatarsal bones were, in fact, so elongated as to resemble those of wading birds." From the Post-pliocene of New Jersey. The remains are mostly in the museum of Yale College.

MELEAGRIS CELER Marsh.

American Journal of Science, iv, 261. Oct., 1872.

A species much smaller than the foregoing, but with legs of slender proportions. Also from the Post-pliocene of New Jersey, and preserved in the Yale museum.

GRALLATORES.

GRUS HAYDENII Marsh.

American Journal of Science, xlix, 214. March, 1870.

A species about as large as the Sandhill Crane (*G. canadensis*). From the Pliocene of Nebraska. Remains preserved in the museum of the Philadelphia Academy.

GRUS PROAVUS Marsh.

American Journal of Science, iv, 261. Oct., 1872.

This species was nearly as large as a Sandhill Crane. The remains representing it were found in the Post-pliocene of New Jersey, and are now in the Yale museum.

ALETORNIS NOBILIS Marsh.

American Journal of Science, iv, 256. Oct., 1872.

Nearly as large as the preceding species. Found in the Eocene deposits of Wyoming, and now in the museum of Yale College.

ALETORNIS PERNIX Marsh.

American Journal of Science, iv, 256. Oct., 1872.

About half the size of the above, and from the same locality. Also in the Yale museum.

ALETORNIS VENUSTUS Marsh.

American Journal of Science, iv, 257. Oct., 1872.

A smaller species, about as large as a Curlew (*Numenius*). From the same locality, and likewise in the Yale museum.

ALETORNIS GRACILIS Marsh.

American Journal of Science, iv, 258. Oct., 1872.

A bird about the size of a Woodcock (*Philohela minor*). From the same formation and locality, and now preserved in the museum of Yale College.

ALETORNIS BELLUS Marsh.

American Journal of Science, iv, 258. Oct., 1872.

A still smaller species, probably belonging to a different genus. From the same locality, and also in the Yale museum.

TELMATORNIS PRISCUS Marsh.

American Journal of Science, xlix, 210. March, 1870.

A species about as large as the King Rail (*Rallus elegans*), and probably allied to the *Rallidæ*. From the Cretaceous formation. Found near Hornerstown, New Jersey, and preserved in the Yale museum.

TELMATORNIS AFFINIS Marsh.

American Journal of Science, xlix, 211. March, 1870.

A somewhat smaller species from the same formation and locality. Also in the museum at Yale.

PALÆOTRINGA LITTORALIS Marsh.

American Journal of Science, xlix, 208. March, 1870.

About equalling a Curlew in size. The remains were found in the Cretaceous green-sand, at the above mentioned locality, and are now preserved at Yale.

PALÆOTRINGA VETUS Marsh.

American Journal of Science, xlix, 209. March, 1870.

A smaller species, from the same formation, found at Arneytown, New Jersey. The known remains are in the Philadelphia Academy.

PALÆOTRINGA VAGANS Marsh.

American Journal of Science, iii, 365. May, 1872.

Intermediate in size between the two preceding species. Discovered in the same formation, near Hornerstown, New Jersey; now in the museum of Yale College.

NATATORES.

SULA LOXOSTYLA Cope.

Transactions of the American Philosophical Society, xiv, 236. Dec., 1870.

A species not so large as the common Gannet (*S. bassana*). From the Miocene of North Carolina. Remains preserved in Prof. Cope's collection.

GRACULUS IDAHENSIS Marsh.

American Journal of Science, xlix, 216. March, 1870.

A typical Cormorant, rather smaller than *G. carbo*. From the Pliocene of Idaho. Most of the known remains are deposited in the Yale museum.

GRACULAVUS VELOX Marsh.

American Journal of Science, iii, 363. May, 1872.

This bird was related to the Cormorants, and was rather smaller than *Graculus carbo*. The remains were found in the green-sand of the Cretaceous formation, near Hornerstown, New Jersey, and are now at Yale College.

GRACULAVUS PUMILUS Marsh.
American Journal of Science, iii, 364. May, 1872.

A smaller species, from the same formation and locality. The remains are in the Yale museum.

GRACULAVUS ANCEPS Marsh.
American Journal of Science, iii, 364. May, 1872.

Apparently a species of Cormorant, about as large as *Graculus violaceus*. From the Cretaceous of Western Kansas. Remains in the Yale College museum.

ICHTHYORNIS DISPAR Marsh.
American Journal of Science, iv, appendix, 344. Oct., 1872.

A bird about as large as a pigeon, and differing from all known birds in having *biconcave vertebrae*. The remains were found in the Cretaceous shale of Kansas, and are in the museum of Yale College.

PUFFINUS CONRADII Marsh.
American Journal of Science, xlix, 212. March, 1870.

A shearwater about the size of *P. cinereus*. From the Miocene of Maryland, and now preserved in the museum of the Philadelphia Academy.

CATARRACTES ANTIQUUS Marsh.
American Journal of Science, xlix, 213. March, 1870.

A Guillemot rather larger than the Common Murre (*Lomvia troile*). From the Miocene of North Carolina. Deposited in the Philadelphia Academy.

CATARRACTES AFFINIS Marsh.
American Journal of Science, iv, 259. Oct., 1872.

A species about as large as the preceding, and nearly related. From the Post-pliocene of Maine. The original specimen is in the Philadelphia Academy.

HESPERORNIS REGALIS Marsh.
American Journal of Science, iii, 360. May, 1872.

This bird was a gigantic Diver, related to the Loons (*Colymbidae*). The skeleton measured about five feet nine inches in length. The known remains were found in the upper Cretaceous shale of Western Kansas, and are now in the Yale museum.

LAORNIS EDVARDSIANUS Marsh.
American Journal of Science, xlix, 206. March, 1870.

This species was nearly as large as a Swan. The remains were discovered in the Middle Marl bed, of Cretaceous age, at Birmingham, New Jersey, and are now in the museum of Yale College.

ADDITIONS AND CORRECTIONS.

INTRODUCTION, *passim*. *For* Order Scansores *read* zygodactyle birds.

P. 9, § 19.—The *Phœnicopteridæ* were not considered as belonging to *Lamellirostres* when the fourth sentence of this paragraph was penned.

P. 22, last two lines. *For* 117 *read* 123; *for* 177 *read* 176; *dele* 154; *insert* 149, 151, 189, 222-3-4, 230.

P. 30, last line but one. *For* no *read* a. (This important error is also repeated in fig 6, where the phalanx in question is omitted.)

P. 35, eighth line. *After* in *insert* nearly.

P. 38, end of ninth line from bottom. *After* Strisores *insert* of some authors.

P. 39, tenth line. *After* no *insert* perfect.

P. 47, § 81, tenth line. *After* (307) *insert* and a species of Accipiter (156).

P. 49, § 86, fourth line. *After* It *insert* when present. Next line, *after* 2t, *insert* when developed.

P. 55, third line. *After* belongs to the *read*: family Picidæ, of the sub-order Pici, of the order Picariæ.

P. 58. Among "abbreviations used" *insert*; —l. c., locus citatus—the place (of a work) just cited. op. cit., opus citatum—the work just cited.

P. 59. Among "works referred to" *insert*; — Sw. and RICH., F. B.-A. ii. *Swainson*, W., and *Richardson*, J.: Fauna Boreali-Americana. Vol. ii. 4to. 1831.

P. 61. *After* Hydrochelidon, *for* 292 *read* 293; *after* Haliplana, *for* 293 *read* 292.

P. 63. *For* GLOTTIS 215, SYMPHEMIA 214 and RHYACOPHILUS 216, *read* TOTANUS 214-6.

P. 63. *For* FULIX 260, and AYTHYA 261, *read* FULIGULA 260-1.

P. 75. *Curve-billed Thrush.* Specimens lately received indicate that the Arizona bird constitutes a variety of *H. curvirostris*: the following is a better description than that given in the text. — Var. *palmeri* RIDGW. Ms. Above, grayish-brown, nearly uniform; wing coverts and quills with slight whitish edging, the edge of the wing itself white; tail feathers with slight whitish tips; below, a paler shade of the color of the upper parts, the throat quite whitish, the crissum slightly rufescent, the breast and belly with obscure dark gray spots on the grayish-white ground; no obvious maxillary streaks, but vague speckling on the cheeks; bill black; feet blackish-brown; bill 1¼; wing 4¼; tail 5; tarsus 1¼; middle toe and claw 1⅜. (Described from 61589, Mus. Smiths. Inst., Tucson, Arizona, Bendire.)

P. 77. *Kennicott's Sylvia.* Add to the quotation: TRISTRAM, Ibis, 1871, 231.

P. 85. *Allied Creeper Wren.* In all probability distinct from the preceding species.

P. 87. *Alaskan Wren.* May be best treated as a variety of the Winter Wren; and this last may be considered as *Anorthura troglodytes* var. *hyemalis*.

P. 122. *Plumbeous Vireo.* Additional material shows that most probably this is a variety of *V. solitarius*, as intimated in the text.

P. 129. *For* Genus CURVIROSTRA Scopoli *read* Genus LOXIA Linnæus. The Red Crossbill may be considered as var. *americana* of the European *Loxia curvirostra*.

(351)

P. 130. *Gray-crowned Finch.* It is hardly necessary to recognize by name more than one variety of this bird—"campestris" being referred to *tephrocotis* proper, and "littoralis" to var. *griseinucha.*

P. 135. *Baird's Bunting.* As very strongly hinted in the text, the supposed specimens of *Centronyx bairdii* from Massachusetts are not this species at all, but a *Passerculus,* apparently new. (*P. princeps* MAYNARD, Am. Nat. vi, 1872, p. 637). Although perfectly aware of this at time of writing, I refrained from anticipating publication of the fact. I venture to foretell, that a second specimen of "Centronyx" will never be found.

P. 136. *St. Lucas Sparrow.* Doubtless only a variety of *P. rostratus.*

P. 140. Good authority contends for the specific validity of *Peucæa cassinii,* but I am not prepared to yield my position.

P. 147. It may be as well to allow *Passerella townsendii* to stand as a species, until its intergradation with *iliaca* is proven. *P. schistacea* goes with *townsendii* as a slight variety.

P. 174, first line. For features *read* feathers.

P. 183. *Vaux's Swift.* I am more inclined to doubt its validity.

P. 186. *Linné Hummingbird.* The implication is, that the specimen accredited to Massachusetts came from a dealer's stock, in exchange for a specimen of *T. colubris* spoilt in stuffing.

P. 207. *Ferruginous Owl.* To the extralimital specimens described, *add:*—No. 61585, Mus. Smiths. Inst., from Tucson, Arizona, since transmitted to me by Lt. C. Bendire, U. S. A. It is the specimen of which some fragments furnished my note in the American Naturalist, as quoted in the text.

P. 213. *Gyrfalcon.* The specimens from the Mackenzie's river region, noticed by Baird (*l. c.*) under name of *F. sacer,* have since been determined by Prof. Newton to be indistinguishable from ordinary var. *islandicus.* I omitted to state, that var. *gyrfalco* is a N. European form, not recognized, I believe, from this country. The name *sacer* has priority over all the others as the specific designation.

P. 222. *Cathartes burrovianus,* there is reason to believe, may be a valid species; it does *not,* however, occur within our limits.

P. 248. *After* Genus PHALAROPUS *insert* Brisson.

P. 270. *Before* GRUIDÆ *insert:*— SUBORDER ALECTORIDES. CRANES, RAILS AND OTHER ALLIES. SEE p. 241.

INDEX AND GLOSSARY.

www.ingramcontent.com/pod-product-compliance
Lightning Source LLC
Chambersburg PA
CBHW030905270326
41929CB00008B/583